A SEARCH FOR ANTIBACTERIAL AGENTS

Edited by **Varaprasad Bobbarala**

A Search for Antibacterial Agents
Edited by Varaprasad Bobbarala

Published by InTech

Janeza Trdine 9, 51000 Rijeka, Croatia

Publishing Process Manager Vedran Greblo

Technical Editor Teodora Smiljanic

Cover Designer InTech Design Team

Additional hard copies can be obtained from orders@intechopen.com

A Search for Antibacterial Agents, Edited by Varaprasad Bobbarala
p. cm.
ISBN 978-953-51-0724-8

For EU product safety concerns:
IN TECH d.o.o., Prolaz Marije Krucifikse Kozulić 3, 51000 Rijeka, Croatia,
info@intechopen.com or visit our website at intechopen.com.

Meet the editor

Dr. Varaprasad Bobbarala received his Ph.D. in Faculty of Science from Andhra University in 2008 under the direction of Professor K. Chandrasekhara Naidu and Professor G. Seshagiri Rao. Specialized in Biochemistry, Medicinal chemistry and Microbiology. He has published over 90 original research articles, reviews, book chapters, and edited three books. He is currently Editor In-Chief of International Journal of Bioassays (ISSN: 2278-778X), Associate Editor and member of the editorial boards as well as the reviewer of dozens of high-impact international periodicals. Dr. B. Varaprasad, previously served as the Chief Scientist of Research and Development (R & D) at Krisani Innovations Pvt. Ltd., before his current role as the Chief Scientist and Technical Director of Research and Development of Adhya Biosciences Pvt. Ltd., India. He is actively engaged in scientific research in the areas of Antimicrobial Resistance, Drug Discovery, Isolation of Bio-active metabolites and bio-efficacy studies.

Contents

Preface XI

Chapter 1 **Relationships Between Chemical Structure and Activity of Triterpenes Against Gram-Positive and Gram-Negative Bacteria 1**
A. G. Pacheco, A. F. C. Alcântara, V. G. C. Abreu and G. M. Corrêa

Chapter 2 **Flower Shaped Silver Nanostructures: An Efficient Bacteria Exterminator 25**
Subash Chandra Sahu, Barada Kanta Mishra and Bikash Kumar Jena

Chapter 3 **Lutamide, a New Ceramide Isolated from the Leaves of *Ficus lutea* 41**
Herve Martial Poumale Poumale

Chapter 4 **Antibacterial Modification of Textiles Using Nanotechnology 47**
Moustafa M. G. Fouda

Chapter 5 **Metal Complexes as Antimicrobial Agents 73**
Marcela Rizzotto

Chapter 6 **Dendrimers as Antibacterial Agents 89**
Metin Tülü and Ali Serol Ertürk

Chapter 7 **Selected Factors Determining the Content of Lactoferrin, Lysozyme and Immunoglobulins G in Bovine Milk 107**
Jolanta Król, Aneta Brodziak, Zygmunt Litwińczuk and Joanna Barłowska

Chapter 8 **The New About Congenital Antimicrobial Defense of Some Epithelial Tissues – Vaginal Mucosa and Hair 125**
Arzumanian Vera, Malbakhova Ekaterina and Vartanova Nune

Chapter 9 **Novel Anti-Microbial Peptides of *Xenorhabdus* Origin Against Multidrug Resistant Plant Pathogens 147**
András Fodor, Mária Hevesi, Andrea Máthé-Fodor, Jozsef Racsko and Joseph A. Hogan

Chapter 10 **Metal Complexes as Prospective Antibacterial Agents 197**
Joshua A. Obaleye, Adedibu C. Tella and Mercy O. Bamigboye

Chapter 11 **Bacteriostatic Agents 219**
Marzieh Rezaei, Majid Komijani and Seyed Morteza Javadirad

Chapter 12 **New Improved Quinlone Derivatives Against Infection 235**
Urooj Haroon, M. Hashim Zuberi, M. Saeed Arayne and Najma Sultana

Chapter 13 **Biocompatibility and Antimicrobial Activity of Some Quaternized Polysulfones 249**
Silvia Ioan and Anca Filimon

Chapter 14 **The Design of Bacteria Strain Selective Antimicrobial Peptides Based on the Incorporation of Unnatural Amino Acids 275**
Amanda L. Russell, David Klapper, Antoine H. Srouji, Jayendra B. Bhonsl, Richard Borschel, Allen Mueller and Rickey P. Hicks

Chapter 15 **Synthesis, Spectral, Magnetic, Thermal and Antimicrobial Studies on Symmetrically Substituted 2, 9, 16, 23-tetra-phenyliminophthalocyanine Complexes 305**
M. H. Moinuddin Khan, K. R. Venugopala Reddy and J. Keshavayya

Chapter 16 **Antisense Antibacterials: From Proof-Of-Concept to Therapeutic Perspectives 319**
Hui Bai and Xiaoxing Luo

Preface

This book contains precisely referenced chapters, emphasizing antibacterial agents with clinical practicality and alternatives to synthetic antibacterial agents through detailed reviews of diseases and their control using alternative approaches. The book aims at explaining bacterial diseases and their control via synthetic drugs replaced by chemicals obtained from different natural resources which present a future direction in the pharmaceutical industry. The book attempts to present emerging low cost and environmentally friendly drugs that are free from side effects studied in the overlapping disciplines of medicinal chemistry, biochemistry, microbiology and pharmacology.

Varaprasad Bobbarala
Chief Scientist at Krisani Biosciences
India

1

Relationships Between Chemical Structure and Activity of Triterpenes Against Gram-Positive and Gram-Negative Bacteria

A. G. Pacheco*, A. F. C. Alcântara, V. G. C. Abreu and G. M. Corrêa
Departamento de Química,
Universidade Federal de Minas Gerais, Belo Horizonte, Minas Gerais,
Brazil

1. Introduction

Bacteria are non-chlorophyllated unicellular organisms that reproduce by fission and do not present nuclear envelope. Gram´s stain is a staining technique used to classify bacteria based on the different characteristic of their cell walls. Gram-positive or Gram-negative bacteria are determined by the amount and location of peptidoglycan in the cell wall, exhibiting different chemical compositions and structures, cell-wall permeabilities, physiologies, metabolisms, and pathogenicities.

Microbial diseases present a significant clinical interest because some species of bacteria are more virulent than other ones and show alteration in sensibility to the conventional antimicrobial drugs, mainly species of the genera *Staphylococcus, Pseudomonas, Enterococcus,* and *Pneumococcus*. The extensive use of the penicillin since the Second World War promoted the appearance of the first strains of penicillin-resistant Gram-positive bacteria (Silveira et al., 2006). Vancomicin and methicillin showed a large spectrum of bactericidal actions against many Gram-positive bacteria. However, some strains also presented resistance to these compounds, as observed to the drugs vancomycin-resistant *Enterococcus* (VRE) and methicillin-resistant *Staphylococcus aureus* (MRSA), respectively. As a consequence, the resistance that pathogenic microorganisms build against antibiotics has stimulated the search of new antimicrobial drugs (Al-Fatimi et al., 2007; Rahman et al., 2002).

In the last few decades, the ethnobotanical search has been the subject of very intense pharmacological studies about drug discovery as potential sources of new compounds of therapeutic value in the treatment of bacterial diseases (Matu & Staden, 2003). The importance of secondary metabolites for the antimicrobial activity has been observed to triterpenoid compounds (Geyid et al., 2005). The triterpenes are widely distributed in the plant and animal kingdoms and occur in either a free state or in a combined form, mainly in the form of esters and glycosides (Ikan, 1991). Triterpenes present a carbon skeleton based on six isoprene units, being biosynthetically derived from the squalene, which may usually yield the pentacyclic triterpenes with six-membered rings. These pentacyclic triterpenes (PCTTs) present a basic skeleton which provides a large amount of derivative structures because different positions on

* Corresponding Author

their skeleton may be substituted. As result, there are at least 4000 known PCTTs (Dzubak et al., 2006), exhibiting a large spectrum of biological activities (James & Dubery, 2009). Some classes of triterpenes present other skeleton, such as fernane- and lupane-type triterpenes.

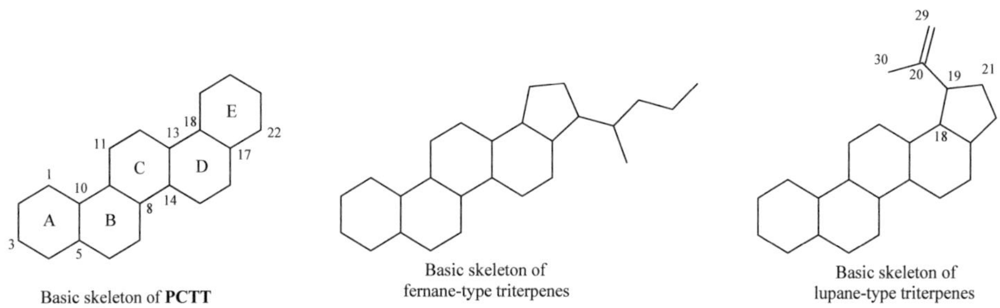

The literature describes the isolation of triterpenes from the vegetal species which exhibit bactericidal activity (Katerere et al., 2003; Sunitha et al., 2001; Ryu et al., 2000; Yun et al., 1999). Table 1 shows the most recent studies relating plant that exhibit bactericidal activity and contain triterpenes. The activity against Gram-negative bacteria has been few studied in relation to Gram-positive ones. The Gram-positive bacteria more studied are *S. aureus*, *B. subtilis*, *B. cereus*, and *S. faecalis* (24, 11, 7, and 6 occurrences, respectively). On the other hand, the Gram-negative bacteria more studied are *P. aeroginosa*, *E. coli*, *K. pneumoniae*, and *S. typhi* (15, 13, 9, and 6 occurrences, respectively).

Species	Isolated compound	Activity against Gram-positive bacteria	Activity against Gram-negative bacteria	Ref.
Abies sachalinensis	Triterpenes	*Bacillus subtilis* and *Staphylococcus aureus*	-	Gao et al., 2008
Acacia mellifera	Triterpenes	*S. aureus*	-	Mutai et al., 2009
Alstonia macrophylla	Triterpenes and steroids	*S. aureus*, *Staphylococcus saprophyticus*, and *Streptococcus faecalis*	*Escherichia coli* and *Proteus mirabilis*	Chattopadhyay et al., 2001
Austroplenckia populnea	Triterpenes	*S. aureus*	-	Miranda et al., 2009
Aquilaria agallocha	Triterpenes, alkaloids, anthraquinones, and tannins	*Bacillus brevis* and *B. subtilis*	*Pseudomonas aeruginosa* and *Shigella flexneri*	Dash et al., 2008
Azadirachta indica	Triterpenes, glycosides, and fatty acids	*Micrococcus luteus* and *S. aureus*	*P. aeruginosa* and *Proteus vulgaris*	Khan et al., 2010
Azima tetracantha	Triterpenes, steroids, and tannins	*S. aureus* and *B. subtilis*	*E. coli*, *Klebsiella pneumoniae*, and *P. aeruginosa*	Ekbote et al., 2010
Calophyllum inophyllum	Triterpenes	*S. aureus*	-	Yimdjo et al., 2004
Cardiospermum helicacabum	Triterpenes, steroids, sugars, alkaloids, phenols, saponins, aminoacids, and tannins	*B. subtilis*	*P. aeruginosa* and *Salmonella typhi*	Viji et al., 2010
Cedrus deodara	Triterpenes, alkaloids, steroids, flavonoids, tannins, phenolic compounds, and	*Bacillus cereus*, *E. faecalis*, and *S. aureus*	*E. coli*, *K. pneumoniae*, and *P. aeruginosa*	Devmurari, 2010

Table 1. Vegetal species that exhibit bactericidal activity and contain triterpenes

Species	Isolated compound	Activity against Gram-positive bacteria	Activity against Gram-negative bacteria	Ref.
Commiphora glandulosa	Triterpenes	*B. subtilis, Clostridium perfringens*, and *S. aureus*	-	Motlhanka et al., 2010
Dendrophthoe falcata	Triterpenes, steroids, tannins, and glycosides	*B. cereus, B. subtilis, M. luteus, S. aureus, Staphylococcus epidermidis*, and *Streptococcus pneumoniae,*	*Enterobacter aerogenes, E. coli, K. pneumoniae, P. aeruginosa, Serratia marcescens*, and *S. typhi*	Pattanayak et al., 2008
Dichrostachys cinerea	Triterpenes and steroids	*B. subtilis* and *S. aureus*	*E. coli* and *P. aeruginosa*	Eisa et al., 2000
Drynaria quercifolia	Triterpenes, coumarins, flavones, lignans, saponins, and steroids	*B. subtilis* and *S. aureus*	*E. coli, K. pneumoniae, P. aeruginosa*, and *S. typhi*	Ramesh et al., 2001
Elaeodendron schlechteranum	Triterpenes	*B. cereus, B. subtilis*, and *S. aureus*	-	Maregesi et al., 2010
Ficus ovata	Triterpenes	*B. cereus, S. aureus*, and *S. faecalis*	*Citrobacter freundii, E. coli, K. pneumoniae, P. aeruginosa*, and *S. typhi*	Kuete et al., 2009
Finlaysonia obovata	Triterpenes	*S. aureus*	*E. coli* and *P. aeruginosa*	Mishra & Sree, 2007
Galium mexicanum	Triterpenes, saponins, flavonoids, sesquiterpene lactones, and glucosides	*S. aureus methicillin-resistant* (MRSA)	-	Bolivar et al., 2011
Garcinia gummicutta	Triterpenes, alkaloids, steroids, oils, catechins, and phenolics	*B. subtilis* and *S. aureus*	*Aeromonas hydrophila, K. pneumoniae, P. aeruginosa*, and *S. typhi*	Maridass et al., 2010
Leucas aspera	Triterpenes	*S. pneumoniae*	*E. coli*	Mangathayaru et al., 2005
Miconia ligustroides	Triterpenes	*B. cereus*	-	Cunha et al., 2010
Mirabilis jalapa	Terpenes and flavonoids	*B. cereus, E. faecalis*, and *M. luteus*	*E. coli, K. pneumoniae*, and *P. aeruginosa*	Hajji et al., 2010
Moringa oleifera	Triterpenes, alkaloids, flavonoids, sesquiterpenes, lactones, diterpenes, and naphtoquinones	*E. faecalis* and *S. aureus*	*Aeromonas caviae* and *Vibrio arahaemolyticus*	Peixoto et al., 2011
Mussaenda macrophylla	Triterpenes	-	*Porphyromonas gengivalis*	Kim et al., 1999
Phyllanthus simplex	Triterpenes, steroids, lignans, flavonoids, glycosides, and phenolic compounds	*S. aureus*	*E. coli, P. aeruginosa*, and *S. flexneri*	Chouhan & Singh, 2010
Psidium guajava	Triterpenes, tannins, and flavonoids	*B. subtilis* and *S. aureus*	*E. coli* and *P. aeruginosa*	Sanches et al., 2005
Pulicaria dysenterica	Triterpenes and steroids	*B. cereus* and *S. aureus*	*Vibrio cholera*	Nickavar & Mojab, 2003
Tridesmostemon omphalocarpoides	Triterpenes	*S. aureus* and *S. faecalis*	*E. coli, K. pneumoniae, P. vulgaris, Shigella dysenteriae*, and *S. typhi*	Kuete et al., 2006
Triumfetta rhomboidea	Triterpenes, Steroids, flavonoids, tannin, and phenolic compounds	*B. cereus, E. faecalis*, and *S. aureus*	*E. coli, K. pneumoniae*, and *P. aeruginosa*	Devmurari et al., 2010
Vochysia divergens	Triterpenes	*S. aureus*	-	Hess et al., 1995

Table 1. Vegetal species that exhibit bactericidal activity and contain triterpenes (contd.)

Some plants exhibit a broad spectrum of activity against both Gram-positive and Gram-negative bacteria and contain other chemical classes, such as coumarins, flavonoids, phenolic compounds, and alkaloids. However, there is an expressive quantity of vegetal species that only triterpenes were isolated, suggesting an intrinsic relationship between this chemical class and the bactericidal activity of these plants. Thus, the present work provides an extensive search in original and review articles addressing the bactericidal activity of triterpenes, which may inspire new biomedical applications, considering atom economy, the synthesis of environmentally benign products without producing toxic by-products, the use of renewable sources of raw materials, and the search for processes with maximal efficiency of energy. To systematization of the results, it was considered that the biological activities are related to the presence of functionalized sites on the chemical structure of each triterpene. Obviously the obtained data do not make them possible the comparison of the intensity of bactericidal activities among the active triterpenes. Moreover, many triterpenes were tested against few species of bacteria, and as a consequence this work only records biological positive test.

Table 2 shows the bactericidal activity of oleanane-type triterpenes isolated from vegetal species and fungi (Compounds **1** to **43** shown in Figure 1). In the case of Gram-positive bacteria, oleananes with different functionalizations exhibit activity against *S. aureus* and a relationship between chemical structure and bactericidal activity could not established. The oleananes **6**, **20**, **21**, **35**, and **36** exhibit activity against *E. faecalis*. All these compounds present functional groups on the alpha side of the triterpene skeleton (hydroxyl group at C-1 and oxygenated group at C-20 or C-16). Compounds **1** to **5**, and **42** exhibit activity against *M. luteus* and present carboxyl group at C-17 or C-20 and oxygenated group at C-3. The presence of a functional group at C-17 is an important criterion to the activity against *B. subtilis*, except compounds **29** and **43**, which are carboxyl group funcionalized at other positions (i.e. C-3 and C-20, respectively). The activity against *S. mutans* is exhibited by the compounds **14**, **15**, **17**, **18**, and **24**, which present oxygenated group at C-3 and carboxyl group at C-17. Few oleanane-type triterpenes were tested against *S. pneumoniae* and *B. pumilus*, and as a consequence, relationships between chemical structure and activity against these Gram-positive bacteria were not possible.

Considering the Gram-negative bacteria, Table 2 shows many oleananes active against *E. coli*. These compounds present different functional groups at the oleanane skeleton, but all them present oxygenated group at C-3. Compounds **13-16**, **19**, **26**, **28-38**, and **43** exhibit activities against *S. typhi* and only present oxygenated group at C-3 in common. The activity against *S. sonnei* is registered for the compounds **7**, **8**, **10**, and **13**, which present carboxyl group at C-17 and oxygenated group at C-3. Similarly, the activity against *P. gingivalis* is registered for the compounds **14**, **15**, **18**, **24**, and **25**, which present carboxyl group at C-17 and oxygenated group at C-3. Only two compounds exhibited activity against *P. fluorencens* (**11** and **12**) and both the oleananes present hydroxyl group at C-19 on the alpha-side of the skeleton. Few oleananes were tested against *V. cholera*, *S. dysenteriae*, *S. flexneri*, *S. boydii*, *P. aeruginosa*, and *C. pneumoniae* and relationships between chemical structure and activity against these Gram-negative bacteria were not possible.

Figure 2 shows the ursane-type triterpenes with bactericidal activity isolated from vegetal species. For the Gram-positive bacteria, the ursanes active against *S. aureus* and *B. subtilis* present oxygenated group at C-3 in common. Few compounds exhibited positive tests against *S. epidermidis*, *A. viscosus*, *M. luteus*, *S. mutans*, *C. perfrigens*, *S. faecalis*, and *B. cereus*.

In the case of Gram-negative bacteria, the ursanes active against *E. coli* present an oxygenated group at C-3 in common. The ursanes active against *S. sonnei, S. flexneri, B. typhi, K. pneumonae,* and *P. aeroginosa* concomitantly present oxygenated groups at C-3 and C-17.

Figure 3 shows the lupane-, friedelane-, and fernane-type triterpenes with bactericidal activity isolated from vegetal species. Friedelin (compound **68**) exhibits the largest spectrum of activities against Gram-positive bacteria (*Bacillus megaterium, Bacillus stearothermophilus, S aureus,* and *S. faecalis*) and Gram-negative bacteria (*C. freundi, E. aerogenes, Enterococcus cloacae, K. pneumoniae, Morganella morganii, P. aeruginosa, P. mirabilis, P. vulgaris, S. dysenterie, S. flexneri,* and *S. typhi, Salmonella typhimurium*). This compound only presents functionalization at C-3 (carbonyl group at position C-3 on the triterpene skeleton). As a consequence, the position C-3 could be considered as a strategic position to bactericidal activity of all triterpenes above-mentioned. However, the fernanes **82-84** do not present functional groups at C-3, but exhibit activity against *M. tuberculosis.*

The compounds shown in the Figures 4 and 5 are miscellaneous-types of triterpenes isolated from vegetal species or obtained from hemi-synthesis which exhibit bactericidal activity. The variety of their chemical structures does not permit to establish relationships with the bactericidal activities showed in the Tables 2 and 3. However, among the triterpenes shown in the Figures 1 to 5 and Tables 2 and 3, 90% of them exhibit activity against Gram-positive bacteria and 60% of them exhibit activity against Gram-negative bacteria. These results indicate higher resistance of Gram-negative Bacteria to the triterpenes.

Compound	Vegetal species	Activity against Gram-positive bacteria	Activity against Gram-negative bacteria	Ref.
2α-Hydroxy-3-oxoolean-12-en-30-oic acid (**1**)	*Dillenia papuana*	*B. subtilis* and *M. luteus*	*E. coli*	Nick et al., 1994
Olean-1,12-dien-29-oic acid, 3-oxo (**2**)	*Dillenia papuana*	*B. subtilis* and *M. luteus*	*E. coli*	Nick et al., 1994
lα-Hydroxy-3-oxoolean-12-en-30-oic acid (**3**)	*Dillenia papuana*	*B. subtilis* and *M. luteus*	*E. coli*	Nick et al., 1994
2-Oxo-3β-hydroxyolean-12-en-30-oic acid (**4**)	*Dillenia papuana*	*B. subtilis* and *M. luteus*	*E. coli*	Nick et al., 1994
Olean-12-en-1,3-dihydroxy (**5**)	*Dillenia papuana*	*B. subtilis* and *M. luteus*	*E. coli*	Nick et al., 1994
3,30-Dihydroxyl-12-oleanen-22-one (**6**)	*Cambretum imberbe*	*E. faecalis* and *S. aureus*	*E. coli*	Angeh et al., 2007; Katerere et al., 2003
Arjulonic acid (**7**)	*Syzygium guineense*	*B. subtilis*	*E. coli* and *Shigella sonnei*	Djoukeng et al., 2005
Terminolic acid (**8**)	*Syzygium guineense*	*B. subtilis*	*E. coli* and *S. sonnei*	Djoukeng et al., 2005
2α,3β,24-Trihydroxyolean-12-en-28-oic acid (**9**)	*Planchonia careya*	MRSA	*Enterococcus vancomicin-resistant* (VRE)	McRae et al., 2008
2,3,23-Trihydroxy-(2α,3β,4α) olean-11-en-28 oic acid (**10**)	*Syzygium guineense*	*B. subtilis*	*E. coli* and *S. sonnei*	Djoukeng et al., 2005

Table 2. Bactericidal activity of triterpenes isolated from vegetal species and fungi

Compound	Vegetal species	Activity against Gram-positive bacteria	Activity against Gram-negative bacteria	Ref.
Arjungenin (**11**)	*Planchonia careya*		*Pseudomonas fluorencens*	McRae et al., 2008
Arjunic acid (**12**)	*Terminalia arjuna*		*P. fluorencens*	Sun et al., 2008
3-Acetyl aleuritolic acid (**13**)	*Spirostacheps africana*	*S. aureus*	*E. coli, Shigella boydii, S. dysenteriae, S. flexneri, S. sonnei, S. typhi,* and *V. cholera*	Mathabe et al., 2008
Oleanolic acid (**14**)	*Periplaca laevigata*	*Spretococcus mutans* and *S. aureus*	*E. coli, P. gingivalis,* and *S. typhi*	Hichri et al., 2003
Oleanolic acid acetate (**15**)	*Periplaca laevigata*	*S. mutans* and *S. aureus*	*E. coli, P. aeruginosa, P. gingivalis,* and *S .typhi*	Hichri et al., 2003
Maslinic acid acetate (**16**)	*Periplaca laevigata*	*S. aureus*	*E. coli, P. aeruginosa,* and *S. typhi*	Hichri et al., 2003
Methyl 3-acetyloleanolic acid (**17**)	*Vitis vinifera*	*S. mutans*		Rivero-Cruz et al., 2008
Methyl oleanolic acid (**18**)	*Vitis vinifera*	*S. mutans*	*P.gingivalis*	Rivero-Cruz et al., 2008
Oleanolic acid 28-*O*-[*β*-*D*-glucopyranosyl] Ester (**19**)	*Drypetes paxii*	*S. aureus*	*E. coli* and *S. typhi*	Chiozem et al., 2009
1*α*,3*β*-Dihydroxyolean-12-en-29-oic acid (**20**)	*Cambretum imberbe*	*E. faecalis* and *S. aureus*	*E. coli*	Angeh et al., 2007; Katerere et al., 2003
1*α*,3*β*-Hydroxyimberbic-acid-23-*O*-*β*-*L*-4-acetylrhamnopyranoside (**21**)	*Cambretum imberbe*	*E. faecalis* and *S. aureus*	-	Angeh et al., 2007; Katerere et al., 2003
1,3,24-Trihydroxyl-12-olean-29-oic acid (**22**)	*Cambretum imberbe*	*S. aureus*	*E. coli*	Angeh et al., 2007; Katerere et al., 2003
1*α*,23-Dihydroxy-12-oleanen-29-oic acid-3*β*-*O*-2,4-diacetyl-*L*-rhamnopyranoside (**23**)	*Cambretum imberbe*	*S. aureus*	*E. coli*	Angeh et al., 2007; Katerere et al., 2003
3-*O*-(30,30-dimethylsuccinyl)-oleanolic acid (**24**)	*Vitis vinifera*	*S. mutans*	*P. gingivalis*	Rivero-Cruz et al., 2008
3-*O*-(20,20-dimethylsuccinyl)oleanolic acids (**25**)	*Vitis vinifera*	-	*P. gingivalis*	Rivero-Cruz et al., 2008
3*β*,6*R*,13*β*-Trihydroxyolean-7-one (**26**)	*Camellia sinensis*	*S. aureus*	*E. coli, S. dysenteriae,* and *S. typhi*	Ling et al., 2010

Table 2. Bactericidal activity of triterpenes isolated from vegetal species and fungi (contd.)

Compound	Vegetal species	Activity against Gram-positive bacteria	Activity against Gram-negative bacteria	Ref.
18*α*-Oleanane-3*β*-ol,19*β*,28-epoxy (**27**)		-	*Chlamydia pneumoniae*	Dehaen et al., 2011
9*β*,25-cyclo-3*β*-*O*-(*β*-*D*-glucopyranosyl)-echynocystic acid (**28**)	*Syniplocos panicrelata*	*B. subtilis* and *S. aureus*	*E. coli* and *P. aeruginosa*	Semwal et al., 2011
3-Oxoolean-l,12-dien-30-oic acid (**29**)	*Dellenia papuana*	*B. subtilis*	*E. coli*	Nick et al., 1994
3*R*-Hydroxyolean-12-en-27-oic acid (**30**)	*Aceriphyllum rossii*	MRSA, quinolone resistance *S. aureus* (QRSA), and *S. aureus*	-	Zheng et al., 2008
3*β*-Hydroxyolean-12-en-27-oic acid (**31**)	*Aceriphyllum rossii*	MRSA, QRSA, and *S. aureus*	-	Zheng et al., 2008
Aceriphyllic acid A (**32**)	*Aceriphyllum rossii*	MRSA, QRSA, and *S. aureus*	-	Zheng et al., 2008
Methyl ester of aceriphyllic acid A (**33**)	*Aceriphyllum rossii*	MRSA, QRSA, and *S. aureus*	-	Zheng et al., 2008
22*α*-Acetyl-16*α*,21*β*-dihydroxyoleanane-13*β*:28-olide-3-*O*-[*β*-glucopyranosyl-(1'''→6')][6''-*O*-coumaroylglucopyranosyl-(1''→2')]-*β*-glucopyranoside (**34**)	*Maesa lanceolata*	*S. aureus*	-	Manguro et al., 2011
16*α*,22*α*-Diacetyl-21*β*-angeloyloleanane-13*β*:28-olide-3*β*-*O*-[*β*-glucopyranosyl-(1''→2')][*β*-glucopyranosyl-(1'''→4')]-*β*-glucopyranoside (**35**)	*Maesa lanceolata*	*B. subtilis, E. faecalis, S. aureus,* and *S. pneumoniae*	*E. coli, P. aeruginosa,* and *V. cholera*	Manguro et al., 2011
16*α*,22*α*,28-Trihydroxy-21*β*-angeloylolean-12-ene-3*β*-*O*-[*α*-rhamnopyranosyl-(1'''→6'')][*β*-glucopyranosyl-(1''→2')]-*β*-xylopyranoside(**36**)	*Maesa lanceolata*	*E. faecalis* and *S. pneumoniae*	*S. typhi* and *V. cholera*	Manguro et al., 2011
16*α*,28-dihydroxy-22*α*-acetyl-21*β*-angeloylolean-12-ene-3-*O*-[*β*-galactopyranosyl-(1''→2')][*α*-rhamnopyranosyl-(1'''→4')]-*α*-arabinopyranoside (**37**)	*Maesa lanceolata*	*B. subtilis*	*S. typhi* and *V. cholera*	Manguro et al., 2011
Chikusetsusaponin IVa methyl Ester (**38**)	*Drypetes laciniata*	-	*E. coli* and *S. typhi*	Fannang et al., 2011
3*β*-[(*α*-*L*-Arabinopyranosyl)-oxy]olean-12-en-28-oic acid (**39**)	*Clematis ganpiniana*	*B. subtilis*	-	Ding et al., 2009
Hederagenin-3*β*-*O*-*α*-*L*-arabinopyranoside (**40**)	*Clematis ganpiniana*	*Bacillus pumilus* and *B. subtilis*	-	Ding et al., 2009
3*β*-*O*-*α*-*L*-Rhamnopyranosyl-(1→2)-*α*-*L*-arabinopyranosyl oleanolic acid (**41**)	*Clematis ganpiniana*	*B. pumilus* and *B. subtilis*	*E. coli*	Ding et al., 2009
α-Hederin (**42**)	*Clematis ganpiniana*	*B. pumilus, B. subtilis, M. luteus,* and *S. aureus*	*E. coli* and *S. dysenteriae*	Ding et al., 2009
5,6(11)-Oleanadien-3*β*-ethan-3-oate (**43**)	*Rhododendron campanulatum*	*B. subtilis* and *S. aureus*	*E. coli, K. pneumoniae,* and *S. typhi*	Tantry et al., 2011
Asiatic acid (**44**)	*Syzygium guineense*	*B. subtillis*	*E. coli* and *S. sonnei*	Djoukeng et al., 2005
Hydroxyasiatic acid (**45**)	*Syzygium guineense*	*B. subtilis*	*E. coli* and *S. sonnei*	Djoukeng et al., 2005

Table 2. Bactericidal activity of triterpenes isolated from vegetal species and fungi (contd.)

Compound	Vegetal species	Activity against Gram-positive bacteria	Activity against Gram-negative bacteria	Ref.
Eleganene-A (**46**)	*Myricana elegans*	*B. subtilis* and *S. aureus*	*S. flexneri* and *S. typhi*	Ahmad et al., 2008
Eleganene-B (**47**)	*Myricana elegans*	*B. subtilis* and *S. aureus*	*E. coli, P. aeruginosa, S. flexneri*, and *S. typhi*	Ahmad et al., 2008
($2\alpha,3\beta$)-2,3,23-Trihydroxy-13,28-epoxyurs-11-en-28-one (**48**)	*Eucalyptus camaldulensis*	*S. aureus* and *S. epidermidis*	*E. coli, K. pneumonae*, and *P. aeruginosa*	Tsiri et al., 2008
Ilexgenin A (**49**)	*Ilex hainanensis*	*Actinomyces viscosus* and *S. mutans*	-	Chen et al., 2011
Rotundic acid (**50**)	*Ilex integra*	*B. subtilis, M. luteus*, and *S. aureus*	*P. aeruginosa*	Haraguchi et al., 1999
Ursolic acid (**51**)	*Geum rivale*	*S. aureus*	*E. coli* and *P. aeruginosa*	Panizzi et al., 2000
$1\beta,2\beta,3\beta$–Trihydroxy-urs-12-ene-23-oic-rhamnoside (**52**)	*Commiphora glandulosa*	*B. subtilis, C. perfringens*, and *S. aureus*	-	Montlhanka et al., 2010
Erythrodiol (**53**)	*Myricana elegans*	*B. subtilis*	*P. aeruginosa* and *S. flexneri*	Ahmad et al., 2008
Corosolic acid (**54**)	*Myricana elegans*	*B. subtilis*	*P. aeruginosa* and *S. flexneri*	Ahmad et al., 2008
$1\beta,3\beta$-Dihydroxyurs-12-en-27-oic acid (**55**)	*Carophora coronata*	*B. subtilis* and MRSA	-	Khera et al., 2003
22β-Acetyl lantoic acid (**56**)	*Lantana camara*	*S. aureus*	*E. coli, P. aeruginosa*, and *S. typhi*	Barre et al., 1997
Lantic acid (**57**)	*Lantana camara*	*B. cereus, B. subtilis, M. luteus, S. aureus*, and *S. faecalis*	*E. coli*	Saleh et al., 1999
22β-Acetoxylantic acid (**58**)	*Lantana Camara*	*S. aureus*	*E. coli, P. aeruginosa*, and *S. typhi*	Barre et al., 1997
Taraxast-20-ene-3β-ol (**59**)	*Saussurea petrovii*	*B. subtilis* and *S. aureus*	*E. coli*	Daí et al., 2001
Taraxast-20(30)ene-$3\beta,21\alpha$-diol (**60**)	*Saussurea petrovii*	*B. subtilis* and *S. aureus*	*E. coli*	Daí et al., 2001
$20\alpha,21\alpha$-Epoxy-taraxastane-$3\beta,22\alpha$-diol (**61**)	*Saussurea petrovii*	*B. subtilis* and *S. aureus*	*E. coli*	Daí et al., 2001
Taraxast-20-ene-3β-ol (**62**)	*Saussurea petrovii*	*B. subtilis* and *S. aureus*	*E. coli*	Daí et al., 2001
Taraxast-20-ene-3 β,30-diol (**63**)	*Saussurea petrovii*	*B. subtilis* and *S. aureus*	*E. coli*	Daí et al., 2001
20(29)-Lupene-3β-isoferulate (**64**)	*Euclea natalensis*	*B. pumilus*	-	Weigenand et al., 2004
Lupeol (**65**)	*Curtisia dentata*	*B. subtilis* and *S. aureus*	*E. coli* and *P. aeruginosa*	Shai et al., 2008
Betulinic acid (**66**)	*Curtisia dentata*	*B. subtilis* and *S. aureus*	*E. coli* and *P. aeruginosa*	Shai et al., 2008

Table 2. Bactericidal activity of triterpenes isolated from vegetal species and fungi (contd.)

Compound	Vegetal species	Activity against Gram-positive bacteria	Activity against Gram-negative bacteria	Ref.
Betulin (**67**)	*Myricana elegans*	-	*C. pneumoniae*	Dehaen et al., 2011; Ahmad et al., 2008
Friedelin (**68**)	*Visnia rubescens*	*Bacillus megaterium*, *Bacillus stearothermophilus*, *S aureus*, and *S. faecalis*	*C. freundi*, *E. aerogenes*, *Enterococcus cloacae*, *K. pneumoniae*, *Morganella morganii*, *P. aeruginosa*, *P. mirabilis*, *P. vulgaris*, *S. dysenterie*, *S. flexneri*, and *S. typhi*, *Salmonella typhimurium*	Tamokou et al., 2009; Kuete et al., 2009, 2007, 2006
3-Oxo-friedelan-20α-oic acid (**69**)	*Maytenus sinegalensis*	*B. subtilis* and *S. aureus*	*E. coli*, *K. pneumoniae*, and *S. flexneri*	Lindsey et al., 2003; Lindsey et al., 2006
3β-Hydroxyfriedelane-7,12,22-trione (**70**)	*Drypetes laciniata*	-	*E. coli*, *P. aeruginosa*, and *S. typhi*	Fannang et al., 2011
12α-Hydroxyfriedelane-3,15-dione (**71**)	*Drypetes paxii*	*S. aureus*		Chiozem et al., 2009
Friedelanol (**72**)	*Visnia rubescens*	*S. aureus*	*P. aeruginosa* and *S. typhi*	Angeh et al., 2007; Katerere et al., 2003
3β-Hydroxyfriedelan-25-al (**73**)	*Drypetes paxii*	*S. aureus*	-	Chiozem et al., 2009
3-Hydroxy-2,24-dioxo-3-friedelen-29-oic acid (**74**)	*Elaeodendron schlechteranum*	*B. cereus* and *S. aureus*	-	Maregesi et al., 2010
22β-Hydroxytingenone (**75**)	*Elaeodendron schlechteranum*	*B. cereus* and *S. aureus*	-	Maregesi et al., 2010
2,3,7-Trihydroxy-6-oxo-1,3,5(10),7-tetraene-24-nor-friedelane-29-oic acid methyl ester (**76**)	*Crossopetalum gaumeri*	*B. cereus*, *M. luteus*, and *S. epidermidis*	-	Ankli et al., 2000
Zeylasterone (**77**)	*Maytenus blepharodes*	*S. aureus*	-	Léon et al., 2010
Dimethylzeylasterone (**78**)	*Maytenus blepharodes*	*S. aureus*	-	Léon et al., 2010
Zeylasteral (**79**)	*Maytenus blepharodes*	*S. aureus*	-	Léon et al., 2010
Dimethylzeylasteral (**80**)	*Maytenus blepharodes*	*S. aureus*	-	Léon et al., 2010
30-Ethyl-2α,16α-dihydroxy-3β-*O*-(β-*D*-glucopyranosyl)-hopan-24-oic acid (**81**)	*Syniplocos panicrelata*	*B. subtilis* and *S. aureus*	*E. coli* and *P. aeruginosa*	Semwal et al., 2011
Hopan-27-al-6β,11R,22-triol (**82**)	*Conoideocrella tenuis* (*fungus*)	-	*Mycobacterium tuberculosis*	Isaka et al., 2011
A'-Neogammacerane-6,11,22,27-tetrol (**83**)	*Conoideocrella tenuis* (*fungus*)	-	*M. tuberculosis*	Isaka et al., 2011

Table 2. Bactericidal activity of triterpenes isolated from vegetal species and fungi (contd.)

Compound	Vegetal species	Activity against Gram-positive bacteria	Activity against Gram-negative bacteria	Ref.
Hopane-6β,7β,22-triol (**84**)	*Conoideocrella tenuis (fungus)*	-	*M. tuberculosis*	Isaka et al., 2011
Dysoxyhainic acid G (**85**)	*Dysoxylum hainanense*	*B. subtilis, M. luteus,* and *S. epidermidis*	-	He et al., 2011
20-Epikoetjapic acid (**86**)	*Osyris lanceolata*	*B. subtilis* and *S. aureus*	*E. coli* and *P. aeruginosa*	Yeboah et al., 2010
Dysoxyhainic acid J (**87**)	*Dysoxylum hainanense*	*B. subtilis* and *S. epidermidis*	-	He et al., 2011
(9,11),(18,19)-Disecoolean-12-en-28-oic acid (**88**)	*Ficus benjamina*	*B. subtilis* and *S. aureus*	*E. coli* and *S.typhimurium*	Parveen et al., 2009
2-Chrysene acetic acid, 9-carboxy-1,2,3,4,4a,4b,5,6,6a,7,8,9,10,10a,12,12a-hexadecahydro-α,α,1,4a,4b,6a,9-heptamethyl-1-(2-oxoethyl),2-methyl ester (**89**)	*Dillenia papuana*	*B. subtilis* and *M. luteus*	*E. coli*	Nick et al., 1994
Polyporenic acid C (**90**)	*Fomitopsis rosea (fungus)*	*S. aureus*	-	Popova et al., 2009
Dysoxyhainic acid I (**91**)	*Dysoxylum hainanense*	*B. subtilis* and *S. epidermidis*	-	He et al., 2011
3α-Hydroxy-24-methylene-23-oxolanost-8-en-26-carboxylic acid (**92**)	*Fomitopsis rosea (fungus)*	*S. aureus*	-	Popova et al., 2009
3α-Carboxyacetoxyquercinic acid (**93**)	*Fomitopsis rosea (fungus)*	*S. aureus*	-	Popova et al., 2009
3α-Oxepanoquercinic acid C (**94**)	*Fomitopsis rosea (fungus)*	*S. aureus*	-	Popova et al., 2009
Lamesticumin F (**95**)	*Lansium domesticum*	*B. cereus* and *B. subtilis*	-	Dong et al., 2011
3α-(3'Butylcarboxyacetoxy)oxepanoquercinic acid C (**96**)	*Fomitopsis rosea (fungus)*	*S. aureus*	-	Popova et al., 2009
Helvolic acid (**97**)	*Pichia guilliermondii (fungus)*	*B. subtilis, S. aureus,* and *Staphylococcus haemolyticus*	*Agrobacterium tumifaciens, E. coli, Pseudomonas lachrymans, Ralstonia solanacearum,* and *Xanthomonas vesicatoria*	Zhao et al., 2010
5α,8α-Epidioxi-24(ξ)-methylcholesta-6,22-diene-3β-ol (**98**)	*Fomitopsis rosea (fungus)*	*S. aureus*		Popova et al., 2009
1,3,16β-yl-Phenypropylacetate-lanostan-5,11,14,16,23,25-hexen-22-one (**99**)	*Stachyterphita jamaicensis*	*S. aureus* and *S. faecalis*	*E. coli* and *P. aeruginosa* -	Maregesi et al., 2010
Dysoxyhainic acid H (**100**)	*Dysoxylum hainanense*	*B. subtilis* and *M. luteus*	-	He et al., 2011
3β-*O*-*cis*-*p*-Coumaroyltormentic acid (**101**)	*Planchonia careya*	*S. aureus*	VRE	McRae et al., 2008
3β-*O*-*trans*-*p*-Coumaroyltormentic acid (**102**)	*Planchonia careya*	*S. aureus*	VRE	McRae et al., 2008

Table 2. Bactericidal activity of triterpenes isolated from vegetal species and fungi (contd.)

Compound	Vegetal species	Activity against Gram-positive bacteria	Activity against Gram-negative bacteria	Ref.
Lamesticumin C (**103**)	*Lansium domesticum*	*B. cereus, B. subtilis, M. luteus, S. epidermidis, S. aureus,* and *Streptococcus pyogenes*	-	Dong et al., 2011
Lamesticumin D (**104**)	*Lansium domesticum*	*B. cereus* and *B. subtilis*	-	Dong et al., 2011
Lamesticumin B (**105**)	*Lansium domesticum*	*B. cereus, B. subtilis, M. luteus, S. aureus, S. epidermidis,* and *S. pyogenes*	-	Dong et al., 2011
Lamesticumin E (**106**)	*Lansium domesticum*	*B. cereus* and *B. subtilis*	-	Dong et al., 2011
Lansic acid 3-ethyl Ester (**107**)	*Lansium domesticum*	*B. cereus, B. subtilis, M. luteus, S. aureus, S. epidermidis,* and *S. pyogenes*	-	Dong et al., 2011
Ethyl lansiolate (**108**)	*Lansium domesticum*	*B. cereus, B. subtilis, M. luteus, S. aureus, S. epidermidis,* and *S. pyogenes*	-	Dong et al., 2011
Lamesticumin A (**109**)	*Lansium domesticum*	*B. cereus, B. subtilis, M. luteus, S. aureus, S. epidermidis,* and *S. pyogenes*	-	Dong et al., 2011
3-Cyclohexene-1-propanoic acid,2-[2-[(1*S*,2*R*,3*R*)-2-(3-ethoxy-3-oxopropyl)-3-(1-hydroxy-1-methylethyl)-2-methyl-6-methylenecyclohexyl]ethyl]-1,3-dimethyl-6-(1-methylethenyl) (**110**)	*Lansium domesticum*	*B. cereus, B. subtilis, M. luteus, S. aureus, S. epidermidis,* and *S. pyogenes*	-	Dong et al., 2011

Table 2. Bactericidal activity of triterpenes isolated from vegetal species and fungi (contd.)

Compound	Activity against Gram-positive bacteria	Activity against Gram-negative bacteria	Ref.
β-D-Galactosideo methyl oleanolate (**111**)	*S. aureus*	-	Takechi & Tanaka, 1992
β-D-Xilosideo methyl oleanolate (**112**)	*S. aureus*	-	Takechi & Tanaka, 1992
β-D-Fucosideo methyl oleanolate (**113**)	*S. aureus*	-	Takechi & Tanaka, 1992
β-L-Fucosideo methyl oleanolate (**114**)	*S. aureus*	-	Takechi & Tanaka, 1992
β-Maltosideo methyl oleanolate (**115**)	*S. aureus*	-	Takechi & Tanaka, 1992
β-Maltotriosídeo methyl oleanolate (**116**)	*S. aureus*	-	Takechi & Tanaka, 1992
Oleanolic acid acetate (**117**)	*S. aureus*	*E. coli* and *P. aeruginosa*	Hichri et al., 2003

Table 3. Bactericidal activity of triterpene derivatives

Compound	Activity against Gram-positive bacteria	Activity against Gram-negative bacteria	Ref.
3β-*O*-Acetate β-amyrin (**118**)	*S. aureus*	*E. coli* and *P. aeruginosa*	Hichri et al., 2003
2β,3β-Dihydroxy-ll-oxooleana-12,18-dien-30-oic acid (**119**)	*B. subtilis*	*Erwinia* sp.	Pitzele, 1974
2β,3α-Dihydroxy- ll -oxooleana-12,18-dien-30-oic acid (**120**)	-	*Erwinia* sp.	Pitzele, 1974
2β,3α-Dihydroxy- ll -oxo-l8β-olean-12-en-30-oic acid (**121**)	*B. subtilis*	-	Pitzele, 1974
2β,3β-Dihydroxy-ll-oxo-18β-olean-12-en-30-oic acid (**122**)	-	*Erwinia* sp.	Pitzele, 1974
2β,3β-Diacetoxy-ll-oxo-18β-olean-12-en-30-oic acid (**123**)	-	*Erwinia* sp.	Pitzele, 1974
3β-Acetyl-11-oxooleanolic acid (**124**)	*S. aureus*	*E. coli, P. aeruginosa,* and *S. typhimurium*	Hichri et al., 2003
Methyl 2β,3α-dihydroxy-18β-olean-12-en-30-oate (**125**)	-	*Erwinia* sp.	Pitzele, 1974
1α-Bromo-2,3-dioxo-18β-olean-12-en-30-oic acid (**126**)	-	*Erwinia sp.*	Pitzele, 1974
3β-*O*-Nicotinoyl-20-(4-methylpiperazin-1-yl)carbonyl-11-oxoolean-12(13)-ene (**127**)	*S. aureus*	-	Kazakovaa et al., 2010
N-3-pyridinacetyloleanolic amide (**128**)	*S. aureus*	*E. coli, P. aeruginosa,* and *S. typhimurium*	Hichri et al., 2003
3β-Hydroxyolean-12-en-28-carboxydiethylphosphonate (**129**)	*S. aureus*	*E. coli, P. aeruginosa,* and *S. typhimurium*	Hichri et al., 2003
3β-Acetoxy-12α-hydroxyoleanan-13β,28-olide (**130**)	-	*S. typhimurium*	Hichri et al., 2003
Oleanan-28-oic acid, 3β,13-dihydroxy-12-oxo-, γ-lactone, acetate (**131**)	*S. aureus*	*E. coli, P. aeruginosa,* and *S. typhimurium*	Hichri et al., 2003
β-Gentiobiosideo methyl ursolate (**132**)	*S. aureus*	-	Takechi & Tanaka, 1993
β-Maltotriosídeo methyl ursolate (**133**)	*S. aureus*	-	Takechi & Tanaka, 1993
Urs-12-ene-28-carboxy-3β–dodecanoate (**134**)	*Bacillus sphaericus, B. subtilis,* and *S. aureus*	*Pseudomonas syringae*	Mallavadhani et al., 2004
Urs-12-ene-28-carboxy-3β–tetradecanoate (**135**)	*Bacillus sphaericus, B. subtilis,* and *S. aureus*	*P. syringae*	Mallavadhani et al., 2004
Urs-12-ene-28-carboxy-3β–hexadecanoate (**136**)	*Bacillus sphaericus, B. subtilis,* and *S. aureus*	*E. coli* and *P. syringae*	Mallavadhani et al., 2004
Urs-12-ene-28-carboxy-3β–octadecanoate (**137**)	*Bacillus sphaericus, B. subtilis,* and *S. aureus*	*E. coli* and *P. syringae*	Mallavadhani et al., 2004
3-Oxo-17-(4-methylpiperazin-1-yl)carbonyloursan-12(13)-ene (**138**)	*S. aureus*	-	Kazakovaa et al., 2010
2-Furfurylidenebetulonic acid (**139**)	*S. aureus*	-	Kazakovaa et al., 2010
(4-Methylpiperazin-1-yl)amide betulonic (**140**)	*S. aureus*	-	Kloos & Zein, 1993
Betulin dioxime (**141**)	-	*C. pneumoniae*	Kloos & Zein, 1993
Umbellatin α (**142**)	*B. cereus* and *B. subtilis*	-	Gonzalez et al., 1992

Table 3. Bactericidal activity of triterpene derivatives (contd.)

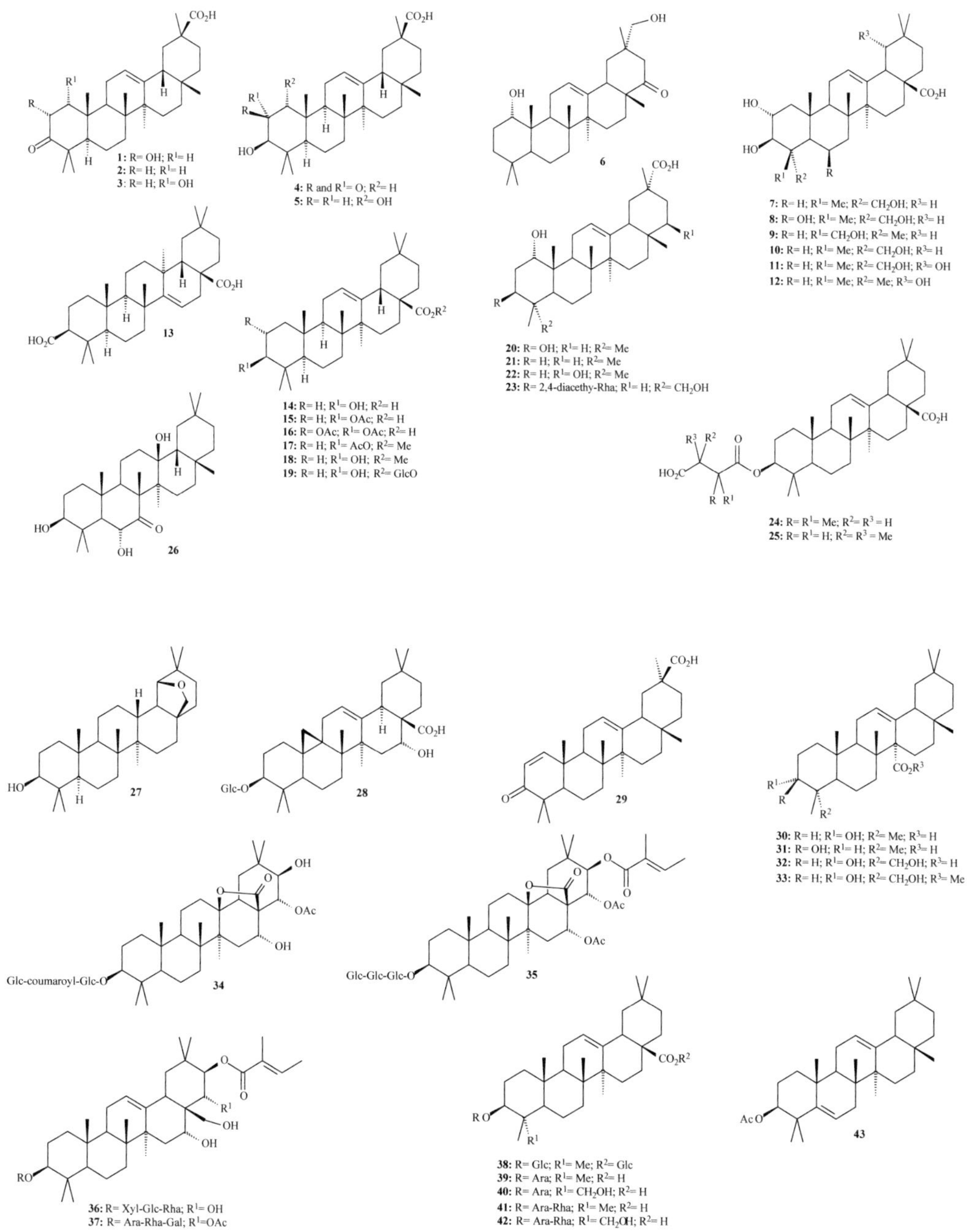

Fig. 1. Oleanane-type triterpenes with bactericidal activity isolated from vegetal species.

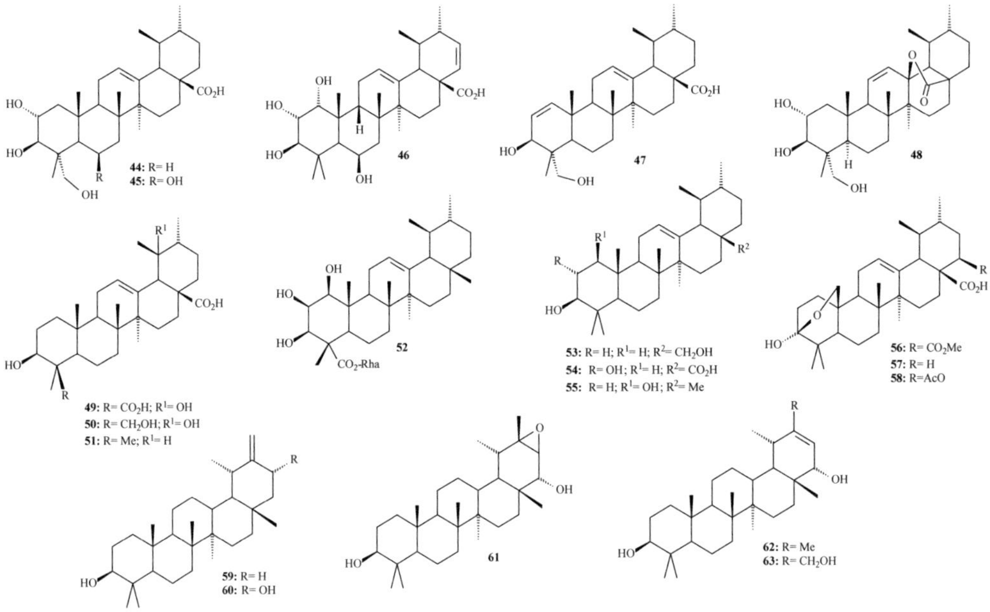

Fig. 2. Ursane-type triterpenes with bactericidal activity isolated from vegetal species.

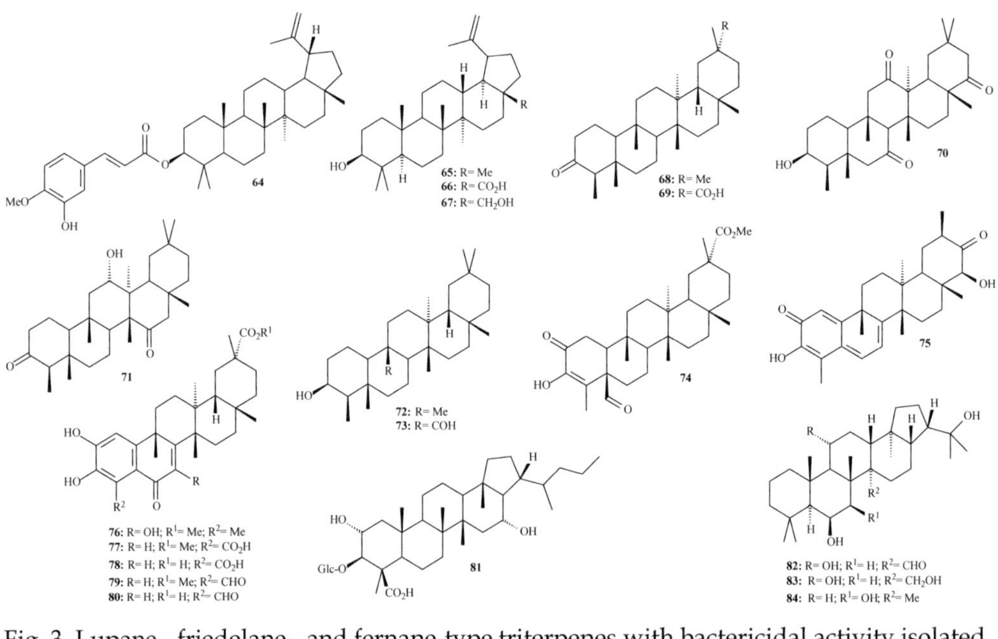

Fig. 3. Lupane-, friedelane-, and fernane-type triterpenes with bactericidal activity isolated from vegetal species.

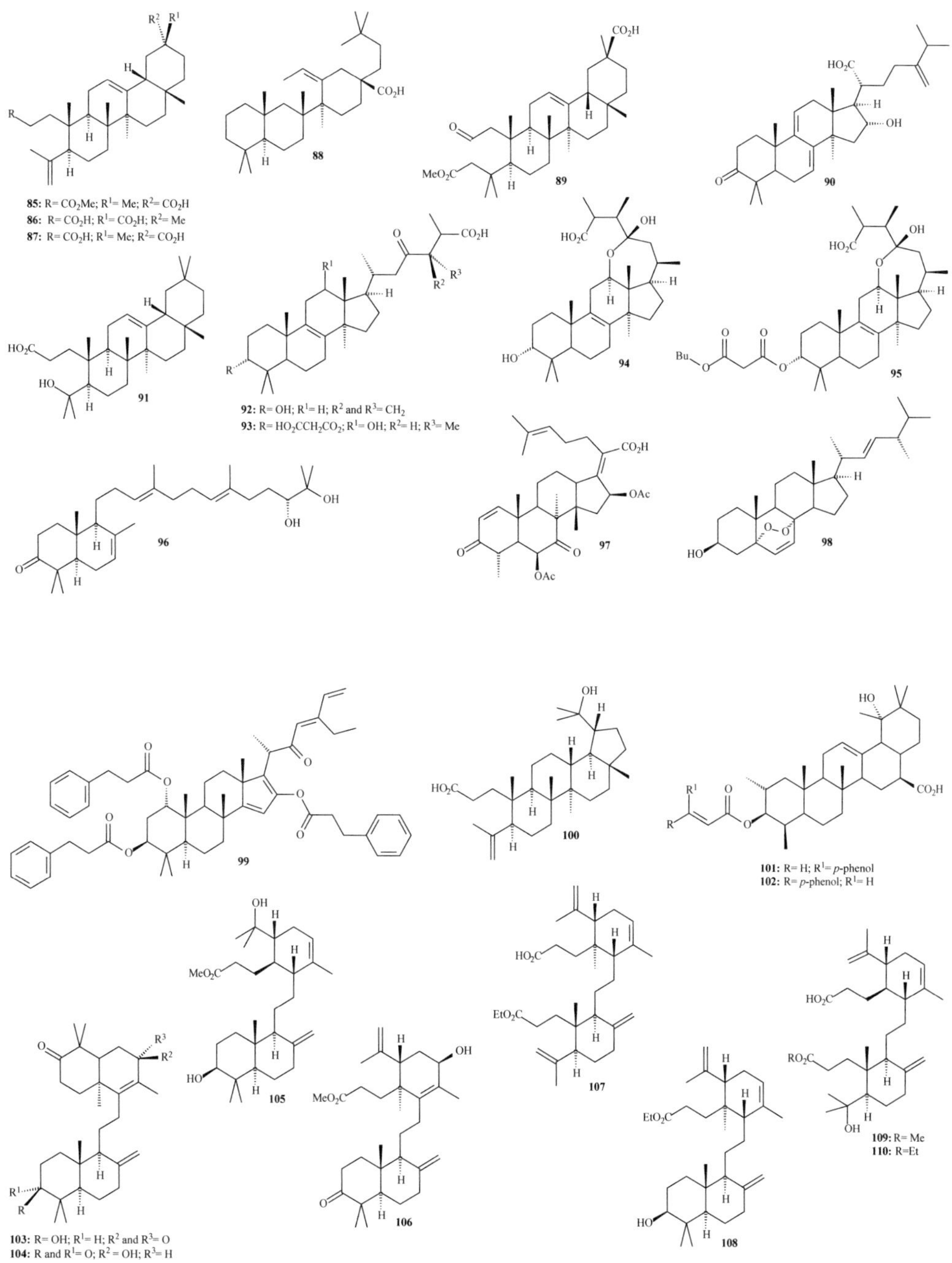

Fig. 4. Miscellaneous types of triterpenes with bactericidal activity isolated from vegetal species.

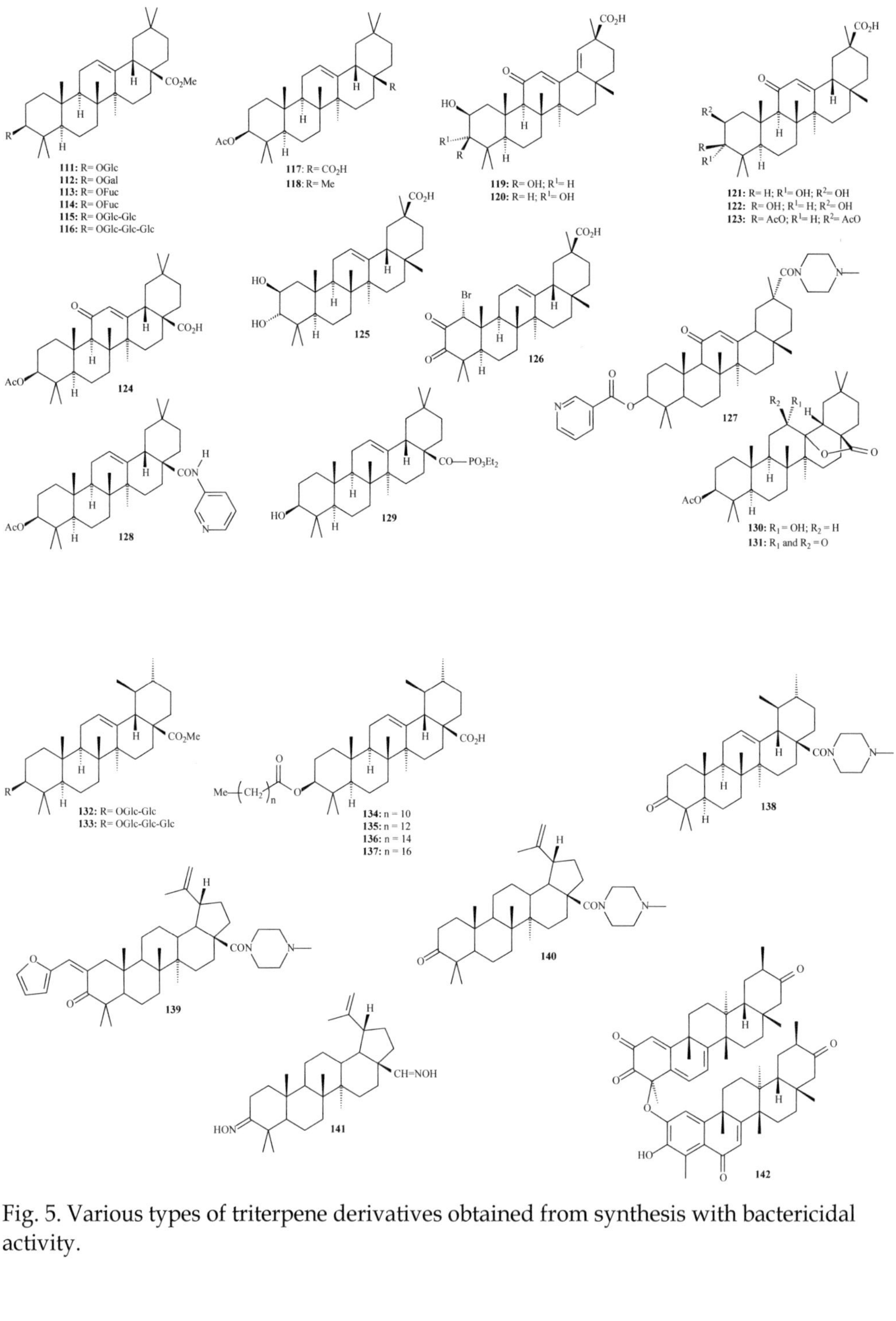

Fig. 5. Various types of triterpene derivatives obtained from synthesis with bactericidal activity.

In conclusion, the general analysis of the relationships between chemical structure and activity of triterpenes against Gram-positive and Gram-negative bacteria indicates that the antibacterial activity of the triterpene may be related to the presence of an oxygenated group at C-3, since 95% of the bactericidal triterpenes present this functionality. This site is represented by hydroxyl, carbonyl, glycosideo, esther (mainly acethyl), or hydroxylimine (compound **141**). The bactericidal activity is also influenced by the chemical structure of the substituent group. Glycoside derivatives usually exhibit higher activity, mainly for 1→6 type bonding in relation to 1→4 type one (Takechi & Tanaka, 1993). The activity is increased for the triterpenes containing free hydroxyl group at C-3, mainly on the beta-side. In fact, the activity usually decreased when the position C-3 is an ester derivative (Abreu et al., 2011). The conversion of the carboxyl group at C-17 on the beta-side to a lactone at C-13 and C-17 increases the bactericidal activity (Hichri et al., 2003).

Moreover, the bactericidal activity attributed to the C-3 site is not influenced by the steric effects, because very active compounds contain groups that present large volumes at C-3, such as compounds **23-25**, **28**, **36-42**, **64**, **81**, **99**, **101**, **102**, **111-116**, **127**, and **134-137**. A carboxyl group at C-17 on the beta side is also important — 78% and 81% of the triterpenes active against Gram-positive and Gram-negative bacteria, respectively, present this functional group. The same analysis can be made for the compounds containing functionality at C-20 on the alpha- or beta-side.

The majority of the active triterpenes presents π-bonding at positions C-5, C-6, C-9, C-11, C-12, and C-13 (i.e., $\Delta^{5,6}$ $\Delta^{9,11}$, and $\Delta^{12,13}$, respectively), few of them present $\Delta^{20,30}$ and $\Delta^{20,21}$, and π-bondings are absent in few active triterpenes. The bactericidal activities are mainly related to functional groups at the rings A and E of the triterpene skeleton. Considering a great quantity of active triterpenes containing π-bonding at the ring C, it may be proposed that this functionalization is also important to the bactericidal activity.

2. References

Abreu, V.G.C.; Takahashi, J.A.; Duarte, L.P.; Piló-Veloso, D.; Junior, P.A.S., Alves, R.O.; Romanha, A.J.; Alcântara, A.F. C. (2011). Evaluation of the bactericidal and trypanocidal activities of triterpenes isolated from the leaves, stems, and flowers of *Lychnophora pinaster*. *Brazilian Journal of Pharmacognosy*, Vol. 21, No. 4, pp. 615-621.

Ahmad, W.; Ahkhan, S.; Muhammadzeeshan; Obaidullah; Nisar, M.; Shaheen, F. & Ahamad, M. (2008). New antibacterial pentacyclic triterpenes from Myricaria elegans Royle. (tamariscineae). *Journal of Enzyme Inhibition and Medicinal Chemistry*, Vol. 2, pp. 1023–1027.

Al-Fatimi, M.; Wurster, M.; Schroder, G.; Lindequist, U. (2007). Antioxidant, antimicrobial and cytotoxic activities of selected medicinal plants from Yemen. *Journal of Ethnopharmacology*, Vol. 111, No. 3, pp. 657–666.

Angeh, J.E.; Huang, X.; Sattler, I.; Swan, G.E.; Dahse, H.; H¨artl, A.; Eloff, J.N. (2007). Antimicrobial and anti-inflammatory activity of four known and one new triterpenoid from *Combretum imberbe* (Combretaceae). *Journal of Ethnopharmacology*, Vol. 110, pp. 56–60.

Ankli, A.; Heilmanna, J.; Heinrich, M.; Sticher, O. (2000). Cytotoxic cardenolides and antibacterial terpenoids from *Crossopetalum gaumeri*. *Phytochemistry*, Vol. 54, pp. 531-537.

Barre, J.T.; Bowden, B.F.; Coll, J.; Jesus, J.; La Fuente, V.E.; Janairo, G.; Ragasa, C.Y. (1997). Bioactive Triterpene from *Lantana camara*. *Phytochemistry*, Vol. 45, No. 2, pp. 321-324.

Bolivar, P.; Cruz-Paredesa, C.; Hernández, L.R.; Juárez, Z.N.; Sánchez-Arreola, E.; Av-Gay, Y.; Bach, H. (2011). Antimicrobial, anti-inflammatory, antiparasitic, and cytotoxic activities of Galium mexicanum. *Journal of Ethnopharmacology*, Vol. 137, No. 1, pp. 141-147.

Chattopadhyay, D.; Maiti, K.; Kundu, A.P.; Chakraborty, M.S.; Bhadra, R.; Mandal, C.; Mandal, A.B. (2001). Antimicrobial activity of *Alstonia macrophylla*: a folklore of bay islands. *Journal of Ethnopharmacology*, Vol. 77, No. 1, pp. 49–55.

Chen, X.Q.; Zan, K.; Yang, J.; Liu, X.X.; Mao, Q.; Zhang, L.; Lai, M.X.; Wang, Q. (2011). Quantitative analysis of triterpenoids in different parts of *Ilex hainanensis*, *Ilex stewardii* and *Ilex pubescens* using HPLC–ELSD and HPLC–MSn and antibacterial activity. *Food Chemistry*, Vol. 126, pp. 1454–1459.

Chiozem, D.D.; Trinh-Van-Dufat, H.; Wansi, J.D.; Mbazoa Djama, C.; Fannang, V.S.; Seguin, E.; Tillequin, F.; Wandji, J. (2009). New Friedelane Triterpenoids with Antimicrobial Activity from the Stems of *Drypetes paxii*. *Chemical & Pharmaceutical Bulletin*, Vol. 57, No. 10, pp. 1119—1122.

Chouhan, H.; Singh, S.K. (2010). Antibacterial activity of *phyllantus simplex* linn. *Pharmacologyonline*, Vol. 3, pp. 169-172.

Cunha, W.R.; Matos, G.X.; Souza, M.G.M.; Tozatti, M.G.; Silva, M.L.A.; Martins, C.H.G.; Silva, R.; Filho A.A.S. (2010). Evaluation of the antibacterial activity of the methylene chloride extract of Miconia ligustroides, isolated triterpene acids, and ursolic acid derivatives. *Pharmaceutical Biology*, Vol. 48 No. 2, pp. 166–169.

Daí, J.; Zhao, C.; Zhang, Q.; Liu, Z.; Zheng, R.; Yang, L. (2001). Taraxastane-type triterpenoids from Saussurea petrovii. *Phytochemistry*, Vol. 58, No. 7, pp. 1107–1111.

Dash, M.; Patra, J.K.; Panda, P.P. (2008). Phytochemical and antimicrobial screening of extracts of Aquilaria agallocha Roxb. *African Journal of Biotechnology*, Vol. 7, No. 20, pp. 3531-3534.

Dehaen, W.; Mashentseva, A.A.; Seitembetov, T.S. (2011). Allobetulin and Its Derivatives: Synthesis and Biological Activity. *Molecules*, Vol. 16, pp. 2443-2466.

Devmurari, V.P. (2010). Antibacterial evaluation of ethanolic extract of *Cedrus deodara* Wood. *Archives of Applied Science Research*, Vol. 2, No. 2, pp. 179-183.

Devmurari, V.P.; Ghodasara, T.J.; Jivani, N.P. (2010). AntibacteriaL Activity and Phytochemical Study of Ethanolic Extract of Triumfetta rhomboidea Jacq. *International Journal of PharmTech Research*. Vol. 2 No. 1, pp. 1182-186.

Djoukeng, J.D.; Abou-Mansour, E.; Tabacchi, R.; Tapondjou, A.L.; Boudab, H.; Lontsi, D. (2005). Antibacterial triterpenes from *Syzygium guineense* (Myrtaceae). *Journal of Ethnopharmacology*, Vol. 101, pp. 283–286.

Ding, Q.; Yang, L.X.; Yang, H.W.; Jianga, C.; Wang, Y.F.; Wanga, S. (2009). Cytotoxic and antibacterial triterpenoids derivatives from *Clematis ganpiniana*. *Journal of Ethnopharmacology*, Vol. 126, pp. 382–385

Dong, S.H.; Zhang, C.R.; Dong, L.; Wu, Y.; Yue, J.M. (2011). Onoceranoid-Type Triterpenoids from *Lansium domesticum*. *Journal Natural Products*, Vol. 74, pp. 1042–1048.

Dzubak, P.; Hajduch, M.; Vydra, D.; Hustova, A.; Kvasnica, M.; Biedermann, D.; Markova, L.; Urban, M.; Sarek, J. (2006). Pharmacological activities of natural triterpenoids and their therapeutic implications. *Natural Products Reports*, Vol. 23, pp. 394–411.

Eisa, M.M.; Almagboul, A.Z.; Omer, M.E.A.; Elegami, A.A. (2000). Antibacterial activity of *DichrostachysCinérea Fitoterapia* Vol. 71, No. 3, pp. 324-327.

Ekbote, M.T.; Ramesh, C.K.; Mahmood, R. (2010). Evaluation of Anthelmintic and Antimicrobial activities of Azima tetracantha Lam. *International Journal of Pharmaceutical Sciences*, Vol. 2, No. 1 pp. 375-381.

Fannang, S.V.; Kuete, V.; Djama, C.M.; Dongfack, M.D.J.; Wansi, J.D.; Tillequin, F.; Seguin, E.; Chosson, E.; Wandji, J. (2011). A new friedelane triterpenoid and saponin with moderate antimicrobial activity from the stems of *Drypetes laciniata*. *Chinese Chemical Letters*, Vol. 22, pp. 171–174.

Gao, H.Y.; Wu, L.J.; Nakane, T.; Shirota, O.; Kuroyanagi, M. (2008). Novel Lanostane and Rearranged Lanostane-Type Triterpenoids from Abies sachalinensis—II— *Chemical & Pharmaceutical Bulletin*,Vol. 56, No. 4, pp. 554—558.

Geyid, A.; Abebe, D.; Debella, A.; Makonnen, Z.; Aberra, F.; Teka, F.; Kebede, T.; Urga, K.; Yersaw, K.; Biza, T.; Mariam, B.H.; Guta, M. (2005). Screening of some medicinal plants of Ethiopia for their anti-microbial properties and chemical profiles. *Journal of Ethnopharmacology*, Vol. 97, No. 3, pp. 421–427.

Gonzalez, A.G.; Jimenez, J.S.; Moujir, L.; Ravelo, A.G.; Luis, J.G.; Bazzochi, I.L.; Gutierrez, A. M. (1992) Two new triterpene dimers from Celastraceae, their partial synthesis and antimicrobial activity. *Tetrahedron*, Vol. 48, No. 4, pp. 769-774.

Hajji, M.; Jarraya, R.; Lassoued, I.; Masmoudi, O.; Damak, M.; Nasri, M. (2010). GC/MS and LC/MS analysis, and antioxidant and antimicrobial activities of various solvent extracts from *Mirabilis jalapa* tubers. *Process Biochemistry*, Vol. 45, No. 9, pp. 1486–1493.

Haraguchi, H.; Kataoka, S.; Okamoto, S.; Hanafi, M.; Shibata, K. (1999). Antimicrobial Triterpenes from *Ilex integra* and the Mechanism of Antifungal Action. *Phytotherapy research*, Vol. 13, pp. 151–156.

He, X.F.; Wang, X.N.; Yin, S.; Dong, L.; Yue, J.M. (2011). Ring A-seco triterpenoids with antibacterial activity from *Dysoxylum hainanense*. *Bioorganic & Medicinal Chemistry Letters*, Vol. 21, pp. 125–129.

Hess, S.C.; Brum, R.L.; Honda, N.K.; Cruz, A.B.; Moretto, E.; Cruz, R.B.; Messana, I.; Ferrari, F.; Filho, V.C.; Yunes, R.A. (1995). Antibacterial activity and phytochemical analysis of *Vochysia divergens* (Vochysiaceae) *Journal of Ethnopharmacology*, Vol. 47, No. 2, pp. 97-100.

Hichri, F.; Ben Jannet, H.; Cheriaa, J.; Jegham, S.; Mighri, Z. (2003). Antibacterial activities of a few prepared derivatives of oleanolic acid and of other natural triterpenic compounds. *Comptes Rendus Chimie*, Vol. 6, pp. 473-483.

Ikan, R. 1991. *Natural Products. A Laboratory Guide.* Academic Press, New York.

Isaka, M.; Palasarn, S.; Supothina, S.; Komwijit, S.; Luangsa-ard, J.J. (2011). Bioactive Compounds from the Scale Insect Pathogenic Fungus *Conoideocrella tenuis* BCC 18627. *Journal Natural Products*, Vol. 74, pp. 782–789

James, U.J.T.; Dubery, I.A. (2009). Pentacyclic Triterpenoids from the Medicinal Herb, *Centella asiatica* (L.) Urban *Molecules,*Vol. 14, pp. 3922–3941.

Katerere, D.R.; Grev, A.I.; Nash, R.J.; Waigh, R.D. (2003). Antimicrobial activity of pentacyclic triterpenes isolated from African Combretaceae. *Phytochemistry*, Vol. 63, No. 1, pp. 81– 88.

Kazakovaa, O.B.; Giniyatullinaa, G.V.; Tolstikova, G.A.; Medvedevaa, N.I.; Utkinab, T.M.; Kartashovab, O.L. (2010). Synthesis, Modification, and Antimicrobial Activity of the N_Methylpiperazinyl Amides of Triterpenic Acids. *Russian Journal of Bioorganic Chemistry*, Vol. 36, No. 3, pp. 383-38915. Khan, I.; Srikakolupu, S.R.; Darsipudi, S.; Gotteti S.D.; Amaranadh, H. (2010). Phytochemical studies and screening of leaf extracts of *Azadirachta indica* for its anti-microbial activity against dental pathogens. *Archives of Applied Science Research*, Vol. 2 No. 2, PP. 246-250.

Khan, I.; Srikakolupu, S.R.; Darsipudi, S.; Gotteti S.D.; Amaranadh, H. (2010). Phytochemical studies and screening of leaf extracts of *Azadirachta indica* for its anti-microbial activity against dental pathogens. *Archives of Applied Science Research*, Vol. 2 No. 2, PP. 246-250.

Khera, S.; Woldemichael, G.M.; Singh, M.P.; Suarez, E.; Timmermann, B. N. (2003). A Novel Antibacterial Iridoid and Triterpene from *Caiophora coronata*. *Journal Natural Products*, Vol. 66, pp. 1628-1631.

Kim, N.C, Desjardins, A.E, Wu, C.D, Kinghorn, A.D. (1999). Activity of Triterpenoid Glycosides from the Root Bark of *Mussaenda macrophylla* against Two Oral Pathogens. *Journal of Natural Products*, Vol. 62, No. 10, pp. 1379-1384.

Kloos, H.; Zein, A.H. 1993. *The Ecology of Health and Diseases in Ethiopia.* West View Press Inc., Washington DC.

Kuete, V.; Nana, F.; Ngameni, B.; Mbaveng, A.T.; Keumedjio, F.; Ngadjui, B.T. (2009). Antimicrobial activity of the crude extract, fractions and compounds from stem bark of *Ficus ovata* (Moraceae). *Journal of Ethnopharmacology*, Vol. 124, No. 3, pp. 556–561.

Kuete, V.; Nguemeving, J. R.; Beng, V. P.; Azebaze, A. G. B.; Etoa, F. X.; Meyer, M.; Bodo, B.; Nkengfack, A. E. (2007). Antimicrobial activity of the methanolic extracts and compounds from *Vismia laurentii* De Wild (Guttiferae). *Journal of Ethnopharmacology*, Vol. 109, pp. 372–379.

Kuete, V.; Tangmouo, J.G.; Beng, V.P.; Ngounou, F.N.; Lontsi, D. (2006). Antimicrobial activity of the methanolic extract from the stem bark of tridesmostemon omphalocarpoides (Sapotaceae). *Journal of Ethnopharmacology*, Vol. 104, No. 1, pp. 5–11.

Léon, L.; López, M.R.; Moujir, L. (2010). Antibacterial properties of zeylasterone, a triterpenoid isolated from *Maytenus blepharodes*, against *Staphylococcus aureus*. *Microbiological Research*, Vol. 165, pp. 617–626.

Lindsey, K.L.; Budesinsky, M.; Kohout, L. & Staden, J.V. (2003). Antibacterial activity of maytenonic acid isolated from the root-bark of Maytenus senegalensis. *South African Journal of Botany*, Vol. 72, pp. 473–477.

Ling, T.J.; Wan, X.C.; Ling, W.W.; Zhang, Z.Z.; Xia, T.; Li, D.X.; Hou, R.Y. (2010). New Triterpenoids and Other Constituents from a Special Microbial-Fermented Tea; Fuzhuan Brick Tea. *Journal of Agricultural and Food Chemistry*, Vol. 58, pp. 4945–4950.

Mallavadhani, U.V.; Mahapatra, A.; Jamil, K.; Srinivasa, P. (2004) Antimicrobial Activity of Some Pentacyclic Triterpenes and Their Synthesized 3-O-Lipophilic Chains. *Biological Pharmaceutical Bulletin*, Vol. 27, No. 10, pp. 1576-1579.

Mangathayaru, K.; Lakshmikant, J.; Sundar, N.S.; Swapna, R.; Grace, X.F.; Vasantha, J. (2005). Antimicrobial activity of Leucas aspera flowers. *Fitoterapia*, Vol. 76, No. 7, pp. 752–754.

Manguro, L.O.A.; Midiwo, J.O.; Tietze, L.F.; Haod, P. (2011). Triterpene saponins of *Maesa lanceolata* leaves. *Arkivoc*, Vol. ii, pp. 172-198.

McRae, J.M.; Yang, Q.; Crawford, R.J.; Palombo, E.A. (2008). Antibacterial compounds from *Planchonia careya* leaf extracts. *Journal of Ethnopharmacology*, Vol. 116, pp. 554–560.

Maregesi, S.M.; Hermans, N.; Dhooghe, L.; Cimanga, K.; Ferreira, D.; Pannecouque, C.; Berghe, D.A.V.; Cos, P.; Maes, L.; Vlietinck, A.J.; Apers, S.; Pieters, L. (2010). Phytochemical and biological investigations of *Elaeodendron schlechteranum*. *Journal of Ethnopharmacology*, Vol.129, pp. 319–326.

Maridass, M.; Ramesh, U.; Raju, G. (2010). Evaluation of Phytochemical, Pharmacognostical and Antibacterial Activity of Garcinia Gummicutta Leaves. *Pharmacologyonline*, Vol. 1, pp. 832-837.

Mathabe, M.C.; Hussein, A.A.; Nikolova, R.V.; Basson, A.E.; Meyer, J.J.M.; Lall, N. (2008). Antibacterial activities and cytotoxicity of terpenoids isolated from *Spirostachys Africana*. *Journal of Ethnopharmacology*, Vol. 116, pp. 194–197.

Matu, E.N.; Staden, J.V. (2003). Antibacterial and anti-inflammatory activities of some plants used for medicinal purposes in Kenya. Vol. 87, pp. 35–41.

Miranda, R.R.S.; Duarte, L.P.; Silva, G.D.F.; Vieira Filho, A.S.; Carvalho, P.B.; Messas, A.C. (2009). Evaluation of antibacterial activity of "Mangabarana" Austroplenckia populnea Reissek (Celastraceae). *Brazilian Journal of Pharmacognosy*, Vol. 19, No. 2a, pp. 370-375.

Mishra, P.M.; Sree, A. (2007). Antibacterial activity and GCMs analysis of the extract of the leaves of Finlaysonia obovata (A Mangrove Plant). *Asian journal of Plant Sciences*, Vol. 6, pp. 168-172.

Motlhanka, D.; Houghton, P.; Miljkovic-Brake, A.; Habtemariam, S. (2010). A novel pentacyclic triterpene glycoside from a resin of *Commiphora glandulosa* from Botswana. *African Journal of Pharmacy and Pharmacology*, Vol. 4, No. 8, pp. 549-554.

Mutai, C.; Bii, C.; Vagias, C.; Abatis, D.; Roussis, V. (2009). Antimicrobial activity of *Acacia mellifera* extracts and lupane triterpenes. *Journal of Ethnopharmacology*, Vol. 123, No. 1, pp. 143–148.

Nick, A.; Wright, O.; Sticher, O. (1994). Antibacterial Triterpenoid Acids from *Dillenia papuana*. *Journal of Natural Products*, Vol. 57, pp. 1245-1250.

Nickavar, B.; Mojab, F. (2003). Antibacterial activity of *Pulicaria dysenterica* Extracts. *Fitoterapia*, Vol. 74, No. 4, pp. 390–393.

Panizzi, L.; Catalano, S.; Miarelli, C.; Cioni, P.L.; Campeol, E. (2000). In vitro antimicrobial activity of extracts and isolated constituents of Geum rivale. *Phytotherapy Research*, Vol. 14, No. 7, pp. 561–563.

Parveen, M.; Ghalib, R.M.; Mehdi, S.H.; Rehman, S.Z.; Ali, M. (2009). A new triterpenoid from the leaves of Ficus benjamina (var. comosa). *Natural Products Research*, Vol. 23, No. 8, pp. 729-736.

Pattanayak, S.P.; Sunita, P. (2008). Wound healing, anti-microbial and antioxidant potential of *Dendrophthoe falcata* (L.f) Ettingsh. *Journal of Ethnopharmacology*, Vol. 120, No. 2, pp. 241–247.

Peixoto, J.R.O.; Silva, G.C.; Costa, R.A.; Fontenelle, J.L.S.; Vieira, G.H.F.; Filho, A.A.F.; Vieira, R.H.S.F. (2011). In vitro antibacterial effect of aqueous and ethanolic Moringa leaf extracts. *Asian Pacific Journal of Tropical Medicine*, Vol. 4, No. 3, pp. 201-204.

Pitzele, B.S. (1974) Synthesis of 2-Oxygenated Glycyrrhetic Acid Derivatives. *Journal of Medicinal Chemistry*, Vol. 17, No. 2, pp. 191-194.

Popova, M.; Trusheva, B.; Gyosheva, M.; Tsvetkova, I.; Bankova, V. (2009). Antibacterial triterpenes from the threatened wood-decay fungus *Fomitopsis rosea*. *Fitoterapia*, Vol. 80, pp. 263–266.

Ramesh, N.; Viswanathan, M.B.; Saraswathy, A.; Balakrishna, K.; Brindha, P.; Lakshmanaperumalsamy P. (2001). Phytochemical and antimicrobial studies on *Drynaria quercifolia*. *Fitoterapia* Vol. 72, No. 8, pp. 934-936.

Rahman, A.U.; Zareen, S.; Choudhary, M.I.; Ngounou, F.N.; Yasin, A.; Parvez, M. (2002). Terminalin A, a novel triterpenoid from Terminalia glaucescens. *Tetrahedron Letters*, Vol. 43, No. 35, pp. 6233–6236.

Rivero-Cruz, J.F.; Zhu, M.; Kinghorn, A.D.; Wua, C.D. (2008). Antimicrobial constituents of Thompson seedless raisins (*Vitis vinifera*) against selected oral pathogens. *Phytochemistry Letters*, Vol. 1, pp.151–154.

Ryu, S.Y.; Oak, M.H.; Yoon, S.K.; Cho, D.I.; Yoo, G.S.; Kim, T.S. (2000). Antiallergic and anti-inflammatory triterpenes from the herb of Prunella vulgaris. *Planta Medica*, Vol. 66, No. 4, pp. 358- 360.

Saleh, M.; Kamel, A.; Li, X.; Swaray, J. (1999). Antibacterial Triterpenoids isolated from *Lantana camara*. *Pharmaceutical Biology*, Vol. 37, pp. 63–66.

Sanches, N.R.; Cortez, D.A.G.; Schiavini, M.S.; Nakamura, C.V.; Dias-Filho, B.P. (2005). An Evaluation of Antibacterial Activities of *Psidium guajava* (L.) *Brazilian Archives of Biology and Technology*, Vol. 48, No. 3, pp. 429-436.

Semwal, R.B.; Semwal, D.K.; Semwal, R.; Singh, R.; Rawat, M.S.M. (2011). Chemical constituents from the stem bark of *Symplocos paniculata* Thunb. with antimicrobial, analgesic and anti-inflammatory activities. *Journal of Ethnopharmacology*, Vol. 135, pp. 78–87.

Shai, L.J.; McGaw, L.J.; Aderogba, M.A.; Mdee, L.K.; Eloff, J.N. (2008). Four pentacyclic triterpenoids with antifungal and antibacterial activity from *Curtisia dentata* (Burm.). *Journal of Ethnopharmacology*, Vol. 119, pp. 238–244.

Silveira, G.P.; Nome, F.; Gesser, J.C.; Sá, M.M.; Terenzi, H. (2006). Recent achievements to combat bacterial resistence. *Quimica Nova*, Vol. 29, No. 4, pp. 844-855.

Sun, F.Y.; Chen, X.P.; Wang, J.H. L.Q.; Yang, S.R.; Du, G.H. (2008). Arjunic Acid, a Strong Free Radical Scavenger from *Terminalia arjuna*. *American Journal of Chinese Medicine*, Vol. 36, No. 1, pp. 197–207.

Sunitha, S.; Nagaraj, M.; Varalakshmi, P. (2001). Hepatoprotective effect of lupeol and lupeol linoleate on tissue antioxidant defence system in cadmium induced hepatotoxicity in rats. *Fitoterapia*, Vol. 72, No. 5, pp. 516– 523.

Takechi, M., Tanaka, Y. (1992). Structure-activity relationships of synthetic methyl oleanolate glycosides. *Phytochemistry*, Vol.31, No.11, pp. 3789-3791.

Takechi, M.; Tanaka, Y. (1993) Structure-activity relationships of synthetic methyl ursolate glycosides. *Phytochemistry*, Vol. 34, No.3, pp. 675-677.

Tamokou, J.D.D.; Tala, M.F.; Wabo, H.K.; Kuiate, J.R.; Tane, P. (2009). Antimicrobial activities of methanol extract and compounds from stem bark of *Vismia rubescens*. *Journal of Ethnopharmacology*, Vol. 124, pp. 571–575.

Tantry, M.A.; Khan, R.; Akbar, S.; Dar, A.R.; Shawl, A.S.; Alam, M.S. (2011). An unusual bioactive oleanane triterpenoid from *Rhododendron campanulatum* D. Don. *Chinese Chemical Letters*, Vol. 22, No. 5, pp. 575–579.

Tsiri, D.; Aligiannis, N.; Graikoub, K.; Spyropoulosa, C.; Chinou, I. (2008). Triterpenoids from *Eucalyptus camaldulensis* Dehnh. Tissue Cultures. *Helvetica Chimica Acta*, Vol. 91, pp. 2110-2114.

Viji, M.; Sathiya, M.; Murugesan, S. (2010). Phytochemical analysis and antibacterial activity of medicinal plant cardiospermum helicacabum linn. *Pharmacologyonline*, Vol. 2, pp. 445-456.

Yeboah, E.M.O.; Majinda, R.R.T.; Kadziola, A.; Muller, A. (2010). Dihydro-α-agarofuran Sesquiterpenes and Pentacyclic Triterpenoids from the Root Bark of *Osyris lanceolata*. *Journal Natural Products*, Vol.73, pp. 1151–1155.

Yimdjo, M.C.; Azebaze, A.G.; Nkengfack, A.E.; Meyer A.M, Bodo, B.; Fomum Z.T. (2004). Antimicrobial and cytotoxic agents from Calophyllum inophyllum. *Phytochemistry*, Vol. 65, No. 20, pp. 2789–2795.

Yun, B.S.; Ryoo, I.J.; Lee, I.K.; Park, K.H.; Choung, D.H, Han, K.H. (1999). Two bioactive pentacyclic triterpene esters from the root bark of *Hibiscus syriacus*. *Journal of Natural Products*, Vol. 62, No. 5, pp. 764– 766.

Weigenand, O.; Hussein, A.A.; Lall, N.; Meyer, J.J.M. (2004). Antibacterial Activity of Naphthoquinones and Triterpenoids from *Euclea natalensis* Root Bark. *Journal Natural Products*, Vol. 67, pp. 1936-1938.

Zhao, J.; Mou, Y.; Shan, T.; Li, S.; Zhou, L. Wang, M.; Wang, J. (2010). Antimicrobial Metabolites from the Endophytic Fungus *Pichia guilliermondii* Isolated from *Paris polyphylla* var. yunnanensis. *Molecules*, Vol. 15, pp. 7961-7970.

Zheng, C.J.; Sohn, M.J.; Kim, D.Y.; Yu, H.E.; Kim, W.G. (2008). Olean-27-carboxylic Acid-Type Triterpenes with Potent Antibacterial Activity from *Aceriphyllum rossii*. *Journal of Agricultural and Food Chemistry*, Vol. 56, pp. 11752–11756.

2

Flower Shaped Silver Nanostructures: An Efficient Bacteria Exterminator

Subash Chandra Sahu, Barada Kanta Mishra and Bikash Kumar Jena*
Institute of Minerals and Materials Technology, Bhubaneswar, Orissa India

1. Introduction

Materials in nanoscale range have dominated several areas of engineering science and technology. In the area of biotechnology nano materials have been used increasingly in biomedical analysis, fabrication of biosensors/biointerface, clinical diagnosis and therapy, drug delivery and so on (Cao, 2008; Mirkin et al., 1996; Alivasatos, 2004; Jena and Raj, 2006, 2011; Xiao et al., 2003; Sengupta et al., 2005; Wu et al., 2003; Brigger et al.,2002; Cui et al., 2001; Gao et al., 2004; Rosi and Mirkin, 2005). Particular interest has been focused on nanostructured noble metal particles for biotechnology application because of their biocompatibility, lower toxicity and higher affinity with wide range of biomolecules (Hazarika et al., 2004; Bae et al., 2005). Therefore, different processes have been adopted for tuning the surfaces of metal nanoparticles to explore the selective binding and monitoring of specific targets in biological sensing (Wang, 2005; Wang et al., 2006; Kneipp et al., 2006). Nanoparticles attached to biomolecules act as an artificial receptor and control the cellular processes for various biological applications such as inhibition of enzymatic activity, transcription regulations, etc (De et al., 2008).

The noble metal nanoparticles are known to display plasmon resonance in the visible region (Kreibing and Volmer, 1995). Optical properties of metal nanoparticles originate from the surface plasmon resonance have attracted substantial interest in biotechnology for diagnosis and sensing (Alivasatos, 2004; Gao et al., 2004; Nam et al., 2003; Xiang et al., 2006; Katz and Willner, 2004). The colour change due to the assembly of Au or Ag nanoparticles developed the colorimetric method for many applications. These colorimetric strategies has been utilised to develop DNA detection methods to study the affinity and specificity of DNA-DNA interactions (Mirkin et al., 1996; Nam et al., 2003; Stoeva et al., 2006; Storhoff et al., 1998; Han et al. 2006). The sensing methodology is based on the colour change due to assembly of Au nanoparticle-linked complementary strands formed by the specific binding of targeted strands. Kanaras et al. demonstrated the use of Au nanoparticle DNA interaction for determination of the enzymatic cleavage of DNA (Kanaras et al. 2007). This process is designed to monitor the enzymatic cleavage activity based on the wave length shifting of Au nanoparticles. Jena and Raj have developed an optical method for the sensing of biomedically important polyionic drugs, protamine and heparin based on the

* Corresponding Author

assembly/disassembly of Au nanoparticles (Jena and Raj, 2008). A colorimetric sensor based on Au nanoparticle-apatamer has been documented for sensing of small molecules and protein (Wang et al., 2008; Liu and Lu, 2006, 2004). Dong and co-workers developed a colorimetric sensor using Au nanoparticle for monitoring α-thrombin based on the specific interaction of aptamer with α-thrombin (Wei et al. 2007).

The wide application of metal nanoparticles in biological process has been motivated to produce nanoparticles of biofiendly in nature. Therefore, synthesis of nanoparticle under green and benign processes were adopted to produce toxic free nanoparticles (Zhang et al., 2007; Mohanpuria et al., 2008; Gill et al., 2007; Goodman et al. 2004; Jin et al. 2007; Singh and Nalwa, 2007; Kannan et al. 2006; Kattumuri et al. 2007). Current nanotechnology research makes great effort to develop bio-friendly methods for production of noble nanostructured materials. For instance, several synthetic methodologies have been explored for the production of metal nanoparticles using vitamins, plant extracts, biomolecules and etc (Nadagauda and Verma, 2006; Shankar et al. 2004; Gardea-Torresdey et al. 1999; Zhang et al., 2006). Shukla et al. explored a green method for synthesis of Au nanoparticles using soybean which are biocompatible in nature (Shukla et al., 2008).

It has been explored that nano-structured materials with novel shapes exhibit more unique physical and chemical properties (Jena et al., 2011). The dramatic enhancement in nanostructure properties has been achieved on tuning the shape in contrast to the size, because different crystal surfaces have different surface atom densities and electronic structures leading to different physical and chemical properties (Jena et al., 2011). Therefore the nanostructured particles with unusual shape possess wider biological and medical applications in comparison with their common spherical counter part. These anisotropic nanomaterials have potential application in signal amplification in bioanalysis and biodiagnosis technology (Grunes et al. 2002; Yu et al. 2003; Turner, 2000; Haes and Van Duyne, 2002). Particle shape has been recognised as an important attribute compared to the size and can be engineered for drug delivery (Mitragotri, 2009). It has been explored that by mimicking the distinctive shapes of bacteria, fungi, and blood cells could improve the ability of nanoparticles to deliver drugs to diseased cells in the body (Mitragotri, 2009). Wijaya et al. have utilized nanoparticles having bone and capsule like shapes for controlled and selective release of drugs (Wijaya et al. 2009). Ray and co-workers have been successfully utilised Au nanorods for screening HIV-1 viral DNA sequencing (Darbha et al., 2008). As far as the particle shape is concerned, very little is explored particularly with respect to structural property relationship of particles and their biological applications. Therefore, the current development is more focused on producing nanosized materials of well-defined morphologies with improved properties.

A substantial amount of research has been carried out for the synthesis of Ag nanostructures owing to their wide and potential applications in antibacterial, antimicrobial, SERS and so forth (Wijaya et al., 2009; Braun et al., 2007; Gupta and Silver, 1998; Schultz et al., 2000). The antibacterial activity of Ag nanostructures has been studied and well documented in the literature. The antibacterial activity of silver has been widely exploited in healing the cuts and wounds (Fox Jr. et al., 1974). It has been effectively implemented in surgical procedures to prevent bacterial infections (Bosetti et al., 2002; Alt et al., 2004). The extensive application of Ag nanoparticles as a successful antibacterial agents are due to their effective bacteria extermination capacity even at very low concentrations (Banerjee et al. 2010). It has been observed that the bacteria and microbes are less likely to build resistance against silver as they usually do against antibiotics (Banerjee et al. 2010). Several studies and explanation has been documented regarding the biocidal activity of Ag nanoparticles. The general perception is that the Ag nanoparticles are prone to bind the sulphur groups of the protein present in bacterial cell wall and open the permeability of the cell membrane and damage the cell wall. Few documents explained that the Ag^+ ions present on the surface of Ag nanoparticles have the actual biocidal properties for cell destructions. The Ag^+ ions are capable of entering to the bacterial cells and get reduced to their elemental states. The cellular process attempts to remove these reduced Ag from the cell interior and eventually hampers the cellular process leading to cell death. It has been explored that the Ag nanostructures target the bacterial membrane, destabilize the plasma membrane potential and ultimately damage the bacterial cell. Chattopadhyay and co-workers have established that the small Ag nanoparticles having sizes less than 10 nm make pores on the cell wall and release the cytoplasmic content of the cell and cause cell death (Gogoi, et al., 2006). Morones et al. explored that antibacterial activity depends on the size of Ag nanostructures (Morones et al., 2005). Song and co-workers explored the shape dependent biocidal activity of Ag nanostructures towards gram negative bacterium, E. coli (Pal et al. 2007).

Many synthetic processes have been developed for spherical shaped Ag nanoparticles and their promising applications. However, more emphasis has been paid for different shaped Ag nanostructures. Various synthetic protocols have been documented for the production of silver nanoparticles with different shapes (Sun and Xia, 2002a, 2002b ; Jin et al., 2001; Lofton and Singmund, 2005; Pastoriza-Santos and Liz-marjan, 2002; Bera and Raj, 2010; Nicewarner-Pena et al. 2001; Kim et al. 2004; Yu and Yam, 2004; Im et al. 2005; Hao et al. 2002; Chen et al. 2005, 2002; Jiang et al. 2006; Ducamp-Sanuesa et al. 1992; Sun et al. 2002; Chen and Carroll, 2002; Xue et al. 2008). Xia and co-workers adopted the polyol process to synthesize a variety of different shaped Ag nanostructures by controlling the concentration of capping agent PVP and precursor, Ag NO_3 in solution (Wiley et al. 2006, 2005, 2004; Sun et al., 2003). Templates, polymers and Surfactants have been mainly used to get anisotropic nanostructures. Though these methods are successful to produce well-defined silver nanostructures, the improved chemical and physical properties of these nanostructure particles are hindered by these strong surface protecting agents. Therefore, substantial interest has been focused to explore sterile, nontoxic and environmentally safe protocols for facile synthesis of different shaped silver nanoparticles.

In this investigation, the unique flower shaped Ag nanostructures were synthesized adopting bio-inspired approach and the antibacterial activities investigated. We observed that the antibacterial properties can be improved by shaping the Ag nanostructures from spherical to flower-like morphology.

2. Materials and methods

2.1 Materials

$AgNO_3$, Rutin hydrate (RT), were obtained from Himedia, India. All other chemicals not mentioned here are of analytical grade and used as received from the suppliers. Carbon coated copper grids were obtained from SPI supplies, USA. All the solutions were prepared with deionised water (18 Ωm) obtained from Millipore system. Glasswares used in this investigation were well cleaned with freshly prepared aqua regia, then rinsed thoroughly with water and dried prior to use (Caution: aqua regia is a powerful oxidizing agent and it should be handled with extreme care).

2.2 Instrumentation

A variety of characterization equipment was employed in order to study the as-synthesised nanostructured material in a comprehensive way. TEM images were obtained from FEI, TECHNAI G2 transmission electron microscope operating at 200 kV. The specimens were prepared by dropping 2 μl of colloidal solution onto carbon coated copper grids. The UV-visible spectra of the colloidal solutions were recorded on a Shimadzu UV-1700 spectrophotometer. X-ray diffraction analysis was carried out with X'pert PRO (Pan Analytica) X-ray diffraction unit using Ni filtered CuKα (λ = 1.54 Å) radiation. X-ray photoelectron spectroscopic analyses were carried out on Kratos Axis 165 X-ray Photoelectron Spectrometer. The X-ray gun was operated at 15 kV and 20 mA and high-resolution spectra were collected using 80 eV pass energy, respectively, with Mg Kα 1253.6 eV radiation. The data were obtained with an acquisition time of 121 seconds.

2.3 Synthesis of flower shaped Ag nanostructures (AgNFs)

In a typical synthesis, 10 ml aqueous solution of $AgNO_3$ (1 mM) was stirred for 2-3 min. and then 0.2 ml of Rutin (15 mM) was added. The stirring was continued for 30 minute and the resulting nanocolloid was stored at 4°C (Jena et al. 2009).

2.4 Synthesis of spherical Ag nanostructures

Citrate stabilized nanoparticles were synthesized according to the previous report (Link and El-Sayed, 1999; Mulvany, 1996) by slight modification. In brief, Ag nanocolloids are prepared by reducing $AgNO_3$ (1 mM) with sodium citrate (0.3 mM) in aqueous medium at boiling temperature.

2.5 Determination of antibacterial activity

Streptococcus Faecalis (SF), Pseudomonas Aeruginosa (PA) and Escherichia coli (EC) bacterias are used in this investigation. The Stocks were created by passing the original reference organisms once through the Muller-Hinton Broth (MHB) and plating on Muller-Hinton agar (MHA) plates for bacterial organisms. For inoculum preparations, bacteria were sub-cultured in Brain-Heart-Infusion (BHI) at 35 °C for 24 hr. The optical density of each culture was measured at 550 nm. The agar well diffusion method was used to determine the antibacterial activity of as synthesized AgNFs. The media used were Muller-Hinton agar (Himedia) for the bacteria under study. The nutrient agar plates were swabbed with cultured bacteria. A total of 2 mm diameter wells were punched into the agar and filled with 100 μl of AgNF colloids standard antibiotics (Ciprofloxacin, Gentamicin, Penicillin, and Chloroamphenicol) used as positive control. The plates were incubated at 35 °C for 24 h for bacteria pathogen. The antibacterial activity was calculated by measuring the inhibition zone diameter. The experiments were repeated thrice and the average values of antibacterial activity were calculated.

3. Result and discussions

3.1 Characterization of Ag nanoparticles

The UV-visible absorption spectra of the as-synthesized Ag nanoparticles were recorded. The absorption spectra of Ag nanoparticles shows its plasmon absorption bands at 438 nm with a small shoulder peak at round 600 nm (Fig. 1). It has been explored that the surface plasmon band of metal nanoparticle depends on the shape, size and its surrounding medium. The anisotropic nanoparticles are known to exhibit characteristic bands corresponding to transverse and longitudinal plasmon absorption. This sort of spectra observed for Ag nanoparticles is attributed to the formation of anisotropic nanostructures rather than spherical. In order to confirm for the same, the transmission electron microscope (TEM) measurements has been carried out. Interestingly TEM image shows the formation of flower-like shape (Fig 2). The Ag nano flowers (AgNFs) have an average size in between 40 to 60 nm. Point should be noted here that present report describes a bio-friendly process for rapid synthesis of flower-like Ag nanocrystals without using any template, polymer and surfactant at room temperature. Fig 2 (F) is the selected area electron diffraction (SAED) pattern of the AgNFs; the clear spots and ordered arrangement reveals that the particles are of single crystalline in nature. For further investigation of surface morphology, the structural analyses of AgNFs were carried out by high-resolution TEM measurements. Fig 2 (D) shows the typical HRTEM of a branched AgNFs. The lattice fringe spacing of the AgNFs was measured to be 0.235 nm revealing that the growth of the nanostructure occurs preferentially on (111) planes of a face centred cubic lattice of Ag. A precise investigation and thorough analysis of AgNFs reveals the nanoparticles consist of twin boundaries. The HRTEM measurement thus exposed that the AgNFs are predominant of Ag(111) lattice planes along with the presence of twin boundaries.

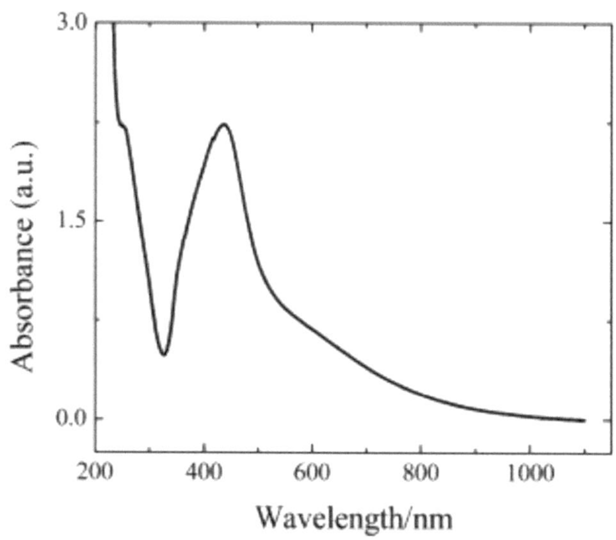

Fig. 1. UV-visible spectrum of Ag nanostructures. Reprinted with permission from (Jena et al., 2009), copyright 2009, American Chemical Society.

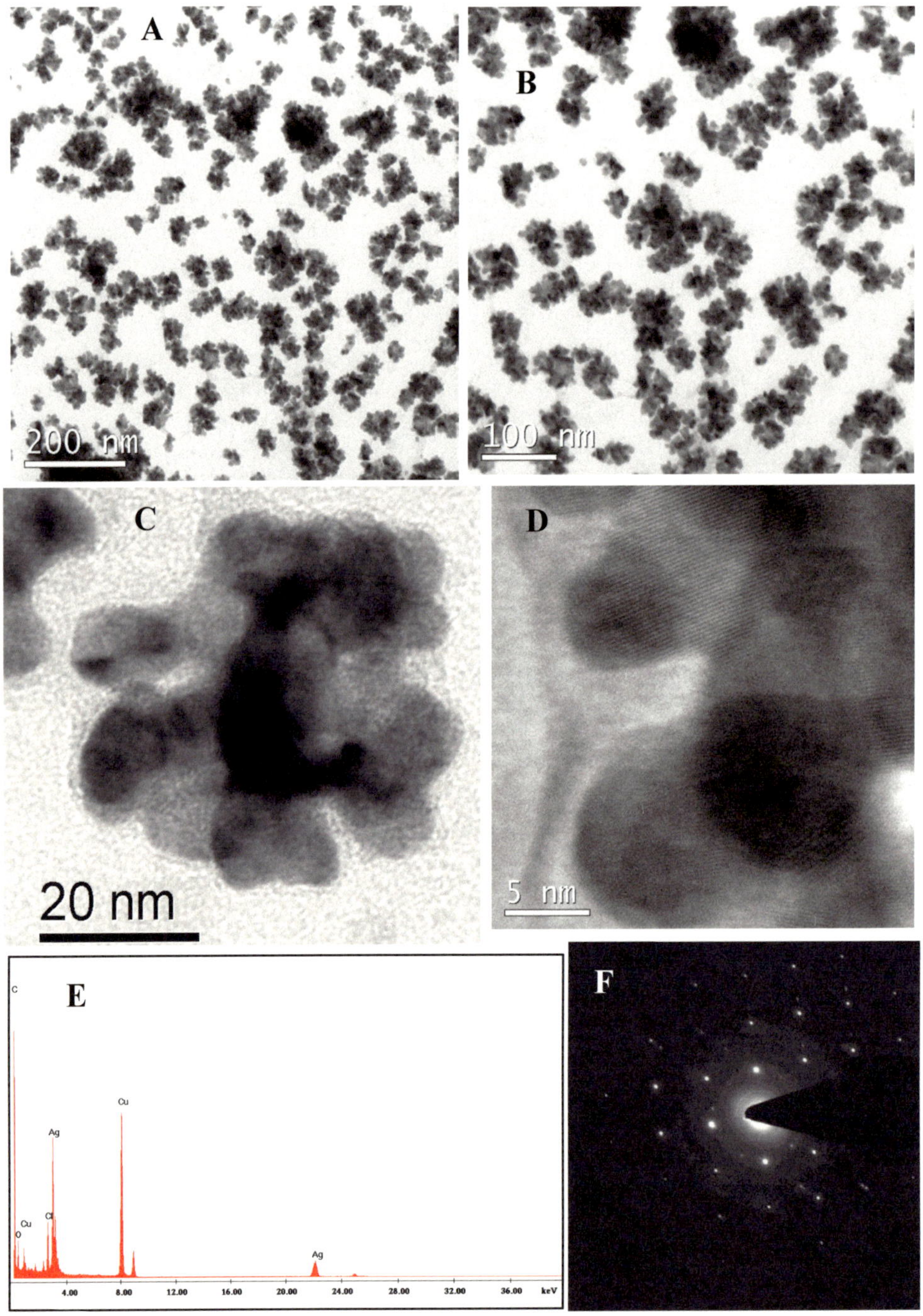

Fig. 2. TEM images of AgNFs induced by rutin

The energy dispersive spectrum confirmed that the AgNFs consist of only Ag (Fig. 2E). Interestingly, the XRD spectrum of AgNFs shows only one peak corresponding to Ag(111) (data not shown). Further the XPS characterization is made to confirm the elemental status. The XPS analysis clearly shows that the prepared AgNFs consist of elemental Ag (data not shown).

The growth of nanostructured particles is highly perceptive to the concentration of rutin. A facile and fruitful synthesis of AgNFs has been achieved at room temperature after manipulating and controlling the concentrations of rutin. By slightly changing the concentration of rutin, nanostructured particles with different morphology were obtained. First, we have examined the influence of concentration of rutin at a fixed concentration of the precursor (1 mM). The TEM images of Ag nanoparticles at different concentrations of rutin have measured (Fig. 3A-C). At higher concentration of rutin, we observed spherical nanoparticles. But at low concentration of rutin we obtained anisotropic nanoparticles. These results indicate the shape and morphology of the nanostructured particles greatly depends on the concentration of rutin. The growth of nanostructured particles is highly sensitive to the concentration of precursor. When the concentration of Ag^+ increases, we obtained nanoparticles with different morphology (Fig. 3D, E). Thus, it is cleared that the size and shape evolution of Ag nanoparticles are highly sensitive to the concentration of precursor as well as reducing/capping agent at normal conditions.

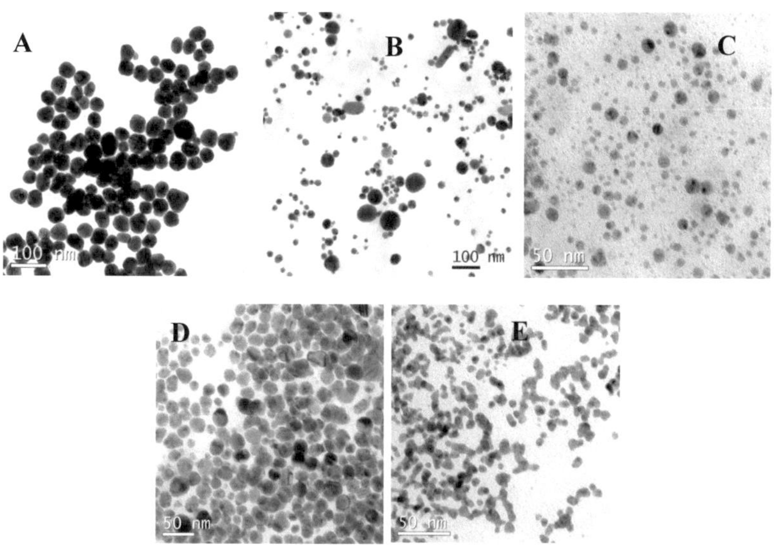

Fig. 3. TEM images of rutin-induced formation of Ag nanoparticles: A-C at fixed concentration (1mM) of Ag^+; [rutin] (A) 3, (B) 1, and (C) 0.5 mM and D, E at fixed concentration (0.3 mM) of rutin. [Ag+]: (A) 0.5 mM and (B) 2mM. Reprinted with permission from (Jena et al., 2009), copyright 2009, American Chemical Society.

The mechanism for the formation of flower shaped silver nanoparticles, were traced out by TEM measurements at different intervals of time over the entire period of formation of nanoparticles. It has been found that at initial stage of the reaction, small spherical shaped nanoparticles were formed and in due course of time, these small nanoparticles led to formation of initial anisotropic nanoparticles of short branches. Thereafter, a further nucleation and assembly process took place for the crystals to grow into a final flower shaped nanoparticles with many branches (Fig. 4).

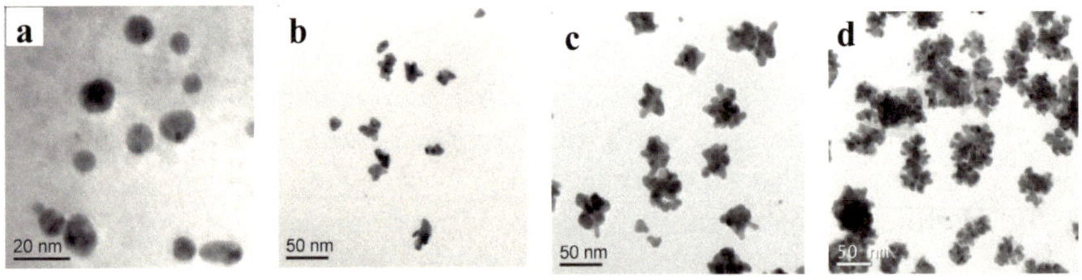

Fig. 4. TEM measurements showing the growth stages of AgNFs at different time intervals of the reaction. (a) 5 min (b) 10 min and (c) 20 min and (d) 30 min.

The formation of AgNFs is attributed to the reduction of Ag+ by rutin. It is well documented in the literature that rutin has the tendency to donate electrons by carbonilation of -OH groups at 3', 4' positions favouring a stable quinonic resonant structure (Fig 5A inset). Thus the formation of elemental Ag^0 is possible due to transfer of electrons from rutin to Ag^+ ions. The UV-visible spectra of rutin shows three absorption peak X, Y, and Z at 398, 320 and 270 nm respectively (Fig. 4 A (a)). The absorption peak X corresponds to B ring position and the intensity of the peak is due to the -OH groups at 3', 4' positions. It was observed that the intensity of peak at X diminishes on subsequent oxidation of rutin to its stable quinonic structure. As a matter of fact, He et al. have shown similar pattern on spectro-electrochemical oxidation of rutin (He et al 2007). In order to confirm the above fact, absorption spectra of supernatant of Ag nanoparticles were recorded. Surprisingly, a decrease in the intensity of X band with slight increase in the intensity of Y and Z bands were observed (Fig. 4Ab). This observation is attributed to the oxidation of rutin by carbonilation of -OH groups at 3', 4' positions to form a stable quinonic resonant structure simultaneously reducing Ag^+ to Ag^0 which undergoes nucleation in due course to produce Ag nanostructures (fig. 5 B).

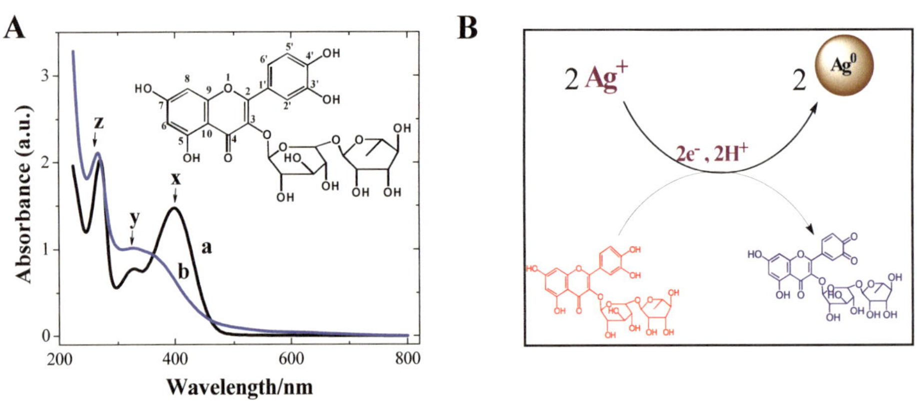

Fig. 5. (A) UV-visible spectra of 0.3 mM rutin (a) and its oxidation product (b). Inset shows the structure of rutin. (B) Scheme showing the possible mechanism towards reduction of Ag+ by rutin.

3.2 Antibacterial application of AgNFs

The antibacterial efficiency of AgNFs was examined against three representative microorganisms, Pseudomonas Aeruginosa (gram negative), S. Faecalis (gram positive), and Escherichia Coli (Gram-negative) of clinical interest. It is well known that E. coli is the most characterized bacterium and can cause gastroenteritis, urinary tract infections and neonatal meningitis; P. Aeruginosa causes chronic respiratory infections in individuals with cystic fibrosis and cancer; and S. Faecalis can cause endocarditic as well as bladder prostate and epididymal infections. Fig. 6X demonstrates the antibacterial activity of AgNFs that shows a clear zone of inhibition after 24 h incubation of the plate at 35 °C. The antibacterial efficacies of AgNFs are compared with the standard antibiotics (Fig. 6Y). Most diagnostically, the AgNFs found to show their potential antibacterial activity against Pseudomonas Aeruginosa, S. Faecalis and Echerichia Coli. Further, a certain amount of colloidal AgNFs was added into the bacterial broth medium and incubated for 24 h. It was found that the coagulation of the bacterial medium was absolutely converted to a transparent medium (Fig. 6Z). Hence, it harmonizes the dramatic antibacterial activity of colloid AgNFs.

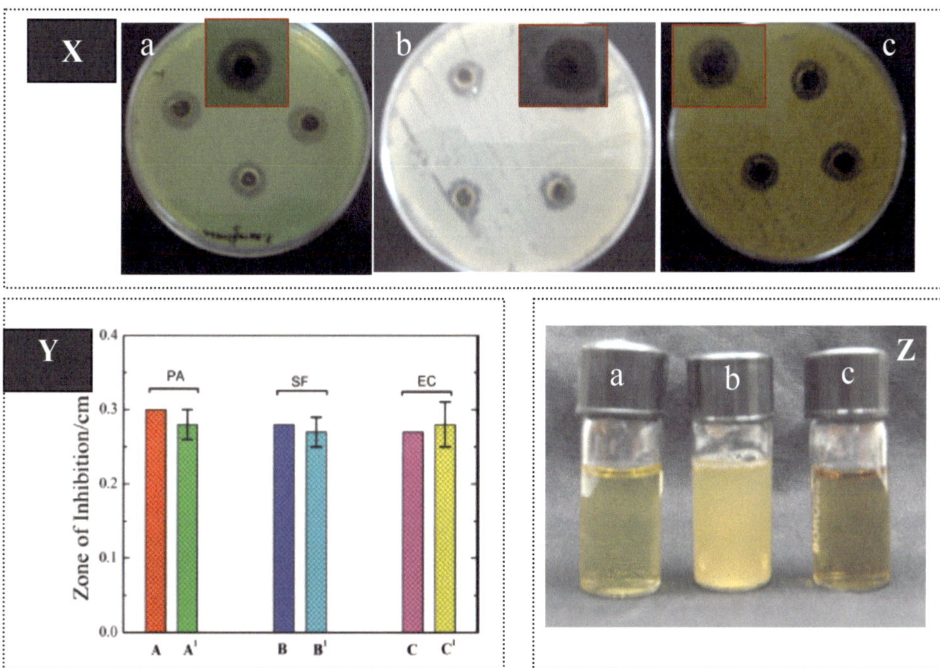

Fig. 6. (X) Diffusion disc showing the antibacterial activity of AgNFs towards (a) P. Aeruginosa (b) S. Faecalis and (c) E. Coli. (Y) Plots showing the antibacterial activity of AgNFs (A', B', C') with respect to the standard antibiotics, Gentamicin (A, C) and Penicillin (B). PA: Pseudomonas Aeruginosa; SF: S. Faecalis; EC: Escherichia Coli. (C) Photographic image showing the antibacterial activity of AgNFs (A) growth medium, (Z) after growth of bacteria (E. Coli) and (c) bacteria solution in presence of AgNFs after 24 h. Reprinted with permission from (Jena et al., 2009), copyright 2009, American Chemical Society.

3.3 Comparative bacterial inhibition studies

A Comparative bacterial inhibition studies were made with different Ag nanoparticles synthesized using rutin and citrate. The citrate stabilized nanoparticles have an average size of 30-40 nm. The antibacterial efficacies of four different nanoparticles were investigated against S. Faecalis bacterium (Fig. 7). As it can be seen (Fig. 7A), the inhibition zone is high in case of AgNFs as compared to other nanoparticles. So it can be inferred that the shape might have played a significant role for their potential antibacterial activity. The effective antibacterial activity of AgNFs can be ascribed to its higher tendency to react with sulphur and phosphorous containing compounds in the bacterial cell leading to bacterial death. It is also worthwhile to mention that the AgNF has higher antibacterial property which makes it a promising candidate material for clinical and industrial applications.

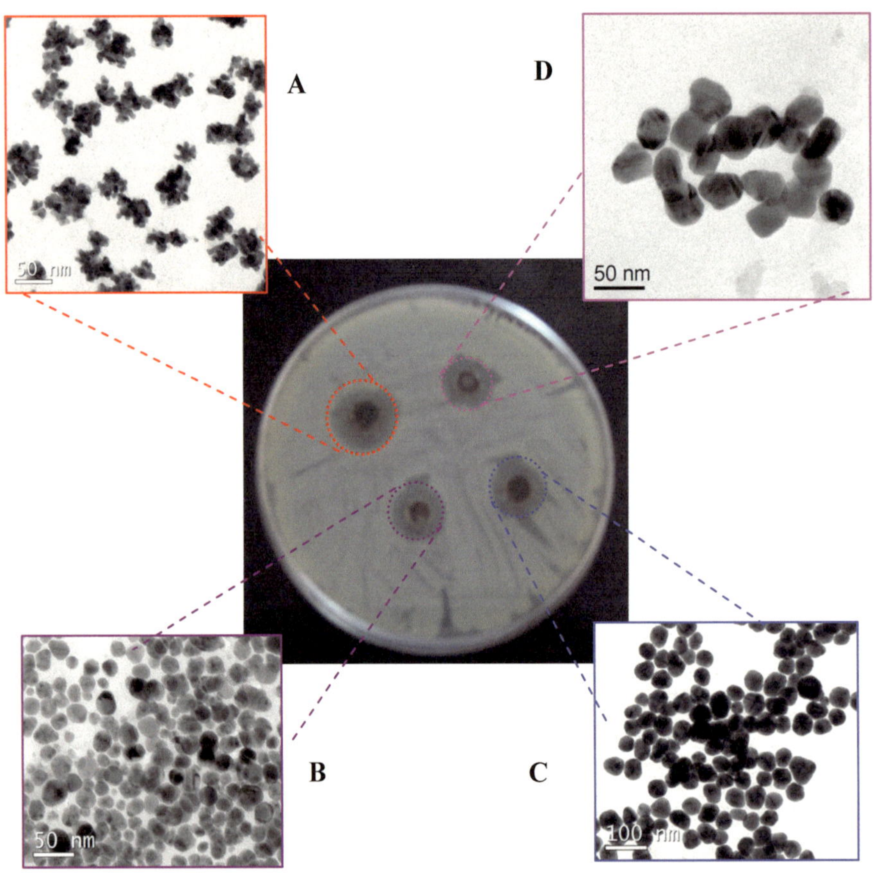

Fig. 7. Diffusion disc showing the antibacterial efficacies of four different shapes of nanoparticles induced by rutin (A, B, C) and citrate (D) against S. Faecalis bacterium. Reprinted with permission from (Jena et al., 2009), copyright 2009, American Chemical Society.

4. Conclusion

We have developed a bioinspired and environmental friendly procedure for rapid room temperature synthesis of flower like Ag nanoparticles circumventing the extra controls and additives. The as prepared AgNFs are potentially capable of attracting diverse biological/clinical applications. The AgNFs shows antibacterial activity towards E. Coli, P. Aeruginosa and S. Faecalis bacterium that are of clinical interest. The AgNFs show an improved potential antibacterial activity compared to its spherical counterparts. The shape of present Ag nanostructure plays a vital role to tune the improved antibacterial properties.

5. Acknowledgment

We are grateful to Dr. Mohan Rao, Indian Institute of Chemical Technology Hyderabad, India for XPS measurement. We thank Institute of Life Sciences and Institute of Technical Education and Research, Bhubaneswar for providing bacteria. SCS thanks CSIR-UGC for fellowship and we acknowledge financial support from CSIR, India.

6. References

Alivisatos, P. (2004). The use of nanocrystals in biological detection. *Nat. Biotechnol.* Vol.22, No.1, pp.47-52.

Alt, V.; Bechert, T.; Steinrucke, P.; Wagener, M.; Seidel, P.; Dingeldein, E.; Domann, E.; Schnettler, R. (2004). An in vitro assessment of the antibacterial properties and cytotoxicity of nanoparticulate silver bone cement. *Biomaterials*. Vol.25, No.18, pp.4383- 4391.

Bae, A. H.; Numata, M.; Hasegawa, T.; Li, C.; Kaneko, K.; Sakurai, K.; Shinkai, S. (2005). 1 D Arrangement of Au Nanoparticles by the Helical Structure of SchizoPhyllan:A Encounter of a Natural Product with Inorganic Compounds. *Angew. Chem., Int. Ed.* Vol.44, No.13, pp.2030-2033.

Banerjee, M.;Mallick, S.;Paul, A.; Chattopadhyay, A.; Ghosh, S. S. (2010). Heightened Reactive Oxygen Species Generation in the Antimicrobial Activity of a Three Component Iodinated Chitosan−Silver Nanoparticle Composite. *Langmuir*. Vol.26, No.8, pp.5901−5908.

Bera, R. K.; Raj, C. R. (2010). Enzyme-Cofactor-Assisted Photochemical Synthesis of Ag Nanostructures and Shape-Dependent optical Sensing of Hg(II)ions. *Chem. Mater.* Vol.22, No.15, pp.4505-4511.

Bosetti, M.; Masse, A.; Tobin, E.; Cannas, M. (2002). Silver coated materials for external fixation devices: in vitro biocompatibility and genotoxicity. *Biomaterials*.Vol. 23, No. 3, pp.887-892.

Braun, G.; Lee, S. J.; Dante, M.; Nguyen, T. Q.; Moskovits, M.; Reich, N. (2007). Surface-Enhanced Raman Spectroscopy for DNA detection by Nanoparticle Assembly onto Smooth Metal Films. *J. Am. Chem. Soc.* Vol.129, No.20, pp.6378-6379.

Brigger, I.; Dubernet, C.; Couvreur, P. (2002). Nanoparticles in cancer therapy and diagnosis. *Adv. Drug. Deliver. Rev.* Vol.54, No.5, pp.631-651.

Cao, Y. C. (2008). Nanomaterials for biomedical applications. *Nanomedicines.* Vol.3, No.4, pp. 467-469.

Chen, S.; Carroll, D. L. (2002). Synthesis and characterization of Truncated Triangular SilverNanoplates. *Nano lett.* Vol.2, No.9, pp.1003-1007.

Chen, S. H.; Fan, Z. Y.;Carroll, D. L. (2002). Silver Nanodisks: synthesis,Characterization,and Self-Assembly. *J. Phys. Chem. B.* Vol.106, No.42, pp.10777-10781.

Chen, Y. B.; Chen, L.; Wu, L. (2005). Structure-Controlled Solventless Thermolytic Synthesis of Uniform Silver Nanodisks. *Inorgan. Chem.* Vol. 44, No.26, pp.9817-9822.

Cui, Y.; Wei, Q. Q.; Park, H. K.; Lieber, C. M. (2001). Nanowire Nanosensors for Highly Sensitive and Selective Detection of Biological and Chemical Species. *Science* Vol.293, No.5533, pp.1289-1292.

Darbha, G. K.; Rai, U. S.; Singh, A. K.; Ray, P. C. (2008). Gold-nanorod-based sensing of sequence specific HIV-1 virus DNA by using hyper rayleigh scattering spectroscopy. *Chem. Eur. J.* Vol.14, No.13, pp.3896-3903.

De, M.; Ghosh, P. S.; Rotello, V. M. (2008). Applications of Nanoparticles in Biology. *Adv. Mater.* Vol. 20, No.22, pp.4225–4241.

Ducamp-Sanguesa, C.; Herrera-Urbina, R.; Figlarz, M. (1992). Synthesis and characterization of fine and monodisperse silver particles of uniform shape. *J. Solid State Chem.* Vol.100, No.2, pp.272-280.

Fox, C. L. J.; Modak, S. M. (1974). Mechanism of silver sulfadiazine action on burn wound infections. *Antimicrobial Agents and Chemotherapy*, Vol.5, No.6, pp.582-588.

Gao, X.; Cui, Y.; Levenson, R. M.; Chung, L.W. K.; Nie, S. (2004). In vivo cancer targeting and imaging with semiconductor quantum dots. *Nat. Biotechnol.* Vol.22, No.8, pp. 969-976.

Gardea-Torresdey, J. L.; Tiemann, K. J.; Gomez, G.; Dokken, K.; Tehuacanero, S.; Jose-Yacaman, M. (1999). Gold nanoparticles obtained by bio-precipitation from gold(III) solutions. *J. Nanopart. Res.* Vol.1, No.3, pp.397-404.

Gill, S.; Lobenberg, R.; Ku, T.; Azarmi, S.; Roa, W.; Prenner, E. J. (2007).Nanoparticles: Characteristics, Mechanisms of Action, and Toxicity in Pulmonary Drug Delivery—A Review .*J. Biomed.Nanotechnol.* Vol.3, No.2, pp.107-119.

Gogoi, S. K.; Gopinath, P.; Paul, A.; Ramesh, A.; Ghosh, S. S.;Chattopadhyay, A. (2006). Green Fluorescent Protein-Expressing Escherichia coli as a Model System for Investigating the Antimicrobial Activities of Silver Nanoparticles. *Langmuir*, Vol.22, No.22, pp.9322–9328.

Goodman, C. M.; McCusker, C. D.; Yilmaz, T.; Rotello,V. M. (2004). Toxicity of Gold Nanoparticles Functionalized with Cationic and Anionic Side Chains. *Bioconjugate Chem.* Vol.15, No.4, pp.897-900.

Grunes, J.; Zhu, J.; Anderson, E. A.; Somorjai, G. A. (2002). Ethylene Hydrogenation over Platinum Nanoparticle Array Model Catalysts Fabricated by Electron Beam Lithography: Determination of Active Metal Surface Area. *J. Phys. Chem. B*, Vol. 106, No.44, pp.11463-11468.

Gupta, A.; Silver, S. (1998). Molecular Genetics:Silver as biocide:Will resistance become a problem? *Nat. Biotechnol.* Vol.16, No.10, pp.888.

Haes, A. J.; Van Duyne, R. P. A. (2002). A Nanoscale Optical Biosensor: Sensitivity and Selectivity of an Approach Based on the Localized Surface Plasmon Resonance Spectroscopy of Triangular Silver Nanoparticles. *J. Am. Chem. Soc*, Vol. 124, No.35, pp.10596-10604.

Han, M.S .; Lytton-Jean, A. K. R.; Oh, B. K.; Heo, J.; Mirkin, C. A. (2006). Colorimetric Screening of DNA-Binding Molecules with Gold Nanoparticle Probes. *Angew. Chem. Int. Ed.* 2006, Vol. 45, No.11, pp.1807–1810.

Hao, E.; Kelly, K. L.; Hupp, J. T.; Schatz, G. C. (2002). Synthesisof Silver Nanodisks Using Polystyrene Mesospheres as Templates. *J. Am. Chem. Soc.* Vo.124, No.51, pp. 15182-15183.

Hazarika, P.; Ceyhan, B.; Niemeyer, C. M. (2004). Reversible Switching of DNA–Gold Nanoparticle Aggregation. *Angew. Chem., Int. Ed.* Vol.43, No.47, pp.6469- 6471.

He, J. –B.; Wang, Y.; Deng, N.; Zha, Z. –G.; Lin, X. –Q. (2007). Cyclic voltammograms obtained from the optical signals: Study of the successive electro-oxidations of rutin. *Electrochimica Acta,* Vol.52 No.24. pp.6665–6672.

Im, S. H.; Lee, Y. T.; Wiley, B.; Xia, Y. N. (2005). Large-Scale Synthesis of silver nanocubes: the Role of HCl in Promoting Cube Perfection and monodispersity. *Angew. Chem. Int. Ed.* Vol. 44, No.14, pp. 2154-2157.

Jena, B. K.; Mishra, B. K.;Bohider, S. (2009). Synthesis of Branched Ag Nanoflowers Based on a Bioinspired Technique: Their Surface Enhanced Raman Scattering and Antibacterial Activity. *J. Phys. Chem. C* . Vol.113, No.33, pp.14753-14758.

Jena, B. K.; Raj, C. R. (2006). Electrochemical biosensor based on integrated assembly of dehydrogenase enzymes and gold nanoparticles. *Anal. Chem.* Vol.78, No.18, pp. 6332-6339.

Jena, B. K.; Raj, C. R. (2008). Optical sensing of biomedically important polyionic drugs using nano-sized gold particles. *Biosens. Bioelectron.* Vol.23, No.8, pp.1285-1290.

Jena, B. K.; Sahu, S. C.; Satpati, B.; Sahu, R. K.; Behera, D.; Mohanty, S. (2011). A facile approach for morphosynthesis of Pd nanoelectrocatalysts. *Chem. Commun.*,Vol. 47, pp.3796-3798.

Jiang, X.; Zeng, Q.; Yu, A. (2006). A self-seeding coreduction method for shape controll of silver nanoplates. *Nanotechnology,* Vol. 17, No.19, pp.4929-4935.

Jin, R. C.; Cao, Y. W.; Mirkin, C. A.; Kelly, K. L.; Schatz, G. C.; Zheng, J. G. (2001). photoinduced Conversion of Silver nanospheres to Nanoprisms. *Science.* Vol.294, No.5548, pp.1901-1903.

Jin, Y. H.; Kannan, S.; Wu, M.; Zhao, J. X. J. (2007). Toxicity of Luminescent Silica Nanoparticles to Living Cells. *Chem. Res. Toxicol.* Vol.20, No.8, pp.1126-1133.

Kanaras, A. G.; Wang, Z.; Brust, M.; Cosstick, R.; Bates, A. D.(2007). Enzymatic Disassembly of DNA–Gold Nanostructures . *Small.* Vol. 3, No.4, pp.590-594.

Kannan, R.; Rahing, V.; Cutler, C.; Pandrapragada, R.; Katti, K. K.; Kattumuri, V.; Robertson, J. D.; Casteel, S. J.; Jurisson, S.; Smith, C.; Boote, E.; Katti, K. V. (2006). Nanocompatible Chemistry toward Fabrication of Target-Specific Gold Nanoparticles. *J. Am. Chem. Soc.* Vol.128, No.35, pp.11342-11343.

Katz, E.; Willner, I. (2004). Integrated Nanoparticle–Biomolecule Hybrid Systems: Synthesis, Properties, and Applications ,*Angew. Chem. Int. Ed.* Vol.43, No.45, pp.6042-6108.

Kattumuri, V.; Katti, K.; Bhaskaran, S.; Boote, E. J.; Casteel, S. W.; Fent, G. M.; Chandrasekhar, M.; Kannan, R; Katti, K. V. (2007). Gum Arabic as a Phytochemical Construct for the Stabilization of Gold Nanoparticles: In Vivo Pharmacokinetics and X-ray-Contrast-Imaging Studies. *Small.* Vol.3, No.2, pp.333-341.

Kim, F.; Connor, S.; Song, H.; Kuykendall, T.; Yang, P. D. (2004). Platonic Gold Nanocrystals. *Angew. Chem., Int. Ed.* Vol. 43, No.28, pp. 3673-3677.

Kneipp, K.; Kneipp, H.; Bohr, H. G. (2006). Single Molecule SERS Spectroscopy. *Series Topics in Applied Physic.* Vol.103,pp.261-277.

Kreibig, U.;Volmer, M. (1995), Optical properties of Metal Clusters. *Springer Series in Material Science.* Vol.25, Pages-532

Link, S.; El-Sayed, M. A. (1999). Spectral Properties and Relaxation Dynamics of Surface Plasmon Electronic Oscillations in Gold and Silver Nanodots and Nanorods. *J. Phys. Chem. B* . Vol.103, No.40, pp.8410-8426.

Lofton, C.; Sigmund, W. (2005). mechanisms controlling Crystal Habits of Gold and silver colloids. *Adv. Funct. Mater.* Vol. 15, No.7, pp.1197-1208.

Liu, J. W.; Lu, Y. (2004). Adenosine-Dependent Assembly of Aptazyme-Functionalized Gold Nanoparticles and Its Application as a Colorimetric Biosensor. *Anal. Chem.* Vol.76, No.6 ,pp.1627-1632.

Liu, J. W.;Lu,Y. (2006). Fast Colorimetric Sensing of Adenosine and Cocaine Based on a General Sensor Design Involving Aptamers and Nanoparticles. *Angew. Chem. Int. Ed.*, Vol. 45, No.1, pp.90-94.

Mirkin, C. A.; Letsinger, R. L.; Mucic, R. C.; Storhoff, J. J. (1996). A DNA- based method for rationally assembling nanoparticles into macroscopic materials. *Nature.* Vol.382, No.6592, pp.607-609.

Mohanpuria, P.; Rana, N. K.; Yadav, K. (2008). Biosynthesis of nanoparticles: technological concepts and future applications. *J Nanopart Res.* Vol.10, No.3, pp.507-517.

Mitragotri, S. (2009). In Drug Delivery, Shape Does Matter.*Pharm. Res.*, Vol.26, No.1, pp.232-234

Morones, J. R.; Elechiguerra, J. L.; Camacho, A.; Holt, K.; Kouri, J. B.; Ramirez, J. T.; Yacaman, M.J. (2005). The bactericidal effect of silver nanoparticles. *Nanotechnology.* Vol. 16, No.10, pp.2346-2353.

Mulvaney, P.(1996). Surface Plasmon Spectroscopy of Nanosized Metal Nanoparticles. *Langmuir.* Vol. 12, No.3, pp.788-800.

Nadagouda, M. N.; Varma, R. S.(2006). Green and controlled synthesis of gold and platinum nanomaterials using vitamin B2: density-assisted self-assembly of nanospheres, wires and rods. *Green Chemistry.* Vol.8, No.6, pp.516–518.

Nam, J. M.; Thaxton, C. S.; Mirkin, C. A. (2003). Nanoparticle-Based Bio-Bar Codes for the Ultrasensitive Detection of Proteins. *Science.* Vol.301, No.5641, pp.1884-1886.

Nicewarner-Peña, S. R.; Freeman, R. G.; Reiss, B. D.; He, L.; Pena, D. J.; Walton, I. D.; Cromer, R.; Keating, C. D.; Natan, M. J. (2001). Submicrometer Metallic Barcodes. *Science.* Vol. 294, No.5540, pp.137-141.

Pal, S.; Tak, Y. K.; Song, J. M. (2007). Does the Antibacterial Activity of Silver Nanoparticles Depend on the Shape of the Nanoparticle? A Study of the Gram-Negative Bacterium Escherichia coli. *Applied and environmental microbiology,* Vol.73, No. 6 pp.1712–1720

Pastoriza-Santos, I.; Liz-marjan, L. M. (2002). Synthesis of Silver Nanoprisms in DMF. *Nano lett.* Vol.2, No.8,pp.903-905.

Rosi, N. L.; Mirkin, C. A. (2005). Nanostructures in Biodiagnostics. *Chem. Rev.,* Vol. 105, No.4, pp.1547-1562.

Schultz, S.; Smith, D. R.; Mock, J. J.; Schultz, D. A. (2000). Single-target molecule detection with nonbleaching with multicolor optical immunolabels. *Proc. Natl. Acad. Sci. U.S.A.* Vol. 97, No.3, pp.996-1001.

Sengupta, S.; Eavarone, D.; Capila, I.; Zhao, G. L.; Watson, N.; Kiziltepe, T.; Sasisekharan, R. (2005). Temporal targeting of tumor cells and neovasculature with a nanoscale delivery system. *Nature.* Vol. 436, No.7050, pp. 568-572.

Shankar, S. S.; Rai, A.; Ankamwar, B.; Singh, A.; Ahmad, A.; Sastry, M. (2004). Biological synthesis of triangular gold nanoprisms. *Nat. Mater*, Vol.3, No.7,pp.482-488.

Shukla, R.; Nune, S. K.; Chanda, N.; Katti, K.; Mekapothula, S.; Kulkarni, R. R.; Welshons, W. V.; Kannan, R.; Katti, K. V. (2008). Soybeans as a phytochemical reservoir for the production and stabilization of biocompatible gold nanoparticles. *Small,* Vol.4, No.9, pp.1425-1436.

Singh, S.; Nalwa, H. S. (2007). Nanotechnology and Health Safety-Toxicity and Risk Assessments of Nanostructured Materials on Human Health. *J. Nanosci. Nanotechnol.* Vol.7, No.9, pp.3048-3070.

Stoeva, S. I.; Lee, J. S.; Smith, J. E.; Rosen, S. T.; Mirkin, C. A.(2006). Multiplexed Detection of Protein Cancer Markers with Biobarcoded Nanoparticle Probes. *J. Am. Chem. Soc.*, Vol. 128, No.26, pp.8378–8379.

Storhoff, J. J.; Elghanian, R.; Mucic, R. C.; Mirkin, C. A.; Letsinger, R. L.; (1998). One-Pot Colorimetric Differentiation of Polynucleotides with Single Base Imperfections Using Gold Nanoparticle Probes. *J. Am.Chem. Soc.*,Vol.120, No.9, pp.1959–1964;

Sun, Y.;Mayers, B.;Xia, Y. (2003). Transformation of Silver Nanospheres into Nanobelts and Triangular Nanoplates through a Thermal Process. *Nano Letters.* Vol.3, No.5, pp.675–679.

Sun, Y.; Xia, Y. (2002). Lage-Scale Synthesis of Uniform Silver Nanowires Through a Soft, Self-Seeding, Polyol Process. *Adv. Mater.*, Vol. 14, No.11, pp. 833-837.

Sun, Y.; Xia, Y. (2002). Shape-Controlled Synthesis of Gold and Silver Nanoparticles. *Science* Vol. 298, No.5601, pp. 2176-2179.

Sun, Y.; Yin, Y.; Mayers, B. T.; Herricks, T.; Xia, Y. (2002). Uniform silver Nanowires synthesis by reducing $AgNO_3$ with Ethylene Glycol in the Presence of Seeds and Poly (vinyl Pyrrolidone). *Chem. Mater.* Vol. 14, No.11, pp.4736-4745.

Turner, A. P. F. (2000). Biosensors--Sense and Sensitivity. *Science.* Vol.290, No.5495, pp.1315-1317.

Wang, C.; Ma, Z.; Wang, T.; Su, Z. (2006). Synthesis, Assembly, and Biofunctionalization of Silica-Coated Gold Nanorods for Colorimetric Biosensing. *Adv. Funct. Mater.* Vol. 16, No.13, pp.1673-1678.

Wang, J. (2005). Nanomaterial-Based Amplified Transduction of Biomolecular Interactions. *Small.* Vol.1, No.11, pp.1036–1043.

Wang, Y.; Li, D.; Ren, W.; Liu, Z.; Dong, S.; Wang, E. (2008). Ultrasensitive colorimetric detection of protein by aptamer–Au nanoparticles conjugates based on a dot-blot assay. *Chem. Commun.*, pp.2520-2522.

Wei, H.; Li, B.; Li, J.; Wang, E.; Dong, S.; (2007) Simple and sensitive aptamer-based colorimetric sensing of protein using unmodified gold nanoparticle probes. *Chem. Commun.*, pp.3735-3737

Wijaya, A.; Schaffer, S. B.; Pallares, I. G.; Hamad-Schifferli, K. (2009). Selective Release of Multiple DNA Oligonucleotides from Gold Nanorods. *ACS Nano.* Vol.3, No.1, pp.80-86.

Wiley, B.; Sun, Y.; Mayers, B.; Xia, Y. (2005). Shape-Controlled Synthesis of Metal Nanostructures: The Case of Silver. *Chem. Euro. J*, Vol.11, No.2, pp.454-463.

Wiley, B.;Im, S. H.; Li, Z. Y.;McLellan, J.; Siekkinen, A.; Xia, Y. (2006). Maneuvering the Surface Plasmon Resonance of Silver Nanostructures through Shape-Controlled Synthesis. *J. Phys. Chem. B*. Vol.110, No.32, pp.15666–15675.

Wiley, B.; Herricks, T.; Sun, Y.; Xia, Y. (2004). Polyol Synthesis of Silver Nanoparticles: Use of Chloride and Oxygen to Promote the Formation of Single-Crystal, Truncated Cubes and Tetrahedrons. *Nano Letters*. Vol. 4, No.9, pp.1733–1739.

Wu, X.; Liu, H.; Liu, J.; Haley, K. N.; Treadway, J. A.; Larson, J. P.; Ge, E.; Peale, F.; Bruchez., M. P. (2003).Immunofluorescent labeling of cancer marker Her2 and other ellular targets with semiconductor quantum dots. *Nat.Biotechnol.* Vol.21, No.1, pp.41-46.

Xiang, J.; Lu, W.; Hu, Y.; Wu,Y.; Yan, H.; Lieber, C. M.(2006). Ge/Si nanowire heterostructures as high-performance field-effect transistors. *Nature,* Vol.441, No.7092, pp.489-493.

Xiao, Y.; Patolsky, F.; Katz, E.; Hainfeld, J. F.; Willner, I. (2003). Plugging into Enzymes: Nanowirings of Redox Enzymes by a Gold Nanoparticle. *Science.* Vol.299, No.5614, pp.1877-1881.

Xue, C.; Mtraux, G.; Millstone, J. E.; Mirkin, C. A. (2008). Mechanistic Study of Photomediated Triangular Silver Nanoprism Growth. *J. Am. Chem. Soc* Vol. 130, No.26, pp. 8337-8344.

Yu, D.; Yam, V. W.(2004). Controlled Synthesis of Monodisperse silver Nanocubes in Water. *J. Am. Chem. Soc.,* Vol. 126, No.41, pp.13200-13201.

Yu, L.; Banerjee, I. A.; Matsui, H. (2003). Direct Growth of Shape-Controlled Nanocrystals on Nanotubes via Biological Recognition. *J. Am. Chem. Soc.*,Vol.125, No.48, pp.14837-14840.

Zhang, G.; Keita, B.; Dolbecq, A.; Mialane, P.; Secheresse, F.; Miserque, F. (2007). Green Chemistry-Type One-Step Synthesis of Silver Nanostructures Based on MoV–MoVI Mixed-Valence Polyoxometalates. *Chem. Mater.* Vol.19, No.24, pp.5821-6058.

Zhang, L.; Shen, Y.; Xia, A.; Li, S.; Jin, B.; Zhang, Q. (2006). One-Step Synthesis of Monodisperse Silver Nanoparticles beneath Vitamin E Langmuir Monolayers. *J. Phys. Chem. B*. Vol.110, No.13, pp.6615–6620.

3

Lutamide, a New Ceramide Isolated from the Leaves of *Ficus lutea*

Herve Martial Poumale Poumale
Department of Organic Chemistry, Faculty of Science, University of Yaoundé I, Yaoundé, Cameroon

1. Introduction

The Moraceae family consists of about 50 genera and nearly 1400 species including important group such as *Artocarpus*, *Morus* and *Ficus*.[1] The genus *Ficus* consists of trees and shrubs that possess latex-like material within their vasculatures, affording protection and self-healing for physical assaults.[2] A number of *Ficus* species are used as food and for medicinal properties in traditional Chinese medicine especially amongst people where these species grow.[3] *Ficus benjamina* is used as ornamental plant in University of Yaounde I, Cameroon.[4] Previous phytochemical studies on the wood of *Ficus lutea* resulted in the isolation of benjaminamide (**2**), *β*-amyrin, *β*-amyrin acetate, lupeol, betulinic acid, *β*-sitosterol glucoside and lutaoside.[5] The strong antioxidant and antibacterial activities exhibited by this genus[6] in addition to the search for the chemical constituents of Cameroonian medicinal plants[7] justified further attempts to isolate and identify active compounds. The few differences between the secondary metabolites isolated from the wood and the leaves of *F. lutea* are may be related to the real specific differences or more probably to a geographic or environmental influence on biosynthesis.

2. Results and discussion

The leaves of *F. lutea* were extracted with MeOH during 30 hours. The extract was submitted to repeated column chromatography to afford benjaminamide (**2**), betulinic acid, 9,19-cycloart-25-ene-3β,24-diol, vitexin as well as one new ceramide (**1a**). The ^{1}H and ^{13}C NMR, and MS of the known compounds were consistent with those reported in the literature.

Lutamide (**1a**) was obtained as an amorphous solid. The molecular formula $C_{34}H_{64}NO_4$ was determined by HRFABMS at *m/z* 550.48348 $[M-H]^-$ (Calcd. 550.48351). The IR spectrum of **1a** indicated absorption bands at ν 3405 cm^{-1} (OH), and strong absorption bands for a secondary amide at ν 1639 and 1590 cm^{-1}. These data were supported by the signals at δ 52.5 and 174.8 in ^{13}C NMR spectrum which confirm the presence of C-N and C=O, respectively. The ^{1}H and ^{13}C NMR spectral data (table 1) of **1a** indicated the presence of an amide linkage, two long chain aliphatic moieties, suggesting the sphingolipid (glycolipid) nature of the molecule. 1D and 2D NMR spectral data of **1a** were nearly superimposable to that of lutaoside (**1c**) which was further isolated from the wood extract of this plant.[5] A careful comparison of the spectra data of **1a** and lutaoside (**1c**) let to the conclusion that, the structure of lutamide (**1a**) was (*2R*)-2-hydroxy-*N*-((*2S,3R,5E,12E*)-1,3-dihydroxyoctadeca-5,12-dien-2-yl)hexadecanamide (**1a**), which is reported here for the first time.

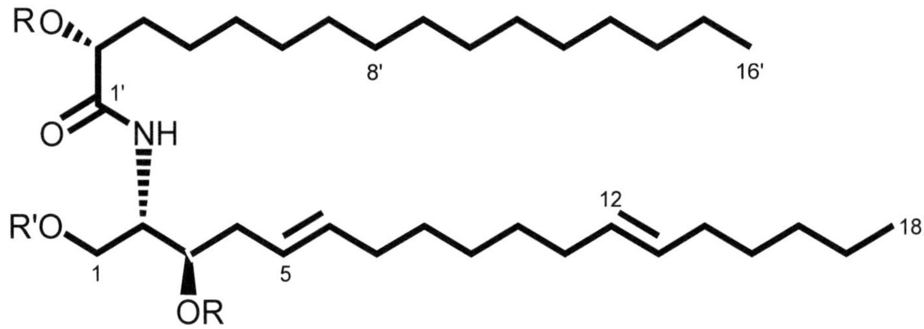

1a: R = R′ = H; **1b**: R = R′ = CH_3-C=O; **1c**: R = H, R′ = *β*-D-glucopyranosyl

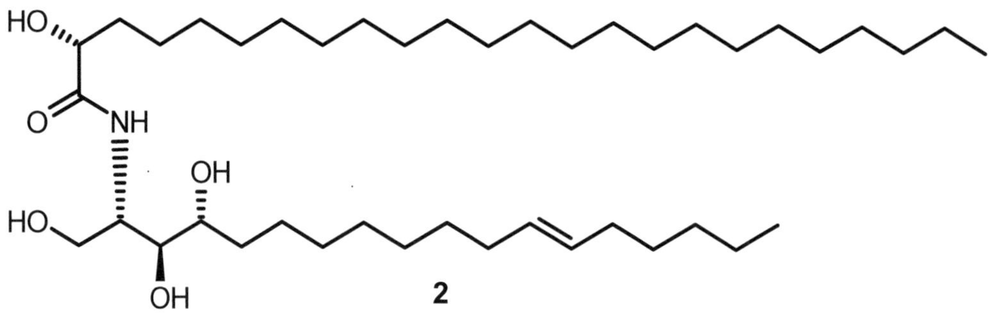

2

Acetylation of compound **1a** gave **1b** ($C_{40}H_{70}NO_7$; m/z 676.51509 $[M-H]^-$; Calcd. 676.51520).

Position	δ_C	δ_H
1*a*	68.2 (t)	4.58 (dd, 10.7; 4.9)
1*β*	68.2 (t)	4.40 (dd, 10.7; 4.3)
2	52.5 (d)	5.18 (m)
3	75.7 (d)	4.39 (m)
4	128.5$^{\varepsilon}$ (d)	5.50$^{\zeta}$ (dd, 15.4; 5.3)
5	128.1$^{\varepsilon}$ (d)	5.10$^{\zeta}$ (dt, 15.4; 4.7)

Position	δ_C	δ_H
6	33.5 (t)	2.25 (m)
7-10, 15-17	24.0-26.5 (t each)	1.27 (br s)
11	33.0 (t)	2.11 (m)
12	130.0^{ε} (d)	5.50^{ζ} (dd, 15.0; 4.8)
13	129.0^{ε} (d)	5.44^{ζ} (dd, 15.0; 4.6)
14	32.1 (t)	1.99 (m)
18	13.0 (q)	0.90 (t, 6.4)
NH	-	8.50 (d, 8.0)
1′	174.8 (s)	-
2′	73.1 (d)	4.20 (t, 7.3)
3′	31.0 (t)	1.78 (m)
4′-14′	27.8-29.0 (t each)	1.27 (br s)
15′	21.9 (t)	1.70 (m)
16′	12.7 (q)	0.89 (t, 6.4)

Multiplicities and coupling constants in Hz are given in parentheses
Resonances with the same superscripts (ε, ζ) in the same column may be interchanged.

Table 1. ^{1}H (400 MHz, C_5D_5N, 30 °C, TMS) and ^{13}C (100 MHz, C_5D_5N) NMR data of lutamide (**1a**)

The antifungal and antibacterial activities of compounds **1a, 1b** and **2** were determined using the agar diffusion method with 8 mm paper disks loaded with 40 μg of each compound (See Table 2). Compound **1a** and **1b** exhibited *in vitro* good antimicrobial activity against *Mucor miehi* and *Bacillus subtilis* compared to the nystatin as reference.

Micro-organisms tested	**Sample**			
	1a	**1b**	**2**	**Nystatin**
Chlorella vulgaris	10	11	-	-
Scenedesmus subspicatus	13	10	10	-
Chlorella vulgaris	11	9	11	-
Mucor miehei	15	15	13	15
Bacillus subtilis	16	15	14	14
Candida albicans	12	13	13	15
Streptomyces viridochromogenes	-		-	14

Diameter of inhibition zone in mm. Nystatin was used as reference and the experiments were repeated 3 times.

Table 2. Antimicrobial activity of compounds **1a, 1b** and **2**

3. Experimental section

3.1 Materials and method

Melting point is uncorrected and was obtained with a micro melting point apparatus (Yanaco, Tokyo-Japan). Optical rotation values were measured with a Horiba SEPA-300 polarimeter, and IR spectra were recorded with JASCO J-20A spectrophotometer. ^{1}H and ^{13}C NMR spectra were acquired with a Jeol EX-400 spectrometer. Chemical shifts are given on a δ (ppm) scale with TMS as an internal standard. Mass spectra were obtained with a Jeol JMS-700 instrument. Column chromatography was conducted on silica gel 60 (Kanto Chemical Co., Inc., Japan), Sephadex LH-20 (Pharmacia, Sweden) and ODS (Fuji Silysia, Japan). TLC analysis was carried out by using precoated silica gel plates (Merck), and the spots were detected by spraying with H_2SO_4/10% vanillin and then heating. Flash chromatography was carried out on silica gel (230-400 mesh). R_f values were measured on Polygram SIL G/UV254 (Macherey-Nagel & Co.).

3.2 Plant material

The leaves of Ficus lutea Vahl were collected in July 2008 at Kribi, South Cameroon. A voucher specimen (Ref. N°. 3471/SRFK) has been deposited in the National Herbarium, Yaoundé, Cameroon.

3.3 Extraction and isolation

The powdered leaves of *Ficus lutea* (2 Kg) were soaked in 10 l of MeOH during 30 hours at room temperature. Solvent was removed under reduced pressure and 60 g of organic extract were obtained. Part of this dark-green residue (58 g) was subjected to vacuum liquid chromatography (VLC) on silica gel and eluted with pure *n*-hexane (Fraction A), followed by mixture of *n*-hexane/ethyl acetate in incremental steps 50%, 100% (Fractions B, C respectively) and finally 10% of the mixture of ethyl acetate/methanol (Fraction D). Four main fractions (A-D) were obtained and, basis of analytical TLC, fractions C and D were combined.

Fraction A (7 g) gave mainly betulinic acid (53.0 mg)[9] and vitexin (11 mg).[10]

Fractions B (6 g) were chromatographed on silica gel and eluted with a mixture of *n*-hexane/ethyl acetate of increasing polarity to yield 54 fractions (ca. 100 ml each). Fractions 1-32 (2 g), subjected to column chromatography over silica gel, yielded mainly 9,19-cycloart-25-ene-3β,24-diol (33 mg)[11] while benjaminamide (**2**, 5 mg) was obtained in fractions 33-54 (3 g) eluted with $CHCl_3$/MeOH (6:1).

Fraction C and D (21 g) was passed through a Sephadex LH-20 column and subjected to silica gel column chromatography and preparative TLC to afford benjaminamide (**2**, 19 mg) and lutamide (**1a**, 34 mg).

*Lutamide or (2R)-2-hydroxy-N-((2S,3R,5E,12E)-1,3-dihydroxyoctadeca-5,12-dien-2-yl)hexadecanamide (**1a**)*: Amorphous powder. – R_f = 0.44 (CH_2Cl_2/10% MeOH). – $[\alpha]_D^{25}$ +19 (*c* 0.6, MeOH). – IR (Film): ν = 3405 (OH), 3201 (NH), 2914, 2853, 1639, 1590, 1418, 1217, 1177, 1078, 1057, 1039, 890 cm^{-1}. – ^{1}H NMR (400 MHz, C_5D_5N, 30 °C, TMS) and – ^{13}C NMR (100 MHz, C_5D_5N): see Table 1. - FABMS: *m/z* 550 $[M-H]^-$. - HRFABMS: *m/z* 550.48348 $[M-H]^-$ (Calcd. 550.48351 for $C_{34}H_{64}NO_4$, $[M-H]^-$).

3.4 Acetylation of lutamide (1a)

Lutamide (**1a**, 6.0 mg) was dissolved in pyridine (1.5 mL) and Ac_2O (1.2 mL). The solution was stirred for 10 hours at 50 °C. The usual work-up gave three-acetoxylutaoside (**1b**) (4.1 mg, 84 %) as an amorphous solid with R_f = 0.94 ($CHCl_3$/10% MeOH). - $[\alpha]_D^{25}$ +23 (*c* 0.9, Pyridine). - IR (Film): ν = 3203 (NH), 2905, 2843, 1653, 1579, 1463, 1217, 1100, 886 cm^{-1}. - ^1H NMR (400 MHz, C_5D_5N, 30 °C, TMS): δ = 0.87 (t, *J* = 6.0 Hz, 6 H, H-18, H-16′), 1.20-1.30 (br s); 1.48-1.55 (m, 4 H, H-3′, H-15′), 1.90 (m, 8 H, H-4, H-7, H-11, H-14), 2.05, 2.06, 2.12 (s, 3 H each, CH_3-C=O), 5.17 (dd, *J* = 11.0, 3.5 Hz, 1 H, H-1a), 5.25 (dd, *J* = 11.0, 5.0 Hz, 1 H, H-1b), 5.42-5.48 (m, 4 H, H-5, H-6, H-12, H-13), 5.51 (m, 1 H, H-2′), 5.53 (m, 1 H, H-3), 8.49 (d, *J* = 8.0 Hz, 1 H, NH). - FABMS: *m/z* 676 $[M-H]^-$. - HRFABMS: *m/z* 676.51509 $[M-H]^-$ (Calcd. 676.51520 for $C_{40}H_{70}NO_7$, $[M-H]^-$).

4. Antimicrobial assay

Agar diffusion tests were performed in the usual manner[8] with *Bacillus subtilis* and *Escherichia coli* (on peptone agar), *Staphylococcus aureus* (Bacto nutrient broth), *Streptomyces viridochromogenes* (M Test agar), the fungi *Mucor miehei* and *Candida albicans* (Sabouraud agar), and three microalgae (*Chlorella vulgaris, Chlorella sorokiniana* and *Scenedesmus subspicatus*).

Compounds were dissolved in an azeotrope chloroform/MeOH (87:13) and 40 μg pro paper disks (Ø 8 mm) were impregnated with each using a 100 μl syringe, dried for 1 h under sterile conditions and placed on the pre-made agar test plates. Bacteria and fungi plates were kept in an incubator at 37 °C for 12 h, micro algae plates for three days at room temperature in a day light incubator. The diameter of inhibition zones was measured.

5. References

[1] K. Venkataraman, *Phytochemistry* 1972, *11*, 1571-1586.

[2] C. C. Berg, *Experientia* 1989, *45*, 605-611.

[3] E. P. Lansky, H. M. Paavilainen, A. D. Pawlus, R. A. Newman, *J. Ethnopharmacol.* 2008, *119*, 195-213.

[4] C. C. F. Simo, S. F. Kouam, H. M. P. Poumale, I. K. Simo, B. T. Ngadui, I. R. Green, K. Krohn, *Biochem. Syst. Ecol.* 2008, *36*, 238-243.

[5] H. M. P. Poumale, A. V. B. S. Djoumessi, B. Ngameni, L. P. Sandjo, B. T. Ngadjui, Y. Shiono, *Acta Chim. Slov.* 2011, *58*, 81–86.

[6] J. B. Harborne, H. Baxter, *Phytochemical Dictionary*; 2nd ed. Taylor and Francis, London, 1999.

[7] B. T. Ngadjui, B. M. Abegaz, *Studies in Natural Products Chemistry, Bioactive Natural Products*; Atta-ur-Rahman Ed., Elsevier: Oxford, 2003; Vol. 29, pp. 761-805.

[8] H. M. P. Poumale, R. T. Kengap, J. C. Tchouankeu, F. Keumedjio, H. Laatsch, B. T. Ngadjui, *Z. Naturforsch.* 2008, *63b*, 1335-1338.

[9] C. Gauthier, J. Legault, S. Rondeau, A. Pichette, *Tetrahedron Lett.* 2009, *50*, 988-991.

[10] R. Heinsbroek, J. V. Brederode, G. V. Nigtevecht, J. Maas, J. Kamsteeg, E. Besson, J. Chopin, *Phytochemistry* 1980, *19*, 1935-1937.

[11] I. Smith-Kielland, J. M. Dornish, K. E. Malterud, G. Hvistendahl, C. Romming, O. C. Bockman, P. Kolsaker, Y. Stenstrom, A. Nordal, *Planta Madica* 1996, *62*, 322-325.

4

Antibacterial Modification of Textiles Using Nanotechnology

Moustafa M. G. Fouda
*Petrochemical Research Chair,
Department of Chemistry, College of Science, King Saud University,
KSA*

1. Introduction

This chapter is undertaken with a view to survey important scientific research and developmental works pertaining to antibacterial modification of textiles using nanotechnology as a new means to achieve such textiles. Inevitably, conventional antimicrobial agents and their applications to textiles are reported. This is followed by a focus on inorganic nanostructured materials that acquire good antibacterial activity and application of these materials to the textiles. Evaluation of the antibacterial efficacy is described. An outlook which envisions the importance of using nanotechnology in the antibacterial finishing of textiles is also outlined.

2. History

During World War II, when cotton fabrics were used extensively for tentage, tarpaulins and truck covers, these fabrics needed to be protected from rotting caused by microbial attack. This was particularly a problem in the South Pacific campaigns, where much of the fighting took place under jungle like conditions. During the early 1940 s, the US army Quartermaster Crops collected and compiled data on fungi, yeast and algae isolated from textiles in tropical and subtropical areas throughout the world. Cotton duck, webbing and other military fabrics were treated with mixtures of chlorinated waxes, copper and antimony salts that stiffened the fabrics and gave them a peculiar odour. At the time, potential polluting effects of the application of, these materials and toxicity-related issue were not a major consideration.

After World War II, and as late as the mid-to-late 1950.s fungicides used on cotton fabrics were compounds such as 8-hydroxygiunoline salts, copper naphthenate, copper ammonium fluoride and chlorinated phenals. As the government and industrial firms became more aware of the environmental and workplace hazards these compounds caused. Alternative products were sought. A considerable amount of work was done by the Southern Regional Research Laboratory of the US Department of Agriculture, the Institute of Textile Technology (ITT) and some of the ITT.s member mills to chemically modify cotton to improve its resistance to rotting and improve other properties by acetylation and cyanoethylation of cotton. These treatments had limited industry acceptance because of relatively high cost and loss of fabric strength in processing. In addition, the growing use of man-made fibres such as nylon, acrylics and polyester, which have inherent resistance to microbial decomposition, came into wider use to replace cotton in many industrial fabrics [2].

3. Introduction

Clothing and textile materials are good media for growth of microorganisms such as bacteria, fungi. According to recent reports, microorganisms could survive on fabric materials for more than 90 days in a hospital environment. Such a high survival rate of pathogens on medically used textiles may contribute to transmissions of diseases in hospitals. As a means to reduce bacterial population in healthcare settings and possibly to cut pathogenic infections caused by the textile materials, utilization of antimicrobial textiles in healthcare facilities is considered to be a potential solution[3].

Textile surface modification provides a way to impart new and diverse properties to textiles while retaining comfort and mechanical strength. Currently, functional finishes on textile fabrics are of critical importance to improve textile products with multifunctional properties. Textile finishes can be divided into aesthetic and functional finishes. Aesthetic finishes are finishes used to modify the appearance or hand of a fiber or fabric. They can alter the texture, luster, or drape of a textile material. Mechanical and chemical processes may be used to impart an aesthetic finish, this type of finishing with a greater emphasis being placed on mechanical processes. Many different chemicals and processes are used in the finishing of textile materials. Functional Finishes that alter fiber or fabric performance, maintenance, durability, safety, and environmental resistance can be considered as functional finishes. Functional finishes are generally applied specifically to alter properties related to care, comfort, and durability. Most functional fabric properties are imparted by using chemical and wet processing methods [4, 5]. Some common functional finishes are listed below:

Antimicrobial, Antistatic, Durable press, Flame resistant/retardant, Soil release/resistant, Water proof/repellent, UV Protection, Self cleaning, Wrinkle recovery. This chapter will focus on the antimicrobial finishes of textiles. The driving force behind the chemical finishing of cotton during the next 10 years is anticipated to comprise several factors. Of these factors, mention is made of the following: (i) chemical finishes which maximize the added value; (ii) chemical finishes which are friendly with the environment; (iii) methods which are convenient for application, and (iv) the need for better quality and minimum use of water and energy [6].

In recent years, antimicrobial finishing of textiles has become extremely important in the production of protective, decorative and technical textile products. This has provided opportunities to expand the use of such textiles to different applications in the textile, pharmaceutical, medical, engineering, agricultural, and food industries [7]. Antimicrobial finishing of textiles protects users from pathogenic or odor-generating microorganisms, which can cause medical and hygienic problems, and protects textiles from undesirable aesthetic changes or damage caused by rotting, which can result in reduced functionality. As a consequence of their importance, the number of different antimicrobial agents suitable for textile application on the market has increased dramatically. These antimicrobial agents differ in their chemical structure, effectiveness, method of application, and influence on people and the environment as well as cost [8-12]. In the literature [9, 10, 12], there are several different classifications of antimicrobial agents according to efficiency, mechanism of antimicrobial activity and washing resistance. According to these studies, products can be divided into biocides and biostats, leaching and bound antimicrobials, controlled-release and barrier-forming agents, and agents of poor and of good washing resistance. In general, the activity of antimicrobial finishes can be biocidal or biostatic. While the biocides (bactericides and fungicides) include agents that kill bacteria and fungi, the biostats (bacteriostats and fungistats) inhibit the microorganisms' growth. The mode of action is directly related to the concentration of the active substance in the textile. The minimum inhibitory concentration (MIC) is required for biostatic activity, but the minimum biocidal concentration (MBC) should be exceeded for biocidal activity.

The majority of antimicrobial agents in the textile industry utilize a controlled-release mechanism [12]. These agents, which are also called, leaching antimicrobials [9], are not chemically bound to the textile fibers and their antimicrobial activity is attributed to their gradual and persistent release from the textile into their surroundings in the presence of moisture, where they act as a poison to a wide spectrum of bacteria and fungi. The antimicrobial efficiency depends directly on the concentration, which should not drop below the MIC. Owing to leaching of the agent into its surroundings, the concentration of the active substance in the textile decreases and gradually falls under the limit of effectiveness. This can induce resistance to these substances in microorganisms; in addition, leaching agents do not withstand repeated laundering. A controlled release mechanism can also be found in agents that are chemically incorporated into the fiber surface, but with an active substance that is leachable in water. The important advantage of these agents over other leaching antimicrobials is that they can be regenerated under appropriate conditions.

The bound antimicrobials [9] include finishes that are chemically bound to the surface of the textile fibers, where they act as a barrier and control microorganisms which come into contact with the fiber surface. Because these agents do not leach into the surroundings of the textile substrate, the probability of microorganisms developing resistance to them is small. Covalent binding of the agent to the textile surface can be ensured if there are enough reactive groups in the agent and the fibers, and if the application process is carried out under suitable conditions. Accordingly, when using bound antimicrobials, the mechanism of chemical binding to the textile surface and the conditions that initiate or catalyze the reaction should be known. Bound antimicrobials are much more resistant to repeated laundering in comparison to leaching agents. However, washing durability of the agent cannot assure its durability of antimicrobial function. The latter could decrease or even expire with the adsorption of dirt, deadly microorganisms or complex formation between the finish and the anionic detergent during laundering. For antimicrobial finishing in the textile industry, it is not only the antimicrobial efficiencies of the agents that are important, the environmental, health and safety aspects of their use must also be taken into account [7].

It should be stressed that the release of finishes from the textile into the surroundings could have negative impacts on living organisms in water because they can affect susceptible bacteria, thereby potentially selecting resistant bacteria. Fortunately, bound antimicrobials are environmentally friendly finishes with no leaching of toxic products into the surroundings. Taking these facts into account, much research has focused on the synthesis of novel antimicrobial agents where leaching antimicrobials have been replaced with bound antimicrobials. Some of the most important examples are presented in this chapter.

4. Conventional Antimicrobial Agents for textiles

4.1 Quaternary ammonium compounds

Cationic surface active agents (cationic surfactants), including particularly quaternary ammonium salts (QASs), are important biocides that for many years have been known to be effective antiseptic and disinfectant agents [8-10],[12, 13]. As antimicrobial agents for textiles, monoammonium and "gemini" or "dimeric" ammonium surfactants (Figure 1) with an alkyl, alkylaryl and perfluorinated hydrocarbon group are used [14, 15]. These are active against a broad spectrum of microorganisms such as Gram-positive and Gram-negative bacteria, fungi and certain types of viruses [16]. The antimicrobial activity of QASs depends on the length of the alkyl chain, the presence of the perfluorinated group and the number of cationic ammonium groups in the molecule. The antimicrobial function arises from attractive interactions between the cationic ammonium group of the QAS and the negatively charged cell membrane of the microbe; these interactions consequently result in the formation of a surfactant–microbe complex. This in turn causes the interruption of all essential functions of the cell membrane and thus the interruption of protein activity [17]. QASs also affect bacterial DNA, causing a loss of multiplication ability [18]. If the long hydrocarbon chain is bonded to the cationic ammonium in the structure of the QAS, two types of interactions between the agent and the microorganism can occur: a polar interaction with the cationic nitrogen of the ammonium group and a non-polar interaction with the hydrophobic chain. Penetration of the hydrophobic group into the microorganism consequently occurs, enabling the alkylammonium group to physically interrupt all key cell functions.

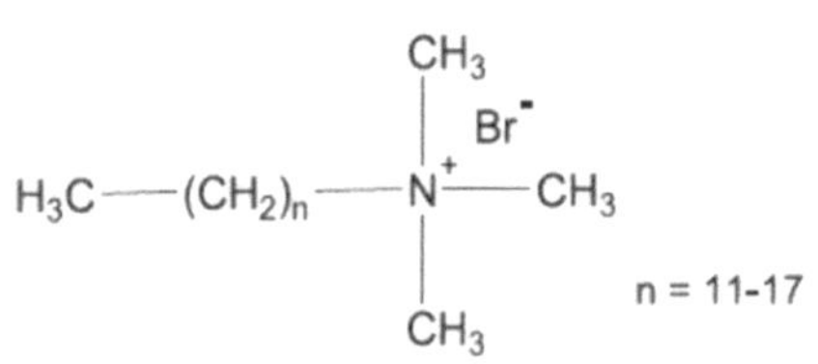

(A)

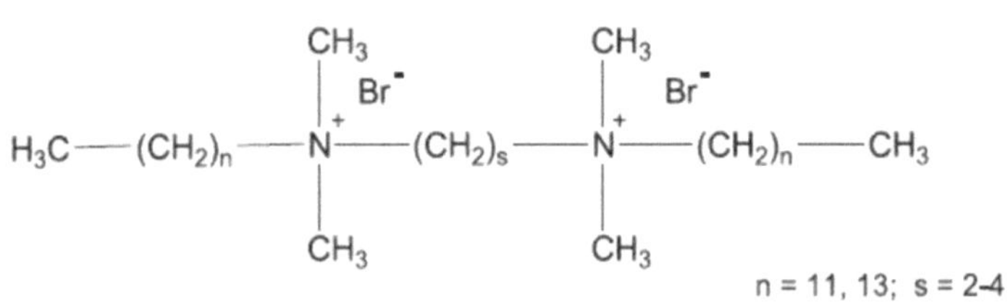

(B)

Fig. 1. Chemical structure of monoquaternary ammonium salt, alkyltrimethylammonium bromide (A), and diquaternary ammonium salt,alkanediyl-α,ω-bis(dimethylalkylammonium bromide) (B).

Despite many positive properties, QASs have an inherent weakness: leaching from the textile. There are no reactive functional groups in the structure of the QAS to allow its chemical bonding to the fibers. Owing to the lack of physical bonding, leaching of the QAS occurs, resulting in a fast decrease in concentration to below the MIC. In addition, QASs have poor wash durability. To develop new, permanently bonded, non-leaching QAS biocidal groups for textile fibers, contemporary studies have synthesized polymerizable QASs [19-21] with acrylate or methacrylate groups for incorporation in the structure (Figure 2). Such QAS monomers have been named surfactant monomers or "surfmers". Under appropriate conditions, "surfmers" polymerize into a bulk polymer network with a polycationic structure, including side QAS groups chemically bonded to the main polyacrylate chain. The merit of fixed bonding to the textile surface is that the QAS groups can act as a biobarrier and kill microorganisms by contact. Furthermore, the formation of a polymer network on the surface of the fibers strongly increases the durability and wash resistance of the antimicrobial agent.

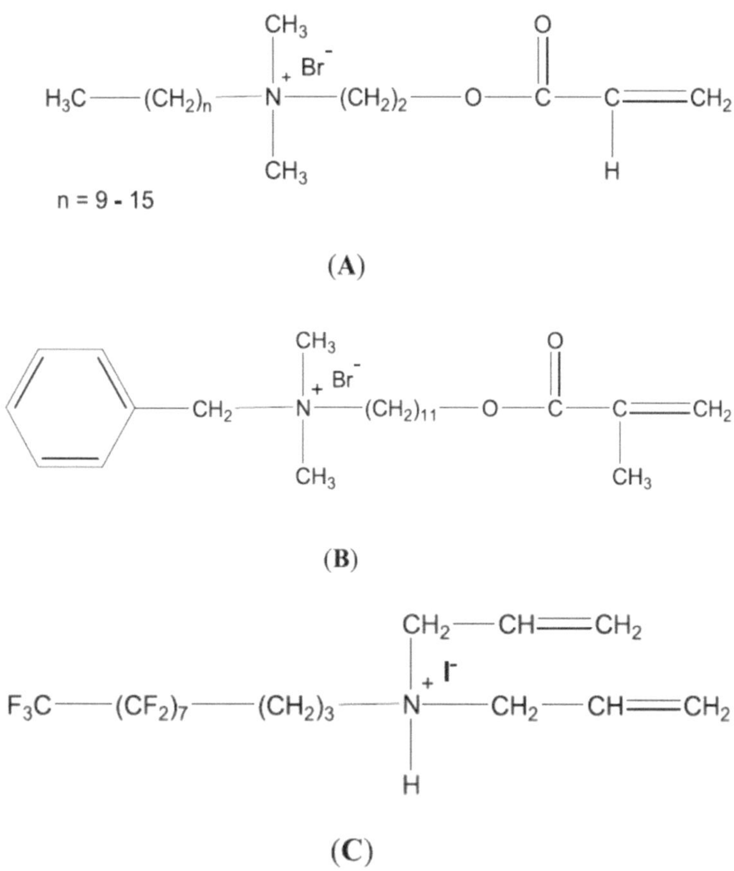

Fig. 2. Chemical structures of various "surfmers": alkyl(2-(acryloyloxy) ethyl)dimethyl ammonium bromide (A) benzyl(11- (acryloyloxy)undecyl)dimethyl ammonium bromide (B) , N- (4,4,5,5,6,6,7,7,8,8,9,9,10,10,11,11- heptadecafluoroundecyl)-N,Ndiallylmethyl ammonium iodide (C)

4.2 N-halamines

N-halamines are heterocyclic organic compounds containing one or two covalent bonds formed between nitrogen and a halogen (N–X), in which the latter is usually chlorine [22]. N–Cl bonds of different stability can be formed by the chlorination of amine, amide or imide groups in dilute sodium hypochlorite. N-halamines are biocides that are active for a broad spectrum of bacteria, fungi and viruses. Their antimicrobial properties are based on the electrophilic substitution of Cl in the N-Cl bond with H; this reaction can be carried out in the presence of water and results in the transfer of Cl+ ions that can bind to acceptor regions on microorganisms. This hinders enzymatic and metabolic processes, leading to the destruction of the microorganisms. As an N–H bond, which does not have antimicrobial properties, is formed in the substitution reaction, further exposure of the agent to dilute sodium hypochlorite is needed for regeneration of its antimicrobial activity [23, 24] .

N-halamines can be applied to various textile surfaces including cellulose [25-27], polyamide [28] and polyester [29] fibers. To increase their effectiveness and the durability of the antimicrobial finish [25], research has been oriented toward synthesis of N-halamide monomers with an incorporated vinyl reactive group (Figure 3) [30, 31] that can polymerize on cellulose fibers under appropriate conditions to form a coating with excellent durability after washing [30].

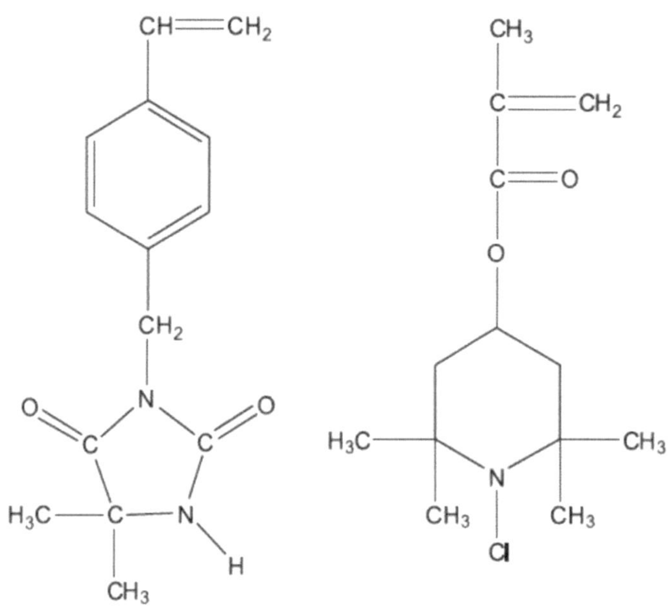

(A) (B)

Fig. 3. Chemical structures of 3-(4'-vinylbenzyl)-5,5-dimethylhydantoin (A) and N-chloro-2,2,6,6-tetramethyl-4-piperidinyl methacrylate (B).

4.3 Chitosan

Chitosan is a deacetylated derivate of chitin, which is a natural polysaccharide mainly derived from the shells of shrimps and other sea crustaceans. Chemically, it can be designated as poly-β-(1→4)-D-glucosamine or poly-(1,4)- 2-amido-deoxy-β-D-glucose (Figure 4) [32]. In addition to its antimicrobial activity, chitosan has some important advantages such as non-toxicity, biocompatibility and biodegradability.

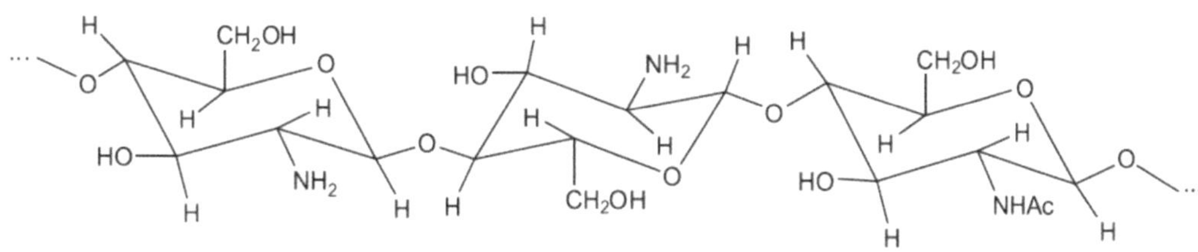

Fig. 4. Chemical structure of chitosan

To provide antimicrobial effect for textiles, chitosan can be used as an additive when spinning antimicrobial fibers [33] and also as a finishing agent [32] for surface modification, mainly of cellulose, cellulose/polyester and wool fibers. Chitosan is positively charged and soluble in acidic to neutral solutions because the amino groups in chitosan have a pKa of ~6.5. Its antimicrobial function arises from its polycationic nature, which is caused by protonation of the amino groups at the C-2 atoms of the glucosamine units; such antimicrobial function is very similar to that determined for QAS. Positively charged amino groups can bind to the negatively charged bacterial surface, resulting in the disruption of the cell membrane and an increase in its permeability. Chitosan can also interact with the DNA of microorganisms to prevent protein synthesis.

The antimicrobial efficiency of chitosan depends on its average molecular weight, degree of deacetylation and the ratio between protonated and unprotonated amino groups in the structure [32]. It is believed that chitosan of a low molecular weight is more antimicrobially active than chitosan oligomers [32]. The efficiency also increases with increased deacetylation, which can exceed 90%. An important disadvantage of chitosan is its weak adhesion to cellulose fibers, resulting in a gradual leaching from the fiber surface with repetitive washing. To enable chitosan to bind strongly to cellulose fibers, various crosslinking agents are used, including mostly polycarboxylic acids (1,2,3, 4-butantetracarboxylic and citric acids) [32, 34] and derivates of imidazolidinone[35]. In the presence of a crosslinking agent, hydroxyl groups of chitosan and cellulose can form covalent bonds with carboxyl groups of polycarboxylic acid in an esterification reaction or with hydroxyl groups of imidazolidinone in an etherification reaction, thus leading to the formation of a crosslink between chitosan and cellulose. This greatly improves durability and wash resistance. In addition, the reactivity of quaternized chitosan has been improved by introducing functional acrylamidomethyl groups to the primary alcohol groups (C-6), which can form covalent bonds with cellulose in alkaline conditions (Figure 5) [36].

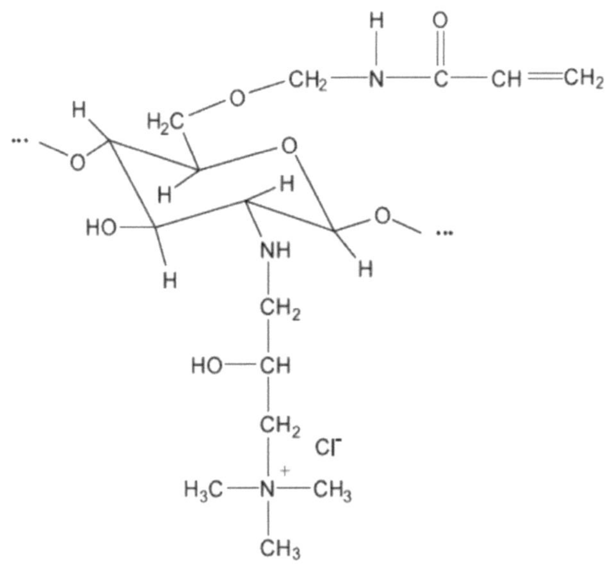

Fig. 5. Chemical structure of reactive O-acrylamidomethyl-N-[(2-hydroxy-3-trimethylammonium)propyl]Chitosan chloride

The chemical binding of chitosan to cellulose fibers can also be achieved by oxidation of cellulose fibers with potassium periodate under acidic conditions to form aldehyde groups, which are allowed to react with the amino groups of chitosan and form a Schiff base (C=N double bond) [37]. Following the model of N-halamine halogenation, some of the amino groups in chitosan have been transformed into an -NHCl structure in the presence of sodium hypochlorite [38]. It has been found that chlorination significantly improves the antimicrobial activity of chitosan.

4.4 Halogenated phenols

Among halogenated phenols, triclosan 5-chloro-2-(2.4-dichlorophenoxy) phenol (Figure 6) [39] is the most widely used biocide; it is present in many contemporary consumer and personal health-care products, detergents and household objects, including textiles and plastics.

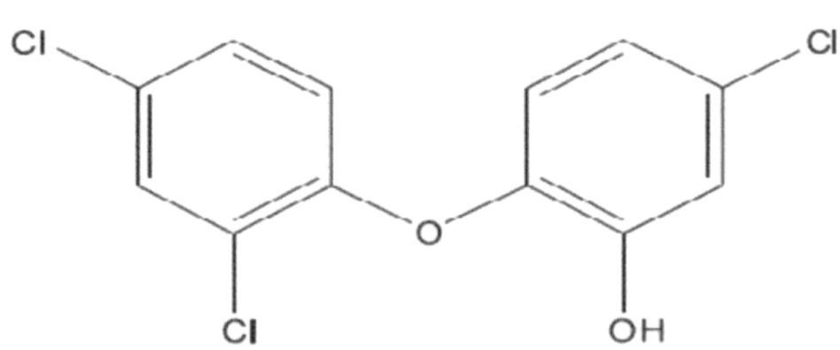

Fig. 6. Chemical structure of triclosan

At bactericidal concentration, triclosan is very effective against a broad range of microorganisms, including antibiotic- resistant bacteria. As the widespread use of triclosan could represent a potential risk in terms of the development of resistant microorganisms [39], strong binding to solid surfaces with subsequent controlled release is important. Triclosan has therefore been applied to cellulose fibers in combination with polycarboxylic acids as crosslinking agents [40]. The application of polycarboxylic acid to fibers previously finished with triclosan enhances the washing durability of the antimicrobial coating.

Novel host–guest complexes including triclosan molecules have been prepared with the use of cationic β-cyclodextrins (Figure 7), which are torus-shaped cyclic oligosaccharides containing six to eight glucose units linked by α-1,4 bonds [41]. Water solubility, stability and antimicrobial activity have been determined for the host–guest complexes. Owing to strong electrostatic attraction, the complexes are adsorbed to the surface of cellulose fibers almost completely. Triclosan has also been encapsulated in biodegradable polylactide as a carrier and used for finishing non-woven textiles [42].

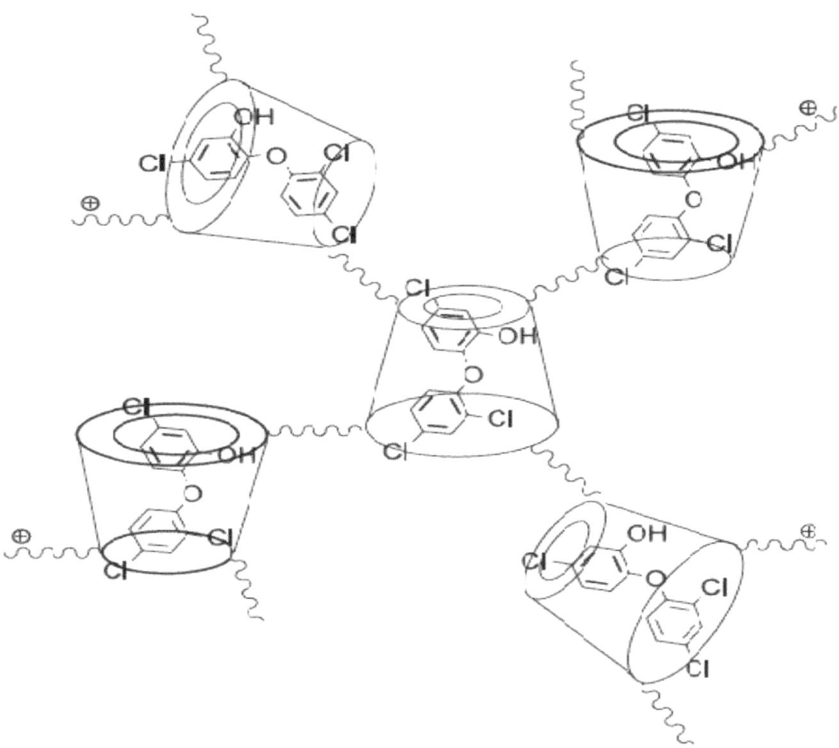

Fig. 7. Schematic presentation of host–guest complexes between cationic β-cyclodextrin and triclosan.

4.5 Polybiguanides

Polybiguanides are polymeric polycationic amines that include cationic biguanide repeat units separated by hydrocarbon chain linkers of identical or dissimilar length. One of the most important antimicrobial agents among them is poly (hexamethylenebiguanide) (PHMB) with an average of 11 biguanide units (Figure 8) [43]; PHMB exhibits much greater antimicrobial activity than corresponding monomeric or dimeric biguanides. PHMB is widely used in medicine as an antiseptic agent, especially for preventing wound infection by antibiotic- resistant bacteria [44]. Owing to its high biocidal activity and low toxicity, it has also attracted attention for the antimicrobial finishing of textiles, mainly for the protection of cellulose fibers [43].

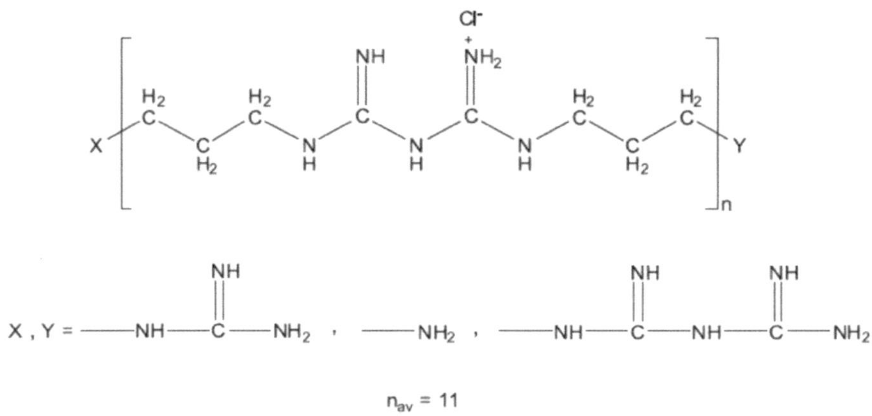

Fig. 8. Chemical structure of Poly (hexamethylenebiguanide). Here n_{av} is the average number of repeat units

The literature [45] indicates that PHMB can bind to the anionic carboxylic groups of cellulose (Figure 9), which are formed through oxidation of glucose rings during pre-treatment processes such as bleaching and mercerizing. At lower concentrations, electrostatic interactions between PHMB and carboxylic groups in the cellulose dominate; however, hydrogen bonding of PHMB with cellulose resulting in multilayer adsorption becomes increasingly dominant as the concentration of PHMB increases. The adsorption of PHMB increases if cellulose fibers are previously dyed with anionic reactive dyes [76, 79], which provide sulfonic acid sites with which the PHMB can react. However, strong PHMG-dye interactions cause a reduction in the antimicrobial activity of PHMB, which is undesirable.

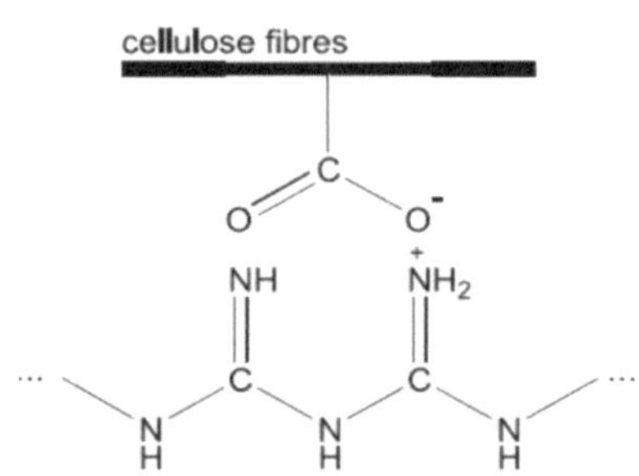

Fig. 9. Binding of poly (hexamethylenebiguanide) to the carboxylic group of cellulose

5. Nanotechnology

In recent years nanotechnology has become one of the most important and exciting forefront fields in physics, chemistry, engineering and biology. It shows great promise for providing us in the near future with many breakthroughs that will change the direction of technological advances in wide range of applications. The prefix nano in the word nanotechnology means one billionth of a meter ($1x10^{-9}$). Nanotechnology deals with various structures of matter having the dimension of the order of a billionth of a meter. Structures on this scale have been shown to have unique and novel functional properties. Based on that principle, many applications of nanotechnology from the simple to the complex have been done. One of these applications is to prepare antimicrobial textiles based on heavy metals in their nanoscale.

Particles at the nanoscale are below the wave length of visible light and therefore, cannot be seen. Consequently they can impart new properties. For example, Ti-nanoparticles are applied for the textile materials in order to develop textile products with UV- protection and self cleaning property. Also Silver nanoparticles are used as antimicrobial agent for wound dressing materials as well as for wound healing. In addition, the production of fibers with diameter less than 100nm is now feasible with the invention of electrospinning process. Electrospinning is a manufacturing new technology capable of producing thin, solid polymer strands from solution by applying a strong electric field to a spinneret with a small capillary orifice. The spun, polymer based nanofibers, can be loaded with different additives which could be metal nanoparticles like silver, drugs or catalysts depending on the required applications. The resulted nanofibers are collected and bundled. These electrospun fibers have high surface area and porous structure, where more than one drug can be encapsulated directly into the fiber. The resulted matrix can be used extensively for medical textiles production with multifunctional properties.

The intersecting fields of study that create this domain of science and engineering perfectly typify the rapid, multidisciplinary advancement of contemporary science and technology. Inorganic materials such as metal and metal oxides have attracted lots of attention over the past decade due to their ability to withstand harsh process conditions. The use of nanoparticles of silver, gold and zinc oxide has been seen as a viable solution to stop infectious diseases due to the antimicrobial properties of these nanoparticles. In view of the textile industry's innovative history, it is no wonder that nanotechnology has found its way into this sector so quickly. Nanotechnology is forecasted as the second industrial evolution in the world. The novel properties and low material consumption amount has attracted global interest across disciplines and industries. The textile sector is no exception.

As stated by the "European Technological Platform for Textiles and Fashion", the textile industry to thrive must improve and reduce the costs of the processes, offer innovative products for traditional markets, develop new products for new markets [46]. Nanotechnology can have an important role to achieve these goals and, in effect, all over the world public and private research institutions and private enterprises are actively engaged in nanotechnology research aimed at applications in the textiles sector. With growth in world population and the spread of disease, the number of antibiotic resistant microorganisms is rising along with the occurrence of infections from these microorganisms. With this increase in health awareness, many people focused their attention on educating and protecting themselves against harmful pathogens. It soon became more important for antimicrobial finished textiles to protect the wearer from bacteria than it was to simply protect the garment from fiber degradation. The need for antimicrobial textiles goes hand-in-hand with the rise in resistant strains of microorganisms. Functional textiles include everything from antimicrobial finished textiles, to durable, or permanent press finished garments, to textiles with self-cleaning properties, and also textiles with nanotechnology [46].

6. Antibacterial modification of textiles using sol-gel technology

The sol-gel process is known as chemical solution deposition, it is a wet-chemical technique widely used in the fields of material science and ceramic engineering. Such methods are used primarily for the fabrication of materials (typically a metal oxide) starting from a chemical solution (or *sol*) that acts as the precursor for an integrated network (or *gel*) of either discrete particles or network polymers. Typical precursors are metal alkoxides and metal chlorides, which undergo various forms of hydrolysis and polycondensation reactions. In this chemical procedure, the 'sol' (or solution) gradually evolves towards the formation of a gel-like diphasic system containing both a liquid phase and solid phase whose morphologies range from discrete particles to continuous polymer networks. In the case of the colloid, the volume fraction of particles (or particle density) may be so low that a significant amount of fluid may need to be removed initially for the gel-like properties to be recognized. This can be accomplished in any number of ways. The simplest method is to allow time for sedimentation to occur, and then pour off the remaining liquid. Centrifugation can also be used to accelerate the process of phase separation. Removal of the remaining liquid (solvent) phase requires a drying process, which is typically accompanied by a significant amount of shrinkage and densification. The rate at which the solvent can be removed is ultimately determined by the distribution of porosity in the gel. The ultimate microstructure of the final component will clearly be strongly influenced by changes imposed upon the structural template during this phase of processing [47].

The precursor sol can be either deposited on a substrate to form a film (e.g., by dip coating or spin coating), cast into a suitable container with the desired shape (e.g., to obtain monolithic ceramics, glasses, fibers, membranes, aerogels), or used to synthesize powders (e.g., microspheres, nanospheres) [47]. The sol-gel approach is a cheap and low-temperature technique that allows for the fine control of the product's chemical composition. Sol-gel derived materials have diverse applications in optics, electronics, energy, space, (bio)sensors, medicine (e.g., controlled drug release), reactive material and separation (e.g., chromatography) technology and in the production of antimicrobial textiles.

Recently, investigations have been conducted to improve properties of textile fabrics by embedding various finishes in sol-gel coating to enhance fabric performance. The following section will focus on different sol-gel applications that create new functional properties in textile fabrics. Textiles coated with sol-gel are reported to impart many important properties, as shown in Figure 10.

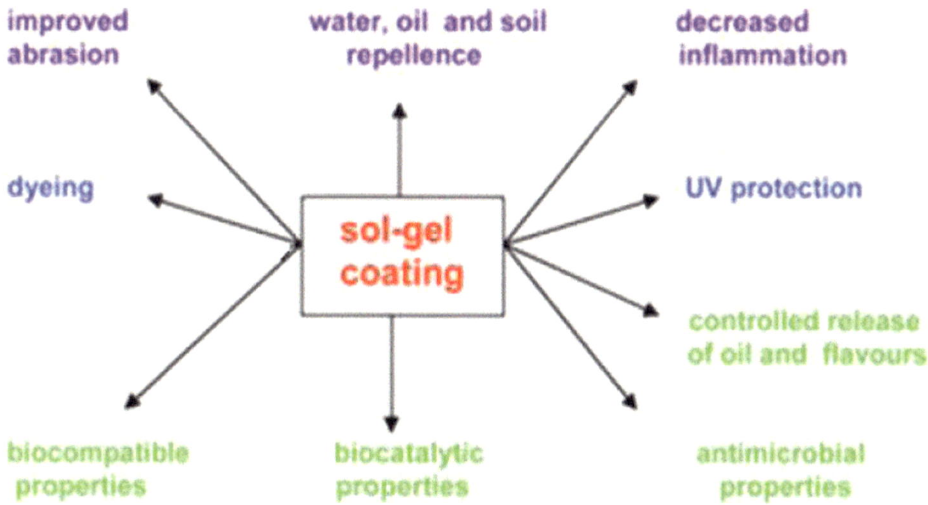

Fig. 10. Different applications of sol-gel coating on textiles [1]

Antimicrobial finishing of cotton textile based on water glass by sol-gel method was investigated by [48]. They prepared the silica sol by acidifying water glass solution with 0.05M H2SO4 solution to pH 11. Antimicrobial treatment was performed by impregnating cotton textile in silica sols and then treated with silver nitrate solution. The antimicrobial activity of the treated cotton fabrics was determined according to AATCC Test method 100-1999. The anti-microbial activity is evaluated by determining the percentage reduction of bacteria count on the sample after exposing the treated fabric to the bacteria.

This result was compared to an untreated control sample by exposing to a bacterial lawn. This antibacterial activity was measured by washing the samples several times after sol-gel treatment. Mahltig et al.,'s (2005) results exhibited that antimicrobial durability of the treated samples increased with increase in water glass content [49]. For the samples without water glass, bacterial reduction % was zero only after 10 washing cycles. But with a water glass content of 2% and 5%, higher silver ions were retained that resulted in more than 99% bacterial 25 reduction even after 20 and 50 cycles of washing. The bacterial reduction % also increases with increase in silver content as well.

Mahltig et al. (2004) investigated antibacterial effect by embedding the biocides in silica coatings. Biocides used in this study were silver nitrate, colloidal Ag, (cetyltrimethylammoniumbromide) CTAB, and octenidine. Silica sol solution was prepared from tetraethoxysilane (TEOS) and 3-glycidyloxypropyltriethoxysilane (GOPTS) [50]. Properties such as antibacterial efficacy wash-out and long-term behavior were analyzed. Their results showed biocidal additives Octenidine and CTAB exhibit high inhibition rate of more than 90% against the fungi Aspergillus niger after 4hr of leaching, whereas colloidal Ag treated samples without sol-gel have the lowest inhibition.

Daoud et al. (2005) coated the cellulosic fibers with titanium oxide nanoparticles, which were obtained from aqueous titania sol. Titanium isopropoxide was hydrolysed and condensed in water to obtain the titania sol coating at low temperature. The stability of titania coating and the antibacterial activity of the coating was analyzed [51]. Treatment with sodium carbonates solutions showed less leaching behavior, which indicated that titania sols were strongly bonded with the cellulosic substrate. They confirmed the same with the help of FESEM images, which shows uniform distribution of the titania sols. The treated fabrics exhibited good antibacterial activity because of the formation of TiO2 surface on the cellulose substrate which prevented the formation of a protective biofilm of adsorbed bacteria.

To fix a QAS on textile fibers, sol-gel technology has also been used in antimicrobial textiles. This enables the formation of a nanocomposite polymer network with an organic-inorganic hybrid structure [52, 53]. Colloidal solutions (sols) have been prepared for this purpose, consisting of mixtures of tetraalkoxysilane (Si(OR)4) and QASs with different structures [54] or organic-inorganic hybrids, including alkyltrialkoxysilanes (Rx-Si(OR)3), with incorporated quaternary ammonium groups (Figure 11) [55, 56]. Alkoxysilanes are sol-gel precursors with alkoxy groups that can hydrolyze in the presence of a catalyst to form silanol (-SiOH) groups, which further condense among each other or with the hydroxyl (-OH) groups of the fibers. The formation of covalent bonds between -SiOH groups of the precursor and -OH groups of the fibers provides increased durability and wash resistance for the nanocomposite network on the finished fibers.

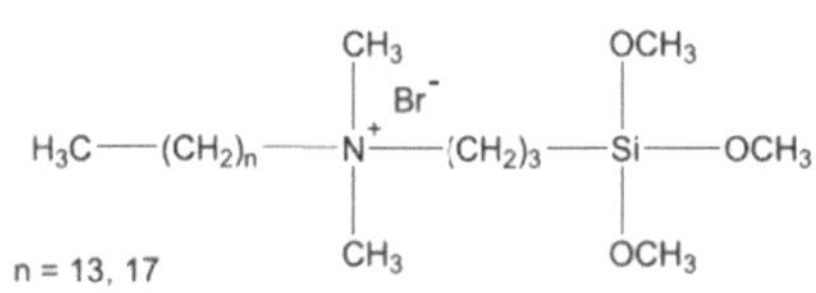

(A)

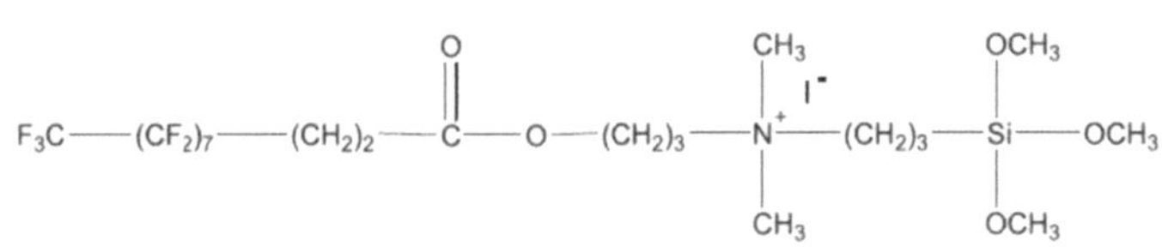

(B)

Fig. 11. Chemical structures of alkyltrialkoxysilanes with incorporated quaternary ammonium groups: alkyl-dimethyl-(3(trimethoxysilyl)- propyl) ammonium bromide (A), perfluorooctylated quaternary ammonium silane coupling agent (B).

Quaternary ammonium functionalized polyhedral *oligomeric* silsesquioxanes (Q-POSS) are also used as sol-gel precursors with antimicrobial properties. These have a cage-like structure of (R*x*-SiO1.5)*n* (*n* = 6, 8, 10, 12,...), where R*x* is a QAS group [57, 58]. The most often used POSS precursors are octasilsesquioxanes (*n* = 8), which have a cubic form with QAS groups bound to silicon atoms. Accordingly, up to eight QAS groups can be incorporated into each molecule (Figure 12), thus enhancing antimicrobial properties. Because Q-POSSs possess a polysiloxane core compatible with the siloxane matrix, they can be used as additives to the polysiloxane coating in combination with polydimethylsiloxane.

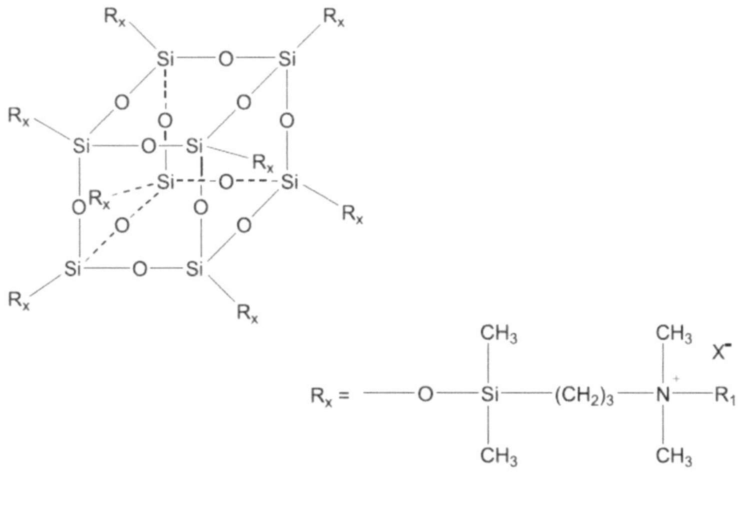

Fig. 12. Schematic presentation of idealized structure of totally quaternized quaternary ammonium Functionalized polyhedral oligomeric silsesquioxanes.

Substituted polycationic polysiloxanes with a pendant QAS (Figure 13A) or imidazolium salt groups (Figure 13B) are also important [59, 60] as bacteriostats on textile fibers. These are mostly copolymers consisting of polydimethylsiloxane, polymethylhydrosiloxane and QAS- or imidazolium- modified polysiloxanes in different molar ratios.

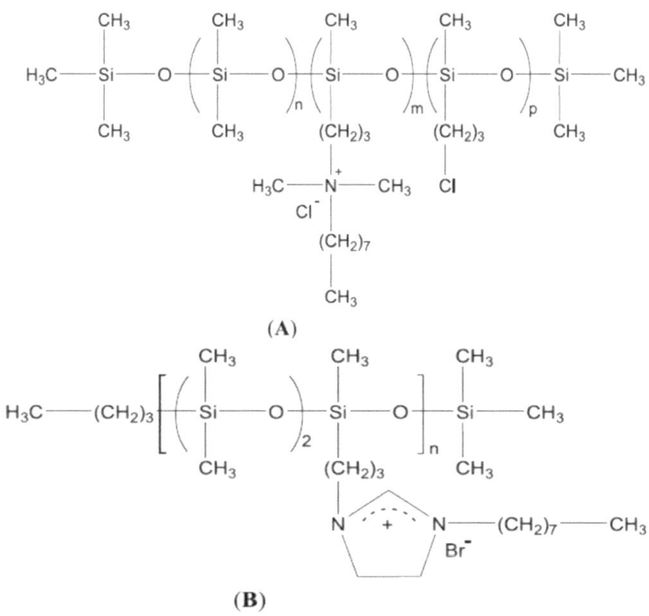

Fig. 13. Substituted polycationic polysiloxane with pendant QAS (A) and imidazolium salt (B) groups.

7. Synthesis of nanoparticles

There are two methods for the production of nanoparticles which is summarized below:

7.1 Top-down technique

The principle behind the top-down approach is to take a bulk piece of the material and then modify it into the wanted nanostructure and subsequent stabilization of the resulting nanosized metal nanoparticles by the addition of colloidal protecting agents. Cutting, grinding and etching are typical fabrication techniques, which have been developed to work on the nano scale. The sizes of the nanostructures which can be produced with top-down techniques are between 10 to 100 nm.

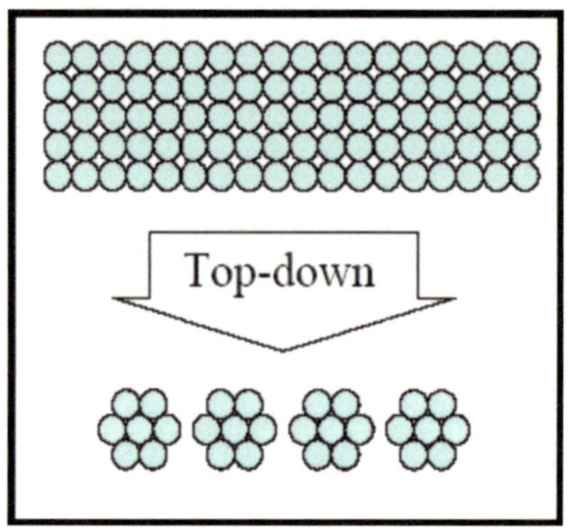

7.2 Bottom-up technique

Bottom-up or self-assembly refers to construction of a structure atom by-atom, molecule-by-molecule or cluster-by-cluster. Colloidal dispersion used in the synthesis of nanoparticles is a good example of a bottom-up approach. An advantage of the bottom-up approach is the better possibilities to obtain nanostructures with less defects and more homogeneous chemical compositions.

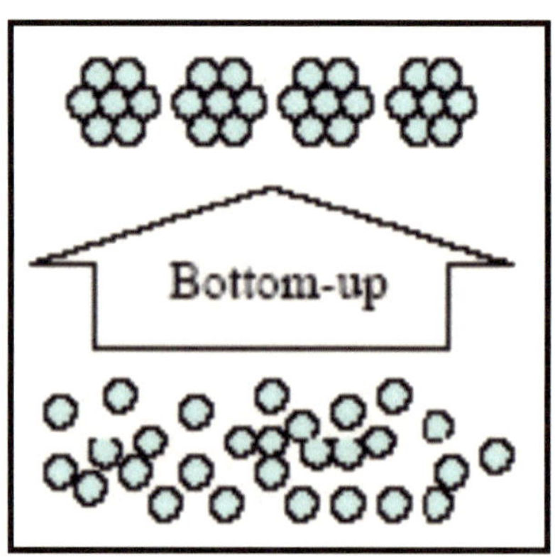

8. Stabilization of nanoparticles

There are two general kinds of stabilization procedures: electrostatic stabilization by the surface adsorbed anions and steric stabilization by the presence of bulky groups.

8.1 Steric stabilization

Steric stabilization can be achieved by the adsorption of large molecules, i. e. polymers, at the surface of the particles [61].

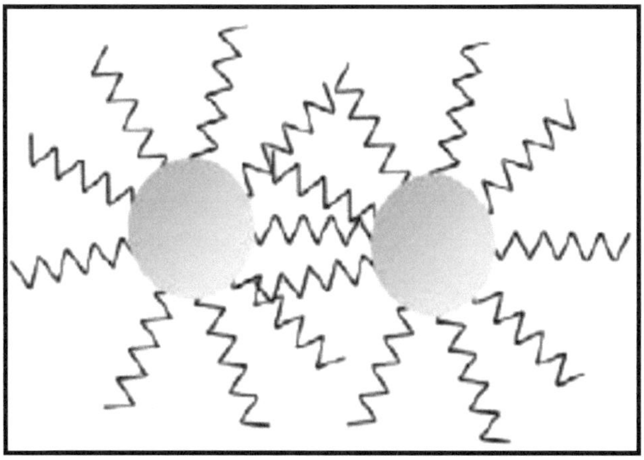

8.2 Electrostatic stabilization

Electrostatic stabilization involves the creation of an electrical double layer arising from ions adsorbed on the surface and associated counter ions that surround the particle [61]. Recently, increasing public concern about hygiene has been driving many investigations for anti-microbial modification of textiles. However, using many anti-microbial agents has been avoided because of their possible harmful or toxic effects. Application of inorganic nano-particles and their nanocomposites would be a good alternative [62]and consequently, they can open up a new opportunity for anti-microbial and multi-functional modification of textiles.

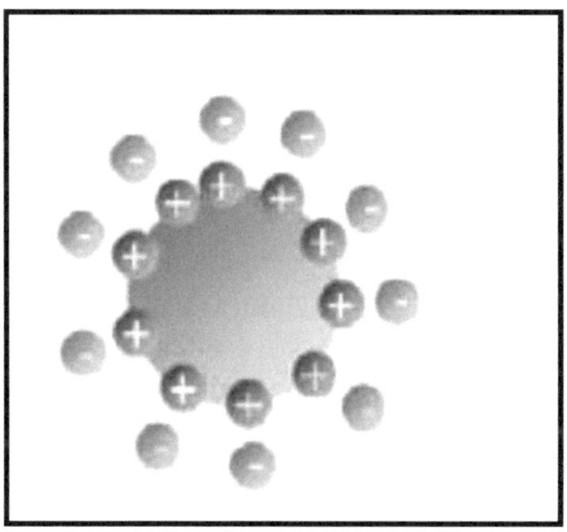

9. Classification of inorganic based nano materials

Nano-structured materials on the basis of inorganic active agents having good potential for anti-microbial activity on textile materials can be categorized in two main groups:

1. Inorganic nano-structured materials and their nanocomposites.
2. Inorganic nano-structured loaded organic carriers.

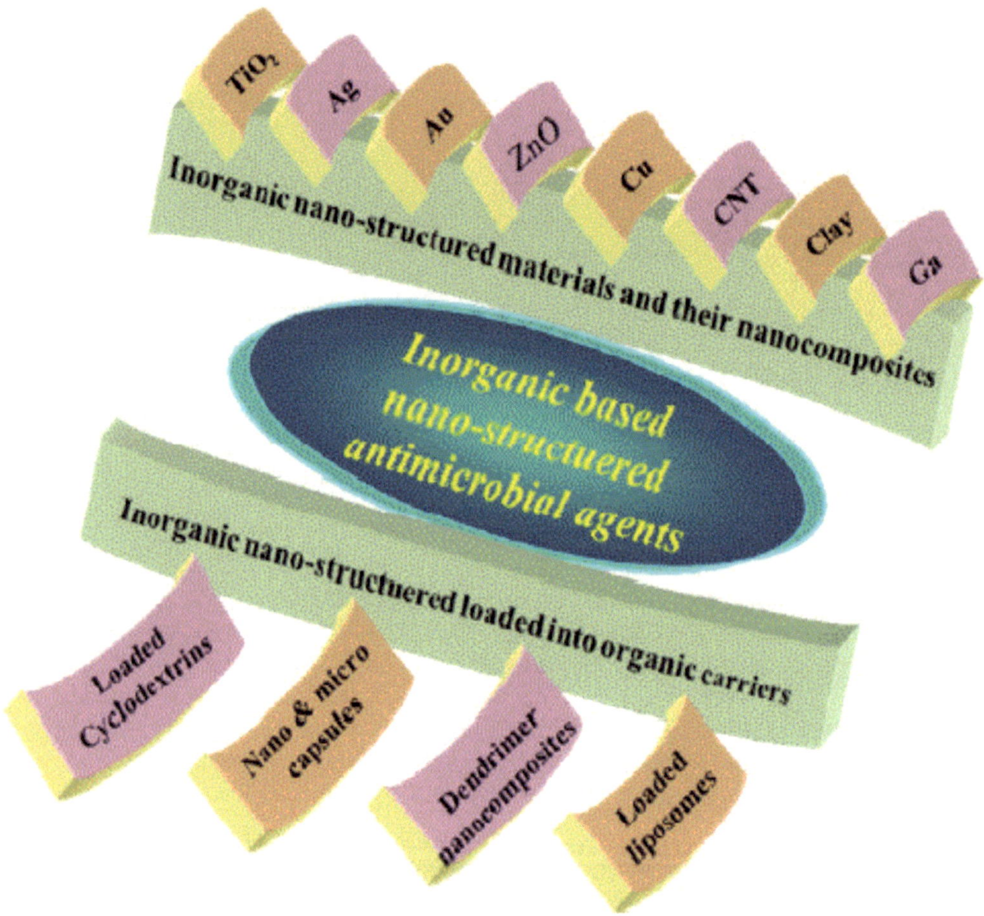

Fig. 14. Classification of inorganic based nano-structured anti-microbial agents.

The inorganic nano-structured materials include titanium dioxide, silver, zinc oxide, copper, gallium, gold nano-particles, carbon nanotubes, nano-layered clay, and their nano-composites. The inorganic nano-structured loaded in organic materials include cyclodextrin loaded with inorganic materials, nano- and micro-capsules having inorganic nano-particles, metallic dendrimer nano-composites and inorganic nano-particles loaded in liposomes. Note that categories such as loaded cyclodextrins, metallic dendrimers nano-composites and loaded liposome can be included in nano-capsules. However, each one has an especial concept, history, architecture and properties. This chapter focus on the inorganic nano-structured materials with good antimicrobial activity potential for textile modification [63].

9.1 TiO2 nanoparticles

Currently, TiO_2 nanoparticles have created a new approach for remarkable applications as an attractive multi-functional material. TiO2 nanoparticles have unique properties such as higher stability, long lasting, safe and broad-spectrum antibiosis. This makes TiO_2 nanoparticles applicable in many fields such as self-cleaning, anti-bacterial agent and UV protecting agent.

9.1.1 Mechanism of action

Titanium dioxide irradiation by light with more energy compared to its band gaps generates electron - hole pairs that induce redox reactions at the surface of the titanium dioxide. Consequently, electrons in TiO2 jump from the valence band to the conduction band, and the electron (e^-) and electric hole (h^+) pairs are formed on the surface of the photo-catalyst. The created negative electrons and oxygen will combine into $O2^-$, the positive electric holes and water will generate hydroxyl radicals. Ultimately, various highly active oxygen species can oxidize organic compounds of cell to carbon dioxide (CO_2) and water (H_2O). Thus, titanium dioxide can decompose common organic matters in the air such as odor molecules, bacteria and viruses.

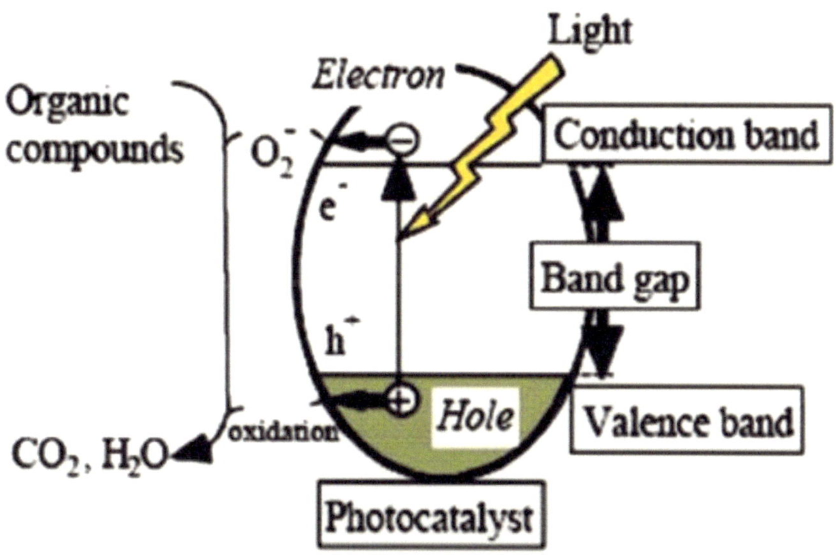

Photocatalysis Mechanism of TiO2

9.2 ZnO nanoparticles

Recently, ZnO has been found highly attractive because of its remarkable application potential in solar cells, sensors, displays, gas sensors, sun-screens, UV absorbers, antireflection coatings, antibacterial, and photo-catalysis. ZnO nano-particles have some advantages, compared to silver nano-particle, such as lower cost, white appearance, and UV-blocking property.

9.3 Silver nanoparticles

Since ancient times among various anti-microbial agents, silver has been most extensively studied and used to fight against infections and prevent spoilage. At present, many researchers have focused on anti-bacterial and multifunctional properties of silver nanoparticles. [64] [65]. Silver is a safer anti-microbial agents in comparison with some organic anti-microbial agents that have been avoided because of the risk of their harmful effects on the human body. Silver has been described as being 'oligodynamic' because of its ability to exert a bactericidal effect on products containing silver, principally due to its anti-microbial activities and low toxicity to human cells. Its therapeutic property has been proven against a broad range of micro-organisms, over 650 disease-causing organisms in the body even at low concentrations.

9.3.1 Mechanism of action

The brief explanation of its antimicrobial mechanism can be explained as follows:

Generally, metal ions destroy or pass through the cell membrane and bond to the –SH group of cellular enzymes. The consequent critical decrease of enzymatic activity causes micro-organism metabolisms change and inhibits their growth, up to the cell's death. The metal ions also catalyze the production of oxygen radicals that oxidize molecular structure of bacteria. The formation of active oxygen occurs according to chemical reaction:

$$H_2O + 1/2\,O_2 \xrightarrow{\text{metal ion}} H_2O_2 \longrightarrow H_2O + (O)$$

Such a mechanism does not need any direct contact between antimicrobial agent and bacteria, because the produced active oxygen diffuses from fiber to the surrounding environment. Silver ions can lead to denaturing of protein and cell death because of their reaction with nucleophilic amino acid residues in proteins, and attach to sulfhydryl, amino, imidazole, phosphate and carboxyl groups of membrane or enzyme proteins. Respiration blocking and cell death also may be caused by forming R–S–S–R bonds.

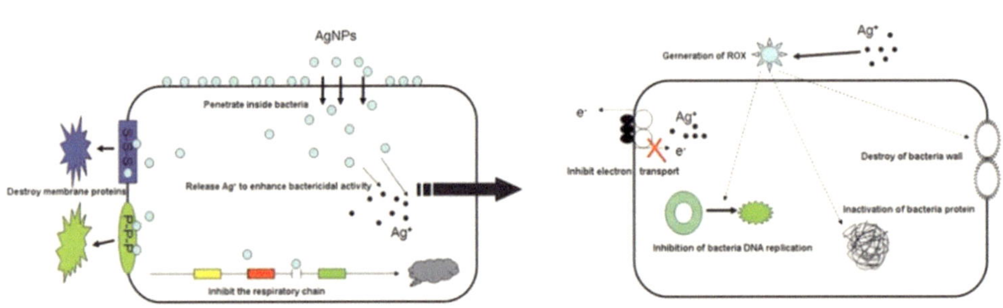

9.4 Gold nanoparticles

Gold nanoparticles are known as a novel biomedical application. Their potent antibacterial effectiveness against acne or scurf and no tolerance to the antibiotic have caused their commercial usage in soap and cosmetic industries. They can remove waste materials from the skin and control sebum.

10. Evaluation of antimicrobial efficacy

A number of test methods have been developed to determine the efficacy of antimicrobial textiles [66]. These methods generally fall into two categories: the agar diffusion test and suspension test.

10.1 Agar diffusion test

Agar diffusion test is a preliminary test to detect the diffusive antimicrobial finish. It is not suitable for non diffusive finishes and textile materials other than fabrics [67]. The agar diffusion tests include AATCC 147-2004 (American Association of Textile Chemists and Colorists), JIS L 1902-2002 (Japanese Industrial Standards) and SN 195920- 1992 (Swiss Norm). They are only qualitative, but are simple to perform and are most suitable when a large number of samples are to be screened for the presence of antimicrobial activity. In these tests, bacterial cells are inoculated on nutrient agar plates over which textile samples are laid for intimate contact. The plates are then incubated at 37°C for 18–24 h and examined for growth of bacteria directly underneath the fabrics and immediately around the edges of the fabrics (zone of inhibition). No bacterial growth directly underneath the fabric sample indicates the presence of antimicrobial activity. The zone of inhibition should not be expected if the antimicrobial agent is firmly attached to the textile (e.g. covalently) which prevents its diffusion into the agar. If the antimicrobial agent can diffuse into the agar, a zone of inhibition becomes apparent and its size provides some indication of the potency of the antimicrobial activity or the release rate of the active agent [67].

10.2 Suspension test

This type of test is exemplified by AATCC 100-2004, JIS L 1902-2002 and SN 195924-1992. These methods provide quantitative values on the antimicrobial finishing, but are more time-consuming than agar diffusion tests. Typically, a small volume (e.g. 1 ml) of bacterial inoculums in a growth media is fully absorbed into fabric samples of appropriate size without leaving any free liquid. This ensures intimate contact between the fabric and the bacteria. After incubating the inoculated fabrics in sealed jars at 37°C or 27°C for up to 24 h, the bacteria in the fabric are eluted and the total number is determined by serial dilution and plating on nutrient agar plates. Antimicrobial activity, expressed as percentage of reduction, is calculated by comparing the size of the initial population with that following the incubation. Appropriate controls, e.g. samples that have gone through the same processing except the antimicrobial finishing, should be included in each experiment to ascertain that the observed decrease in bacterial number is truly due to the antimicrobial finishing.

It should be noted that suspension tests are often performed under artificial conditions that promote bacterial growth (e.g. rich nutrients in the inoculum and saturating moisture in the testing fabrics). The moisture in the tests is also essential for the action of the biocide. As a result, dramatic results are often produced (e.g. >99% bacterial cells are killed during the assays), leading to an overwhelming impression of the efficacy of the antimicrobial ability. However, such conditions are rarely found during the normal use of a textile product. To date, very few studies have examined the antimicrobial effects under normal wearing conditions. To more closely mimic the real-life situation, the JIS L 1902-2002 method recommends the use of bacterial cells suspended in heavily diluted nutrient media to limit nutrient levels.

The ISO (International Organization for Standardization) has developed a test method (ISO 20743) in which bacteria are "printed" onto the surface of textiles without them being in an aqueous suspension [25, 26]. The printed samples are then incubated under humid conditions at 20°C for a specified time (18–24 h) following which the surviving cells are counted. Antimicrobial tests only assess the antimicrobial effectiveness of the treated textiles. Before marketing, the textile products have to pass biocompatibility tests which involve three separate assays: cytotoxicity, sensitization and irritation. These assays are outside the scope of this chapter but are discussed elsewhere [66].

11. Remarks and outlooks

Inorganic and metallic-based nanostructure materials have created a new interesting fields in all sciences for the continuous investigations due to their undeniably unique properties. Their applications have already led to the development of new practical productions. Considering the indubitable role of textiles in human life, these new fields in textile industry have been increasingly welcomed. However, designing new applicable and affordable techniques for manufacturing scale-up production will not only create a new field of study, but meet the expanding human requirements.

12. References

[1] B. Mahltig and H. Bottcher, Modified silica sol coatings for water-repellent textiles. Journal of Sol-Gel Science and Technology, 2003. 27(1): p. 43-52.
[2] T. Ramachandran, K. Rajendrakumar, and R. Rajendran, Antimicrobial Textiles- an Overview. IE (I) Journal—TX, 2004. 84.
[3] L. Qian and G. Sun, Durable and Regenerable Antimicrobial Textiles: Synthesis and Applications of 3-Methylol-2,2,5,5-tetramethylimidazolidin-4-one (MTMIO. Journal of Applied Polymer Science, 2003. 89: p. 2418-2425.
[4] P. B. Hudson, A. C. Clapp, and D. Kness, Joseph's Introductory Textile Science,Edn. 6th. Harcourt Brace Jovanovich College Publishers, New York, 1993.
[5] C. Tomasino, Chemistry and Technology of Fabric Preparation and Finishing. North Carolina State University, Raleigh, NC, 1992.
[6] M. Hashem, R. Refaie, and A. Hebeish, Crosslinking of partially carboxymethylated cotton fabric via cationization. Journal of Cleaner Production, 2005. 13 (9): p. 869-965.
[7] B. Simoncic and B. Tomsic, Structures of Novel Antimicrobial Agents for Textiles - A Review. Textile Research Journal, 2010. 0(0): p. 1-17.
[8] Y. Gao and R. Cranston, Recent Advances in Antimicrobial Treatments of Textile. Text. Res. J., 2008. 78: p. 60–72.
[9] W. D. Schindler and P. J. Hauser, "Chemical Finishing of Textiles" Woodhead Publishing Ltd, Cambridge, 2004.
[10] I. Dring, Anti-microbial, Rotproofing and Hygiene Finishes. In "Textile Finishing", Heywood, D. (Ed.), Society of Dyers and Colourists, Bradford, 2003 p. 351–371.

[11] B. Mahltig, H. Haufe, and H. Böttcher, Functionalization of Textiles by Inorganic Sol-Gel Coatings. J. Mater. Chem., 2005. 15: p. 4385–4398.
[12] T. L. Vigo, Protection of Textiles from Biological Attack, . In "Functional Finishes, Part A, Chemical Processing of Fibres and Fabrics, Handbook of Fiber Science and Technology, , 1983. II" Sello, S. B. (Ed.), Marcel Dekker, Inc., New York,: p. 367–426.
[13] R. Purwar and M. Joshi, Recent Developments in Antimicrobial Finishing of Textiles - A review. AATCC Rev., 2004. 4: p. 22–26.
[14] M. C. Murguia, L. M. Machuca, M. C. Lura, M. I. Cabrera, and R. J. Grau, Synthesis and Properties of Novel Antifungal Gemini Compounds Derived from N-Acetyl Diethanolamines. J. Surfactants Deterg., 2008. 11: p. 223–230.
[15] L. Massi, F. Guittard, R. Levy, and S. Gêribaldi, Enhanced Activity of Fluorinated Quaternary Ammonium Surfactants Against Pseudomonas aeruginosa. Eur. J. Med. Chem., 2009. 44: p. 1615–1622.
[16] B. Ahlström, M. Chelminska-Bertilsson, R. A.Thompson, and L. Edebo, Long-chain Alkanoylcholines, a New Category of Soft Antimicrobial Agents that are Enzymatically Degradable,. Antimicrob. Agents Chemother., 1995. 39: p. 50–55.
[17] P. Gilbert and L. E. Moore, Cationic Antiseptics: Diversity of Action Under a Common Epithet,. J. Appl. Microbiol., 2005. 99: p. 703–715.
[18] M. Marini, M. Bondi, R. Iseppi, M. Toselli, and F. Pilati, Preparation and Antibacterial Activity of Hybrid Materials Containing Quaternary Ammonium Salts via Sol-Gel Process. Eur. Polym. J., 2007. 43: p. 3621-3628.
[19] M. Summers, J. Eastoe, and R. M. Richardson, Concentrated Polymerized Cationic Surfactant Phases. Langmuir, 2003. 19: p. 6357-6362.
[20] H. Shao, L. Jiang, W.D. Meng, and F. L. Qing, Synthesis and Antimicrobial Activity of a Perfluoroalkyl-Containing Quaternary Ammonium Salt. J. Fluorine Chem., 2003. 124: p. 89–91.
[21] L. Caillier, E. T. de Givenchy, R. Levy, Y. Vandenberghe, S. Géribaldi, and F. Guittard, Synthesis and Antimicrobial Properties of Polymerizable Quaternary Ammoniums. Eur. J. Med. Chem., 2009. 44: p. 3201–3208.
[22] G. Sun, T. Y. Chen, W. B. Wheatley, and S. D. Worley, Preparation of Novel Biocidal N-Halamine Polymers. J. Bioact. Compat. Polym., 1995. 10: p. 135–144.
[23] L. Qian and G. Sun, Durable and Regenerable Antimicrobial Textiles: Chlorine Transfer Among Halamine Structures, . Ind. Eng. Chem. Res., 2005. 44: p. 852–856.
[24] K. Barnes, J. Liang, R. Wu, S. D. Worley, J. Lee, R. M. Broughton, and T. S. Huang, Synthesis and Antimicrobial Applications of 5,50-ethylenebis[5-methyl-3-(3-triethoxysilylpropyl) hydantoin]. Biomaterials, 2006. 27: p. 4825–4830.
[25] K. Barnes, J. Liang, R. Wu, S. D. Worley, J. Lee, R. M. Broughton, and T. S. Huang, Synthesis and Antimicrobial Applications of 5,50-ethylenebis[5-methyl-3-(3-triethoxysilylpropyl) hydantoin]. Biomaterials, 2006. 27: p. 4825-4830.
[26] K. Barnes, J. Liang, S. D. Worley, J. Lee, R. M. Broughton, and T. S. Huang, Modification of Silica Gel Cellulose and Polyurethane with a Sterically Hindered N-Halamine Moiety to Produce Antimicrobial Activity, . J. Appl. Polym. Sci., 2007. 105: p. 2306–2313.
[27] X. Ren, H. B. Kocer, S. D. Worley, R. M. Broughton, and T. S. Huang, Rechargeable Biocidal Cellulose: Synthesis and Application of 3-(2,3-dihydroxypropyl)-5,5-dimethylimidazolidine-2,4-dione. Carbohydr. Polym., 2009. 75: p. 683–687.

[28] J. Lin, V. Cammarata, and S. D. Worley, Infrared Characterization of Biocidal Nylon. Polymer, 2001. 42: p. 7903–7906.
[29] X. Ren, H. B. Kocer, L. Kou, S. D. Worley, R. Broughton, M. Tzou, Y. M, and T. S. Huang, Antimicrobial Polyester. J. Appl. Polym. Sci., 2008. 109: p. 2756–2761.
[30] X. Ren, L. Kou, H. B. Kocer, C. Zhu, S. D. Worley, R. M. Broughton, and T. S. Huang, Antimicrobial Coating of an N-Halamine Biocidal Monomer on Cotton Fibers via Admicellar Polymerization. Colloids Surf., 2008. A, 317, 711–716.
[31] A. E. I. Ahmed, J. N. Hay, M. E. Bushell, J. N. Wardell, and G. Cavalli, Biocidal Polymers (I): Preparation and Biological Activity of Some Novel Biocidal Polymers Based on Uramil and its Azo-Dyes. React. Funct. Polym., 2008. 68: p. 248–260.
[32] M. M. G. Fouda, A. El Shafei, S. Sharaf, and A. Hebeish, Microwave curing for producing cotton fabrics with easy care and antibacterial properties. Carbohydrate Polymers, 2009. 77: p. 651–655.
[33] L. Fan, Y. Du, B. Zhang, J. Yang, J. Zhou, and J. F. Kennedy, Preparation and Properties of Alginate/Carboxymethyl Chitosan Blend Fibers. Carbohydr. Polym., 2006. 65: p. 447-452.
[34] M. M. G. Fouda, R. Wittke, D. Knittel, and E. Schollmeyer, Use of Chitosan/Polyamine Biopolymers Based Cotton as a Model System to Prepare Antimicrobial Wound Dressing. Int. J. Diabetes Mellitus, 2009. 1: p. 61–64.
[35] K. S. Huang, W. J. Wu, J. B. Chen, and H. S. Lian, Application of Low-Molecular-Weight Chitosan in Durable Press Finishing. Carbohydr. Polym., 2008. 73: p. 254-260.
[36] S. H. Lim and S. M. Hudson, Synthesis and Antimicrobial Activity of a Water-Soluble Chitosan Derivative with a Fiber-Reactive Group. Carbohydr. Res., 2004. 339: p. 313–319.
[37] Y. Kitkulnumchai, A. Ajavakom, and M. Sukwattanasinitt, Treatment of Oxidized Cellulose Fabric with Chitosan and its Surface Activity Towards Anionic Reactive Dyes. Cellulose, 2008. 15: p. 599–608.
[38] Z. Cao and Y. Sun, N-Halamine-based Chitosan: Preparation, Characterization, and Antimicrobial Function. J. Biomed. Mater. Res., 2007. Part A, 85A,: p. 99–107.
[39] S. P. Yazdankhah, A. A. Scheie, E. A. Høiby, B. T. Lunestad, E. Heir, T. Ø. Fotland, K. Naterstad, and H. Kruse, Triclosan and Antimicrobial Resistance in Bacteria: An Overview. Microb. Drug Resist., 2006. 12: p. 83–90.
[40] M. Orhan, D. Kut, and C. Gunesoglu, Improving the Antibacterial Activity of Cotton Fabrics Finished with Triclosan by the Use of 1,2,3,4-butanetetracarboxylic Acid and Citric Acid. J. Appl. Polym. Sci., 2009. 111: p. 1344–1352
[41] L. Qian, Y. Guan, Z. Ziaee, B. He, A. Zheng, and H. Xiao, Rendering Cellulose Fibers Antimicrobial Using Cationic β-cyclodextrin-based Polymers Included with Antibiotics. Cellulose, 2009. 16: p. 309–317.
[42] B. Goetzendorf-Grabowska, H. Królinkowska, P. Bak, M. Gadzinowska, B. Brycki, and A. Szwajca, Triclosan Encapsulated in Poly(L,L-lactide) as a Carrier of Antibacterial Properties of Textiles. Fibres Text. East. Eur., 2008. 16: p. 102–107.
[43] A. Kawabata and J. A. Taylor, The Effect of Reactive Dyes Upon the Uptake and Antibacterial Efficacy of Poly(hexamethylene biguanide) on Cotton. Part 3: Reduction in the Antibacterial Efficacy of Poly(hexamethylene biguanide) on

Cotton, Dyed with Bis(monochlorotriazinyl) Reactive Dyes. Carbohydr. Polym., 2007. 67: p. 375–389.
[44] K. Moore and D. Gray, Using PHMB Antimicrobial to Prevent Wound Infection. Wounds UK, 2007. 3: p. 96–102.
[45] R. S. Blackburn, A. L. Harvey, L. Kettle, J. D. Payne, and S. J. Russell, Sorption of Poly(hexamethylenebiguanide) on Cellulose: Mechanism of Binding and Molecular Recognition. Langmuir, 2006. 22: p. 5636–5644.
[46] R. Rajendran, C. Balakumar, A. Hasabo, M. Ahammed, S. Jayakumar, K. Vaideki, and E. M. Rajesh, Use of zinc oxide nano particles for production of antimicrobial textiles. International Journal of Engineering, Science and Technology, 2010. 2(1): p. 202-208.
[47] C. J. Brinker and G. W. Scherer, Sol-Gel Science: The Physics and Chemistry of Sol-Gel Processing. Academic Press. ISBN 0121349705, 1990.
[48] Y. Xing, X. Yang, and J. Dai, Antimicrobial finishing of cotton textile based on water glass by sol-gel method. Journal of Sol-Gel Science and Technology, 2007. 43(2): p. 187-192.
[49] B. Mahltig, H. Haufe, and H. Bottcher, Functionalisation of textiles by inorganic sol-gel coatings. Journal of Materials Chemistry, 2005. 15: p. 4385-4398.
[50] B. Mahltig, D. Fiedler, and H. Bottcher, Antimicrobial sol-gel coatings. Journal of Sol-Gel Science and Technology, 2004. 32(1): p. 219-222.
[51] W. A. Daoud, J. H. Xin, and Y. Zhang, Surface functionalization of cellulose fibers with titanium dioxide nanoparticles and their combined bactericidal activities. Surface Science, 2005. 599(1-3): p. 69-75.
[52] B. M. Novak, Hybrid Nanocomposite Materials - Between Inorganic Glasses and Organic Polymers. Adv. Mater., 1993. 5: p. 42–433.
[53] H. S. Nalwa (ed.), "Handbook of Organic-Inorganic Hybrid Materials and Nanocomposites". American Scientific Publisher,Stevenson Ranch, 2003.
[54] X. Wang and C. Wang, The Antibacterial Finish of Cotton via Sols Containing Quaternary Ammonium Salts. J. Sol-Gel Sci. Technol., 2009. 50: p. 15–21.
[55] Z. Li, D. Lee, X. Sheng, R. E. Cohen, and M. F. Rubner, Two-Level Antibacterial Coating with Both Release-Killing and Contact Killing Capabilities. Langmuir, 2006. 22: p. 9820–9823.
[56] M. Yu, G. Gu, W. D. Meng, and F. L. Qing, Superhydrophobic Cotton Fabric Coating Based on a Complex Layer of Silica Nanoparticles and Perfluorooctylated Quaternary Ammonium Silane Coupling Agent. Appl. Surf. Sci., 2007. 253: p. 3669–3673.
[57] J. Chojnowski, W. Fortuniak, P. Ros´ciszewski, W. Werel, J. Lukasiac, W. Kamysz, and R. Halasa, Polysilsesquioxanes and Oligosilsesquioxanes Substituted by Alkylammonium Salts as Antibacterial Biocides, . J. Inorg. Organomet. Polym. Mater., 2006. 16: p. 219–230.
[58] P. Majumdar, E. Lee, N. Gubbins, N. Stafslien, S. J. Daniels, J. Thorson, and C. J. Chisholm, Synthesis and Antimicrobial Activity of Quaternary Ammonium-Functionalized POSS (Q-POSS) and Polysiloxane Coatings Containing Q-POSS,. Polymer, 2009. 50: p. 1124–1133.
[59] P. Majumdar, E. Lee, N. Patel, S. J. Stafslien, J. Daniels, and B. J. Chisholm, Development of Environmentally Friendly, Antifouling Coatings Based on

Tethered Quaternary Ammonium Salts in a Crosslinked Polydimethylsiloxane Matrix, . J. Coat. Technol. Res. , 2008. 5: p. 405–417.

[60] U. Mizerska, W. Fortuniak, J. Chojnowski, R. Halasa, A. Konopacka, and W. Werel, Polysiloxane Cationic Biocides with Imidazolium Salt (ImS) Groups, Synthesis and Antibacterial Properties, . Eur. Polym. J., 2009. 45: p. 779–787.

[61] J. Dutta and H. Hofmann, Encyclopedia of Nanoscience and Nanotechnology; Vol. X; Nalwa, H. S., Ed. American Scientific Publishers: Stevenson Ranch, CA, 2003: p. 1-23.

[62] Q. Chen, X. Shen, and H. Gao, One-step synthesis of silver-poly(4- vinylpyridine) hybrid microgels by ⌋ -irradiation and surfactant-free emulsion polymerization, the photoluminescence characteristics. Colloids and Surfaces A: Physicochemical Engineering Aspects, 2006. 275: p. 45-49

[63] R. Dastjerdi and M. Montazer, A review on the application of inorganic nano-structured materials in the modification of textiles: Focus on anti-microbial properties. Colloids and Surfaces B: Biointerfaces, 2010. 79: p. 5–18.

[64] A. Hebeish, M. A. Ramadan, M. E. El-Naggar, M. H. El-Rafie, Rendering cotton fabrics antibacterial using silver nanoparticles- based finishing formulation. Research Journal of Textiles &Apparel (Accepted for publication), 2009.

[65] A. Hebeish, M. H. El-Rafie, M. A. Ramadan, and M. E. El-Naggar, Investigation into the synthesis and characterization of silver nanoparticles. Research Journal of Textiles & Apparel (accepted for publication) 2009.

[66] D. Hofer, Antimicrobial Textiles—Evaluation of their Effectiveness and Safety in "Biofunctional Textiles and the Skin", Hipler, U. C., and Elsner, P. (eds), Karger, Basel, 2006. 42–50.

[67] Y. Gao and R. Cranston, Recent Advances in Antimicrobial Treatments of Textiles. Textile Research Journal, 2010. 78(1): p. 60-72.

5

Metal Complexes as Antimicrobial Agents

Marcela Rizzotto
Faculty of Biochemistry and Pharmacy, National University of Rosario, Argentina

1. Introduction

The treatment of infectious diseases still remains an important and challenging problem because of a combination of factors including emerging infectious diseases and the increasing number of multi-drug resistant microbial pathogens. In spite of a large number of antibiotics and chemotherapeutics available for medical use, at the same time the emergence of old and new antibiotic resistance created in the last decades revealed a substantial medical need for new classes of antimicrobial agents. There is a real perceived need for the discovery of new compounds endowed with antimicrobial activity, possibly acting through mechanism of action, which is distinct from those of well-known classes of antimicrobial agents to which many clinically relevant pathogens are now resistant.

Due to the outbreak of infectious diseases caused by different pathogenic bacteria and the development of antibiotic resistance, researchers are searching for new antibacterial agents. Therefore, new antimicrobial agents and nanotechnological materials have to be synthesized for the treatment of resistant bacterial diseases.

Historically, medicinal inorganic chemistry is rich in metal- or metalloid-based drugs, including Paul Erlich's organoarsenic compound for the treatment of syphilis, antiarthritic gold preparations, and diagnostic agents for magnetic resonance imaging (Gd, Mn, Fe) among others.

Some metals have been used as drugs and diagnostic agents to treat a variety of diseases and conditions. Platinum compounds, cisplatin (cis-$[Pt(NH_3)_2Cl_2]$), carboplatin and oxaliplatin are among the most widely used cancer therapeutic agents. Gold drugs, myocrisin and auranofin are used for the treatment of rheumatoid arthritis. Another important aspect of medicinal inorganic chemistry is the development of radiopharmaceuticals and diagnostic agents. A technetium radiopharmaceutical, cardiolite supplies 99mTc, which is selectively taken up by myocardial tissue and is used to image the heart. $_{186}Re/_{188}Re$ has been identified as important radionuclides with therapeutic potential. The use of lanthanides and transition metals (Gd, Fe, Mn) as paramagnetic contrast agents for magnetic resonance imaging is becoming more exciting with the development of new complexes having the ability to target specific tissues and physiological states.

The field of bioinorganic chemistry, which deals with the study of role of metal complexes in biological systems, has opened a new horizon for scientific research in coordination compounds. A large number of compounds are important from the biological point of view.

Some metals are essential for biological functions and are found in enzymes and cofactors required for various processes. For example, hemoglobin in red blood cells contains an iron porphyrin complex, which is used for oxygen transport and storage in the body. Chlorophyll in green plants, which is responsible for photosynthetic process, contains a magnesium porphyrin complex. Cobalt is found in the coenzyme B12, which is essential for the transfer of alkyl groups from one molecule to another in biological systems. Metals such as copper, zinc, iron and manganese are incorporated into catalytic proteins (the metalloenzymes), which facilitate a multitude of chemical reactions needed for life. Today medicinal inorganic chemistry remains a field of great promise with many challenges. The potential for a major expansion of chemical diversity into new structural and reactivity motifs of high therapeutic impact is unquestionable.

Biological metal ions play key roles in the structural organization and activation of certain enzymes, which are involved in the transfer of genetic information from DNA, leading to the synthesis of specific proteins. Transition metal complexes have attracted attentions of inorganic, metallo-organic as well as bio-inorganic chemists because of their extensive applications in wide ranging areas from material to biological sciences.

Some chelating agents have been developed for metal intoxication, e.g., water soluble phosphine chelating agents are designed for chelating metals such as technetium, rhenium, platinum and gold. Many organic compounds used in medicine do not have a purely organic mode of action; some are activated or biotransformed by metal ions including metalloenzyme, others have a direct or indirect effect on metal ion metabolism. The pharmacological activities of these metal compounds depend on the metal ion, its ligands and the structure of the compounds. These factors are responsible for reaching them at the proper target site in the body. It is known that certain metal ions penetrate into bacteria and inactivate their enzymes, or some metal ions can generate hydrogen peroxide, thus killing bacteria.

Biologically relevant metal complexes have several requirements in terms of their synthetic design. First, a biologically active metal complex should have a sufficiently high thermodynamic stability to deliver the metal to the active site. The metal-ligand binding should be hydrolytically stable. The kinetics with which the metal ion undergoes ligation or deligation reactions is of great importance. The molecular weight of the metal complex is also critical. The compounds of low molecular weight with neutral charge and some water solubility are soluble in almost any medium and may slip through biological membranes by passive diffusion.

Generally, drug combinations have proven to be an essential feature of antimicrobial treatment due to a number of important considerations: (i) they increase activity through the use of compounds with synergistic or additive activity; (ii) they thwart drug resistance; (iii) they decrease required doses, reducing both cost and the chances of toxic side effects; (iv) they increase the spectrum of activity.

Various biological aspects of the metal based drugs/ligands entirely depend on the ease of cleaving the bond between the metal ion and the ligand. As a consequence, it is essential to understand the relationship between ligand and the metal in biological systems. Several metal complexes are known to accelerate the drug action and the efficacy of the organic therapeutic agent. The efficacy of the various organic therapeutic agents can often be enhanced upon coordination with a suitable metal ion. The pharmacological activity of metal complexes is highly dependent on the nature of the metal ions and the donor sequence of the ligands because different ligands exhibit different biological properties. There is a real perceived need for the discovery of new compounds endowed with antimicrobial activities. The newly prepared compounds should be more effective and possibly act through a distinct mechanism from those of well-known classes of antimicrobial agents to which many clinically relevant pathogens are now resistant.

2. Metal complexes of sulfonamides

Sulfonamides were the first effective chemotherapeutic agents employed systematically for the prevention and cure of bacterial infections in humans. Among the many and so different families of organic–inorganic chemicals being currently investigated today because of their applications, sulfonamides and their N-derivatives are one of the outstanding groups. Sulfonamides represent an important class of medicinally important compounds which are extensively used as antibacterial agent. It interferes with PABA (p-aminobenzoic acid) in the biosynthesis of tetrahydrofolic acid, which is a basic growth factor essential for the metabolic process of bacteria.

N-Substituted sulfonamides are still among the most widely used antibacterial agents in the world, mainly because of their low cost, low toxicity, and excellent activity against bacterial diseases.

Many activities apart from carbonic anhydrase have been recently reviewed that include endotelin antagonism, anti-inflammatory, tubular transport inhibition, insulin release and saluretic activity. It is well documented that toxicological and pharmacological properties are enhanced when sulfonamides are administered in the form of their metal complexes

In 2006 the synthesis, characterization and comparative biological study of a series of antibacterial copper complexes with heterocyclic sulfonamides (L) were reported. Two kinds of complexes were obtained with the stoichiometries $[Cu(L)_2]\cdot H_2O$ and $[Cu(L)_2(H_2O)_4]\cdot nH_2O$, which were characterized by infrared and electronic spectroscopies. The antimicrobial activity was evaluated for all the synthesized complexes and ligands using the agar dilution test. The results showed that the complexes with five-membered heterocyclic rings were more active than the free sulfonamides while the pyrimidine, pyridine and pyridazine complexes had similar or less activity than the free ligands. In order to find an explanation for this behavior lipophilicity and superoxide dismutase-like activity were tested, showing that the $[Cu(sulfamethoxazol)_2(H_2O)_4]\cdot 3H_2O$ presented the highest antimicrobial potency and a superoxide dismutase-like activity comparable with pharmacological active compounds. In spite of the fact that the different species were added in the agar as a suspension, due to their low solubility, all the compounds were active against S. aureus and E. coli. They acted as antibacterial agents with different behaviors. $[Cu(sulfadiazine)_2]\cdot H_2O$, $[Cu(sulfamerazine)_2]\cdot H_2O$ and $[Cu(sulfapyridine)_2]\cdot H_2O$ were less effcient than the corresponding sulfonamides, while $[Cu(sulfamethoxypyridazine)_2]\cdot H_2O$ had the same microbiological activity. On the contrary, $[Cu(sulfisoxazole)_2(H_2O)_4]\cdot 2H_2O$, $[Cu(sulfamethoxazole)_2(H_2O)_4]\cdot 3H_2O$, $[Cu(sulfamethoxazole)_2]\cdot H_2O$ and $[Cu(sulfamethizole)_2]\cdot H_2O$ were more efficient than the free sulfonamides (MIC from 4 to 32 μg/mL). In this last group all the ligands have a five membered heterocycle (isoxazole and diazomethyzole) and all the corresponding complexes coordinate through the heterocyclic N. None of the copper sulfate dilutions, used as controls, inhibited the growth of bacteria. Taking into account the previous knowledge, it could be suggest that one reason for the higher activity of the last four complexes may be due to higher lipophilicity in relation with free sulfonamides. The complexes with five-membered heterocyclic rings showed more activity than free ligands, in particular $[Cu(sulfamethoxazole)_2(H_2O)_4]\cdot 3H_2O$ provided the highest antimicrobial potency (4 μg/mL against *Staphylococcus aureus* ATCC 29213, *S. aureus* and *Escherichia coli* (from patient exudates).

In 2007, tri- and di-positive metal complexes of the sulfa-drugs, Schiff base namely 2-thiophene carboxaldehyde-sulfametrole (HL) and its have been synthesized and characterized. The proposed general formulae of the prepared complexes are $[M_2X_4(HL)(H_2O)_4]$ (whereM = Mn(II), Co(II), Ni(II), Cu(II), Zn(II) and Cd(II), X= Cl, $[Fe_2Cl_6(HL)(H_2O)_2]$, $[(FeSO_4)_2(HL)(H_2O)_4]$ and $[(UO_2)_2(HL)(NO_3)_4]\cdot H2O$. Chloramphenicol and Grisofluvine were used as reference compounds for antbacterial and antifungal activities, respectively. *Escherichia coli, Salmonella typhi, Staphylococcus aureus* and *Bacillus subtillus* (bacteria) or *Aspergillus terreus* and *Aspergillus flavus* (Fungi) were used as the test organisms. Results have been recorded in the form of inhibition zones (diameter, cm). It can be seen that the antibacterial and antifungal of HL ligand are more or similar to that of the standards. Further the antibacterial and antifungal actions of HL ligand may be significantly enhanced on chelation with transition metal ions. Also, complexes showed more or same antibacterial activity and moderate antifungal activity comparable to that of the standards. So, the results suggest that metallation increases activity.

Based on the significant biological and pharmacological properties that the indole moiety possesses, a new class of such compounds was reported by combining the chemistry of sulfonamides with indole-3-carbaldehyde and to explore their biological activities with the aim of obtaining more potent antibacterial and/or antifungal compounds. Synthesis of seven new indolenyl sulfonamides, have been prepared by the condensation reaction of indole-3-carboxaldehyde with different sulfonamides such as, sulphanilamide, sulfaguanidine, sulfathiazole, sulfamethoxazole, sulfisoxazole, sulfadiazine and sulfamethazine. These synthesized compounds have been used as potential ligands for complexation with some selective divalent transition metal ions (cobalt, copper, nickel & zinc). Structure of the synthesized ligands has been deduced from their physical, analytical (elemental analyses) and spectral (IR, 1H NMR and 13C NMR & UV–vis) data. All the compounds have also been assayed for their in vitro antibacterial and antifungal activities examining six species of pathogenic bacteria (*Escherichia coli, Shigella flexneri, Pseudomonas aeruginosa, Salmonella typhi, Staphylococcus aureus* and *Bacillus subtilis*) and six of fungi (*Trichophyton longifusus, Candida albicans, Aspergillus flavus, Microsporum canis, Fusarium soloni* and *Candida glabrata*). Antibacterial and antifungal results showed that all the compounds showed significant antibacterial activity whereas most of the compounds displayed good antifungal activity (MIC from 1.3 to 0.65 x 10^{-4} M)

In 2010 a new series of sulfonamide derived Schiff bases has been synthesized by a condensation reaction of various sulfonamides with aromatic aldehydes. The so obtained sulfonamide were further investigated for their chelation and biological properties with first row d-transition metal ions [cobalt(II), copper(II), nickel(II) and zinc(II)]. An octahedral geometry has been suggested for all the complexes. The ligands and their metal complexes have been screened for in vitro antibacterial, antifungal and cytotoxic properties. The result of these studies have revealed that all compounds showed moderate to significant antibacterial activity against one or more bacterial strains and good antifungal activity against various fungal strains. All the synthesized compounds exhibited varying degree of inhibitory effect on the growth of different tested strains. A significant activity was observed by all the compounds against *E. coli*. Antibacterial activity is overall enhanced after complexation of the ligands. However, the Zinc (II) complexes ($[Zn(L1\text{-}H)_2(H_2O)_2]$: $C_{38}H_{36}N_8O_8S_2\,Br_2Zn$) and ($[Zn(L2\text{-}H)_2(H_2O)_2]$: $C_{34}H_{28}N_8O_8S_2Br_2\,Zn$) of both the ligands were observed to be the most active against all strains (MIC was in the range of 3.204×10^{-8} to 1.341×10^{-7} M). The antifungal screening of all compounds was carried out against *T. longifusus, C. albican, A. .avus, M. canis, F. solani and C. glaberata* fungal strains according to the literature protocol. Majority of the synthesized compounds showed good antifungal activity against different fungal strains. Compounds $[Zn(L1\text{-}H)_2(H_2O)_2]$, $[Ni(L2\text{-}H)_2(H_2O)_2]$ and $[Zn(L2\text{-}H)_2(H_2O)_2]$ showed excellent while all other compounds showed moderate to excellent activities against various fungal strains. The results of inhibition were compared with the results of inhibition with the standard drugs miconazole and amphotericin B.

The reaction between phthalylsulfathiazolate (PST) and cobalt(II) aqueous solutions leads to a stable complex compound, $[Co^{II}(PST)(H_2O)_4]\cdot 2H_2O$ (Co-PST). Reflectance diffuse spectrum is in agreement with a distorted octahedral environment of the Co(II) ion. Vibrational FTIR and Raman spectroscopic data reveal that the ligand would be doubly deprotonated. Spectroscopic and chemical data let us suggest that the Nthiazolic and the Nsulfonamide atoms could be the binding sites for the Co(II) ion to the phthalylsulfathiazole moiety. The following strains from the American Type Culture Collection (ATCC), Rockville, MD, USA, Malbrán Institute (MI), Pasteur Institute (PI) and from the Laboratorio de Microbiologı´a (LM, Facultad de Ciencias Médicas, Universidad Nacional de Cuyo, Mendoza, Argentina) were used: Gram-negative bacteria: *Escherichia coli* ATCC 25922, LM1-*Escherichia coli*, LM2-*Escherichia coli*, *Pseudomonas aeruginosa* ATCC 27853 and Gram-positive bacteria: LM-*Staphylococcus aureus*, *Staphylococcus aureus* methicillin-sensitive ATCC 29213 and *Staphylococcus aureus* methicillin-resistant ATCC 43300 were used for the antibacterial evaluation. Bacteria were grown on Müeller–Hinton agar medium. Co-PST showed antibacterial activity similar to the ligand (20-30 μg/mL). Activity against *Candida albicans*, if moderate (125 μg/mL), was better than the ligand one (>250 μg/mL). Co-PST did not show direct mutagenicity with the Ames test in the range of assay doses nor hemolytic effects to human erythrocytes in vitro at concentrations in which it is active. The phytotoxicity of the complex, evaluated with the Allium test, was similar to the phthalylsulfathiazole one in the whole tested range.

More recently (2011), copper(II) and nickel(II) complexes of sulfadimethoxine, sulfadiazine, sulfamerazine and sulfamethazine were synthesized and characterized by single-crystal X-ray diffraction and electrochemistry. Structural inspections showed that the antibacterial entity of ligands remains non-coordinated to metal ions in the complex high-lighting the fact that in each cluster, antiseptic activity of the metal has been associated to the antibiotic activity of the ligand. The antibacterial activity of the complex is as important as the ligands one with the addition of antiseptic activity via the incorporation of copper ions. Bacterial strains tested were *Staphylococcus aureus, Escherichia coli, Klebsiella pneumoniae, Enterococcus faecalis, Pseudomonas aeruginosa*. The MICs of *E. faecalis* and *P. aeruginosa* are equal to or greater than 128 μg/mL. The MICs of *E. coli, S. aureus* and *K. pneumoniae* are between 32 and 128 μg/mL. All the results obtained with the complexes are thus at least identical or greater than those observed for the free ligands.

3. Antibacteria in materials

Bacteria can grown in different materials that are in close contact with humans, foods, etc., so, it is very important to control this matter in order to prevent risk of infections. The three following investigations are examples in this way.

The growth of bacteria on cellulosic textile is one of its inherent properties. Infection by bacteria causes cross-infection by pathogens, development of odour, staining and loss of the performance properties of textile, so application of antimicrobial finishing agents are necessary for many textiles such as hygienic, medical textiles and odour free sport wear. The antimicrobial function can be incorporated into textile either by chemical finishing of fabrics with biocidal agents or by physical incorporation of the agents into fibers. An area of polymer research that presents great current interest is that of the development of polymers with antimicrobial activities, generally known as polymeric biocides. In the area of health care and hygienic applications, biocidal polymers may be incorporated into fibers, or possibly extruded into fibers themselves, and used for contact disinfectants in many biomedical applications. One method of achieving antimicrobial polymers is to add an organic or inorganic biocide to the polymers during processing of the material.The antimicrobial study emphasise that Cu/oxidized polyvinyl pyridine (PVP) and Ag/oxidized PVP have retarded the growth of bacteria significantly, and Ag/oxidized PVP has a far better biocidal activity. The antibacterial activity of both metal ions survived after washing 10 times.

Jute, a lignocellulosic natural fiber, has 58-68% a-cellulose, 12-14% lignin and 21-24% hemicellulose as a major constituent. Traditionally and due to its inherent coarseness property, jute is being used as a low cost packaging material and also to some extent for producing floor covering and decorative items. A study was carried out using chitosan-metal complex aiming to impart the jute fabric antimicrobial properties. In this regards, Ag^+, Zn^{2+} and Zr^{2+} ions were allowed separately to form a complex with chitosan. It has been found that, jute fabrics treated with chitosan-metal complex show better antimicrobial properties than those fabrics treated with either chitosan or metal salt separately. Moreover, the jute fabrics treated with chitosan-Zn complex have higher antimicrobial properties compared with those samples treated with chitosan-Zr or chitosan-Ag complexes. It has been found that, jute fabrics treated with chitosan-metal complex show better antimicrobial properties than those fabrics treated with either chitosan or metal salt separately. Moreover, the jute fabrics treated with chitosan-Zn complex have higher antimicrobial properties compared with those samples treated with chitosan-Zr or chitosan-Ag complexes.

Antimicrobial ceramics (ACs) are becoming increasingly important because of their wide range of applications, including fabrics, building materials, cosmetics, electrical appliances, etc. The antimicrobial ceramics (AC) based on hydroxyapatite (HA) were made in a wet chemical process with additions of $AgNO_3$, $Cu(NO_3)_2 \cdot 3H_2O$ and $Zn(NO_3)_2 \cdot 6H_2O$. The aerobic *Escherichia coli* was used in the study. An obvious antimicrobial effect against *E. coli* was observed in Ag(I) AC. In contrast to Ag(I) AC, it was difficult to ascertain any bactericidal effect in the case of Cu(II) and Zn(II) AC. This suggests that Ag(I) dissolved out and reacted with *E. coli* and inactivated the *E. coli* metabolism, thus inhibiting its growth.

4. Cu, Co, Ni, Zn

Many biologically active compounds used as drugs possess modified pharmacological and toxicological potentials when administered in the form of metal based compounds. Various metal ions potentially and commonly used are cobalt, copper, nickel and zinc because of forming low molecular weight complexes and therefore, prove to be more beneficial against several diseases.

The antibiotic activity of metal complexes of N-methylthioformohydroxarnic acid against gram-negative *Escherichia coli* and gram-positive *Staphylococcus aureus* was investigated. The kinetically labile, square-planar, divalent (Cu, Ni, and Pd) and octahedral, trivalent (Fe, Co, and Cr) complexes displayed activity (0.5 to 5 μM against *S. aureus*), whereas the more inert platinum(II) or rhodium(III) complex displayed no activity, or activity only at elevated concentrations. The free ligand did not suppress the growth of the above organisms, and the sulfur atom of the ligand in its metal complexes appears crucial for activity.

Neutral thiabendazole (TBZH) when uncoordinated to a metal centre is a poor anti-Candida agent and has very little chemotherapeutic potential. Complexes of Cu(II), in which the TBZH is present as a neutral chelating ligand, are all potent anti-candida agents with activity comparable to the prescription drug ketoconazole. Coordination of neutral TBZH to a copper centre in two of these complexes resulted in a significant increase in its chemotherapeutic potential.

Synthesis and antimicrobial activity of new metal [Co(II), Ni(II), Zn(II)] complexes from 2-(1′-hydroxynaphthyl)benzoxazoles have been described in 2007. The cobalt complex $C_{34}H_{20}CoN_2O_4$ showed significant antifungal activity (MIC 6.25-12.5 μg/mL).

Interaction of norfloxacin (Nor) and ofloxacin (Ofl) with copper(II) and copper(II)/phenanthroline has been studied in aqueous solution and the stability constants of the binary complexes Cu(II)/fluoroquinolone and of the ternary complexes Cu(II)/phenanthroline/fluoroquinolone have been determined by potentiometry and UV-vis spectrophotometry. The stability constants for the binary and ternary complexes of norfloxacin were always higher than those found for ofloxacin and comparing the values obtained for the binary and ternary species (log$K$) it is possible to conclude that the ternary complexes are more stable than the binary ones, suggesting that an interaction occurs between the ligands in the ternary complexes. From the distribution diagrams it is possible to state that at physiological pH 7.4, the copper ternary complexes, are the main species in solution not only at the concentration used to determined the stability constants but also at the minimum inhibitory concentration. The antibacterial activity of these complexes, in different bacterial strains, was determined, at physiological pH, and the results obtain show that these complexes may be good candidates as metalloantibiotics. (MIC against *Escherichia coli* ATCC 25922: Cu.Ofl. 0.015 and Cu.Ofl.Phe: 0.03 μg/mL). This work shows that copper(II)/phenanthroline complexes with fluoroquinolones are very stable, at physiological pH and they seem to be a good approach for the development of drugs with similar activity against bacteria but with the possibility of lowering their level of resistance.

A novel copper(II) complex of the fluoroquinolone antibacterial drug *N*-propyl-norfloxacin (Hpr-norf) in the presence of 1,10-phenanthroline (Phen) has been synthesized, characterized and studied its biological properties as antitumor antibiotic and antimicrobial agent. The antimicrobial activity of the complex has been tested, revealing an increased potency in comparison to the free Hpr-norf. (MIC: 4–16 μg/mL)

A few mixed ligand transition metal carbodithioate complexes of the general formula [M(4-MPipzcdt)$_x$(phen)$_y$]Y (M = Mn(II), Co(II), Zn(II); 4-MPipzcdt = 4-methylpiperazine-1-carbodithioate; phen = 1,10-phenanthroline; x = 1 and y = 2 when Y = Cl; x = 2 and y = 1 when Y = nil) were synthesized and screened for their antimicrobial activity against *Candida albicans, Escherichia coli, Pseudomonas aeruginosa, Staphylococcus aureus and Enterococcus faecalis* by disk diffusion method. All the complexes exhibited prominent antimicrobial activity against tested pathogenic strains with the MIC values in the range <8-512 μg/mL. The complexes [Mn(4-MPipzcdt)$_2$(phen)] and [Co(4-MPipzcdt)(phen)$_2$]Cl inhibited the growth of *Candida albicans* at a concentration as low as 8 μg/mL. The complexes were also evaluated for their toxicity towards human transformed rhabdomyosarcoma cells (RD cells). Moderate cell viability of the RD cells was exhibited against the metal complexes.

Mixed-ligand ternary transition metal (i.e., Cu, Co and Ni) complexes bearing iminodiacetic acid ligand and 1,10-phenanthroline co-ligand, were synthesized and characterized from spectral methods. The complexes usually adopt a distorted octahedral geometry around the metal ion. Among the synthesized ternary complexes, copper and cobalt complexes showed remarkable antibacterial and antifungal activities (6.25 to 12.5 μg/mL).

Transition metal complexes of the type [M(L)$_2$] and those containing monodentate phosphines of the type [M(L)$_2$(PPh$_3$)] (M= Ni, Co, Cu and Zn; L = cyclohexylamine-N-dithiocarbamate; PPh3 = triphenylphosphine) have been synthesized. The spectral studies in all compounds revealed that the coordination of metals occurs via the sulphur atom of the dithiocarbamate ligand in a bidentate fashion. The ligand and their metal complexes were screened in vitro for their antibacterial activity against *Escherichia coli, Staphylococcus aureus, Salmonella typhi, Enterococcus faecalis, Pseudomonas aeruginosa* and *Bacillus cereus* and antifungal activities against *Aspergillus flavus, Aspergillus carbonarius, Aspergillus niger* and *Aspergillus fumigatus*. The metal complexes exhibited higher antimicrobial activity than the parent ligands. Generally, the zinc complexes were effective against the growth of bacteria with Zn(L)$_2$ displaying broad spectrum bacteriocidal activity at concentrations of 50 μg/mL; and Ni(L)$_2$ was more effective against the growth of fungi at concentrations of 100–400μg/mL under laboratory conditions.

Condensation of o-acetoacetylphenol and 1,2-diaminopropane in 1:1 molar ratio under condition of high dilution yielded the mono-condensed dibasic Schiff base ligand with a N_2O_2 donors. Reactions of the ligand with metal salts yielded mono- and homo-bi-nuclear complexes formulated as [M(HL)], where M = Co(II), Ni(II) and Cu(II), [M$_2$(L)Cl(H$_2$O)$_2$]·½H$_2$O, where M: Co(II) and Ni(II) and [Cu(H$_2$L)Cl]. The mononuclear Ni(II) complex, [Ni(HL)], was used to synthesize homo- and hetero-bi- and tri-nuclear complexes with the molecular formulae [Ni$_2$(L)Cl(H$_2$O)$_2$], [Ni$_2$(L)$_2$FeCl(H$_2$O)]·H$_2$O and [Ni$_2$(HL)$_2$CoCl$_2$]. The structures of the complexes were characterized by various techniques such as elemental and thermal analyses, IR, ^{1}H and ^{13}C NMR, mass and electronic spectra as well as conductivity and magnetic moment measurements. Square-planar and octahedral geometries are suggested for the Cu(II), Co(II) and Ni(II) complexes. The Schiff base and its metal complexes were evaluated for antimicrobial activity against Gram positive bacteria (*Staphylococcus aureus*), Gram negative bacteria (*Escherichia coli*) and fungi (*Candida albicans* and *Aspergillus flavus*). The ligand and some of its complexes were found to be biologically active. Results recorded revealed that among all the compounds $Zn(L)_2$ was effective at lower concentration (50µg/mL) and also exhibit bactericidal activity against all the bacteria. On the other hand $Co(L)_2$ was the compound with the weakest activity against bacteria and mostly static except with *P. aeruginosa*.

The transition metal (Fe, Co, Ni or Cu) ternary complexes containing pyridine-2,6-dicarboxylate dianion as primary ligand and 4-picoline as an auxiliary ligand were synthesized and characterized from spectral methods. The spectral data are consistent with a distorted octahedral geometry. The iron, copper and cobalt complexes showed remarkable antibacterial and antifungal activities while nickel complex showed these activities up to less extent. The antimicrobial activities investigated against *Escherichia coli, Bacillus subtilis, Staphylococcus aureus, Salmonella typhymurium, Candida albicans, Aspergillus fumigatus* and *Penicillium marneffei* showed significant activities (6.25 to 12.5 µg/mL).

Equilibrium studies on the ternary complex systems involving ampicillin (amp) as ligand (A) and imidazole containing ligands viz., imidazole (Him), benzimidazole (Hbim), histamine (Hist) and histidine (His) as ligands (B) at 37 °C show the presence of CuABH, CuAB and $CuAB_2$. The antimicrobial activity of Cu(II)-amp(A) and Cu(II)-amp(A)-Him/Hist/His(B) complexes were tested against bacteria *S. typhi*, yeast *S. cerevisae* and fungi *L. theobrome* and *F. oxysporum*. The bacteria and yeast were tested by "agar diffusion method" and the fungal activity have been tested using "potato dextrose agar method" using DMF as control. The zone inhibition against the growth of bacteria, yeast and fungi for the binary and ternary complexes show that the inhibition zone of ternary complexes (6.7-9.1 mm) is higher than the binary complex and control (3.6-7.9). On chelation, the polarity of Cu(II) ion will be reduced to a greater extent due to the overlap of ligand orbital and partial sharing of the positive charge of the Cu(II) ion with donor groups. Further, it increases the delocalization of π-electrons over the whole chelate ring and enhances lipophilicity of the complexes. This increased lipophilicity enhances the penetration of the complexes into lipid membranes and blocking of the metal binding sites in the enzymes of microorganism. These complexes also disturb the respiration process of the cell and thus block the synthesis of protein that restricts further growth of the organism.

The straightforward condensation of 3-acetyl-2-one indol and hydrazinecarbothioamide to yield the novel Schiff base ligand has been reported. Its flexible back bone, together with the presence of N, S and O donor atoms, renders this compound interesting for studying its coordination behaviour with transition metals ion. In this work some complexes with copper, nickel and cobalt have been characterized . The spectral data indicate that ligand behaves as a neutral pentadentate ligand, with two different coordinating sites, one provided by a nitrogen and an oxygen donor atoms and one by the C=N, C=O and C=S groups, each one accommodating a metal ion. Antimicrobial study reveals that, metal complexes have more biological activity than free ligand. Complex $HLNi_2(OAc)_4(H_2O)_3$ shows best antimicrobial activity against all microorganism (MIC: 10-12 µg/mL, while Imipenem = standard drug: 6-8 µg/mL, ligand: 65-125 µg/mL).

The antibactericidal activity of the cupric chloride, fluoroquinolones and its complexes were tested against two Gram(+) *S. aureus, B. subtilis*, and three Gram(−) *S. marcescens, E. coli* and *P. aeruginosa* organisms using double dilution method. An acceptable reason for this increase in bactericidal activity may be considered in the light of Overtone's concept and chelation theory. According to Overtone's concept of cell permeability, the lipid membrane that surrounds cell favors the passage of only lipid soluble materials so that liposolubility is an important factor which controls bactericidal activity. This increased lipophilicity enhances the penetration of the complexes into lipid membranes and blocks the metal binding sites in bacterial enzymes. The antimicrobial activity of all complexes against five microorganisms (MIC from 0.36 to 2.07 μg/mL) is much higher than metal salt. The complex shows better antimicrobial activity than the metal salt ≈ 3000 μg/mL), another quinilones (1.1-5.7 μg/mL), and enrofloxacin (1.4-3.9 μg/mL).

The study provides useful information about the nature of bonding in zinc–thione complexes. We have shown that thiones react with $ZnCl_2$ to form the complexes of the type, L_2ZnCl_2 in which the ligands coordinate in the thione form in solution as well as in the solid state. Antimicrobial activities of the complexes were evaluated by minimum inhibitory concentration and the results showed that some complexes exhibited significant activities against gram-negative bacteria (*Escherichia coli, Pseudomonas aeruginosa*: $[Zn(Dmtu)_2Cl_2]$ 50-40 μg/mL) and yeasts ($[Zn(Tmtu)_2Cl_2]$ 80 μg/mL against *Saccharomyces cerevisiae*). However, moderate activity was observed against molds (*Aspergillus niger, Penicillium citrinum*).

5. Silver antimicrobial agents

Silver and its compounds have long been used as antimicrobial agents in medicine. Silver is active at low concentrations and has a low toxicity. Silver sulfadiazine is a widely used broad-spectrum antibiotic ointment, effective against a broad range of bacteria and some yeasts. It is used to prevent and treat skin infections on the areas of burnt skin. Silver complexes of oxygen donor ligands such as, $[Ag(hino)]_2$ (where hino = 4-isopropyltopolone) and water-soluble silver(I) complexes of 2-pyrollidone-2-carboxylic acid displayed wide ranging and effective activities against some bacteria, yeasts and moulds. It has been found that the silver–oxygen bonding properties rather than the chiral helical or achiral polymer structure play a role in exhibiting antimicrobial activities. The antimicrobial activities of the silver(I)–oxygen bonding complexes are independent of whether the ligand itself possesses antimicrobial activities. Examples of such complexes are $\{[Ag(L\text{-}Hasp)]_2\}$ and $\{[Ag(LHasp)]_2\}_n$, where the ligand showed no activity. Similarly, salicylic acid (salH) did not prevent the growth of a fungal pathogen, while the silver complexes, $[Ag(salH)]_2$ and $[Ag(NH_3)(salH)]_2$ greatly inhibited cell reproduction. These complexes also produced a cytotoxic response against human cancer cells. The biological action of the silver(I) oxygen complexes comes from a weaker bonding property of the Ag–O bond. In the biological system, the ease of ligand replacement of the silver(I) complexes would result in further replacement with biological ligands. The Ag–O bonding complexes can readily undergo ligand replacement with O-, N- or S-donor atoms. The antimicrobial activities by silver(I)–oxygen bonding complexes are due to the silver(I) ion itself, i.e., due to a direct interaction between the silver(I) ion and biological ligands such as protein, enzymes and membrane. The coordinating ligands of the silver(I) complexes play the role of carrier for the silver(I) ion to the biological system. The magnitude of antimicrobial properties of silver complexes is related to the ease with which they participate in ligand exchange reactions. For example, it has been speculated that the weak Ag–O and Ag–N bond strengths might play an important role in exhibiting wider spectrum of antimicrobial and antifungal activities and that the potential target sites for inhibition of bacterial and yeast growth by silver complexes might be the sulfur containing residues of proteins. Generally, Ag–S complexes have been shown to have a narrower spectrum of antibacterial activity than Ag–N complexes but no antifungal activity. In contrast, the compounds with Ag–P bonds have shown no activity against bacteria, yeast or molds. A sulfur coordinated, water soluble silver(I) complex of thiomalic acid showed remarkable antimicrobial activity against some bacteria, yeast and moulds. Although aqueous silver nitrate itself has similar activities for bacteria, the complexation of silver(I) with thiomalate leads to appreciably high activities for some moulds. Antimicrobial activities have been observed for sulfur bonded silver complexes of 2-mercaptonicotinic acid, $[Ag(Hmna)]_6 \cdot 4H_2O$ and 2-mercaptobenzoic acid, $[Ag(Hmba)]_n$ and $[Na\{Ag(Hmba)\} \cdot H_2O]_n$. The spectra of antimicrobial activities observed in Ag–S bonded compounds have so far been narrower than those in the Ag–N bonded compounds such as $[Ag(im)]_n$ (im = imidazole) and $[Ag(triaz)]_n$ (triaz = 1,2,4-triazole). The key factor determining the spectra of antimicrobial activity is the nature of atom coordinated to the silver(I) atom and its bonding properties, (i.e., the ease of ligand replacement), rather than the solid state structure, solubility, charge and degree of polymerization of the complexes

The antibacterial property of silver has been known for thousands of years. Silver nanoparticles have proved to be most effective as they have good antimicrobial efficacy against bacteria, viruses and other eukaryotic microorganisms. Silver is toxic to microorganisms by poisoning respiratory enzymes and components of electron transport system, and it also binds to bacterial surface, altering the membrane function and inhibiting replication. Silver compounds are used as antimicrobial agents in a variety of applications, including coating of catheters, dental resin composites, burn wounds and homeopathic medicine, antimicrobial matter with a minimal risk of toxicity in humans. The synthesis, spectroscopic, thermal and antimicrobial properties and characterization of five new (saccharinato)silver(I) complexes with diverse diamine derivatives, N,N,N′,N′-tetramethylethylenediamine, N,N′-diethylethylenediamine, N,N′-dimethylethylenediamine, N,Ndiethylethylenediamine and 1,3-diamino-2,2-dimethylpropan ligands, namely $[Ag_2(sac)_2(tmen)_2]$ (1), $[Ag_2(sac)_2(deten)_2]$ (2), $[Ag_2(sac)_2(dmen)_2]_n$ (3), [Ag(sac)(N,N-eten)] (4) and $[Ag(sac)(dmpen)]_n$ (5) and Ag(I)–saccharin complex (6) were reported. There are no significant differences among antimicrobial activities of new complexes. Complex 6 is more efficient against all bacteria and yeast than the new complexes except *P. aeruginosa*. While the values of MIC for 6 are in the range of 13.5–27.5 μg/mL, value of MIC for *P. aeruginosa* is 45 μg/mL. But the values of MIC for new complexes (1–5) are in the range of 30–55 μg/mL. Antimicrobial activities of complexes 1–5 and 6 did not show differences between gram (+), gram (-) and eukaryotes in point of effective dose. It is known that there are a disparity between prokaryotes (gram +, gram -) and eukaryotes. But this difference is not important for antimicrobial effect of silver-saccharinate complexes. This suggests that they are potent as broad spectrum topical antimicrobial agents, but they need to be investigated with respect to their toxicity.

Two water-soluble, silver(I) complexes showing a wide spectrum of effective antibacterial and antifungal activities, i.e., $\{[Ag(Hhis)]0.2Et(OH)\}_2$ (**1**; H_2his = L-histidine) and [Ag(Hpyrrld)]2 (**2**; H_2pyrrld = (*S*)-(-)-2-pyrrolidone-5-carboxylic acid) were prepared. Antimicrobial activities of the free ligands, H_2his and H_2pyrrld, were estimated as >1000 μg/mL for bacteria, yeast, and mold and, thus, showed no activity. The Ag^+ ion, as aqueous $AgNO_3$, has shown remarkable activities against Gram-negative bacteria (*E. coli, P. aeruginosa*), moderate activities against one Gram-positive bacteria (*B. subtilis*), and no activity (>1600 μg/mL) against 2 yeasts and 10 molds. The complex **2**, on the contrary, showed remarkable and excellent activities against a wide spectrum of Gram-negative and -positive (*B. subtilis* and *S. aureus*) bacteria and yeast (*C. albicans* and *S. cerevisiae*) and even against many molds except *A. niger* and *A. terreus*. A similar wide spectrum was also obtained in **1**. Of particular note is the fact that in **1** and **2** activities against many molds are observed. The antibacterial and antifungal activities of **1** and **2** were remarkable (7.9-62.5 μg/mL) and comparable to another silver(I)-N-heterocycle complexes (6.3-50 μg/mL).

The alkanol *N*-functionalized silvercarbene complexes Silver(I)-2,6-bis(ethanolimidazolemethyl) pyridine hydroxide and silver(I)-2,6-bis(propanolimidazolemethyl)pyridine hydroxide are soluble in aqueous media. The solubility and stability of silver complexes in chloride solution are key factors that limit the use of silver complexes for in vivo application. The bactericidal activities of the water-soluble silver(I)-carbene complexes were found to be improved over that of silver nitrate, so the use of Ag-C donor (carbene) compounds has demonstrated its potential as a therapeutic agent.

The antimicrobial activity of the bis(*N*-heterocyclic carbene) (NHC) silver acetate complex, was evaluated against a variety of test organisms including a panel of highly resistant opportunistic pathogens recovered, primarily, from the respiratory tract of patients with cystic fibrosis (CF). The silver complex was also found to be a very effective antimicrobial agent when tested on fungi. Against *A. niger* and *S. cerevisiae*, it was found to be effective with a fungicidal MIC values of 13 μg/mL and 4 μg/mL. It shows a fungistatic effect on *C. albicans* with a MIC value of 4 μg/mL. Application of this NHC silver complex to primary cultures of murine respiratory epithelial cells followed by microarray analysis showed minimal gene expression changes at the concentrations effective against respiratory pathogens. Furthermore, methylated caffeine without silver showed some antibacterial and antifungal activity.

Preliminary in vivo toxicity studies demonstrated very low toxicity for both the parent methylated caffeine and the silver complex. Given the water solubility of this silver complex and its low toxicity, it may prove useful as a nebulized therapy in patients colonized with these resistant organisms.

$[Ag(2\text{-amino-3-methylpyridine})_2]NO_3$ (1) and $[Ag(\text{pyridine-2-carboxaldoxime})NO_3]$ (2) were prepared from corresponding ligands and $AgNO_3$ in water/ethanol solutions, and the products were characterized by IR, elemental analysis, NMR, and TGA. The X-ray crystal structures of the two compounds show that the geometry around the silver(I) ion is bent for complex 1 with nitrate as an anion and trigonal planar for complex 2 with nitrate coordinated. The geometries of the complexes are well described by DFT calculations using the ZORA relativistic approach. The compounds were tested against 14 different clinically isolated and four ATCC standard bacteria and yeasts and also compared with 17 commonly used antibiotics. Both 1 and 2 exhibited considerable activity against *S. lutea, M. lutea,* and *S. aureus* (0.6-17.9 μg/mL) and against the yeast *Candida albicans* (2.3 μg/mL), while 2-amino-3-methylpyridine is slightly active and pyridine-2-carboxaldoxime shows no antimicrobial activity. In addition, the interaction of these metal complexes with DNA was investigated. Both 1 and 2 bind to DNA and reduce its electrophoretic mobility with different patterns of migration, while the ligands themselves induce no change.

The two compounds show antibacterial effects against different bacteria and yeast, quite comparable to commercial antibiotics in vitro, but their activity spectrum is different, on both a microgram per milliliter basis and a Ag per milliliter basis.

6. Metal ions from the 5º and 6º periods

Dimers of vancomycin (Van), linked by a rigid metal complex, $[Pt(en)(H_2O)_2]^{2+}$, (en: ethylenediamine) exhibit potent activities (MIC: 0.8 μg/mL, 720 times more potent than that of Van itself) against vancomycin resistant enterococci (VRE). The result suggests that combining metal complexation and receptor/ligand interaction offers a useful method to construct multivalent inhibitors. In summary, metal complex can be used as a new platform to construct multivalent inhibitors, which are as effective as other rigid linkers used for multivalency. One of the concerns on platinum-based complexes is its cytotoxicity. Preliminary study has shown that these *cis*-platin based divalent Vans are not toxic toward mammalian cells. Our future work will examine other metal complex linkers, which may help further elucidate the structural basis of vancomycin resistance, as well as the mechanism of multivalent Vans binding to vancomycinsensitive strains, which has yet to be established.

Cationic gold(I) complexes containing 1-[2-(acridin-9-ylamino)ethyl]-1,3-dimethylthiourea $[AuL(1)]^{n+}$ (where L is Cl-, Br-, SCN-, PEt_3, PPh_3, or 1), derived from a class of analogous platinum(II) antitumor agents, have been synthesized. Unlike platinum, gold does not form permanent adducts with DNA, and its complexes are 2 orders of magnitude less cytotoxic in non-small-cell lung cancer cells than the most active platinum-based agent. Instead, several gold analogues show submicromolar and selective antimicrobial activity against *Mycobacterium tuberculosis* (MIC: 0.49-0.82 μM). In conclusion, the current set of complexes shows considerable potential as relatively nontoxic anti-Mtb agents. Given the urgent need for effective treatment options for multidrug resistant forms of TB, novel gold(I) complexes based on improved prodrug design and deliverymay represent a promising approach to combating this disease.

A series of novel palladium(II) chloride and bromide complexes with three types of quinolinylaminophosphonates have been synthesized and structurally characterized. All organophosphorus ligands contain three potential donor atoms, quinoline nitrogen, amino nitrogen and phosphoryl oxygen, but their coordination behaviour towards palladium(II) ion is different. In complexes either quinoline or both quinoline and amino nitrogens are involved in metal(s) bonding forming mononuclear dihalide adducts either with cis- or trans-configuration as well as dinuclear tetrahalide complexes. Phosphoryl oxygen is not coordinated and is free to be involved in hydrogen bonding, which is the main feature of crystal structures of complexes. The stereochemistry of the complexes, the nature of metal-ligand binding and hydrogen bond interactions are investigated by spectroscopy and X-ray structure analysis. Biological properties of complexes were examined by screening of their ability to inhibit the cancer growth in vitro in a panel of human tumor cell lines and their antimicrobial activity in a wide spectrum of bacterial and fungal strains. While no specific antimicrobial effects of both the free organophosphorus ligands and their palladium(II) halide complexes were noted, the majority of complexes demonstrated cytostatic activity, which was especially pronounced in the case of dipalladium tetrahalide complexes with IC50:10 μM. It may be concluded that palladium complexes of investigated quinolinylaminophosphonates represent an interesting class of new complex compounds from the viewpoint of their physicochemical, structural and biological properties.

Some organotin(IV) complexes containing benzil bis(benzoylhydrazone) and different numbers of bipyridyl units have been synthesised and fully characterized. In all the complexes the bis(hydrazone) ligand is doubly deprotonated and behaves as N_2O_2 tetradentate chelate. The in vitro antimicrobial activity against bacteria and fungi was evaluated. Methyl derivatives are devoid of antimicrobial properties, whereas all butyltin(IV) complexes show a good activity against bacteria that increases with the number of bipyridyl units (12-25 μg/mL).

7. Conclusions

In general, when the antimicrobial activity of metal complexes is concerned, the following five principal factors may be considered:

i. The chelate effect, i.e. bidentate ligands, such as the quinolones, show higher antimicrobial efficiency towards complexes with monodentate ligands
ii. The nature of the ligands
iii. The total charge of the complex; generally the antimicrobial efficiency decreases in the order cationic > neutral > anionic complex
iv. The nature of the counter ion in the case of the ionic complexes
v. The nuclearity of the metal center in the complex; dinuclear centers are more active than mononuclear ones.

The antimicrobial activities of metal complexes depended more on the metal center itself than on the geometry around the metal ion.

The biological activity of new silver(I) complexes is potentially important, and these compounds are developed not only with wound care in mind but some cases of antibiotic resistence, and also with, for example, the treatment of lungs chronically infected with cystic fibrosis.

Because of the broad spectrum activity displayed by some of the tested compounds, it is would be necessary to evaluate the cytotoxicity of these compounds as their applications in the formulation of novel antimicrobial therapeutic drugs seem promising. Useful test in the preclinical phase of medicaments are bioassays like the Ames test and a micronucleus test (*Allium cepa* test for example).

8. References

Beerse et al USA Patent N° US 6,294,186 B1, Sep. 25, 2001

Chohan, Zahid H.; Hazoor A. Shad a, Moulay H. Youssoufi b, Taibi Ben Hadda Some new biologically active metal-based sulfonamide, European Journal of Medicinal Chemistry 45 (2010) 2893-2901

Dai, Hui-Xiong; Antonia F. Stepan, Mark S. Plummer, Yang-Hui Zhang, and Jin-Quan Yu Divergent C–H Functionalizations Directed by Sulfonamide Pharmacophores: Late-Stage Diversification as a Tool for Drug Discovery, J. Am. Chem. Soc., Article ASAP DOI: 10.1021/ja201708f, Publication Date (Web): April 13, 2011,

Efthimiadou Eleni K., Alexandra Karaliota, George Psomas Metal complexes of the third-generation quinolone antimicrobial drug sparfloxacin: Structure and biological evaluation, Journal of Inorganic Biochemistry 104 (2010) 455–466

Higazy, Asha; Mohamed Hashem, Ali ElShafei, Nihal Shaker, Marwa Abdel Hady Development of antimicrobial jute packaging using chitosan and chitosan-metal complex Carbohydrate Polymers, Volume 79, Issue 4, 17 March 2010, Pages 867-874

Kantouch, A.; Atef El-Sayed A Polyvinyl pyridine metal complex as permanent antimicrobial finishing for viscose fabric International Journal of Biological Macromolecules, Volume 43, Issue 5, 1 December 2008, Pages 451-455

Kascatan-Nebioglu, Aysegul, Abdulkareem Melaiye, Khadijah Hindi, Semih Durmus, Matthew J. Panzner, Lisa A. Hogue, Rebekah J. Mallett, Christine E. Hovis, Marvin Coughenour, Seth D. Crosby, Amy Milsted, Daniel L. Ely, Claire A. Tessier, Carolyn L. Cannon, and Wiley J. Youngs Synthesis from Caffeine of a Mixed N-Heterocyclic Carbene−Silver Acetate Complex Active against Resistant Respiratory Pathogens, J. Med. Chem., 2006, 49 (23), pp 6811–6818

Katsarou, Maria E.; Eleni K. Efthimiadou, George Psomas, Alexandra Karaliota and Dionisios Vourloumis Novel Copper(II) Complex of N-Propyl-norfloxacin and 1,10-Phenanthroline with Enhanced Antileukemic and DNA Nuclease Activities J. Med. Chem., 2008, 51 (3), pp 470–478)

Kim, T. N., Q. L. Feng, J. O. Kim, J. Wu, H. Wang, G. C. Chen, F. Z. Cui. Journal of Materials Science: Materials in Medicine 9 (1998) 129-134 Antimicrobial effects of metal ions (Ag(I), Cu(II), Zn(II) in hydroxyapatite

Kozarich, John W. Medicinal Inorganic Chemistry: Promises and Challenges, , Medicinal Inorganic Chemistry, Chapter 2, pp 4–14 ACS Symposium Series, Vol. 903, ISBN13: 9780841238992 eISBN: 9780841220218 Publication Date (Print): August 25, 2005, Copyright © 2005 American Chemical Society

Lippard, S.; Chapter 1Metal Ion Chemistry for Sustaining Life, , In Medicinal Inorganic Chemistry; Sessler, J., et al.; ACS Symposium Series; American Chemical Society: Washington, DC, 2005

M.R. Malik et al., Inorg. Chim. Acta (2011), doi:10.1016/j.ica.2011.06.017

Monti, Laura; Ana Pontoriero Natalia Mosconi, Cecilia Giulidori, Estela Hure, Patricia A. M. Williams, María Victoria Rodríguez, Gabriela Feresin, Darío Campagnoli and Marcela Rizzotto* "Synthesis, characterization and antimicrobial properties of a Co(II)-phthalylsulfathiazolate complex" Biometals (2010), 23:1015-1028 and references cited therein

Morais Leme D, Marin-Morales MA (2009) Allium cepa test in environmental monitoring: a review on its application. Mutat Res 682:71–81

Mortelmans K, Zeiger E (2000) The Ames Salmonella/microsome mutagenicity assay. Mutat Res 455:29–60 28:2187–2195

Nomiya, Kenji *, Kuniaki Onodera, Ken Tsukagoshi, Kouhei Shimada, Akira Yoshizawa, Tada-aki Itoyanagi, Akiyoshi Sugie, Shinichiro Tsuruta, Ryo Sato, Noriko

Chikaraishi Kasuga Inorganica Chimica Acta 362 (2009) 43–55, Syntheses, structures and antimicrobial activities of various metal complexes of hinokitiol

Olar, Rodica; Mihaela Badea, Dana Marinescu, Mariana-Carmen Chifiriuc, Coralia Bleotu, Maria Nicoleta Grecu, Emilia-Elena Iorgulescu, Veronica Lazar; N,N-dimethylbiguanide complexes displaying low cytotoxicity as potential large spectrum antimicrobial agents European Journal of Medicinal Chemistry, Volume 45, Issue 7, July 2010, Pages 3027-3034

Patil, Sangamesh A.; Vinod H. Naik, Ajaykumar D. Kulkarni, Prema S. Badami DNA cleavage, antimicrobial, spectroscopic and fluorescence studies of Co(II), Ni(II) and Cu(II) complexes with SNO donor coumarin Schiff bases Spectrochimica Acta Part A: Molecular and Biomolecular Spectroscopy, Volume 75, Issue 1, January 2010, Pages 347-354

S. Bellú, E. M. Hure, M. Trapé, C. Trossero, G. Molina, C. Drogo, P. A. M. Williams, A. M. Atria, J. C. Muñoz Acevedo, S. Zacchino, M. Sortino, D. Campagnoli and M. Rizzotto* "Synthesis, Structure and Antifungal Properties of Co (II)-sulfathiazolate complexes" Polyhedron 24 (2005) 501-509.

Saeed Ahmad*, Anvarhusein A. Isab, Saqib Ali, Abdul Rahman Al-Arfaj. Perspectives in bioinorganic chemistry of some metal based therapeutic agents (Polyhedron 25 (2006) 1633–1645)

Shebl, Magdy; Saied M.E. Khalil, Saleh A. Ahmed, Hesham A.A. Medien;Synthesis, spectroscopic characterization and antimicrobial activity of mono-, bi- and tri-nuclear metal complexes of a new Schiff base ligand Journal of Molecular Structure, Volume 980, Issues 1-3, 10 September 2010, Pages 39-50

Sultana, N., M. Saeed Arayne, Somia Gul, Sana Shamim;Sparfloxacin–metal complexes as antifungal agents - Their synthesis, characterization and antimicrobial activities Journal of Molecular Structure, Volume 975, Issues 1-3, 30 June 2010, Pages 285-291

Wilkinson, Brendan L.; Alessio Innocenti, Daniela Vullo, Claudiu T. Supuran and Sally-Ann Poulsen Inhibition of Carbonic Anhydrases with Glycosyltriazole Benzene Sulfonamides J. Med. Chem., 2008, 51 (6), pp 1945–1953)

Xing, Bengang; Chun-Wing Yu, Pak-Leung Ho,Kin-Hung Chow, Terence Cheung, Hongwei Gu, Zongwei Cai, and Bing Xu Multivalent Antibiotics via Metal Complexes: Potent Divalent Vancomycins against Vancomycin-Resistant Enterococci J. Med. Chem., 2003, 46 (23), pp 4904–4909

Yesilel, Okan Zafer Gökhan Kastas, Cihan Darcan, Inci Ilker, Hümeyra Pasaoglu, Orhan Büyükgüngör Inorganica Chimica Acta 363 (2010) 1849–1858 Syntheses, thermal analyses, crystal structures and antimicrobial properties of silver(I)-saccharinate complexes with diverse diamine liga

Zafar A. Siddiqi*, M. Shahid, Mohd. Khalid, S. Kumar European Journal of Medicinal Chemistry 44 (2009) 2517–2522, Antimicrobial and SOD activities of novel transition metal ternary complexes of iminodiacetic acid containing a-diimine as auxiliary ligand

Zafar A.; Mohd Khalid, Sarvendra Kumar, M. Shahid, Shabana Noor, Antimicrobial and SOD activities of novel transition metal complexes of pyridine-2,6-dicarboxylic acid containing 4-picoline as auxiliary ligand European Journal of Medicinal Chemistry, Volume 45, Issue 1, January 2010, Pages 264-Siddiqi, 269

6

Dendrimers as Antibacterial Agents

Metin Tülü and Ali Serol Ertürk
Yıldız Technical University
Turkey

1. Introduction

No more than 100 years ago, if a person acquired a bacterial infection, the body had to clear the infection by itself or else the infection would eventually result in death. After penicillin and many other effective antibiotics were discovered, however, that changed. In the decades after penicillin was discovered in 1928, a number of powerful antibiotics were developed. They were used plentifully and often carelessly prescribed needlessly for certain bacterial infections and even for viral infections where they have no effect. Farmers found that animals fed low levels of antibiotics grow faster and are less subject to disease, so thousands of tons of antibiotics were (and still are) added to animal feed. The problem: unlike disinfectants, antibiotics generally act against a single component of a bacterium. Thus, in environments where antibiotics are present, there is great selective pressure toward bacteria that can make the relatively minor mutations needed to render them resistant. Once a single bacterium has developed resistance to an antibiotic, it can be amplified across bacterial species by quick propagation and the tendency to share antibiotic resistance genes with other bacteria. In the last decade, resistance to antibiotics even antibiotics once thought to be "last ditch" treatments has increased remarkably and is continually on the rise. Doctors are finding many once-treatable infections are now deadly (e.g. highly publicized Methicillin-resistant Staphylococcus aureus (MRSA) infections). Large pharmaceutical companies, once major sources of new antibiotics, have exhausted most "easy" targets for new antibiotics and have shifted their research and development focus to long-term, chronic diseases rather than antibiotic discovery to increase profits (Tanner, 2009).

Antibacterial and **antimicrobial are two similar concepts and sometimes they are used interchangeably; however there are some differences between them. Antibacterial:** Anything that destroys bacteria or suppresses their growth or their ability to reproduce. Heat, chemicals such as chlorine, and antibiotic drugs all have antibacterial properties. Many antibacterial products for cleaning and hand washing are sold today. Such products do not reduce the risk for symptoms of viral infectious diseases in otherwise healthy persons. This does not preclude the potential contribution of antibacterial products to reducing symptoms of bacterial diseases in the home (Dorland Medical Dictionary, 2010). The term **antibacterial** is often used synonymously with the term antibiotic(s); today, however, with increased knowledge of the causative agents of various infectious diseases, antibiotic(s) has come to denote a broader range of antimicrobial compounds, including anti-fungal and other compounds.

Antimicrobial is a substance that kills or inhibits the growth of microorganisms (Merriam-Webster Online Dictionary, 2009) such as bacteria, fungi, or protozoans. Antimicrobial drugs either kill microbes (microbiocidal) or prevent the growth of microbes (microbiostatic). Disinfectants are antimicrobial substances used on non-living objects or outside the body (Wikipedia, 2011).

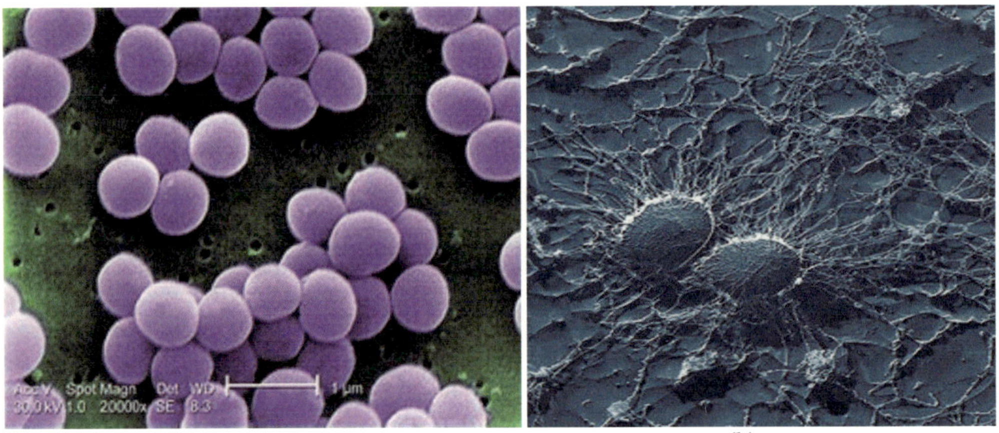

(a) (b)

Fig. 1. a) Scanning electron micrograph of S. aureus, 20,000 times enlargement, and false color added b) Bacterial cells of Staphylococcus aureus, which is one of the causal agents of mastitis in dairy cows. Its large capsule protects the organism from attack by the cow's immunological defenses (Kluytmans et al., 1997).

2. How do antibacterial agents work?

Most antimicrobial agents used for the treatment of bacterial infections may be categorized according to their principal mechanism of action. There are 5 major modes of action: (1) interference with cell wall synthesis, (2) inhibition of protein synthesis, (3) interference with nucleic acid synthesis, (4) inhibition of a metabolic pathway, and (5) Disruption of bacterial membrane structure. (Table 1.)(Neu, 1992). Antibacterial drugs that work by inhibiting bacterial cell wall synthesis include the b-lactams, such as the penicillins, cephalosporins, carbapenems, and monobactams, and the glycopeptides, including vancomycin and teicoplanin (McManus, 1997; Neu, 1992). β-Lactam agents inhibit synthesis of the bacterial cell wall by interfering with the enzymes required for the synthesis of the peptidoglycan layer. Vancomycin and teicoplanin also interfere with cell wall synthesis, but do so by binding to the terminal D-alanine residues of the nascent peptidoglycan chain, thereby preventing the cross-linking steps required for stable cell wall synthesis. Macrolides, aminoglycosides, tetracyclines, chloramphenicol, streptogramins, and oxazolidinones produce their antibacterial effects by inhibiting protein synthesis. Bacterial ribosomes differ in structure from their counterparts in eukaryotic cells. Antibacterial agents take advantage of these differences to selectively inhibit bacterial growth. Macrolides, aminoglycosides, and tetracyclines bind to the 30S subunit of the ribosome, whereas chloramphenicol binds to the 50S subunit. Fluoroquinolones exert their antibacterial effects by disrupting DNA synthesis and causing lethal double-strand DNA breaks during DNA replication (Drlica & Zhao, 1997), whereas sulfonamides and trimethoprim (TMP) block the pathway for folic acid synthesis, which ultimately inhibits DNA synthesis (Petri, 2005; Yao & Moellering, 2003). The common antibacterial drug combination of TMP, a folic acid analogue, plus sulfamethoxazole (SMX) (a sulfonamide) inhibits 2 steps in the enzymatic pathway for bacterial folate synthesis. Disruption of bacterial membrane structure may be a fifth, although less well characterized, mechanism of action. It is postulated that polymyxins exert their inhibitory effects by increasing bacterial membrane permeability, causing leakage of bacterial contents (Storm et al., 1977). The cyclic lipopeptide daptomycin apparently inserts its lipid tail into the bacterial cell membrane (Carpenter & Chambers, 2004), causing membrane depolarization and eventual death of the bacterium.

1.	Interference with cell wall synthesis : - Lactams: penicillins, cephalosporins, carbapenems, monobactams - Glycopeptides: vancomycin, teicoplanin
2.	Protein synthesis inhibition - Bind to 50S ribosomal subunit: macrolides, chloramphenicol, clindamycin, quinupristin-dalfopristin, linezolid - Bind to 30S ribosomal subunit: aminoglycosides, tetracyclines - Bind to bacterial isoleucyl-tRNA synthetase: mupirocin
3.	Interference with nucleic acid synthesis - Inhibit DNA synthesis: fluoroquinolones - Inhibit RNA synthesis: rifampin
4.	Inhibition of metabolic pathway: sulfonamides, folic acid analogues
5.	Disruption of bacterial membrane structure: polymyxins, daptomycin

Table 1. Mechanisms of action of antibacterial agents

3. Methods for antibacterial activity

3.1 Disk diffusion method

Screening for antibacterial and antifungal activities are carried out using sterilized antibiotic discs (6 mm), following the procedure performance standards for Antimicrobial Disk Susceptibility Tests, outlined by the National Committee for Clinical Laboratory Standards NCCLS (Collins, 1989; Villonova, 1993).

According to Disc Diffusion Method; when a filter paper disc impregnated with a chemical is placed on agar the chemical will diffuse from the disc into the agar. This diffusion will place the chemical in the agar only around the disc. The solubility of the chemical and its molecular size will determine the size of the area of chemical infiltration around the disc. If an organism is placed on the agar it will not grow in the area around the disc if it is susceptible to the chemical. This area of no growth around the disc is known as a "zone of inhibition".

Principle; antiseptics, disinfectants and antibiotics are used in different ways to combat microbial growth. Antiseptics are used on living tissue to remove pathogens. Disinfectants are similar in use but are used on inanimate objects. Antibiotics are substances produced by living organisms, such as *Penicillium* or *Bacillus*, that kill or inhibit the growth of other organisms, primarily bacteria. Many antibiotics are chemically altered to reduce toxicity, increase solubility, or give them some other desirable characteristic that they lack in their natural form. Other substances have been developed from plants or dyes and are used like antibiotics. A better term for these substances is antimicrobials, but the term antibiotic is widely used to mean all types of antimicrobial chemotherapy. Many conditions can affect a disc diffusion susceptibility test. When performing these tests certain things are held constant so only the size of the zone of inhibition is variable. Conditions that must be constant from test to test include the agar used, the amount of organism used, the concentration of chemical used, and incubation conditions (time, temperature, and atmosphere). The amount of organism used is standardized using a turbidity standard. This may be a visual approximation using a McFarland standard 0.5 or turbidity may be determined by using a spectrophotometer (optical density of 1.0 at 600 nm). For antibiotic susceptibility testing the antibiotic concentrations are predetermined and commercially available. Each test method has a prescribed media to be used and incubation is to be at 35-37° C in ambient air for 18-24 hours. The disc diffusion method for antibiotic susceptibility testing is the Kirby-Bauer method. The agar used is Meuller-Hinton agar that is rigorously tested for composition and pH. Further the depth of the agar in the plate is a factor to be considered in the disc diffusion method. This method is well documented and standard zones of inhibition have been determined for susceptible and resistant values. There is also a zone of intermediate resistance indicating that some inhibition occurs using this antimicrobial but it may not be sufficient inhibition to eradicate the organism from the body.

The standardized methods for antiseptic and disinfectant testing are more rigorous and more difficult to reproduce in a student laboratory. Two common tests are the Phenol Coefficient Test (a comparison of the effect of the chemical and phenol on several organisms) and the Use Dilution Test (testing the chemical under actual conditions of use). A disc diffusion test can be used to approximate the Use Dilution Test. The chemical under consideration is used to saturate a filter paper disc. This disc is then used to introduce the chemical to the agar for testing. The actual zone sizes have not been standardized as in the Kirby-Bauer method, but a comparison of zone sizes for the same chemical among organisms will provide an approximate effectiveness of the chemical.

3.2 Dilution method

Screening for antibacterial and antifungal activities was carried out by preparing a broth micro dilution, following the procedure outlined in Manual of Clinical Microbial (Jones et al., 1985). The broth dilution method depends upon inoculation at a specific inoculum density of broth media (in tubes or microtitre plates) containing antibiotics at varying levels usually doubling dilutions are used and after incubation, turbidity is recorded either visually or with an automated reader, and the breakpoint concentration established. Microtitre plates or ready-to-use strips are commercially available with antibiotics ready prepared in the wells. A variation on this approach is the agar dilution method where a small volume of suspension is inoculated onto agar containing a particular concentration of antibiotic, when the inoculum has dried the plate is incubated and again examined for zones of growth.

Biological data: Standardized samples of Penicillin-g (blocking the formation of bacterial cell walls, rendering bacteria unable to multiply and spread; Ampicillin (penetrating and preventing the growth of Gram-negative bacteria); Cefotaxime (used against most Gram-negative enteric bacteria); Vancomycin (acting by interfering with the construction cell walls in bacteria), Ofloxacin (entering the bacterial cell and inhibiting DNA-gyrase, which is involved in the production of genetic material, preventing the bacteria from reproducing); Tetracyclines (exerting their antimicrobial effect the inhibition of protein synthesis; Nystatin (binding to sterols in the fungal cellular membrane altering the permeability to allow leakage of the cellular contents and destroying the fungus); Ketoconazole (inhibiting the growth of fungal organisms by interfering with the formation of the fungal cell wall) and Clotrimazole (interfering with their cell membranes and causing essential constituents of the fungal cells leakage). Mueller Hinton media, Nutrient Broth and Malt Extract Broth are purchased from Difco and yeast extracts is obtained from Oxoid.

4. Dendrimers as antibacterial or antimicrobial agents

High molecular surface functional group concentration of dendrimers can dominate antibacterial properties to the interacting molecule. If the end or surface groups of dendrimers are functionalized with biologically active antimicrobial groups, we might expect an increase in antimicrobial activity of the dendrimers depending on the high molecular antimicrobial surface functional group increase on the surface of the molecule.

The target for antimicrobials must be selected carefully. This is because of the fact that bulkier dendrimers may not be able to penetrate the cell membrane barrier and may have difficulty reaching the target site for the anticipated antimicrobial action (Chen & Cooper, 2002).

Biocides immobilized on dendrimers can be more effective if the target sites are cell walls and/ or membranes. It has been shown that small quaternary ammonium compounds exert their antimicrobial action by disrupting and disintegrating the cell membrane (Ghosh, 1988; Kourai et al., 1980; Panarin et al., 1985). Converting functional end groups of dendrimer to ammonium salts, dendrimer biocides can be synthesized. These dendrimer biocides have been shown to be more potent than their small molecule counterparts as they bear high local density active groups. Thus, dendrimer biocides may be very beneficial in terms of activity, localization in specific organs, reduced toxicity, and increased duration of action (Donaruma, 1978, 1980).

As the bacteria are negatively charged and dendrimer biocides have high positive charge density, electrostatic interactions bring them into contact with each other. Depending on the concentration of the dendrimer biocides, membrane permeability can be slightly changed or denaturized. At the end, high concentrations of dendrimer biocides can lead to complete disintegration of the bacterial membrane causing to a bactericidal effect (Neu, 1992).

By adding water soluble functional end groups to dendrimers, water soluble dendrimers are obtained. When these dendrimers interacted with bacteriostatic weak water soluble or insoluble antibiotics, the antibacterial properties antibiotics can be altered, especially improved, and also clinical applications can be experimentally observed by conducting studies. Bacterial inflections remain major causes of mortality in hospitals all around the world. Against these bacterial inflections including enteric and urinary tract, and respiratory tract, sulfonamides are widely used. They are preferred due to ease of administration and wide spectrum of anti-bacterial activity. However, the clinical use of sulfonamides is limited due to their extremely low solubility in water, rapid elimination in blood, low level of association to plasma proteins and several side effects. Microbiology studies showed that PAMAM dendrimers could increase water solubility and the anti bacterial activity of a kind of sulfonamide, Sulfamethoxazole (SMZ)(Fig. 2.) (a 4- or 8 fold increase in the antibacterial activity of SMZ in dendrimer solution compared to pure SMZ dissolved in dimethyl sulfoxide (DMSO) or 0.01 M NaOH solution) (Ma, et al., 2007). Likewise, Quinolones (Fig. 3.) which are expanding class of clinically established potent antibiotics are not freely soluble in water. Microbiological studies of the quinolones (nadifloxacin and prulifloxacin) showed that strong antimicrobial activities of nadifloxacin and prolifloxacin were still significantly increased in the presence of PAMAM dendrimers and also their water solubility increased (Cheng et al., 2007).

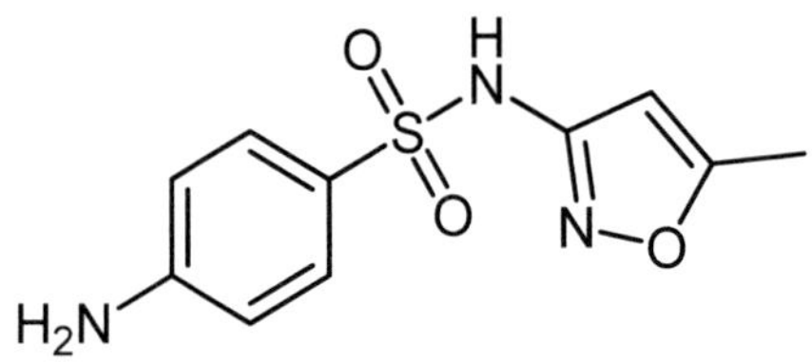

Fig. 2. Structure of Sulfamethoxazole (SMZ) 4-amino-*N*-(5-methylisoxazol-3-yl)-benzenesulfonamide

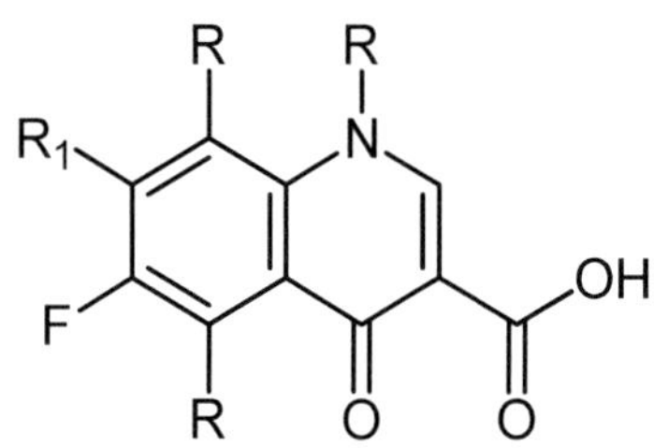

Fig. 3. Essential structure of all quinolone antibiotics: R_1 is usually piperazine; if the connection contains fluorine (F), it is a fluoroquinolone.

Highly branched dendritic structures and unique properties of dendrimers have attracted great interest in the recent studies of antimicrobial activities of dendrimers and their derivatives (Chen & Cooper, 2000). In most cases, Dendrimers have the ability of carrying biologically active agents by encapsulating them in the interior or generally, attaching them on the periphery of dendrimers. For example, most dendrimers displaying antimicrobial activities are terminated with antimicrobial agents, including ferrocene (Abd-Elzaher & Ali, 2006), quaternary ammonium (Chen et al., 2000; Chen & Cooper, 2000, 2002), boron complexes (De Queiroz, 2006), carbohydrates (Rojo, 2004), and peptides (Bernstein et al., 2003, Bourne et al., 2000; Janiszewska et al., 2006; Klajnert et al., 2006; Pini et al., 2005; Sechi et al., 2006).

Poly (amido amine) PAMAM (e.g, the generation 3 (G3) PAMAM in Fig. 4.) dendrimers are the most extensively studied dendrimers. PAMAM dendrimers with a wide variety of functional groups at the periphery are commercially available. Some of the dendrimers having terminal amino groups are shown as they are having low toxicity to eukaryotic cells (Malik et al., 2000; Roberts et al., 1996). Modification of the amino groups of the PAMAM dendrimers with poly (ethylene glycol) (PEG) or lauroly chains further improves the biocompatibility (Jevprasesphant et al., 2003a, 2003b; Luo et al., 2002). As the number of PEG or lauroyl chains increases, the cytotoxicity of PAMAM to human colon adenocarcinoma cells decreases (Jevprasesphant et al., 2003). Shielding of the positive charges of the protonated amino groups on the exterior of the dendrimer by the PEG or lauroyl chains is likely the reason for reduced cytotoxicity (Jevprasesphant et al., 2003). Due to their excellent biocompatibility, PEG-modified PAMAM dendrimers have been used as carriers of imaging agents and pharmaceuticals, including antimicrobial agents such as penicillin V and silver (Aymonier et al., 2002; Balogh et al., 2001; Bielinska et al., 1996; Svenson & Tomalia, 2005; Yang & Lopina, 2003).

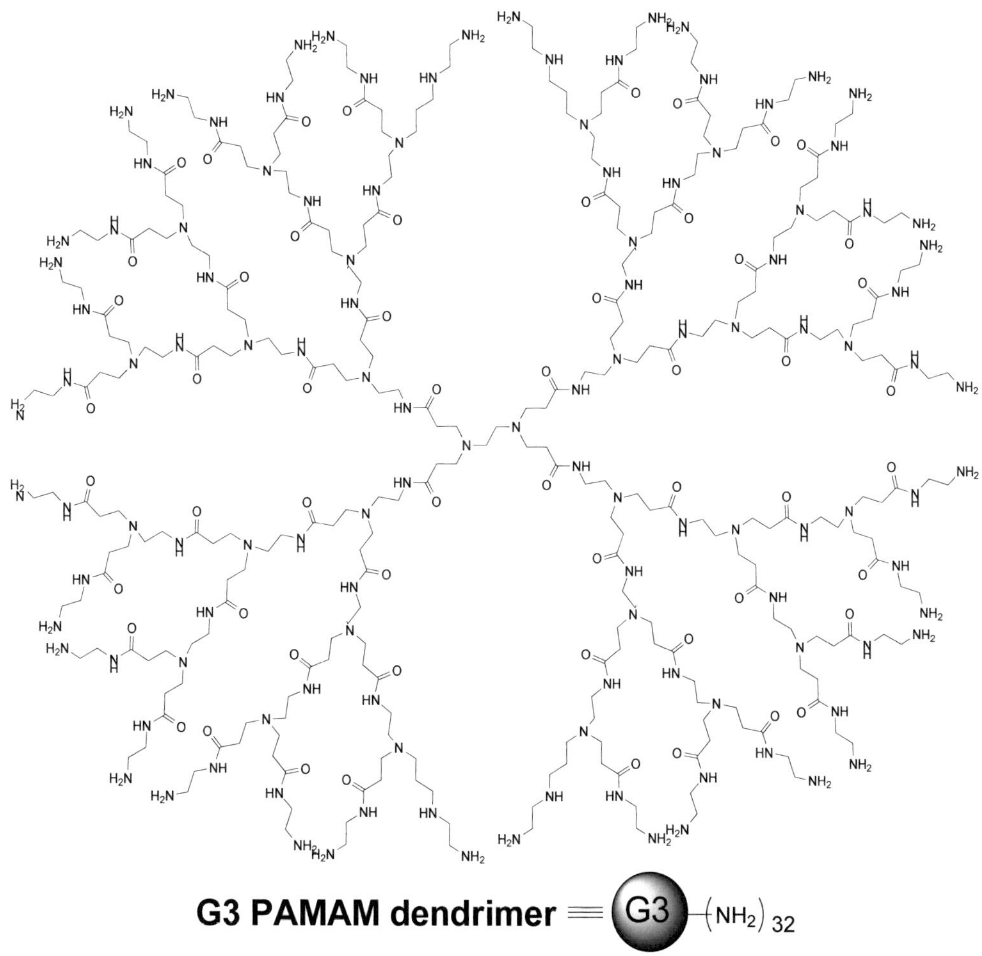

Fig. 4. Structure of the G3 PAMAM dendrimer possessing ~32 amino groups at the periphery.

Even if mostly PAMAM dendrimers stated as the carriers or the modifiers of the antibacterial agents, it has been shown that PAMAM dendrimers themselves show antibacterial properties and they are highly toxic to some bacteria. They are also suggested PAMAM dendrimers as the antimicrobial agents because of the fact that PEG-coated PAMAM derivatives exhibited low toxicity to human corneal epithelial cells (Calabretta et al., 2007). Epidemiologic studies performed in Europe countries report an increase in infectious diseases and bacterial resistance to antibiotics. This has resulted in intensive world wide search for natural linear antimicrobial peptides and their derivatives. Antimicrobial peptides are 10 to 50 amino-acid-long peptide antibiotics that have been recently discovered in many living organism including humans. These peptides are an important part of the innate defense system. They are an important part of the innate defense system, possessing a high potency and broad spectrum of activity against prokaryotic cells with only a minor impact on eukaryotic cells. Such properties raised some hope that natural antimicrobial peptides and their synthetic analogs may be adaptable for use in vivo as new generation antibiotics (Lequin et al., 2003; Papo & Shai, 2003; Powers & Hancock, 2003).

Active structures of linear peptides can be modeled by application of dendrimer chemistry. Dendrimers have nanoscopic dimensions having surface functional groups. Dendrimers can be synthesized by using organic chemistry methods and they can be indicated as the well suited compounds for biotechnological and biochemical applications. This is because of the fact that they can have multivalent nature, unambiguous composition, reliability and versatility in their synthesis. Eventually, antibacterial properties between two structurally different class of molecules- linear peptides and dendrimers can be transferred to attribute active conformation of amphiplic dendrimeric peptides (Janiszewska et al., 2007).

There is a significant global need for new antibacterial and alternative mechanisms of action given the rise in resistance among bacteria (MacDougall & Polk, 2005; Mah & O'Toole, 2005). Of the various known antibacterial agent classes, amphiphilic compounds act through perturbation and disruption of the prokaryotic membrane (Denyer, 1995). It has been hypothesized that amphiphilic anionic dendrimers may exhibit antibacterial activity with minimal eukaryotic cell cytotoxicity, since dendrimers with terminal anionic charges are generally noncytotoxic and have low toxicity in zebrafish whole animal development studies (Heiden, 2007). On the other hand, cationic dendrimers, some of which have antibacterial properties if the positive charge is properly shielded (Chen & Cooper, 2002), have repeatedly shown cytotoxicity against a variety of eukaryotic cell lines (Gurdag et al., 2006; Hong et al, 2004).

Apart from the cytotoxicity studies about the antimicrobial activities of the dendrimer integrated molecules, dendrimer studies also have come to a major interesting point in terms of many oral care products. Most of these products, especially toothpastes include Triclosan (TCN) (Fig. 5.) in order to prevent the bacterial growth called dental plague on the surface of the teeth. PAMAM dendrimers are of interest because they preferred in strategy in formulation to increase the delivery efficiency of antibacterial. TCN has broad spectrum antimicrobial activity against many types of Gram positive and Gram-negative non-sporulating bacteria, some fungi and yeasts (Brading & Marsh, 2003). For an antibacterial to be effective when delivered from an oral care product it should be efficiently retained and subsequently released at the site of interest. Dendrimers are good candidates for delivery systems as they can be modified by the addition of molecules to their surface groups. Active molecules either can be encapsulated or conjugated to surface groups. The surface groups also can be modified to enable specific targeting of the dendrimer carrier (Sampathkumar & Yarema, 2005). TCN is not an ionic molecule, itself. In theory, cationic dendrimers may have intrinsic mucoadhesive properties for use in the oral cavity, as mucin (which covers oral epithelia) is negatively charged (Hao & Heng, 2003), thus creating an electrostatic attraction between mucus and dendrimer. An agent such as TCN, when encapsulated in the dendrimer architecture may then be slowly released into the oral cavity, potentially increasing efficacy (Gardiner et al., 2008).

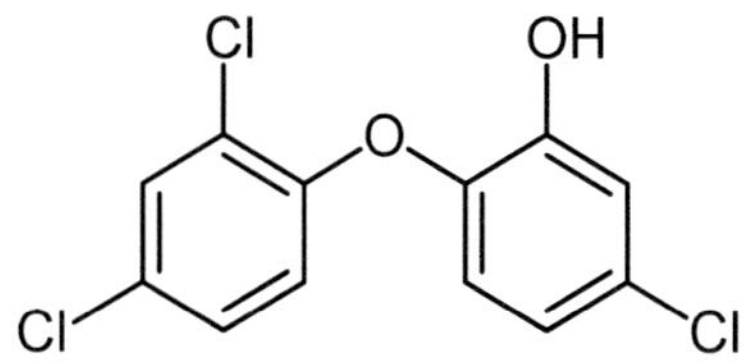

Fig. 5. Structure of Triclosan

Metal containing dendritic nanoparticles called as the Metallodendrimers. They are incorporated with metal atoms. Silver has known since ancient times as a very effective antimicrobial agent. Silver containing compounds and materials have been routinely used to prevent attack of a broad spectrum of microorganisms on prostheses (Gosheger et al., 2004), catheters (Semuel & Guggenbichler, 2004), vascular grafts (Strathmann & Wingender, 2004), human skin (Lee & Jeong, 2005), also used in medicine to reduce infection in burnt treatment (Parikh et al., 2005; Ulkur et al., 2005), arthroplasty (Alt, 2004). However they exhibit low toxicity to mammalian cells. The antibacterial activity of silver nanoparticles was tested against *Bacillus subtilis* and *Staphylococcus aureus* bacteria at different concentrations by using the diffusion disc technique. The results have showed that the antibacterial activity increases with the increase of concentration of the active agent (Mahapatra & Karak, 2008). An example demonstration of Nano-scaled silver interaction with bacteria cell can be shown from Fig. 6.

The scope of Metallodendrimers has been developing by the interest of the antibacterial studies conducted by the application of them to different type of materials as in interaction. Silver complexes of dendrimers have a broad spectrum of preventing variety of microorganism attacks and so they are preferred as antibacterial study materials. For example, it has been shown that Dendrimers have been used as vehicle to develop the antimicrobial properties of textile fabrics by modifying (PAMAM) G3 dendrimer to provide antimicrobial properties. By accomplishing this, metal nanoparticles $AgNO_3$-PAMAM (G3) complex as well as a MesoSilver-PAMAM (G3) complex has been formed and applied to Cotton/Nylon blend fabric. SEM analysis has shown that Dispersion of the silver nanoparticles onto the fabric (see Fig. 7. and Fig. 8.) was well and treated fabric against *Staphylo-coccus aureus* exhibited significantly biocide activities for each type of modified dendrimers (Ghosh et al., 2010).

Dispersion of organic/ inorganic hybrid materials could be utilized to form regular thin film coatings with antibacterial effects by using dendritic-polymer templates. The antibacterial activity of the coating films based on the hyper branched core/shell type hybrids and closely associated with the silver ions release of the films. The molecular architecture of the core/shell type hybrids used as hosts for silver nanoparticles can provide an improved control of the silver ion release and therefore an adjustment of the antibacterial affect specific to its application. This can be succeeded by the utility of the controllable generation structure of dendrimer nanoparticles (Gladitz et al., 2009). It has been shown that dendrimer films effectively inhibited the colonization of the Gramnegative bacteria *Pseudomonas aeruginosa* (strain PAO1) and, to a lesser extent, the Gram-positive bacteria *Staphylococcus aureus* (SA). Moreover, The antibacterial activity of the films was maintained even after storage of the samples in Phosphate Buffered Saline (PBS) for up to 30 days (Wang et al., 2011).

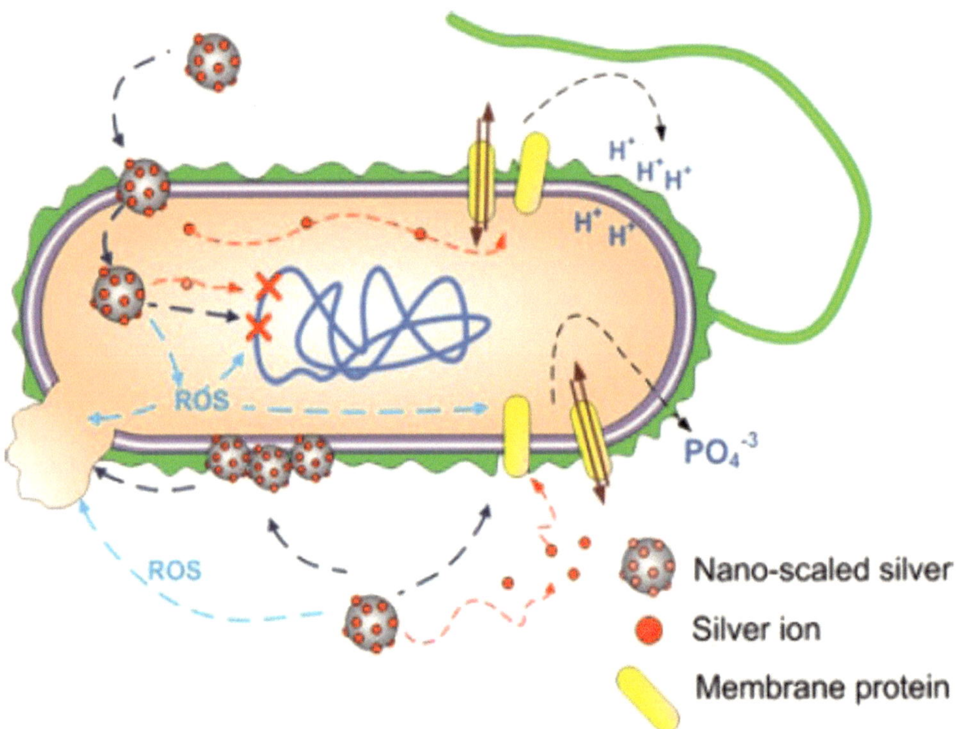

Fig. 6. Diagram summarizing nano-scaled silver interaction with bacterial cells. Nano-scaled silver may (1) release silver ions and generate reactive oxygen species (ROS); (2) interact with membrane proteins affecting their correct function; (3) accumulate in the cell membrane affecting membrane permeability; and (4) enter into the cell where it can generate ROS, release silver ions, and affect DNA. Generated ROS may also affect DNA, cell membrane, and membrane proteins, and silver ion release will likely affect DNA and membrane proteins (Marambio-Jones & Hoek, 2010).

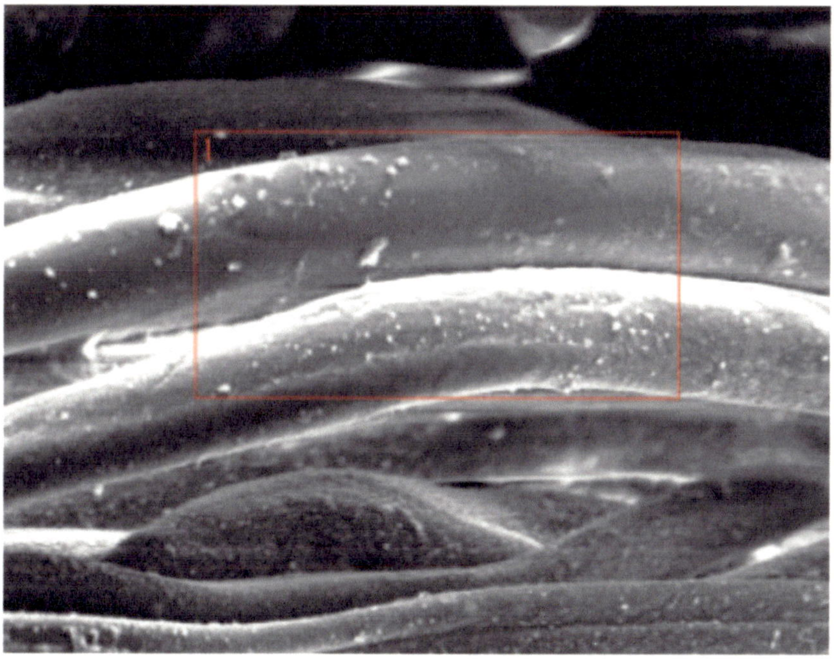

Fig. 7. SEM image of the fabric treated with silver/ dendrimers complex at 1200 magnification (10.0 kV and 10 mm distance) (Ghosh et al., 2010)

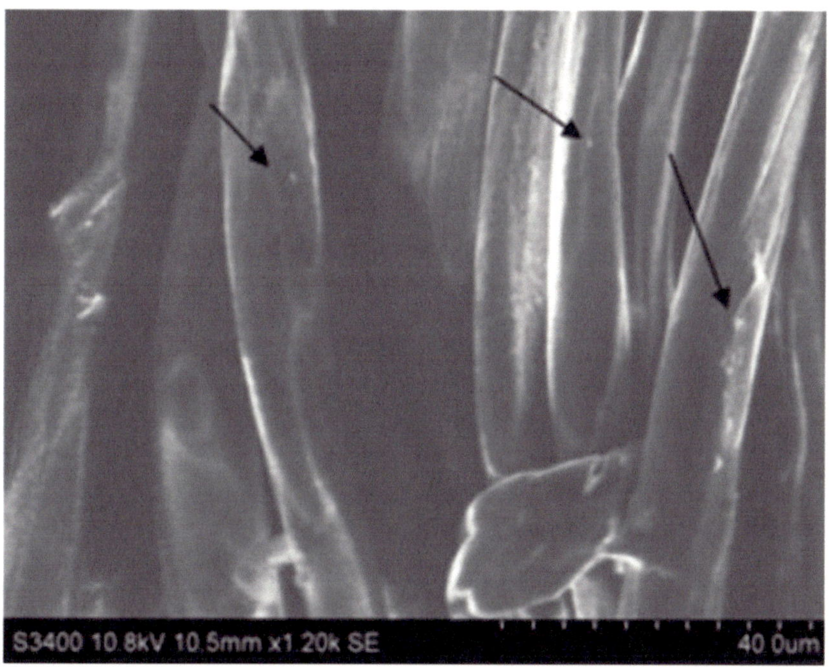

Fig. 8. SEM image of MesoSilver–dendrimer composite treated-fabric (1200 mag) (Ghosh et al., 2010)

5. Conclusion

It is clear that bacteria will continue to develop resistance to currently available antibacterial drugs by either new mutations or the exchange of genetic information, that is, putting old resistance genes into new hosts. In many healthcare facilities around the world, bacterial pathogens that Express multiple resistance mechanisms are becoming the norm, complicating treatment and increasing both human morbidity and financial costs. Prudent use of antibacterial drugs using the appropriate drug at the appropriate dosage and for the appropriate duration is one important means of reducing the selective pressure that helps resistant organisms emerge. The other vital aspect of controlling the spread of multidrug resistant organisms is providing sufficient personnel and resources for infection control in all healthcare facilities. New antibacterial agents with different mechanisms of action are also needed. It is difficult to outsmart organisms that have had several billion years to learn how to adapt to hostile environments, such as those containing antimicrobial agents. Yet, with sufficient efforts to use antimicrobial agents wisely, thereby preventing the emergence of resistant organisms, and strict attention to infection control guidelines to contain the spread of resistant organisms when they develop, we should be able to stay at least 1 step ahead of the next resistant plague.

6. Acknowledgment

We gratefully thank the Yıldız Technical University Project Office (2011-01-02-KAP04 & 2011-01-02-KAP05) and EU Lifelong Learning Programme* (Webgentech Project number: 2010-1-TR1-LEO05-16728) as well as the numerous students, postdoctoral associates, colleagues, and collaborators for their input assistance, and hard work through the now nearly two and one-half decades of dendrimers and fractal constructs.

7. References

Abd-Elzaher, M. M., Ali, I.A.I. (2006). Preparation, characterization and biological studies of some novel ferrocenyl compounds. *Appl. Organomet. Chem.*, Vol.20, Issue.2, (February 2006), pp. 107-111, ISSN 0268-2605

Alt, V., Bechert, T., Steinrucke, P., Wagener, M., Seidel, P., Dingeldein, E., Domann, U., & Schnettler, R. (2004). An in vitro assessment of the antibacterial properties and cytotoxicity of nanoparticulate silver bone cement. *Biomaterials*, Vol.25, Issue.18 (August 2004), pp. 4383-4391, ISSN 0142-9612

Aymonier, C., Schlotterbeck, U., Antonietti, L., Zacharias, P., Thomann, R., Tiller, J. C., & Mecking, S. (2002). Hybrids of silver nanoparticles with amphiphilic hyperbranched macromolecules exhibiting antimicrobial properties. *Chem. Commun.*, Issue.24, (November 2002), pp. 3018- 3019

* Decision N°1720/2006/EC of the European Parliament and of the Council of 15/11/2006 establishing an action programme in the field of lifelong learning, published in the Official Journal of the EU N°L327/45 on 24/11/2006.

Balogh, L., Swanson, D.R., Tomalia, D.A., Hagnauer, G.L., & McManus, A.T. (2001). Dendrimer–Silver Complexes and Nanocomposites as Antimicrobial Agents. *Nano Lett.*, Vol.1, Issue.1, (January 2001), pp. 18-21, ISSN 530-698

Bernstein, D. I., Stanberry, L. R., Sacks, S., Ayisi, N. K., Gong, Y. H., Ireland, J., Mumper, R. J., Holan, G., Matthews, B., McCarthy, T., & Bournel, N. (2003). Evaluations of Unformulated and Formulated Dendrimer-Based Microbicide Candidates in Mouse and Guinea Pig Models of Genital Herpes. *Antimicrob. Agents Chemother.*, Vol.47, No.12, (December 2003), 47, pp. 3784- 3788, ISSN 1098-6596

Bielinska, A., Kukowska-Latallo, J. F., Johnson, J., Tomalia, D. A., & Baker, J. R. (1996). Regulation of in vitro gene expression using antisense oligonucleotides or antisense expression plasmids transfected using starburst PAMAM dendrimers. *Nucleic Acids Res.*,Vol.24, No.11, (June 1996), pp. 2176-2182, ISSN 0305-1048

Bourne, N., Stanberry, L. R., Kern, E. R., Holan, G., Matthews, B., & Bernstein, D. I.. (2000). Dendrimers, a New Class of Candidate Topical Microbicides with Activity against Herpes Simplex Virus Infection. *Antimicrob. Agents Chemother.*, Vol.44, No.9, (September 2000), pp. 2471-2474, ISSN 1098-6596

Brading, M.G., Marsh, P.D. (2003). The oral environment: The challenge for antimicrobials in oral care products. *IntDent J, Vol. 53, Suppl.*1, (December 2003), pp. 353–362, ISSN 0020-6539

Calabretta, M.K., Kumar, A., McDermott, A.M., & Cai, C. (2007). Antibacterial Activities of Poly(amidoamine) Dendrimers Terminated with Amino and Poly(ethylene glycol) Groups. *Biomacromolecules*, Vol.8, Issue.6, (June 2007), pp. 1807-1811, ISSN 1525-7797

Carpenter, C.F., Chambers, H.F. (2004). Daptomycin: another novel agent for treating infections due to drug-resistant gram-positive pathogens. *Clinical Infectious Diseases*, Vol.38, Issue 7, (March 2004), pp. 994 -1000, ISSN 1058-4838

Chen, C.Z., Cooper, S.L. (2002). Interactions between dendrimer biocides and bacterial membranes. *Biomaterials* , Vol.23, No.16, (January 2002), pp. 3359-3368, ISSN 0142-9612

Chen, C.Z.S., Beck-Tan, N.C., Dhurjati, P., van Dyk, T.K., LaRossa, R.A., & Cooper, S.L. (2000). Quaternary Ammonium Functionalized Poly(propylene imine) Dendrimers as Effective Antimicrobials: Structure–Activity Studies. *Biomacromolecules*, Vol.1, Issue.3, (August 2000), pp. 473- 480, ISSN 1525-7797

Chen, C.Z.S., Cooper, S. L. (2000). Recent Advances in Antimicrobial Dendrimers. *Advanced Materials*, Vol.12, Issue.11, (June 2000), pp. 843-846, ISSN: 1521-4095

Cheng, Y., Qu, H., Ma, M., Xu, Z., Xu ,P., Fang, Y., & Xu, T. (2007). Polyamidoamine (PAMAM) dendrimers as biocompatible carriers of quibolone antimicrobials: An in vitro study. *European Journal of Medicinal Chemistry*, Vol.42, Issue.7, (July 2007), pp. 1032-1038

Collins, C.H., (1989). *Microbiological methods*, (6th ed), Butterworth-Heinemann, ISBN-10: 0750614285, London

De Queiroz, A. A. A., Abraham, G. A., Camillo, M. A. P., Higa, O. Z., Silva, G. S., Fernandez, M. D., & San Roman, J. (2006). Physicochemical and antimicrobial properties of boron-complexed polyglycerol-chitosan dendrimers. *J. Biomater. Sci., Polym. Ed.*, Vol. 17, No.6, (June 2006), pp. 689-707, ISSN 0920-5063

Denyer, S. P. (1995). Mechanisms of action of antibacterial biocides. *Int. Biodeterior. Biodegrad.*,Vol.36, Issue.3-4, (October-December 1995), pp. 227–245, ISSN 0964-8305

Donaruma, L.G., Vogl, O.(1978). *Polymeric drugs*, Academic Press, New York

Donaruma, L.G.,Vogl, O., & Ottenbr ite R.M. (1980). *Anionic polymeric drugs*, Wiley, New York

Dorlands Medical Dictionary. (17.11.2010). Antibacterial, In: *antibactrerial*. 05.09.2011, Available from: <http://www.mercksource.com/pp/us/cns/cns_hl_dorlands_split.jsp?pg=/ppdocs/us/common/dorlands/dorland/one/000005889.htm>

Drlica, K., Zhao, X. (1997). DNA gyrase, topoisomerase IV, and the 4-quinolones. *Microbiol Mol Biol Rev.*, Vol.61, No.3, (September 1997), pp. 377–392, 1092-2172

Gardiner, J., Freeman, S., Matthew, L., Leach, M., Green, A., Alcock, J., & D'Emanuele, A. (2008). PAMAM dendrimers for the delivery of the antibacterial Triclosan. *Journal of Enzyme Inhibition and Medicinal Chemistry*, Vol.23, No.5, (January 2008), pp. 623-628, ISSN 1475-6366

Ghosh, M. (1988). Synthetic macromolecules as potential chemotherapeutic agents. *Polym News*, Vol.13, (1988), pp. 71-77, ISSN 0032-3918

Ghosh, S., Yadav, S., Vasanthan, N., & Sekosan, G. (2010). A Study of Antimicrobial Property of Textile Fabric Trated with Modified Dendrimers. *Journal of Applied Polymer Science*, Vol.115, Issue.2 (January 2010), pp. 716-722, ISSN 1097-4628

Gladitz, M., Reinemann, S., & Radusch, H-J. (2009). Preperation of Silver Nanoparticle Dipersions via a Dendritic-Polymer Template Approach and their use for Antibacterial Surface Treatment. *Macromol. Mater. Eng*, Vol.294, Issue.3, (March 2009), pp. 178–189, ISSN 1439-2054

Gosheger, G., Hardes, J., Ahrens, H., Streitburger, A., Buerger, H., Erren, M., Gunsel, A., Kemper, F.H, Winkelmann, W., & Eiff, C. (2004). Silver-coated megaendoprostheses in a rabbit model—an analysis of the infection rate and toxicological side effects. *Biomaterials*, Vol.25, Issue.24, (November 2004), pp. 5547-5556, ISSN 0142-9612

Gurdag, S., Khandare, J., Stapels, S., Matherly, L. H., & Kannan, R. M. (2006). Activity of Dendrimer−Methotrexate Conjugates on Methotrexate-Sensitive and -Resistant Cell Lines. *Bioconjugate Chem.*,Vol.17, Issue.2, (March 2006), pp. 275– 283, ISSN 1043-1802

Hao, J. , Heng, P.W.S. (2003). Buccal delivery systems. *Drug Dev Ind Pharm.*, Vol.29, No.8, (January 2003), pp. 821–832, ISSN 0363-9045

Heiden, T. C., Dengler, E., Kao, W.J., Heideman, W., & Peterson, R.E. (2007). Developmental toxicity of low generation PAMAM dendrimers in zebrafish. *Toxicology and Applied Pharmacology*, Vol.225, Issue. 1, (November 2007), pp. 70–79, ISSN 0041-008X

Hong, S., Bielinska, A.U., Mecke, A., Keszler, B., Beals, J.L., Shi, X., Balogh, L., Orr, B.G., Baker, J.R., Jr., & Banaszak Holl, M.M. (2004). Interaction of Poly(amidoamine) Dendrimers with Supported Lipid Bilayers and Cells: Hole Formation and the

Relation to Transport. *Bioconjugate Chem*, Vol.15, Issue.4, (July 2004), pp. 774–782, ISSN 1043-1802

Janiszewska, J., Urbanczyk-Lipkowska, Z.(2006) Synthesis, antimicrobial activity and structural studies of low molecular mass lysine dendrimers. *Acta Biochim. Pol.*, Vol.53, No.1, (February 2006), pp. 77-82, ISSN 1734-154X

Janiszewska, J., Urbanczyk-Lipkwska, Z. (2007). Amphiphilic Dendrimeric Peptides as Model Non-Sequential Pharmacophores with Antimicrobial Properties. *Journal of Molecular Microbiology and Biotechnology*, Vol.13, No.4, (September 2007), pp. 220-225

Jevprasesphant, R., Penny, J., Attwood, D., McKeown, N. B., & D'Emanuele, A. (2003). Engineering of Dendrimer Surfaces to Enhance Transepithelial Transport and Reduce Cytotoxicity. *Pharm. Res.*, Vol.20, Issue.10, (October 2003), pp. 1543-1550, ISSN 0724-8741

Jevprasesphant, R., Penny, J., Jalal, R., Attwood, D., McKeown, N. B., & D'Emanuele, A. (2003). The influence of surface modification on the cytotoxicity of PAMAM dendrimers. *Int. J. Pharm.*, Vol.252, Issues.1-2, (February 2003), pp. 263-266, ISSN 0378-5173

Jones, R.N., Barry, A.L., Gaven, T.L., & Washington , J.A. (1985) In: *Manual of clinical microbiology*, Lennette, E.H., Balows, A., & Shadomy, W.J., pp. 972-977, American Society for Microbiology, ISBN-10: 0914826654, Washington

Klajnert, B., Janiszewska, J., Urbanczyk-Lipkowska, Z., Bryszewska, M., Shcharbin, D.,Labieniec, M. (2006). Biological properties of low molecular mass peptide dendrimers. *Int. J. Pharm.*, Vol.309, Issues.1-2, (February 2006), pp. 208-217, ISSN 0378-5173

Kluytmans, J., van Belkum, A., & Verbrugh, H. (1997). Nasal carriage of Staphylococcus aureus: epidemiology, underlying mechanisms, and associated risks. *Clinical Microbiology Reviews*, Vol.10, No.3, (July 1997), pp. 505-520, ISSN 0893-8512

Kourai, H., Horie, T., Takeichi, K., & Shibasaki, I.J. (1980). Antimicrobial activities of amino acid derivatives. *Antibacterial Antifungal Agents*, Vol.8, (1980), pp. 9–17

Lee, J.H, Jeong, S.H. (2005). Bacteriostasis and Skin Innoxiousness of Nanosize Silver Colloids on Textile Fabrics. *Text. Res. J.*, Vol.75, Issue.7, (July 2005), pp. 551-556, ISSN 0040-5175

Lequin, O., Bruston, F., Convert, O., Chassaing, G., & Nicolas, P. (2003). Helical structure of dermaseptin B2 in a membrane-mimetic environment. *Biochemistry*, Vol.42, Issue.34, (September 2003), pp. 10311–10323, ISSN 0006-2960

Luo, D., Haverstick, K., Belcheva, N., Han, E., & Saltzman, W. M. (2002). Efficient Control on Molecular Weight in the Synthesis of Poly(p-xylylene)s via Gilch Polymerization. *Macromolecules*, Vol.35, Issue.9, (April 2002), pp. 3456-3462, ISSN 0024-9297

Ma, M., Cheng, Y., Xu, Z., Xu, P., Qu, H., Fang, Y., Xu, T., & Wen, L. (2007). Evaluation of polyamidoamine (PAMAM) dendrimers as drug carriers of anti-bacterial drugs using sulfamethoxazole (SMZ) as a model drug. *European Journal of Medicinal Chemistry*, Vol.42, Issue.1 (November 2006), pp. 93-98, ISSN 0223-5234

MacDougall, C., Polk, R.E. (2005). Antimicrobial Stewardship Programs in Health Care Systems. *Clin. Microbiol. ReV.*, Vol.18, Issue.4, (October 2005), pp. 638–56, ISSN 0893-8512

Mah, T.-F., O'Toole, G.A. (2001). Mechanisms of biofilm resistance to antimicrobial agents. *Trends Microbiol.*, Vol.9, No.1, (January 2001), pp. 34–39, ISSN 0966-842X

Mahapatra, SS., Karak, N. (2008). Silver nanoparticle in hyperbranchedpolyamine: Synthesis, characterization and antibacrerial activity. *Materials Chemistry and Physics*, Vol.112, Issue.3, (December 2008), pp. 1114–1119, ISSN 0254-0584

Malik, N., Wiwattanapatapee, R., Klopsch, R., Lorenz, K., Frey, H., Weener, J. W., Meijer, E. W., Paulus, W., & Duncan, R. (2000). Dendrimers: Relationship between structure and biocompatibility in vitro, and preliminary studies on the biodistribution of 125I-labelled polyamidoamine dendrimers in vivo. *J. Controlled Release*, Vol.65, Issues.1-2, (March 2000), pp. 133-148, ISSN 0168-3659

Marambio-Jones, C., Hoek, E.M.V. (2010). A review of the antibacterial effects of silver nanoparticles and potential implications for human health and the environment. *J Nanopart Res.*, Vol.12, No.5, (June 2010), pp. 1531–1551, ISSN 1388-0764

McManus, M.C. (1997). Mechanisms of bacterial resistance to antimicrobial agents. *Am J Health Syst Pharm.*, Vol.54, No.12, (June 1997), pp. 1420-1433, ISSN 1079-2082

Merriam-Webster Online Dictionary. (2009). Antimicrobial, In: *antimicrobial.* (05.09.2011), Available from: < http://www.merriam-webster.com/dictionary/Antimicrobial>

Neu, H.C. (1992). The crisis in antibiotic resistance. *Science.*, Vol.257, No.5073, (August 1992), pp. 1064-1073, ISSN 0036-8075

Panarin, E.F., Solovskii, M.V., Zaikina, N.A., & Afinogenov GE. (1985). Biological activity of cationic polyelectrolytes. *Makromol. Chem. Suppl.*, Vol.9, (1985), pp. 25-33.

Papo, N., Shai, Y. (2003). Exploring peptide membrane interaction using surface plasmon resonance: differentiation between pore formation versus membrane disruption by lytic peptides. *Biochemistry*, Vol.42, Issue.2, (January 2003), pp 458–466, ISSN 0006-2960

Parikh, D.V., Fink, T., Rajasekharan, K., Sachinvala, N.D., Sawhney, A.P.S., Calamari, T.A., & Parikh, A.D. (2005). *Text. Res. J.*, Vol.75, Issue.2, (February 2005), pp. 134-138, ISSN 0040-5175

Petri, W.A.J. (2005). Antimicrobial agents: sulfonamides, trimethoprim-sulfamethoxazole, quinolones, and agents for urinary tract infections. In: *Goodman & Gilman's The Pharmacological Basis of Therapeutics*, pp. 1111-1126, McGraw- Hill Professional, ISBN-10: 0071422803 New York.

Pini, A., Giuliani, A., Falciani, C., Runci, Y., Ricci, C., Lelli, B., Malossi, M., Neri, P., Rossolini, G. M., & Bracci, L. (2005). Antimicrobial Activity of Novel Dendrimeric Peptides Obtained by Phage Display Selection and Rational Modification. *Antimicrob. Agents Chemother.*, Vol.49, No.7, (March 2005), pp. 2665-2672, ISSN 0066-4804

Powers, J.P.S., Hancock, R.E.W. (2003). The relationship between peptide structure and antibacterial activity. *Peptides*, Vol.24, Issue.11, (November 2003), pp. 1681–16, ISSN 0196-9781

Roberts, J. C., Bhalgat, M. K., & Zera, R. T. (1996). Preliminary biological evaluation of polyamidoamine (PAMAM) StarburstTM dendrimers. *J. Biomed. Mater. Res. Part A*,Vol.30, Issue.1, (January 1996), pp. 53-65, ISSN 1552-4965

Rojo, J., Delgado, R. (2004). Glycodendritic structures: promising new antiviral drugs, *J. Antimicrob. Chemother.*, Vol.54, No.3, (August 2004), pp. 579- 581, ISSN 0305-7453

Sampathkumar, S-G., Yarema, K.J. (2005). Targeting cancer cells with dendrimers. *Chem & Biol.*, Vol.12, No.1 , (January 2005), pp. 5–6, ISSN 1074-5521

Sechi, M., Casu, F., Campesi, I., Fiori, S., & Mariani, A. (2006). Hyperbranched Molecular Structures with Potential Antiviral Activity: Derivatives of 5,6-Dihydroxyindole-2-Carboxylic Acid. *Molecule*, Vol.11, Issue.12, (December 2006), pp. 968-977, ISSN 1420-3049

Semuel, U., Guggenbichler, J.P. (2004). Prevention of catheter-related infections: the potential of a new nano-silver impregnated catheter. *Int. J. Antimicrob. Agents*, Vol.23, Suupl.1, (March 2004), pp. 75-78, ISSN 0924-8579

Storm, D.R., Rosenthal, K.S., & Swanson, P.E. (1977). Polymyxin and related peptide antibiotics. *Annu Rev Biochem.*, Vol.46, (July 1977), pp. 723–763

Strathmann, M., Wingender, J. (2004). Use of an oxonol dye in combination with confocal laser scanning microscopy to monitor damage to Staphylococcus aureus cells during colonisation of silver-coated vascular grafts. *Int. J. Antimicrob. Agents*, Vol.24, Issue.3, (September 2004), pp. 234-240, ISSN 0924-8579

Svenson, S., Tomalia, D. A. (2005). Dendrimers in biomedical applications—reflections on the field. *AdV. Drug DeliVery ReV.*, Vol.57, Issue.15, (December 2005), pp. 2106-2129, ISSN 0169-409X

Tanner, B. (2009). Antimicrobial Fabrics-Issues and Opportunities in the Era of Antibiotic Resistance, In: *Antibotic Resistance*, November 2009, Available from: <http://www.antimicrobialtestlaboratories.com/antimicrobial_fabrics_and_antibiotic_resistance.pdf>

Ulkur, E., Oncul, O., Karagoz, H., Yeniz, E., & Celikoz, B. (2005). Comparison of silver-coated dressing (Acticoat™), chlorhexidine acetate 0.5% (Bactigrass®), and fusidic acid 2% (Fucidin®) for topical antibacterial effect in methicillin-resistant Staphylococci-contaminated, full-skin thickness rat burn wounds. *Burns*, Vol.31, Issue.7, (November 2005), pp. 874-877, ISSN 0305-4179

Villanova, P.A., (1993). Performance Standards for Antimicrobial Disk Susceptibility Tests, M2-A5, pp. 1-32, Approved Standard NCCLS Publication, USA

Wang, L., Erasquin, U.J., Zhao, M., Ren, L., Zhang, M.Y., Cheng, G.J., Wang, Y., & Cai, C. (2011). Stability, Antimicrobial Activity, and Cytotoxicity of Poly (amidoamine) Dendrimers on Titainium Substrates. *ACS Appl. Mater. Interfaces*, Vol.3, Issue.8, (August 2011), pp. 2885–2894, ISSN 1944-8244

Wikipedia. (05.09.2011). Antimicrobial, In: *antimicrobial.* (05.09.2011), Available from: < http://en.wikipedia.org/wiki/Antimicrobial>

Yang, H., Lopina, S. T. (2003). Penicillin V-conjugated PEG-PAMAM star polymers. *J. Biomater. Sci., Polym. Ed.*, Vol.14, No.10, (October 2003), pp. 1043- 1056

Yao, J., Moellering, R.J. (2003). Antibacterial agents. In: *Manual of Clinical Microbiology*, pp. 1039-1073, ASM Press, ISBN-10: 1555812554, Washington DC.

7

Selected Factors Determining the Content of Lactoferrin, Lysozyme and Immunoglobulins G in Bovine Milk

Jolanta Król[1], Aneta Brodziak[2], Zygmunt Litwińczuk[2] and Joanna Barłowska[1]

[1]*University of Life Sciences in Lublin*
Department of Commodity Science and Processing of Animal Raw Materials
[2]*Department of Breeding and Genetic Resources Conservation of Cattle*
Poland

1. Introduction

According to the statement "prevention is better than cure" the consumers' concern about their health increases, which is reflected in the sales of products supporting the immune system. Such products are made of milk and colostrum, especially the immune proteins isolated from them. These proteins are increasingly used to enrich baby food, dietary medications or high-protein formulas recommended for convalescents and athletes. Furthermore, these proteins are used in pharmacology and cosmetology. They play a major role in the transmission of passive immunity to the offspring and protect the host mammary gland (Gapper et al., 2007; Stelwagen et al., 2009). This group of proteins include immunoglobulins, lactoferrin, lactoperoxidase and lysozyme. Increasingly, these milk components are used by humans for the prophylactic or therapeutic aims.

Content of antibacterial proteins in the diet is one of the factors determining the normal immune response of the organism. Lactoferrin, lysozyme and immunoglobulins are therefore essential components of milk from the consumers' viewpoint. In order to meet the demands of the consumers, the dairy industry should seek high quality raw materials tested for biologically active compounds. It is therefore important to know the factors that determine their content in cow milk. In this study the influence of selected factors on content of lactoferrin, lysozyme and immunoglobulins G in cow milk is discussed.

2. Characteristics of milk proteins with the antibacterial properties

2.1 Immunoglobulins

Very important group of proteins exhibiting antimicrobial activity are the immunoglobulins. These compounds are high molecular globulins, which are present in plasma and body fluids. Depending on the physicochemical structure and biological activity, three major classes of immunoglobulins are distinguished, i.e. IgG, IgM and IgA. IgG dominate (approximately 80 %) in milk of ruminants, while in other mammalian milk, including human, – IgA (approximately 90 %). They determine the specific humoral immunity of a body (El-Loly & Farrag, 2007; Pakkanen & Aalto, 1997). During the process of antigens binding as well as phagocytosis or complement activation these proteins are involved in the destruction of pathogenic microorganisms, i.e. *Escherichia coli, Candida albicans, Clostridium difficile, Shigella flexneri, Streptococcus mutans* and *Helicobacter pylori* (Gapper et al., 2007; Korhonen, 2004). Furthermore, the immunoglobulins block the action of toxins and viruses. In many countries, the specimens on the basis of immunoglobulins are commercially available. They are meant for the livestock, mainly for newborn calves and pigs in order to prevent gastro-intestinal infections (El-Loly, 2007). Other increasingly popular products are also Ig-based products used as food additives (Gapper et al., 2007; Stelwagen et al., 2009; Struff & Sprotte, 2008). It has been shown that the supplementation use is beneficial for immunity system and prevents diseases of the digestive system (Mehra et al., 2005; Struff & Sprotte, 2008). The reduction of diarrhea incidence caused by rotavirus was noted in the infants and children up to four years (El-Loly, 2007; Gapper et al., 2007; Rawal et al., 2008). Clinical studies have confirmed the efficacy of these specimens in the analgesic therapy in patients with fibromyalgia syndrome (Goebel et al., 2008; Struff & Sprotte, 2008).

2.2 Lactoferrin

Lactoferrin is a glycoprotein with a molecular weight of about 80 kDa. Due to the size and construction, it belongs to the transferrin family, which has a specific ability to bind iron (Baker & Barker, 2005; Legrant et al., 2008). It occurs as a single polypeptide chain which consists of about 690 amino acids (Baker & Barker, 2005, Pakkanen & Aalto, 1997). This protein was first isolated from milk in the 60s of the previous century. Lactoferrin also occurs in other secretions such as saliva, tears, semen, bronchial mucous secretion, digestive and genital tract. It is also a component of neutrophil secondary granules from which, during an injury, infection and inflammation, it is released into the blood (Artym, 2010, Baker & Barker, 2005; Garcia-Mantoya, 2011; Małaczewska & Rotkiewicz, 2007). This protein is an essential element of non-specific innate immunity of humans and other mammals (Kruzel et al., 2007; Legrant et al., 2008). By binding and sequestrating of iron, lactoferrin exhibits antibacterial properties against Gram-positive and Gram-negative bacteria, non-capsular and capsular viruses as well as various types of fungi and parasites (Małaczewska & Rotkiewicz, 2007; Orsi, 2004; Steijns & Hooijdonk, 2000; Wakabayashi et al., 2006). This action partially results from the ability of protein to chelate iron (Fe^{3+} ions), thereby removing this element from the environment of microbial growth (Andersen et al., 2003). Other mechanisms of the lactoferrin antimicrobial action include direct destruction of sheaths and disturbance of bacterial cell metabolism, inhibition of the processes of bacterial adherence to body tissues of the host (Hendrixon et al., 2003), inhibition of biofilm formation by some bacteria (Singh et al., 2002) and stimulation of the immune system of host to fight against pathogens (Artym, 2006). Lactoferrin protects the intestinal epithelial cells, and at the same time inhibits the growth of *E. coli* and other pathogenic intestinal bacteria, mainly *Enterobacteriaceae,* while stimulating the growth of useful intestinal microflora of the *Bifidobacterium genus* (Wakabayashi et al., 2006). This is particularly crucial in the case of neonates in whom there is a gradual colonization of the alimentary canal by diverse microflora. The development of normal bacterial flora ensures the efficient digestion, protects against the pathogenic bacteria development and increases immunity (Actor et al., 2009, Griffiths et al., 2003). These factors determined the use of lactoferrin in food for infants (Satue-Gracia et al., 2000; Wakabayashi et al., 2006). An additional advantage of lactoferrin in the fight against bacterial infections is the possibility of increasing a bacteria sensitivity to certain antibiotics (vancomycin, penicillin) and lowering their effective doses. The combination of penicillin with lactoferrin doubles an inhibitory activity of antibiotic against *Staphylococcus aureus* (Diarra et al., 2002). Lactoferrin also shows an antiviral activity (Van der Strate et al., 2001; Zimecki & Artym, 2005). In clinical trials, it proved to be effective in inhibiting of hepatitis C and B type virus infections (Ikeda et al; 2000; Ishii et al., 2003; Okada et al., 2002), herpes (*Herpes simplex virus*) (Andersen et al., 2003; Jenssen et al., 2004), HIV (Berkhout et al., 2002; Semba et al., 1998) and rotavirus, which are the most important etiological factor for acute diseases running with a diarrhea - major cause of mortality of infants and young children in developing countries (Brock, 2002; Superti et al., 1999; Van der Strate et al., 2001). Moreover, the synergistic effect with antiviral drugs, including interferon, acyclovir and cidofovir, has been observed (Andersen et al., 2003; Ishii et al., 2003). This effect allows the reduction of the doses of used drugs, characterized by high toxicity for the organism. Lactoferrin also has antioxidant capacity and prevents the formation of free radicals, as well as regulates the production and release of cytokines and tumor necrosis factor - TNF, secreted by macrophages (Małaczewska & Rotkiewicz, 2007). *In vitro* and *in vivo* research showed that lactoferrin applied direct antitumor effect on melanoma and colon cancer cells (Spadaro et al., 2008). Moreover, the results suggest that lactoferrin accelerates the formation of bone tissue, and therefore can be used in the prevention and treatment of the bone diseases, including osteoporosis. Due to cytoprotective properties, it may be involved in slowing down the development of neurodegenerative diseases, such as Alzheimer's or Parkinson's disease or multiple sclerosis (Blains et al., 2009, Cornish et al., 2004).

Very important properties of lactoferrin is its resistance to heat and proteolytic enzymes (Małaczewska & Rotkiewicz, 2007; Steijns & Hooijdonk, 2000; Wakabayashi et al., 2006). In veterinary medicine, what is particularly promising is the use of lactoferrin in aquaculture as an agent safe for the environment and human health (Małaczewska et al., 2009).

2.3 Lysozyme

Lysozyme (N-acetylmuramide glycanhydrolase, E.C.3.2.1.17) is a low molecular weight (14.4 kDa) enzymatic protein from the hydrolase group. It is widely distributed in nature, occurring in many body fluids and tissues of living organisms (Fox & Kelly, 2005). The highest concentration of the enzyme was found in tears and egg white protein, which is currently the basic source of its obtaining on an industrial scale (Cegielska-Radziejewska et al., 2008; Chiang et al., 2006; Liśnierowski, 2009; Malicki et al., 2003). The relatively large quantities were also noted in human milk (Benkerroum, 2008; Pakkanen & Aalto, 1997; Shah, 2000). Lysozyme is natural defense mechanism of an organism. Its action is based on disintegrating the bacteria by dissolving the polysaccharide-peptide complex (peptidoglycan), which is the main component of the cell wall of numerous bacteria (Masschalck & Michiels, 2003). Under natural conditions, the antibacterial activity of lysozyme (monomer form) is limited to Gram-positive bacteria, and only after modification (despite the reduction of hydrolytic activity) its bactericidal activity extends to Gram-negative bacteria, including many pathogenic bacteria (Benkerroum, 2008; Chang & Li, 2002; Ibrahim et al., 1996; Liśnierowski, 2009; Liśnierowski et al., 2004; Masschalck et al., 2001;

Masschalck & Michiels, 2003). Lysozyme is also one of the mechanisms of non-specific, humoral immune response (Benkerroum, 2008; Montagne et al., 1998; Pakkanen & Aalto, 1997). Antibacterial properties of lysozyme causes a considerable interest in its practical utilization in many sectors of food industries. It is used primarily as an additive to food, showing the preservative properties (Leśnierowski, 2009; Malicki et al., 2003; Proctor and Cunningham, 1988; Rosiak & Kołożyn-Krajewska, 2003). During rennet cheese production lysozyme limits the growth of butyric fermentation bacteria, especially *Clostridium tyrobutyricum*, causing cheese bloating (Danyluk & Kiev, 2001; Proctor & Cunningham, 1988). Lysozyme is also used in medical diagnostics, pharmacology and veterinary medicine. The enzyme has found a wide application in the therapies of viral and bacterial infections, treatment of skin as well as eye diseases, periodontitis, leukemia and cancer (Benkerroum, 2008; Proctor & Cunningham, 1988; Zimecki & Artym, 2005). Lysozyme exerts an antibiotic adjuvant mechanism of action, and therefore it is often called the endogenous antibiotic. It has been shown that the administration of lysozyme enriched milk for the premature has a positive influence on their development and speeds up the fight against an infection (Zimecki & Artym, 2005).

3. Antibacterial proteins content in milk of different animal species

Antibacterial proteins content in milk of different species of livestock is highly variable (table 1). The greatest amount of lactoferrin and lysozyme contains human milk, however, it is a poor source of immunoglobulins G. The milk of mares, similar to human milk, is characterized by a high content of lysozyme, at lower level of lactoferrin. Milk of other livestock species contains considerably less of these proteins, however, in this group of animals the camel and cow milk is distinguished by a higher content of lactoferrin and the ovine milk - lysozyme. The richest source of IgG is the camel milk. It is worth noting that the lactoferrin derived from human milk reveals the lowest antibacterial activity, while the highest from the camel milk (Coness et al., 2008). A changeability within the species is primarily due to the differences in lactation period, feeding regimen, number of analyzed samples, breeds, and methods of analysis.

Specification	Lactoferrin	Lysozyme	Immunoglobulins G
Human	700-2000	100-890	40-54
Cow	80-500	0.37-0.60	100-800
Buffalo	50-320	0.13-0.15	460-1300
Camel	200-728	0.73-5.00	2000
Goat	98-150	0.25	100-400
Ewe	140	1-4	500
Mare	820	400-890	390

Table 1. Average concentrations of lactoferrin, lysozyme and immunoglobulins G in milk of different species (mg/l) (Dračková et al., 2009; El-Hatmi et al., 2007; Konuspayeva et al., 2008; Liu et al., 2009; Pandya & Khan, 2006; Park et al., 2007; Stelwagen et al., 2009; Wheeler & Hodgkinson, 2007)

4. Materials and methods

4.1 Materials

4.1.1 Animals

The studies were conducted throughout four successive years (2006-2009) on milk samples collected from seven different breeds of dairy cows maintained in Poland, i.e. three breeds with an international meaning (Polish Holstein-Friesian, Simental and Jersey) as well as four local breeds (Polish Red, Whitebacks, Polish Black and White and Polish Red and White). The Polish Holstein-Friesian, Simental and Jersey cows were managed under an intensive system in free-stall barns. An animal feeding system, established over both the winter and summer season, was based on a Total Mixed Ration - TMR (corn silage, haylage and feed concentrate). It should be also mentioned that one group of the Simental cows, included into the research, was maintained in the Southern Poland and a conventional feeding system was used. In the summer the animals grazed pasture *ad libitum*, and in the winter they received haylage, hay, and concentrate. However, the local breeds of cows are predominantly managed in small-sized farms in South-Eastern Poland (mountain, submountain or boggy terrains). Owing to small number and genetic distinction, the animals of these breeds are included into the genetic resources conservation programme (Litwińczuk et al., 2006). These cows were housed in a conventional system, i.e. in tie-stall barns. The summer feeding was based on pasture forage, i.e. green fodder comprising grasses and legumes supplemented by hay or straw, while the winter feeding on haylage and hay with fodder beet additive. In all the farms, a feed ration was supplemented with a feed concentrate.

4.1.2 Milk samples collection

Milk samples were collected individually from each cow during trial milking which occurred twice a year, once in the summer period and again in the winter. A total of 3,105 milk samples were examined (table 2).

Breed	Summer season	Winter season
	Intensive system	
Polish Holstein-Friesian	502	539
Simental	215	242
Jersey	184	192
Total	901	973
	Conventional system	
Simental	221	207
Polish Red	159	126
Whitebacks	133	117
Polish Black and White	91	75
Polish Red and White	54	48
Total	658	573

Table 2. Number of milk samples taken to the research

4.2 Chemical analysis

All milk samples were transported to the laboratory of the Department of Commodity Science and Processing of Animal Raw Materials, University of Life Sciences in Lublin (Poland). Each milk sample was analyzed for somatic cell count (SCC) using flow cytometry technology - Somacount 150 apparatus (Bentley, USA). In 2,662 samples the SCC did not exceed 400,000 cells/ml, i.e. the standard accepted by the European Parliament and the Council of 29 April 2004 (EC Regulation No. 853/2004 on the specific hygiene rules for food of animal origin). These milk samples were included solely into the research evaluating an effect of cow breed, age of cows and stage of lactation and feeding system.

Lactoferrin and lysozyme contents were determined using the reversed-phase high-performance liquid chromatography (RP-HPLC) with UV-Vis detector. From each sample of raw milk was taken 50 ml and adjusted to pH 4.6 with 0.1 mol/l HCl, and allowed to stand at room temperature for about one hour to allow for the acid precipitation of caseins. Consequently, whey (7 ml) was taken from each of the samples separately and then centrifuged at 10,000 rpm for 15 min. Finally, whey solutions were filtered through paper quality filter discs (diameter: 125 mm, density: 65 g/m^2, grade: 3 hours (Munktell, Germany)) and 0.20-μm disposable sterile filters (Millipore type GSTF, USA). The supernatants in vials were kept refrigerated until further analysis, and were injected into the chromatograph at the suitable time (in the amount of 20 μl). Protein separation was performed on liquid chromatography ProStar 210 model and UV-Vis ProStar 325 detector (Varian, USA). The measurements were carried out using the water/acetonitrile mobile phase at gradient elution and column NUCLEOSIL 300-5 C18 (Varian, USA) of 250 mm length and 4.6 mm diameter. The mobile phase was solvent A (90% water, 10% acetonitryle) and solvent B (90% acetonitryle, 10% water), purchased from Sigma (Germany). The solvents were filtered through 0.45-μm filters (Millipore, USA) and degassed by using ultrasounds. The total analysis time for a single sample was 35 min at 205 nm wavelength with column temperature of 37°C. The analyses of reference substances were conducted under the same conditions. On the grounds of the obtained chromatograms, using program Star 6.2 Chromatography Workstation (Varian, USA), the qualitative and quantitative identification of each substance was performed followed by their concentration determination. Calibration of the chromatographic system for whey proteins determination was carried out by the external standard method. For this purpose, each protein was calibrated individually by injecting solutions of the standards (20 μl). The standards were purified proteins, i.e. lactoferrin (90 %) from bovine milk and lysozyme (95 %) from hen egg whites, which were purchased from Sigma (Germany). All chemicals were of HPLC analytical grade. Concentrations of lactoferrin and lysozyme solutions ranged from 0 to 200 mg/l and from 0 to 20 μg/l respectively, and were prepared to create the calibration curves. The limits of quantification LOQ (for lactoferrin - 40 mg/l and for lysozyme - 2.8 μg/l) and detection LOD (for lactoferrin - 8.7 mg/l and lysozyme - 0.9 μg/l) were determined. The immunoglobulin G (IgG) levels were established by the aid of radial immunodiffusion technique with Bovine IgG LL tests (The Binding Site, Birmingham, UK). Control samples, provided in the IgG LL set, were used before the test samples.

4.3 Statistical analysis

Data are given as mean ± standard deviation. The obtained results were analyzed by the General Linear Model (GLM) - factorial ANOVA procedures of Statsoft Inc. Statistica ver.6

(Statsoft Inc. 2003). It was done on the grounds of one-way and multi-way analysis of variance with interaction, using Tukey's HSD procedure.

The following factors were taken into consideration:

- cow breed

Seven breeds of cows kept in Poland, i.e. three breeds of international importance (Polish Holstein-Friesian, Jersey and Simmental) and four local breeds, kept only in Poland (Polish Red, Whitebacks, Polish Black and White and Polish Red and White).

- age of cows and stage of lactation

Age classes, mostly noticed as subsequent lactation: I, II, III and IV.

Stage of lactation, i.e. up to 120 days, from 121 to 200 and from 201 to 305 days of lactation.

- feeding system

Two feeding systems solely for the Simental cows were distinguished:

- conventional (in the summer cows grazed the pasture (ad libitum), in the winter cows were fed with haylage and hay),
- intensive (total mixed ration (TMR) feeding system was used throughout the year).

- somatic cell count.

The research material for each breed was split into four groups according to somatic cell count detected in the milk samples: Group I - up to 100,000 cells/ml, Group II - 101,000-400,000 cells/ml, Group III - 401,000-500,000 cells/ml, and Group IV - 501,000-1,000,000 cells/ml. Next, to regulate the SCC distribution, the SCC data were transformed into log 10 SCC for each sample before statistical analysis could take place.

5. Factors affecting the content of lactoferrin, lysozyme and immunoglobulins G

5.1 Cow breed

It has been shown that among the many breeds of cows involved in milk production in Poland, the cows of local breeds produce a raw material with a higher content of antibacterial proteins, such as lactoferrin and immunoglobulins G. The highest concentration of these proteins was found in the milk of Polish Red cows (128.7 and 558.1 mg/l). A slightly lower level of them was established in milk of Polish Red and White (120.9 and 545.6 mg/l) as well as Whitebacks (115.2 and 540.2 mg/l). Significant amounts of lactoferrin and immunoglobulins G were also reported in the milk of Simental cows (116.74 and 579.9 mg/l), managed under an intensive production system. However, the milk obtained from two breeds of cows with the greatest importance in milk production, both in Poland and the world, i.e. Holstein-Friesian and Jersey, was a poorer source of these proteins (table 3). Simultaneously, milk of Jersey cows characterized by a high content of lysozyme (13.02 μg/l). Milk of local cow breeds contains only slightly less lysozyme (from 10.79 to 12.42 μg/l). The comparable to the present research lactoferrin content in raw milk from Swedish farms, i.e. 90 mg/l (70-110 mg/l) was obtained by Lindmark-Månsson et al. (2003). In another Swedish study (Wedholm et al., 2006), both in the milk of Swedish

Holstein cows, and Swedish Red-White, established 120.0 mg/l of lactoferrin. Higher level of this protein (145.66-204.89 mg/l) reported Wielgosz-Groth et al. (2009) in the milk of Polish Holstein-Friesian cows. A substantially lower this protein content (7.30-14.73 mg/l) in milk from Black-White variety cows with 50-75 % share of Holstein-Friesian genes obtained Reklewska et al. (2003). The authors, however, noted the closest to the present study lysozyme concentration. That milk, depending on the feeding system, contained from 12.54 to 16.43 μg/l of lysozyme. Significantly more of this protein (70 μg/l) found Elegamy et al. (1996). Tsuji et al. (1990), analyzing lactoferrin concentration in the colostrum obtained from four breeds of cows in Japan, showed a higher share of this protein in the colostrum produced by Holstein-Friesian and Jersey cows (1.96-2.11 mg/ml), compared to Japanese Brown and Japanese Black (0.40-0.56 mg/ml). Newstead (1976) determined IgG concentration in milk of Jersey and Friesian cows. The author found (compared to the present research) a substantially lower level of this protein in milk of Jersey cows, i.e. 0.32 g/l. While milk gained from Friesian cows contained 0.46 g/l IgG. In the research of Levieux and Ollier (1999) revealed the comparable to the present study concentration of IgG. Krukowski et al. (2006) in the milk from Black and White variety cows with 50-75 % Holstein-Friesian genes reported higher level of these proteins - 628 mg/l. Whereas, in milk from the Polish Holstein-Friesian Black and White variety cows, the differences in IgG content reached over 200 mg/l. Murphy et al. (2005) showed a breed influence on concentration of IgG in colostrum of beef cows (from 75.7 mg/ml for the Limousin to 95.5 mg/ml for Charolais).

Specification	Lactoferrin (mg/l)	Lysozyme (μg/l)	Immunoglobulins G (mg/l)
Polish Holstein-Friesian	88.42^{A} 14.72	8.22^{A} 0.99	423.6^{A} 18.4
Simental	116.74^{BC} 16.54	9.84^{AB} 1.88	579.9^{B} 16.5
Jersey	103.48^{AB} 15.68	13.02^{B} 2.21	508.6^{AB} 29.1
Polish Red	128.7^{C} 17.35	12.17^{B} 5.76	558.1^{B} 22.4
Whitebacks	115.2^{BC} 21.01	12.42^{B} 3.38	540.2^{B} 46.1
Polish Black and White	105.9^{AB} 19.37	10.79^{AB} 4.82	530.2^{B} 29.3
Polish Red and White	120.9^{BC} 19.22	11.51^{AB} 5.23	545.6^{B} 48.3

A, B, C – differences significant at $P \leq 0.01$

Table 3. Effect of breed on the antimicrobial protein content in bovine milk (mean ± standard deviation) (Król et al., 2007, 2010a, 2010b; unpublished data)

5.2 Age of cows and stage of lactation

Age of cows (usually referred to as the subsequent lactation) and stage of lactation are the main physiological factors affecting the productivity and chemical composition of cow milk. It has been shown that the poorest source of analyzed antimicrobial proteins proved to be the milk obtained from primiparous cows (table 4). The lowest level of IgG was found in I lactation (454.8 mg/l). In subsequent lactations share of these compounds increased gradually, with significant differences (P≤0.01) occurred only in the IV lactation. Lysozyme content progressed successively with the lactations, with peaks reported in IV lactation. Whereas, the highest lactoferrin concentration was observed in milk from cows at II and IV lactation, 107.18 and 115.61 mg/l, respectively. In the studies of Levieux & Ollier (1999), similarly to the present research, the primiparous cows produced significantly less IgG, as compared to cows at II-IV lactation (P≤0.05) and older (P≤0.01). The results of Guliński et al. (2006) and Mian-Bin Wu & Xu Yin-Jun (2009) confirmed the higher concentration of these proteins in milk and colostrum produced by the multiparous cows. It was also noted that in the subsequent lactations were significant differences in lactoferrin content (Back & Thompson, 2005; Hagiwara et al., 2003; Liu et al., 2009; Tsuji et al., 1990). Older cows, i.e. in IV lactation, produced milk with significantly higher concentration of this protein, compared to younger ones (Back & Thompson, 2005; Liu et al., 2009; Tsuji et al., 1990). Whereas, Hagiwara et al. (2003) reported significantly lower lactoferrin concentration in V lactation in relation to II (P≤0.01) and III (P≤0.05) lactation.

Specification	Subsequent lactation			
	I	II	III	IV
Lactoferrin (mg/l)	89.39^{A} 17.89	107.18^{B} 18.26	99.50^{AB} 16.32	115.61^{B} 19.33
Lysozyme (μg/l)	7.56^{a} 2.11	9.29^{ab} 1.68	9.87^{b} 2.13	10.28^{b} 2.36
Immunoglobulins G (mg/l)	454.8^{A} 16.4	511.7^{AB} 21.5	513.2^{AB} 19.7	541.1^{B} 25.9

a, b – differences significant at P≤0.05; A, B – differences significant at P≤0.01

Table 4. Effect of subsequent lactation on the antimicrobial protein content in bovine milk (mean ± standard deviation) (Król et al., 2010a)

Antibacterial protein level also underwent significant changes throughout the lactation period (table 5). Milk collected during the first months of lactation was characterized by the lowest lactoferrin and lysozyme content. Significantly more of these proteins amounts were determined in the late stage of lactation, i.e. by 48.11 mg/l (P≤0.01) and 3.12 mg/l (P≤0.05). Milk obtained from this lactation stage also contained the highest immunoglobulins G level (553.8 mg/l). During the first stages of lactation, these proteins content was significantly lower (P≤0.05), i.e. ranged from 70.6 mg/l (early stage) to 78.4 mg/l (middle stage). Concentration of IgG investigated by Caffin et al (1983) was also noted to change (alike in the present study) during the lactation course. In the early (30 days) and the middle stage (150 days) of lactation, there were found comparable values, i.e. 0.37 and 0.38 mg/ml. The significantly higher IgG content was reported in milk obtained in late lactation (270 days) – 0.60 mg/ml. Changes in IgG level in milk during lactation also confirmed the research of Liu et al. (2009). Wielgosz-Groth et al. (2009) showed an increase in the content of lactoferrin in milk of Jersey cows during lactation. Its concentration ranged from 51.91 μg/ml in the first month of lactation to 259.43 μg/ml in the tenth one. The changes in these protein content thought the lactation were also reported by Hiss et al. (2008), who analyzed goat milk. Until 32nd lactation week the lactoferrin level maintained between 10 to 28 mg/ml, then it successively progressed to reach over 100 μg/ml in 44th lactation week. In human milk also stated that the lactoferrin and lysozyme concentration increase in the duration of lactation (Hennart et al., 1991; Montagne et al., 2001). Cheng et al. (2008) found a high correlation coefficient between the content of lactoferrin and lactation stage (r = 0.557).

Specification	Lactation stage		
	<120 days	121-200 days	201-305 days
Lactoferrin (mg/l)	76.12^{A} 15.69	109.50^{AB} 19.58	124.23^{B} 21.36
Lysozyme (μg/l)	8.16^{a} 2.25	9.18^{ab} 2.64	11.28^{b} 3.29
Immunoglobulins G (mg/l)	483.2^{a} 21.3	475.4^{a} 19.8	553.8^{b} 23.6

a, b – differences significant at P≤0.05; A, B – differences significant at P≤0.01

Table 5. Effect of lactation stage on the antimicrobial protein content in bovine milk (mean ± standard deviation) (Król et al., 2010a)

5.3 Feeding system

Among the environmental factors, the feeding system has the major impact on the milk yield and its chemical composition. For example the Simental cows, for which it was possible to distinguished two groups of feeding system, the significant differences in the content of these proteins were showed between the groups (table 6). Milk of cows grazing the pasture characterized by a higher content of lactoferrin (by 27.9 mg/l), lysozyme (by 0.65 mg/l), as well as IgG (by 39.6 mg/l). Higher level of the functional whey proteins in milk of cows grazing the pasture also obtained King et al. (2007) and Reklewska et al. (2003). Different results reported Wielgosz et al. (2009). The authors found lower levels of lactoferrin in the milk of cows kept on the pasture (145.66-148.83 mg/l) in comparison with milk of cows fed in barns (174.63-204.89 mg/l). Turner et al. (2003) also reported higher levels of lactoferrin in milk of cows fed the TMR system in relation to milk of cows grazing the pasture. In the subsequent study (Turner et al., 2007) the content of lactoferrin in milk of cows using the pasture in various degree was compared. Significantly (P≤0.05) higher lactoferrin yield (in g per day) was found in the milk of cows having unrestricted access to pasture (*ad libitum*). Mackle et al. (1999) showed that a pasture supplementing with maize grain and silage led to slightly decreasing of IgG content.

Specification	Feeding system	
	Traditional system	Intensive system
Lactoferrin (mg/l)	127.52^{B} 19.1	110.71^{A} 17.8
Lysozyme (μg/l)	10.90^{b} 2.4	10.25^{a} 2.4
Immunoglobulins G (mg/l)	602.4^{B} 40.1	562.8^{A} 31.6

a, b – differences significant at P≤0.05; A, B – differences significant at P≤0.01

Table 6. Effect of feeding system on the antimicrobial protein content in bovine milk (mean ± standard deviation) (Król et al., 2008; Brodziak, 2011; unpublished data)

5.4 Somatic cell count

Somatic cell count is a commonly recognized indicator of bovine udder health, milk quality and its technological usability. SCC is also one of the criteria for admission to the purchase of milk. According to many authors and bovine quarter producing milk with the SCC over 200,000 cells/ml shows the symptoms of subclinical mastitis. In Poland and other EU countries in accordance with the applicable decree, i.e. Regulation (EC) No. 853/2004, raw cow milk should not contain more than 400,000 somatic cells/ml. SCC has been shown to influence the immunoactive protein content. With the growth of SCC significantly increased the concentration of lactoferrin, lysozyme and IgG (table 7). Milk with the highest number of somatic cells (group IV) contained most of these proteins. In comparison to milk of cows belonging to I group, these differences were for lactoferrin - 26.66 mg/l (i.e. 32.6 % value of group I), lysozyme - 5.14 µg/l (60.2 %) and IgG - 219.4 mg/l (42.6 %). A substantial effect of SCC on immunoactive proteins content was confirmed by relatively high positive values of calculated correlation coefficients. For the content of lactoferrin: $r = 0.65$, for lysozyme $r = 0.63$ and for immunoglobulins G $r = 0.79$. Urech et al. (1999), in the study of quarter milk, indicated similar tendencies when 100,000 cells/ml were recognized as the threshold of somatic cell count. Quarter milk obtained from clinically healthy mammary glands contained an average of 84,000 of somatic cells/ml, whereas the milk from infected glands had 293,000 cells/ml ($P<0.001$). The authors showed the significant growth of lactoferrin (by 0.45 %) in milk from the affected udder. Similarly, Hamann (2002) defined "the gold standard" for a cell count to be up to 100,000 somatic cells/ml. Counts that reached above this point provided evidence of disturbed milk secretion that would lead to a reduced production of daily milk, changes in its chemical composition, and a deterioration of the processing properties of the milk. According to Lindmark-Månsson et al. (2000, 2006) an exceeding 5,000 cells/ml for somatic cell count limit leads to increase of lactoferrin content in milk, and a close relationship between this milk component and the status of udder health has been confirmed by very high correlation coefficients between milk lactoferrin concentration and somatic cell counts ($r = 0.962$ and $r = 0.918$ at $P<0.001$), obtained in two independent studies. A significant increase ($P<0.05$) of lactoferrin content with concurrent progression of a mammary gland disease was also noted in goat and buffalo milk (Leitner et al., 2004; Piccinini et al., 2006). The effect of SCC growth on the content of IgG was reported also by other authors (Caffin et al., 1983; Liu et al., 2009).

Specification	SCC group			
	I	II	III	IV
Lactoferrin (mg/l)	81.84^{a} 17.2	82.56^{a} 18.4	91.07^{ab} 19.6	108.5^{b} 20.2
Lysozyme (µg/l)	8.65^{A} 1.74	9.68^{AB} 1.46	11.54^{AB} 2.05	13.79^{B} 2.68
Immunoglobulins G (mg/l)	514.6^{a} 18.9	520.9^{a} 22.7	587.1^{ab} 35.5	734.0^{b} 54.3

a, b - differences significant at P≤0.05; A, B, C - differences significant at P≤0.01

Table 7. Effect of SCC on the antimicrobial protein content in bovine milk (mean ± standard deviation) (Litwińczuk et al., 2011; unpublished data)

Summarized in table 8 the results of two-way analysis of variances indicate significant influence of the analyzed factors (breed, age of cows, stage of lactation, feeding system, SCC) on the content of individual proteins. For lactoferrin content the significant interactions between breed of cows and stage of lactation as well as breed and SCC were also found. In the case of the following interactions: breed and age of cows as well as age of cows x stage of lactation the significant correlations have been shown for lysozyme content.

Factor	Lactoferrin	Lysozyme	Immunoglobulins G
Breed	0.004	0.010	0.002
Age of cows	0.000	0.000	0.002
Stage of lactation	0.000	0.021	0.030
Feeding system	0.001	0.031	0.008
SCC	0.015	0.000	0.025
Breed x age of cows	0.521	0.015	0.150
Breed x stage of lactation	0.029	0.365	0.131
Breed x SCC	0.000	0.553	0.777
Age of cows x stage of lactation	0.075	0.003	0.778
Age of cows x SCC	0.363	0.470	0.951
Stage of lactation x SCC	0.893	0.560	0.817

Table. 8. Results of one-way and two-way variances analysis for chosen milk proteins (P values) (Litwińczuk et al., 2011; unpublished data)

6. Conclusion

Content of the antibacterial proteins is determined by the genetic factors (breed of cows), environmental (feeding system and SCC) as well as physiological (age of cows and stage of lactation). Significant increase of these proteins concentration could be achieved by a production system changing, i.e. the introduction of pasture.

7. References

Actor, J.K., Hwang, S., Kruzel, M.L. (2009). Lactoferrin as a natural immune modulator. *Curr. Pharm. Des.*, Vol. 15, No. 17, pp. 1956-1973

Andersen, J.H., Jenssen, H., Gutteberg, T.J. (2003). Lactoferrin and lactoferricin inhibit Herpes simplex 1 and 2 infection and exhibit synergy when combined with acyclovir. *Antivir. Res.*, No. 58, pp. 209-215

Artym, J. (2010). Role of lactoferrin in iron administration in the organism. Part II. Antimicrobial and antiinflammatory action by an iron sequestration. *Postepy Hig. Med. Dosw.*, No. 64, 604-616

Back, P.J, Thomson, N.A. (2005). Exploiting cow genotype to increase milk value through production of minor milk components. *Proceedings of the New Zealand Society of Animal Production*, Vol. 65, pp. 53-58

Baker, E.N., Baker, H.M. (2005). Molecular structure, binding properties and dynamics of lactoferrin. *Cell Mol Life Sci.*, Vol. 62, pp. 2531-2539

Benkerroum, N. (2008). Antimicrobial activity of lysozyme with special relevance to milk. *African Journal of Biotechnology*, Vol. 7, No. 25, pp. 4856-4867

Berkhout, B., Van Wamel, J.L., Belijaars, L., Meijer, D.K., Visser, S., Floris, R. (2002). Characterization of the anti-HIV effects of native lactoferrin and other milk proteins and protein-derived peptides. *Antiviral Res.*, Vol. 55, pp. 341-355

Blais, A., Malet, A., Mikogami, T., Martin-Rouas, C., Tomé, D. (2009). Oral bovine lactoferrin improves bone status of ovariectomized mice. *Am. J. Physiol. Endocrinol. Metab.*, No. 296, pp. 1281–1288

Brodziak, A. (2011). Influence of the chosen environmental and genetic factors on the content and functional properties of milk whey proteins. *Doctoral thesis*. University of Life Sciences in Lublin

Caffin, J.P., Poutrel, B., Rainard, P. (1983). Physiological and pathological factors influencing bovine immunoglobulin G_1 concentration in milk. *J. Dairy Sci.*, Vol. 66, No. 10, pp. 2161-2166

Cegielska-Radziejewska, R., Leśnierowski, G., Kijowski, J. (2008). Properties and application of egg white lysozyme and its modified preparations - a review. *Pol. J. Food Nutr. Sci.*, Vol. 58, No. 1, pp. 5-10

Chang, J.Y., Li, L. (2002). The unfolding mechanism and disulfide structures of denatured lysozyme. *FEBS Letters*, No. 511, pp. 73–78

Cheng, J.B., Wang, J.Q., Bu, D.P., Liu, G.L., Zhang, C.G., Wei, H.Y., Zhou, L.Y., Wang, J.Z. (2008). Factors affecting the lactoferrin concentration in bovine milk. *J. Dairy Sci.*, Vol. 91, No. 3, pp. 970-976

Chiang, B.H., Su, C.K., Tsai, G.J., Tsao, G.T. (2006). Egg white lysozyme purification by ultrafiltration and affinity chromatography. *J. Food Sci.*, Vol. 58, pp. 303-306

Conesa, C., Sánchez, L., Rota, C., Pérez, M.D., Calvo, M., Farnaud, S., Evans, R.W. (2008). Isolation of lactoferrin from milk of different species: Calorimetric and antimicrobial studies. *Comparative Biochemistry and Physiology*, Part B, No. 150, pp. 131–139

Cornish, J., Callon, K.E., Naot, D., Palmano, K.P., Banovic, T., Bava, U., Watson, M., Lin, J.M., Tong, P. C., Chen, Q., Chan, V.A., Reid, H.E., Fazzalari, N., Baker, H.M., Baker, E.N., Haggarty N.W., Grey, A.B., Reid, I.R. (2004). Lactoferrin is a potent regulator of bone cell activity and increases bone formation *in vivo*. *Endocrinology*, Vol. 145, No. 9, pp. 4366-4374

Danyluk, B., Kijowski, J. (2001). The effect of lysozyme monomer on the growth of *Clostridium tyrobutyricum*. *Przem. Spoż.*, No. 12, pp. 16-19 (in Polish)

Diarra, M.S., Petitelere, D., Lacasse, P. (2002). Effect of lactoferrin in combination with penicillin on the morphology and the physiology of *Staphylococcus aureus* isolated from bovine mastitis. *J. Dairy Sci.*, Vol. 85, pp.1141-1149

Dračková, M., Borkovcová, I., Janštová, B., Naiserová, M., Přidalová, H., Navrátilová, P., Vorlová, L. (2009). Determination of lactoferrin in goat milk by HPLC method. *Czech J. Food Sci.*, Vol. 27, pp. 102-104

Elagamy, E.I., Ruppanner, R., Ismail, A., Champagne, C.P., Assaf, R. (1996) Purification and characterization of lactoferrin, lactoperoxidase, lysozyme and immunoglobulins from camel's milk. *Inter. Dairy J.*, Vol. 6, No. 2, pp. 129-145

El-Hatmi, H., Girardet, J.M., Gaillard, J.L., Yahyaoui, M.H., Attia, H. (2007). Characterisation of whey proteins of camel (*Camelus dromedarius*) milk and colostrums. *Small Ruminant Research*, Vol. 70, pp. 267-271

El-Loly, M.M. (2007). Bovine milk immunoglobulins in relation to human health. *Inter. J. Dairy Sci.*, Vol. 2, No. 3, pp. 183-195

El-Loly, M.M., Farrag, A.F. (2007). Isolation of immunoglobulin-rich fractions from whey. *Milchwissenschaft*, Vol. 62, No. 2, pp. 199-202

Fox, P.F., Kelly, A.L. (2005). Indigenous enzymes in milk: Overview and historical aspects – Part 2. *Inter. Dairy J.*, No. 16, pp. 517-532

Gapper, L.W., David, E.J., Copestake, D.E.J., Otter, D.E., Indyk, H.E. (2007). Analysis of bovine immunoglobulin G in milk, colostrum and dietary supplements: a review. *Anal. Bioanal. Chem.*, Vol. 389, pp. 93-109

Garcia-Mantoya, I.A., Cendon, T.S., Arevalo-Gallegos, S., Rascon-Cruz, Q. (2011). Lactoferrin a multiple bioactive protein: An overview. *Biochemica et Biophysica Acta*, doi:10.1016/j.bbagen.2011.06.018, pp. 1-11

Goebel, A., Buhner, S., Schedel, R., Lochs, H., Sprotte, G. (2008). Altered intestinal permeability in patients with primary fibromyalgia and in patients with complex regional pain syndrome. *Rheumatology*, Vol. 47, No. 8, pp. 1223-1227

Griffiths, E.A., Duffy, L.C., Schanbacher, E.L., Dryja, D., Leavens, A., Neiswander, R.I., Oiao, H., DiRienzo, D., Ogra, P. (2003). In vitro growth responses of bifidobacteria and enteropatogenes to bovine and human lactoferrin. *Dig. Dis. Sci.*, Vol. 48, pp. 1324-1332

Guliński, P., Niedziałek, G., Salmończyk., E, Górski T. (2006). Immunoglobulin content in colostrum of cows within select genetic and environmental factors (in Polish). *Medycyna Wet.*, Vol. 62, No. 3, pp. 339-342

Hagiwara, S., Kawai, K., Anri, A., Nagahata, H. (2003). Lactoferrin concentrations in milk from normal and subclinical mastitic cows. *J. Vet. Med. Sci.*, Vol. 65, No.3, pp. 319-323

Hamann, J. (2002). Relationship between somatic cell count and milk composition. *Bulletin FIL-IDF*, Vol. 372, pp. 56-59

Hendrixson, D.R., Qiu, J., Shewry, S.C., Fink, D.L., Petty, S., Baker, E.N., Plaut, A.G., St Geme, J.W. (2003). Human milk lactoferrin is a serine protease that cleaves Haemophilus surface proteins at arginine-rich sites. *Mol. Microbiol.*, Vol. 47, pp. 607–617

Hennart, P.F., Brasseur, D.J., Dologne-Desnoeck, J.B., Dramaix, M.M., Robyn, C.E. (1991). Lysozyme, lactoferrin, and secretory immunoglobulin A content in brest milk: influence of duration of lactation, nutrition status, prolactin status, and parity of mather. *Am. J. Clin. Nutr.*, Vol. 53, pp. 32-39

Hiss, S., Meyer, T., Sauerwein, H. (2008). Lacyoferrin concentrations in goat milk throughout lactation. *Small Ruminant Research*, Vol. 80, pp. 87–90

Ibrahim, H.R., Higashiguchi, S., Juneja, L.R., Kim, M., Yamamoto, T., Sugimato, Y., Aoki, T. (1996). Partially unfolded lysozyme at neutral pH agglutinates and kills Gram-negative and Gram-positive bacteria through membrane damage mechanism. *J. Agric. Food Chemistry*, Vol. 44, pp. 3799–3806

Ikeda, M.A., Nozaki, K., Sugiyama, T., Tanaka, A., Naganuma, K., Tanaka, H., Sekihara, K., Shimotohno, M., Saito, Kato. N. (2000). Characterization of antiviral activity of lactoferrin against hepatitis C virus infection in human cultured cells. *Virus Res.*, Vol. 66, pp. 51-63

Ishii, K., Takamura, N., Shinohara, M., Wakui, N., Shin, H., Sumino, Y., Ohmoto, Y., Teraguchi, S., Yamauchi, K. (2003). Long-term follow-up of chronic hepatitis C patients treated with oral lactoferrin for 12 months. *Hepatol. Res.*, Vol. 25, pp. 226-233

Jenssen, H., Andersen, J.H., Uhlin-Hansen, L., Gutteberg, T.J., Rekdal, Ø. (2004). Anti-HSV activity of lactoferricin analogues is only partly related to their affinity for heparan sulfate. *Antivir. Res.*, Vol. 61, pp. 101-109

Konuspayeva, G., Loiseau, G., Levieux, D., Faye, B. (2008). Lactoferrin and immunoglobulin content in camel milk from Bactrian, Dromedary and hybrids in Kazakhstan. *Journal of Camelid Sciences*, No. 1, pp. 54-62

Korhonen, H. (2004). Isolation of immunoglobulins from colostrums. *Bulletin of the IDF*, Vol. 389, pp. 78-84

Korhonen, H., Marnila, P., Gill, H.S. (2000). Milk immunoglobulins and complement factors. *British Journal of Nutrition*, Vol. 84, pp. 75-80

Król, J., Brodziak, A., Litwińczuk, Z., Szwajkowska, M. (2011). Whey proteins use in health promotion. *Żywienie Człowieka i Metabolizm*, Vol. XXXVIII, No. 1, pp. 36-45

Król, J., Litwińczuk, Z., Barłowska, J., Kędzierska-Matysek, M. (2007). Initial results on casein and whey protein content in milk of Polish Red and Whitebacked cows. *Ann. Anim. Sci.*, No. 1, pp. 207-211

Król, J., Litwińczuk, Z., Brodziak, A, Barłowska, J. (2010a). Lactoferrin, lysozyme and immunoglobulin G content in milk of four breeds of cows manager under intensive production system. *Polish Journal of Veterinary Science*, Vol. 13, No. 2, pp. 357-361

Król, J., Litwińczuk, Z., Brodziak, A, Sawicka-Zugaj, W. (2010b). Bioactive protein content in milk from local breeds of cows included in the genetic resources conservation programme. *Ann. Anim. Sci.*, Vol. 10, No. 3, pp. 213-221

Król, J., Litwińczuk, Z., Litwińczuk, A., Brodziak, A. (2008). Content of protein and its fractions in milk of Simmental cows with regard to a rearing technology. *Ann. Anim. Sci.*, Vol. 8, No.1, pp. 57-61

Krukowski, H., Lisowski, A., Różański, P., Polkowska, I. (2006). IgG level in mastitis cows' milk (in Polish). *Roczn Nauk PTZ*, Vol. 2, No. 2, pp. 65-69

Kruzel, M.L., Actor, J.K., Boldogh, I., Zimecki, M. (2007). Lactoferrin in health and disease. *Postepy Hig. Med. Dosw.*, Vol. 61, pp. 261-267

Legrand, D., Pierce, E., Carpentier, M., Mariller, C., Mazurier, J. (2008). Structure and function lactoferrin. *Adv. Exp. Med. Biol.*, Vol. 606, No. 94, pp. 163

Leitner, G., Merin, U., Silanikove, N. (2004). Changes in milk composition as affected by subclinical mastitis in goats. *J. Dairy Sci.*, Vol. 87, pp. 1719-1726

Leśnierowski, G. (2009). New manners of physical-chemical modification of lysozyme. *Nauka Przyroda Technologie*, Vol. 3, No. 4, pp. 1-18

Leśnierowski, G., Cegielska-Radziejewska, R., Kijowski, J. (2004). Thermally and chemical-thermally modified lysozyme and its bacteriostatic activity. *World Poult. Sci. J.*, Vol. 60, pp. 303–310

Levieux, D., Ollier, A. (1999). Bovine immunoglobulin G, β-lactoglobulin, α-lactalbumin and serum albumin in colostrum and milk during the early post partum period. *J. Dairy Res.*, Vol. 66, pp. 421-430

Lindmark-Månsson, H., Sensson, U., Paulsson, M., Alden, G., Frank, B., Johnson, G. (2000). Influence of milk components, somatic cells and supplemental zinc on milk processability. *Inter. Dairy J.*, Vol. 10, pp. 423-433

Lindmark-Månsson, H., Fonden, R., Pettersson, H.E. (2003). Composition of Swedish dairy milk. *Inter. Dairy J.*, Vol. 13, pp. 409-425

Lindmark-Månsson, H., Brånning, C., Alden, G., Paullsson M. (2006). Relationship between somatic cell count, individual leukocyte populations and milk components in bovine udder quarter milk. *Inter. Dairy J.*, Vol. 16, pp. 717-727

Litwińczuk, Z., Chabuz, W., Stanek, P., Sawicka, W. (2006). Genetic potential and reproductive performance of Whitebacks - Polish native breed of cows. *Arch. Tierz., Dummerstorf*, Vol. 49, pp. 289–296

Litwińczuk, Z., Król, J., Brodziak, A, Barłowska, J. (2011). Changes of protein content and its fractions in bovine milk from different cow breeds subject to somatic cell count. *J. Dairy Sci.*, Vol. 94, No. 2, pp. 684-691

Liu, G.L., Wang, J.Q., Bu, D.P., Cheng, J.B., Zhang, C.G., Wei, H.Y., Zhou, L.Y., Zhou, Z.F., Hu, H., Dong, X.L. (2009). Factors affecting the transfer of immunoglobulin G1 into the milk of Holstein cows. *The Veterinary Journal*, Vol. 182, pp. 79-85

Mackle, T.R., Bryant, A.M., Petch, S.F., Hill, J.P., Auldist, M.J. (1999). Variation in the composition of milk protein from pasture-fed dairy cows in the late lactation and the effect of grain and silage supplementation. *New Zealand Journal of Agricultural Research*, Vol. 42, pp. 147-154

Małaczewska, J., Rotkiewicz, Z. (2007). Lactoferrin as a multipotential protein. *Medycyna Wet.*, Vol. 63, No. 2, pp. 136-139

Małaczewska, J., Wójcik, R., Siwicki, A.K. (2009). Possibilities of lactoferrin using as an immune stimulator in fish and shellfish. *Medycyna Wet.*, Vol. 65, No. 2, pp. 95-98

Malicki, A., Jarmoluk, A., Bużewicz, S. (2003). Effect of lysozyme additive on the durability and microbial safety of the sausage in barrier sheath. *Medycyna Wet.*, Vol. 2, No. 2, pp. 29-36

Masschalack, B., Van Houdt, R., Van Haver, E.G., Michiels, C.W. (2001). Inactivation of gram-negative bacteria by lysozyme, denatured lysozyme, and lysozyme-derived peptides under high hydrostatic pressure. *Appl. Environ. Microbiol.*, Vol. 67, No. 1, pp. 339-344

Masschalck, B., Michiels, C.W. (2003). Antimicrobial properties of lysozyme in relation to foodborne vegetative bacteria. *Microbiol. Rev. Crit.*, Vol. 29, No. 3, pp. 191-214

Mehra, R., Marnila, P., Korhonen, H. (2005). Milk immunoglobulins for health promotion. *Inter. Dairy J.*, Vol. 16, pp. 1262-1271

Mian-bin, W., Yin-jun, X. (2009) Isolation and purification of lactoferrin and immunoglobulin G from bovine colostrum with serial cation-anion exchange chromatography. *Biotechnology and Bioprocess Engineering*, Vol. 14, pp. 155-160

Montagne P., Cuillière M.L., Molé C., Béné M.C., Faure G. (1998). Microparticle-enhanced nephelometric immunoassay of lysozyme in milk and other human body fluids. Clinical Chemistry, Vol. 44, pp. 1610-1615

Montagne, P., Cuillière, M.L., Molé, C., Béné, MC., Faure, G. (2001). Changes in lactoferrin and lysozyme levels in human milk during the first twelve weeks of lactation. *Adv. EXp. Med. Biol.*, Vol. 501, pp. 241-247

Murphy, B.M., Drennan, M.J., O'Mara, F.P., Earley, B. (2005). Cow serum and colostrum immunoglobulin (IgG1) concentration of five suckler cow breed types and subsequent immune status of their calves. *Irish Journal of Agricultural and Food Research,* Vol. 44, pp. 205–213

Newstead D.F. (1976) Carotene and immunoglobulin concentrations in the colostrums and milk of pasture-fed cows. *J. Dairy Res.*, Vol. 43, pp. 229-237

Okada, S., Tanaka, K., Sato, T., Ueno, H., Saito, S., Okusaka, T., Sato, K., Yamamoto, S., Kakizoe, T. (2002). Dose-response trial of lactoferrin In patients with chronie hepatitis C. *Jpn. J. Cancer Res.*, Vol. 93, pp. 1063-1069

Orsi, N. (2004). The antimicrobial activity of lactoferrin: current status and perspectives. *Biometals*, Vol. 17, pp. 189–196

Pakkanen, R., Aalto, J. (1997). Growth factors and antimicrobial factors of bovine colostrum. *Inter. Dairy J.*, Vol. 7, pp. 285-297

Pandya, A.J., Khan, M.M.H. (2006). Buffalo milk. In: *Handbook of milk of non-bovine mammals.* Park Y.W., Haenlein G.F.W. (Eds), Iowa, Blackwell Publishing Professional, pp. 195-273

Park, Y.W., Juarez, M., Ramos, M., Haenlein, G.F.W. (2007). Physico-chemical characteristics of goat and sheep milk. *Small Ruminant Research*, Vol. 68, pp. 88–113

Piccinini, R., M. Mirelli, B. Ferri, C. Tripaldi, M. Belotti, V. Dapra, S. Orlandini, and A. Zecconi. 2006. Relationship between cellular and whey components in buffalo milk. *J. Dairy Res.*, Vol. 73, pp. 129-133

Proctor, V.A., Cunningham, F.E. (1988). The chemistry of lysozyme and its use as a food preservative and a pharmaceutical. *Crit. Rev. Food Sci. Nutr.*, Vol. 26, No. 4, pp. 359-395

Rawal, P., Gupta, V., Thapa, B.R. (2008). Role of colostrum in gastrointestinal infections. *Indian J. Pediatr.*, Vol. 75, No. 9, pp. 917-921

Reklewska, B., Bernatowicz, E., Reklewski, Z., Nałęcz-Tarwacka, T., Kuczyńska, B., Zdziarski, K., Oprządek, A. (2003). Biological active proteins content in milk of cows in relations to feeding system and season. *Zesz. Nauk. Przeg. Hod.* 68, 85-98

Satue-Gracia M.T., Frankel E.N., Rangavajhyala N., German J.B. (2000). Lactoferrin in infant formulas: effect on oxidation. *J. Agric. Food Chem.*, Vol. 48, pp. 4984-4990

Semba, R.D., Miotti, P.G., Lan, Y., Chiphangwi, J.G., Hoover, D.R., Dallabetta, G.A., Yang, L.P., Saah, A.J. (1998). Maternal serum lactoferrin and vertical transmission of HIV. *AIDS*, Vol. 12, pp. 331-332

Shah, N.P. (2000). Effects of milk-derived bioactives: an overview. *British Journal of Nutrition*, Vol. 84, Suppl. 1, pp. 3-10

Singh, P.K., Parsek, M.R., Greenberg, E.P., Welsh, M.J. (2002). A component of innate immunity prevents bacterial biofilm development. *Nature*, Vol. 417, pp. 552–555

Spadaro, M., Caorsi, C. Ceruti, P., Varadhachary, A., Forni, G., Pericle, F. Giovarelli, M. (2008). Lactoferrin, a major defense protein of innate immunity, is a novel maturation factor for human dendritic cells. *The FASEB Journal*, Vol. 22, pp. 2747-2757

Steijns, J.M., Hooijdonk, A.C.M. (2000). Occurrence, structure, biochemical properties and technological characteristics of lactoferrin. *British Journal of Nutrition*, Vol. 84, Suppl. 1, pp. 11-17

Stelwagen, K., Carpenter, E., Haigh, B., Hodgkinson, A., Wheeler, T.T. (2009). Immune components of bovine colostrum and milk. *J. Anim Sci.*, Vol. 87, pp. 3-9

Struff, W.G., Sprotte, G. (2008). Bovine colostrum as a biologic in clinical medicine: a review - part II: clinical studies. *Int. J. Clin. Pharmacol. Ther.*, Vol. 46, No 5, pp. 211-225

Sung, K., Khan, S.A., Nawaz, M.S., Cerniglia, C.E., Tamplin, M.L., Philips, R.W., Kelley, L.C. (2011). Lysozyme as a barrier to growth of Bacillus anthracis strain Sterne in liquid egg white, milk and beef. *Food Microbiology*, Vol. 28, pp. 1231-1234

Superti, F., Ammendolia M.G., Valenti P., Seganti L. (1997). Antirotaviral activity of milk proteins: lactoferrin prevents rotavirus infection in the enterocyte-like cell line HT-29. *Med. Microbiol. Immunol.*, Vol. 186, pp. 83-91

Szwajkowska M., Wolanciuk A., Barłowska J., Król J., Litwińczuk Z. (2011) Bovine milk proteins as the source of bioactive peptides influencing the consumers' immune system - a review. *Animal Science Papers and Reports*, Vol. 29, No. 4, pp. 269-280

Tsuji, S, Hirata, Y, Mukai, F, Ohtagaki, S (1990). Comparison of lactoferrin content in colostrum between different cattle breeds. *J. Dairy Sci.*, Vol. 73, pp. 125-128

Turner, S.A., Williamson, J.H., Roche, J.R., Kolver, E.S. (2003). Diet and genotype affect milk lactoferrin concentrations in late lactation. *Proceedings of the New Zealand Society of Animal Production*, Vol. 63, pp. 87-90

Turner, S.A., Thomson, N.A., Auldist, M.J. (2007). Variation of lactoferrin and lactoperoxidase in bovine milk and the impact of level of pasture intake. New *Zealand Journal of Agricultural Research*, Vol. 50, pp. 33-40

Urech, E., Z. Puha, and M. Schallibaum (1999). Changes in milk protein fraction as affected by subclinical mastitis. *J. Dairy Sci.*, Vol. 82, pp. 2402-2411

Van der Strate, B.W.A., Beljaars, L., Molema, G., Harmsen, M.C., Meijer, D.K.F. (2001). Antiviral activities of lactoferrin. *Antiviral Res.*, Vol. 52, pp. 225–239

Wakabayashi, H., Yamauchi, K., Takase, M. (2006). Lactoferrin research, technology and applications. *Inter. Dairy J.*, Vol. 16, pp. 1241-1251

Wedholm, A., Hallén, E., Larsen, L.B., Lindmark-Månsson, H., Karlsson, A.H., Allmere, T. (2006). Comparison of milk protein composition in a Swedish and a Danish dairy herd using reversed phase HPLC. *Acta Agriculturae Scand Section A*, Vol. 56, 8-15

Wheeler, T.T., Hodgkinson, A.J. (2007). Immune components of colostrums and milk - a historical perspective. *Journal Mammary Gland Biol. Neoplasia*, Vol. 12, 237-247

Wielgosz-Groth, Z., Sobczuk-Szul, M., Wroński, M., Rzemieniewski, A. (2009). Season and production level on composition of proteins in milk of Polish Holstein-Friesian. *Biul. Nauk.*, Vol. 30, pp. 135-139

Zimecki, M., Artym, J. (2005). Therapeutic properties of proteins and peptides from colostrum and milk. *Postepy Hig. Med. Dosw.*, Vol. 59, pp. 309-323

8

The New About Congenital Antimicrobial Defense of Some Epithelial Tissues – Vaginal Mucosa and Hair

Arzumanian Vera[1], Malbakhova Ekaterina[2] and Vartanova Nune[1]
[1]*Mechnikov Research Institute for Vaccines and Sera, Moscow*
[2]*Research Center for Obstetrics, Gynecology and Perinathology, Moscow*
Russia

1. Introduction

Antimicrobial peptides (AMP) are a family of more than 500 different substances which protect mucosal and dry epithelial surfaces of all multicellular organisms (Bals, 2000; Zasloff, 2002). They are widely dispersed in nature and active against broad spectrum of Gram-positive and Gram-negative bacteria, yeasts, fungi and enveloped viruses, therefore they called "natural antibiotics". Most of AMP are the cationic peptides very diversed with structure but demonstrate an affinity for the negatively charged phospholipids which are present on the outer surfaces of the cytoplasmic membranes of many microbial species. As far as it is difficult for a microbe to change the phospholipid organization of its membrane, resistance to the AMP occurs at levels that are much lower than those observed for conventional antibiotics. Besides the direct antimicrobial function AMP are the inflammation mediators participated in such different processes as proliferation, immune induction, wound healing, cytokines release, chemotaxis, protease-antiprotease balance, redox homeostasis. Different human epithelial locuses are investigated for availability of AMP - respiratory tract, oral cavity, skin, colon, vaginal tract. However some "blank spots" occur until now, for example, to what extent the role of AMP in the defense of vaginal tract is important, and do the AMP take part in the defense of hair?

2. Antimicrobial peptides as factor of local immunity in vulvovaginal candidosis

Vulvovaginal candidosis (**VVC**) affected women of reproductive age, at that acute VVC , which strike up to 75% of women , and chronic recurrent VVC (up to 20%) are distinguished (Fidel Jr, 2007). Factors promoted the VVC development reputed the following: availability of current diseases (infectional, endocrinopathic, autoimmune etc.); mechanic traums; chemical factors - use of corticosteroids, cytostatics, antibiotics, oral contraceptives etc.; pregnancy.

It is known that in systemic candidosis a main protective role belongs to cellular immunity, namely to polymorphonuclear leucocytes; in mucosal candidosis, except VVC, most significant are Th1-cells, circulated in the blood and local (Fidel Jr, 2007). In VVC most important are reputed the factors of local immunity, among which listed phagocytes and epithelial cells (Nomanbhoy et al, 2002), antibodies (Barousse et al, 2004), cytokines and interferon (Shabashova et al, 2006), concentration of lactate and pH, as well as antimicrobial peptides (AMP) secreted by different leucocytes and epithelial cells (Valore et al, 2002, 2006).

In connection with vaginal protection the following peptides are listed: lactoferrin, calprotectin, secretory leukoprotease inhibitor , cathelicidines, lysozyme and defensins.

In **table 1** features of these substances are summarized: some their chemical properties, localization in human organism, antimicrobial activity spectrum, mechanism of action, and minimal inhibiting concentrations against the *Candida* spp., as well as their concentrations in vaginal secretions of healthy women and in patients with VVC.

Origin of VVC is caused by the decrease of protective function of vaginal epithelium, with combined action of different AMP as the part of the protection. Summarizing the data of table 1 one can conclude that the levels of some AMP somewhat higher in patients with VVC than in healthy women, however researchers till now did not demonstrate exact results, which phase of disease was studied - acute or remission? It is logically to propose that the beginning of acute phase of VVC must be caused by the sudden fall of AMP level, whereas the further development of the disease most likely accompanied with gradual increase of AMP level.

Compare of AMP concentrations, which really can suppress *C. albicans* growth, with AMP levels occurred in vaginal secretions may give the information about some peptide substance. Apparently the most significant may be calprotectin, defensin HNP-1 and in some extent lysozyme. Certain of the AMP known to have a synergistic action, for example, lactoferrin and calprotectin can intensify the effect of each other during the growth inhibition of *C. albicans* (Okutomi et al, 1998).

As far as all of AMP are produced by immune cells it is obviously that decrease of their level must be the result of: 1) decrease of the compounds synthesis in the cells; 2) reduction of quantity of the cells; 3) change the structure of AMP resulting to the loss of activity. Anyway the local cell immunity is primary towards the AMP. Deficit of such effective tools of the first defense line, like AMP, the human organism should compensate by alternative mechanisms of resistance to fungal microflora: activation of phagocytal function and increase of specific immunoglobulines synthesis. Recently was found out that monoclonal antibodies Mab C7 obtained by *C. albicans* mannoprotein not only suppressed the adhesion of yeast cells to different surfaces, but had the direct candidacidal activity (Omaetxebarria et al, 2005). Earlier unknown mechanism of AMP induction in vagina was discovered: cathelicidine hCAP-18, contained at high concentrations, but in inert form in semen plasma, falling in vagina with low pH transformed in its active form (Sorensen et al, 2003).

3. Role of AMP compare to other factors of local immunity in women with vulvovaginal candidosis

Reasons of frequent appearance of VVC during pregnancy and increase of its recurrent form still are not clear. Among provoking factors of VVC local immunity and hormones disturbances are usually listed. However it is still a question which parameter of local immunity is most important and which factors are causal in the process of symbiotic to pathogenic flora and acute to chronic VVC transformation.

Antimicrobial peptide	Chemical features	Cells-producers; locuses	Known activity spectrum	Mechanism of action on *Candida* cells	Concentration provide *in vitro* activity against *C. albicans*	Concentration in vaginal fluid, μg/ml		Reference
						In healthy women	In patients with VVC	
	1	2	3	4	5	6	7	
Lactoferrin	Ferrum containing cationic glycoprotein, 76 – 80 kDa	Neutrophils (one of main peptides); secrets of endocrine glands, mucous membranes and human milk	Protozoa, fungi, bacteria, viruses	Exhaustion of ferric ions in medium; disruption of cytoplasmic membrane integrity; leakage of intracellular reserves of Ca(2+)	20 μg/ml	0.9 ± 0.2 μg/ml	no data	**2,3,4:** Salmon et al, 1997; Samaranayake 2001; Viejo-Diaz, 2004, van der Kraan, 2005 Lupetti et al, 2004 **5:** Bellamy et al, 1993 **6:** Valore et al , 2006
Calprotectine (=leucocyte protein L1)	Calcium and zinc – binding protein; 36,5 kDa, consist from 3 subunits with mol. mass 12,5 kDa	Form about 60% of protein fraction of neutrophils cytosol ; monocytes, macrophages of reactive tissues, squamous epithelium; blood plasma	Fungi and bacteria	Consumption of zinc from medium (competition in zinc ions)	4-32 μg/ml	5 -14 μg/ml	7 - 15 μg/ml	**2,3,5**: Brandtzaeg et al, 1995, **4:** Loomans et al,1998, Lulloff SJ, 2004 **6,7:** Valore et al, 2006
Lysozyme	Enzyme muramidase, 14,5 kDa	Monocytes, polymorphonuclear neutrophil; mucous membranes	Fungi and bacteria	Presumably damage of cell wall and cytoplasmic membrane	10 – 30 μg/ml	0.4 – 3 μg/ml	1.8 – 4.8 μg/ml	**2,3,4,5:** Ibrahim et al, 2001; Samaranayake et al, 1997 -2001 Marquis et al, 1982, 1993 **6,7:** Valore et al, 2006
secretory leukoprotease inhibitor (SLI)	Cationic nonglycosilated high-based acid-stable protein with high percent of cysteine, 11.7 kDa	Neutrophils, macrophages; c secrets of exocrine glands and mucous membranes	Fungi, bacteria, viruses	Presumably damage of cytoplasmic membrane; inhibition of microbial proteases	23 – 175 μg/ml	0,05-0,20 μg/ml	0,04-0,18 μg/ml	**1,2:** Doumas et al, 2005 Tomee et al, 1998; **5**: Tomee et al, 1997; **6,7:** Valore et al., 2006

Table 1. Continued

Antimicrobial peptide	Chemical features	Cells-producers; locuses	Known activity spectrum	Mechanism of action on *Candida* cells	Concentration provide *in vitro* activity against *C. albicans*	Concentration in vaginal fluid, µg/ml		Reference
						In healthy women	In patients with VVC	
	1	2	3	4	5	6	7	
cathelicidines	Amphypatic, cationic, alpha-spiraled peptides	Squamous epithelium and neutrophils of respiratory, gastrointestinal and urogenital tracts	Yeasts, bacteria	Destruction of cytoplasmic membrane				**1:** Moon et al. ,2006 **2,5:** Frohm et al, 1999 **3,4,5**: den Hertog et al, 2005 **6:** Valore et al., 2002
hCAP18	19 kDa				33 µg/ml	no data	no data	
LL-37	4,5 kDa				9 – 90 µg/ml	0,065 – 1,0	no data	
Defensins: **Beta:** HBD-1	Cationic nonglycosilated peptides with 6 cysteine residues, which form 3 intramolecular disulphide bridges of three-chain structure 3,5 – 4,5 kDa	Limphocytes; phagocytes; epitheliocytes of respiratory, gastrointestinal and urogenital tracts	Yeasts, bacteria and viruses	Membranes destruction	> 40 µg/ml	0,015- 0,035 µg/ml	0,010- 0,035 µg/ml	**2:** Taggart et al, 2003 **1:** Bals , 2000; Schneider et al, 2005; **6,7:** Valore et al, 2006 **5:** Feng et al, 2005; **5**: Cullor et al, 1990
HBD-2					4 µg/ml	0,005- 0,040	0,035-0,100	
Alpha: HNP-1-3					10 (only for HNP 1)	1,5 – 5	2.5 – 12.5	
HNP-5					50 µg/ml	0,007-0,025	0,005-0,025	

Table 1. Antimicrobial peptides of vaginal mucosa – biochemical properties.

From a quantity of factors of local immunity in VVC we can mark out few factors with direct antifungal activity: phagocytes (Nomanbhoy et al., 2002), antibodies (Barousse et al., 2004) and antimicrobial peptides (AMP) (Valore et al., 2002, 2006).

Phagocytosis of *C. albicans* is performed by neutrophyles and macrophages, that demonstrated by *in vitro* studies (Vonk et al., 2002). But there are no data on the action of these cells *in vivo*.

Among immunoglobulins most presented in vaginal secretions are IgG and secretory IgA. (Mestecky et al., 2005). But data concerning a relationship between immunoglobulins level and yeast population size are few a number and discrepant (de Carvalho et al., 2003; Kurnatowska et al., 2002; Mestecky et al., 2005).

However phagocytosis and immunoglobulins are common and relatively well investigated parameters of immunity whereas AMP as medical research subjects are insufficiently known.

In context of vaginal epithelium protection lactoferrin, calprotectin, lysozyme, leukoprotease secretory inhibitor, cathelicidins and defensins are mentioned.

The aim of the study was to determine a relationship between yeast population, severity of the disease and some parameters of local immunity in pregnant women with VVC.

3.1 Materials and methods

The study included 45 pregnant women aged 22 to 35 years old, conventionally divided into two main groups by the presence or absence of a chronic process. Each group was divided into two subgroups depending on the phase of the process at the time of the survey. Group has a name: RVVCE - women with recurrent VVC with exacerbation (n=9); RVVCR - women with recurrent VVC in remission (n=10); AVVC - women with primary acute VVC (n=13) ; ASYM - women with the minimum of typical symptoms of VVC at the time of the survey (n=13).

To assess the severity of the VVC the combination of the following symptoms was used: itching, burning sensation, the nature and amount of vaginal discharge, pain during urination, dyspareunia, dermatitis of perianal area, swelling, redness and erosive lesions of the vaginal walls. Each symptom was evaluated on a scale from 0 to 2, where a 0 means the absence of symptoms, 1 - moderate intensity, and for 2 - a vivid manifestation of a symptom.

Material for inoculation of medium was taken from the vaginal fornix with a sterile applicator, which was placed in a test tube with sterile transport medium «Amies» and delivered to the laboratory. Inoculation was carried out by the standard method on glucose-peptone-yeast growth medium containing the antibiotic.

Material for microscopy was collected with a sterile spatula from the vaginal fornix in a sterile container. A small amount (about 5 μl) sample was placed on a glass slide, pressed the coverslip so as to remove excess and to observe a monolayer of cells. Microscopy of samples were carried out at a total magnification of x1750. Efficiency of phagocytosis was evaluated as the ratio between the number of yeast cells, localized within the phagocytes, and the total number of yeast cells in the field of microscope.

Collection of vaginal secretions was carried out by the following method: put a tampon in the vagina «OB Pro comfort » of 10 minutes, after which the tampon was transferred to a plastic column 15x75 mm and eluted with 7 ml of distilled water using vacuum pump. The eluate was filtered through a bacterial membrane filter Millipore with a pore diameter of 0,22μ. The filtrate was freeze-dried and diluted in sterile distilled water so as to obtain a 10-fold concentrated relative to the initial filtrate (VF).

Immunoglobulins was assessed by dot-blot analysis . An antigen used in the analysis was obtained by selective extraction of surface proteins of cells *Candida albicans* (Arzumanian et al, 2000). Antigen at a concentration of 1mg/ml volume of 0.8 μl was applied onto a nitrocellulose membrane ("Whatman") with pore size 0.2 μ, dried and washed with 0,1% Tween-20 in saline, then with distilled water. Then incubated the membrane in the wells with 80 μl of blocking solution - PBS with 10% bovine serum - within 30 minutes, after which the well was added 20 μl of VF and incubated overnight. After incubation, liquid was removed, well washed, as described above, added 100 μl of conjugate solution in blocking solution. In determining the sIgA as a conjugate used mouse monoclonal antibodies to human sIgA conjugated with peroxidase at a dilution of 1:250; in determining IgG - mouse monoclonal antibody to human IgG conjugated with peroxidase at a dilution of 1:5000. After incubation for 1 hour the liquid was removed, the well with the membrane was washed as described above, added 100 μl of a solution containing hydrogen peroxide, TMB and precipitating agent, developed within 15-20 minutes. Results of the analysis were evaluated by 6-point scale from 0 to 5 depending on the intensity of the color spot.

Antifungal activity was determined as follows: two days old test culture of *Candida albicans* (№ 927, collection of Mechnikov Institute) were incubated with aliquots of VF with a temperature of 32^0 C and the ratio 20 μl of VF / 5 μl yeast suspension with concentration of cells 10^5 CFU / ml. From this mixture aliquots were inoculated on agar plates immediately after mixing and after 2 h of incubation. The result was expressed as the percentage of cells killed in the process of incubation.

3.2 Results and discussion

One of the aims of the study was to evaluate the local cellular antifungal immunity based on data fungal populations microscopy. We studied different of morphotypes data of fungal cells directly into the smears of women with VVC: single yeast cells (blastospores), pseudomycelium, true mycelium and blastospores enclosed in phagocytes. Summary data of microscopy of samples, culture tests and features of the VVC course in different groups of patients are presented in **table 1**.

The sum of symptoms characterizing the severity of theVVC for the group RVVCE varied from 5 to 13, whereas for the group RVVCR - from 1 to 9, for the group AVVC - from 2 to 8, for the group ASYM - from 1 to 6. Table 1 shows the median of these values. Obviously that this index correspond to the nature of the VVC. At the same time attention is drawn to the presence of symptoms in the group ASYM. The main contribution in this category belongs to the symptom of "the number and nature of vaginal discharge", namely, all patients in this subgroup had "milky" smears and in 38,5% of them there were abundant. Hereinafter we consider this category of patients as close to normal.

According to the established norm the increased contamination by yeast cells was observed in the two categories of patients - in AVVC and to a greater extent in RVVCE. However, despite the low seeding in the group RVVCR occurred the highest efficiency of phagocytosis (80%). This index was significantly lower in the acute phase, and in patients of ASYM group it was the lowest. Efficiency of phagocytosis in the subgroups varied as follows: in group ASYM from 0 to 50%, in AVVC - 20 ÷ 100%, in RVVCR - 50 ÷ 90%, and in RVVCE - 20 ÷ 100%. The frequency of detection of psevdomycelial cells was maximal in RVVCE and minimal in ASYM. Mycelium was detected much less frequently and only in subgroups with evidence of inflammation. Likewise divided within the groups frequency of positive cultural tests. Namely the presence of viable yeast cells in vaginal secretions in the group RVVCE was found in 100% of patients and varied in the range 1800 - 200,000 CFU/ml.

Compare all the above parameters with the sum of symptoms in the groups studied revealed the presence of a high degree of correlation of all indexes, except the efficiency of phagocytosis. In addition, the correlation coefficient between the frequency of positive cultural tests in the groups and medians of maximal number of yeast cells in the field of microscope was 0.921. There was no relationship between the efficiency of phagocytosis and yeast dissemination (***r*** = 0,194), as well as between the efficiency of phagocytosis and the frequencies of positive cultural tests (***r*** = - 0,134).

The next task of the study was to investigate the participation of local humoral immunity, namely immunoglobulins G and A. It is noteworthy that in all groups studied, more or less often the local IgG-antibodies to antigens of *C. albicans* were found. This is understandable, considering that none of the groups had patients who more or less would not be a carrier of opportunistic yeasts. However, in the acute phases of VVC was a significant increase in frequency of detection of IgG, especially in the group AVVC. The levels of these antibodies varied in all groups from 0 to 5, median values were: for RVVCE – 1, RVVCR – 0, AVVC - 2, ASYM - 0. Immunoglobulins of sIgA class did not occur in the patients group RVVCR, but at the exacerbation of a chronic process, these antibodies were detected in 25% of cases, while AVVC - 50% of cases. Interestingly, even in the group ASYM patients met with sIgA-antibodies. Levels of these immunoglobulins were varied as follows: for RVVCE- 0 to 3 (median 0), for RVVCR - 0; for AVVC - 0 to 3 (median 1), for ASYM - 0 to 2 (median 0).

The absolute values of the levels of IgG and sIgA did not correlate with either the abundance of yeast cells in vaginal secretions, or with frequency of positive cultural tests or with each other. Total of all 45 patients only 3 were detected the presence of both antibodies, while they were patients in the acute phase of the VVC. However, the detection rate of IgG-antibodies in the groups correlated with the medians of contamination of vaginal secretions (r = 0,772), whereas the frequency of detection of sIgA-antibodies was not associated with this index (r = 0,143).

The third objective of the study was determination of antifungal activity of antimicrobial peptides presented in vaginal secretions. We assessed the cumulative effect of AMP on the cells of *C. albicans* test culture (see "Materials and Methods"). It is important that the action of VF on yeast cells had a dose-depend character (**figure 1**). The highest activity against yeast cells possessed preparations obtained from vaginal secretions of ASYM group: spread of activity values was 34,4 ÷ 92,5%. In the group of patients with RVVCR this index ranged from 0 ÷ 63,8%; at RVVCE - 0 ÷ 48,4%; at AVVC - 0 ÷ 26,9%. Medians of these values are shown in table 1. Marked inverse correlation between the medians of antifungal activity and medians of following factors took place: the severity of the VVC (r = - 0,811), sowing

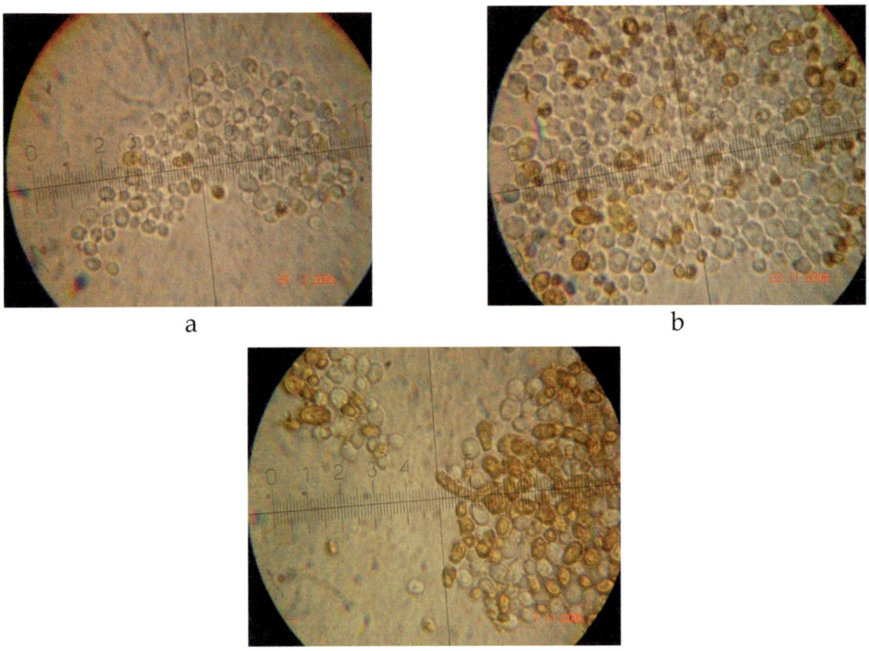

a b c

Fig. 1. Effect of different aliquots of vaginal fluid on the cells of *C. albicans* reference culture (stained by bromocresol purple); a - *C. albicans* incubated with 1 µl of vaginal fluid obtained from woman of ASYM group; b - 20 µl of vaginal fluid; c - 40 µl of vaginal fluid; alive cells – white, dead cells – yellow.

(r = - 0,689), contamination (r = - 0,855), IgG-antibodies (r = - 0,894), sIgA- antibodies (r = - 0,544). No correlation was found between the medians of antifungal activity and efficiency of phagocytosis (r = - 0,117).

Phagocytosis is an important mechanism of protection from opportunistic fungi. Traditionally an evaluation of phagocytosis and intracellular lysis *C. albicans* carry out by neutrophils and macrophages on *in vitro* models of cell cultures (Vonk et al., 2002). Revealed the presence of yeast blastospores not only in the intercellular space of vaginal smears, but within the phagocytes, we decided to estimate the ratio of the number of free and phagocyted yeast cells and use this index as an indicator of the efficiency of phagocytosis in this locus. Similar study with *C. albicans* test culture has previously been carried out (Valore E., 2006), but without analysis of correlation between VVC severity and antimicrobial activity of VF. According to the most recent data most presented in the vaginal secretions is immunoglobulin G (Mestecky et al, 2005). Attempts to establish the relationship between the severity of VVC and levels of immunoglobulins did not give unambiguous results (de Carvalho et al, 2003; Kurnatowska , Magnowski, 2002; Mestecky et al, 2005). Nevertheless, it is useful to compare the level and frequency of detection of secretory immunoglobulin in chronic and acute VVC with other parameters of local immunity.

Summarizing the data of correlation analysis of the studied parameters, we can conclude that for the category of patients, which is close to normal (ASYM), the most important part of local immunity is the total antifungal activity of soluble components of the vaginal secretions (**table**

1, 2). It is clear that at high antifungal activity of AMP the functioning of phagocytes and immunoglobulins is not necessary. Primary acute process (AVVC) caused by low antifungal activity, but it is characterized by significant activation of phagocytes and immunoglobulins. Chronization of acute process (RVVCE) also accompanied by a low antifungal activity and increased phagocytosis, but less frequent detection of immunoglobulins than in AVVC. Based on data of microscopy and cultural tests can be concluded that the acute phase of chronic VVC is always accompanied by the greatest severity of symptoms, an abundance of blastospores and often filamentous elements. These data are consistent with the results of studies conducted on large samples of patients with VVC (2861 pers.), where it is shown that the presence of hyphae of the mycelium is a marker of disease severity (Demirezen , Beksac , 2004). In the remission stage of chronic process (RVVCR) was noted the relatively high antifungal activity, a high level of efficiency of phagocytosis and the lowest level of immunoglobulins.

Group of patients	**n**	**Severity of VVC, (median)**	**Maximal amount of yeast cells in one field (median)**	**Frequency of positive cultural tests, %**	**Efficiency of phagocytosis, % of phagocyted cells in smears (median)**	**IgG, frequency of detection, %**	**sIgA, frequency of detection, %**	**Antifungal activity of VF, % of killed yeast cells (median)**
RVVCE	9	8	43	100	40	50	25	13,6
RVVCR	10	4	19	0	80	42,9	0	43,2
AVVC	13	5	30	30,8	50	62,5	50	12,5
ASYM	13	3	8	0	15	16,7	33,3	54,1
*r**			*0,968*	*0,982*	*0,057*	*0,588*	*0,062*	*- 0,811*

* - correlation coefficient between certain index and severity of VVC

Table 2. Correlation between severity of VVC and data of microscopy of vaginal smears in different groups of patients

Thus, the direct antifungal activity of the soluble fraction of vaginal secretions, apparently, is the first condition, limiting the increase of fungal population. Reasons for the decrease of this activity may be, on the one hand, reducing the number of AMP, produced by neutrophils and epithelial cells, on the other - change the structure of AMP, leading to the loss of their activity. Human organism is forced to fill up the deficit of AMP through alternative mechanisms - activation of the local phagocytic function and increase the synthesis of specific secretory immunoglobulins. All these conclusions we summarized on the scheme (**table 3**).

GROUPS	**PHAGOCYTOSIS**	**IMMUNOGLOBULIN G**	**ANTIMICROBIAL PEPTIDES**
RVVCE	△	△	▼
RVVCR	△	△	△
AVVC	△	△	▼
ASYM	▼	▼	△

Table 3. Interrelation of investigated parameters of local immunity depending on form of VVC in pregnant women. ▼relatively low index; △ relatively high index

4. Bacterial vaginosis and the local antimicrobial activity in women

The colonization of female genital tract by alien microorganisms is determined by several factors, among which are normally listed the competition with resident microflora, exfoliation of the squamous epithelium, acidic and lactate rich medium. More than 20 years ago the antimicrobial peptides produced by polymorphonuclear neutrophils and epithelial cells were mentioning as a barrier in this locus (Cohen , 1984). Details concerning the role of each of the AMP, as well as contribute to the overall antimicrobial activity in vulvovaginal candidiasis mentioned above (**table 1**). In recent years a detailed study of the spectrum of these peptides in bacterial vaginosis (**BV**) and in the norm was carried out (Valore et al, 2002, 2006). In bacterial vaginosis was showed a reduction in the concentrations of AMP as compared with healthy women. The purpose of this study was to compare antimicrobial activity of vaginal discharge with the severity of bacterial vaginosis, as well as pH and some microbiological parameters.

4.1 Materials and methods

The study was conducted in 53 pregnant women, 44 of them - with bacterial vaginosis, and 9 - without it.

The severity of BV was evaluated by total score used a combination of the following symptoms: the nature and amount of bleeding, specific "fish" odor, pain when urinating, itching, burning, dyspareunia, dermatitis of perianal area, swelling, redness of the vaginal walls. Each symptom was evaluated on a scale from 0 to 2, where a 0 means the absence of symptoms, 1 - moderate intensity, and for 2 – vivid manifestation of a symptom. In accordance with the values of obtained scores all women was divided into subgroups: 1 - group consisted of women without signs of BV, then the subgroup is referred to as the "norm"; 2 - with mild BV (scores of 3 to 8 points); 3 - with moderate BV (9 to 12 points), 4 - with severe BV (13 to 18 points).

Material for microscopy taken with a sterile spatula from the vaginal fornix and placed on a glass slide. After staining the smears by Gram number of lactobacilli, bacteroides, gram-positive cocci and gardnerellas were estimated. Abundance expressed in points on four-point scale, where 0 - absence of this group of microorganisms, 1 - single cells in the visual field, 2 - moderate number of cells, 3 - an abundance of cells. Lactobacilli and bacteroides were combined into a group of "obligatory microflora", and cocci and gardnerellas - a group of " facultative microflora".

Collection of vaginal secretions (VF)was carried out as mentioned above (see chapter 3.1). **Total antimicrobial activity** was determined as follows: cells of 4 days old test culture of *Escherichia coli* (№ 23, a collection of Mechnikov Institute) were incubated with aliquots of VF at a temperature of 32^0 C and a ratio of 40 μl of VF/ 10 μl of bacterial suspension density of 10^4 CFU / ml. Aliquotes of the mixture were seeded on Petri dishes with agar medium immediately after mixing and after 2 h of incubation. The result was expressed as the percentage of cells killed in the process of incubation.

Separation of proteins in samples was carried out by SDS-PAGE in 5-20% gradient polyacrylamide gel (Lambin et al., 1976). Samples were prepared in nondenaturating conditions by mixing 1 volume of sample with 2 volumes of buffer and causing the sample to 40 μl per track. Staining was performed using Coomassie R-250. As molecular weight standards using a mixture of LMW ("Amersham Pharmacia").

Glucose was determined using VO meter "Accu-check Active" "Roche Diagnostics GmbH" (Germany). The result was expressed as mmol / l.

The pH was evaluated using indicator strips, intended for measuring range 3.8 - 6.0 ("LLC Lach-Ner", Czech Republic).

4.2 Results and discussion

Lactic-acid bacteria and bacteroids form a large part of the normal vaginal microflora (up to 10^8 CFU / ml), and are obligate microorganisms inhabiting this ecosystem. At the same time, the Gram-positive cocci, which are represented by entero-, strepto-, and staphylococci, are minor components of the vaginal microbiota (up to 10^3 CFU / ml), which can be distinguished by microscopic study. Even more rare and less desirable are the bacteria of the genus *Gardnerella*. Namely with increase in their population, and population of aerobic cocci is often linked BV. Therefore, the microorganisms identified by smear microscopy, we divided into two main groups - obligate (lactobacilli and bacteroides) and facultative (cocci and gardnerellas). Based on data of microscopy every smear was expessed in points. The results of the assessment in the form of arithmetic means are given in **table 4**. From the table it is followed that the highest scores on the obligatory microflora was observed in the absence of symptoms of BV. In part, index of facultative microorganisms were the lowest in the group "norm", and largest, respectively, in the groups with BV.

In general, in all groups the abundance of obligate microflora was in high inverse correlation relationship with abundance of facultative one (Pearson's correlation coefficient r = - 0,968). At that, the abundance of facultative microflora was directly correlated with the

Group of patients	n	**Severity of BV (medians)**	**Abundance of microflora, (medians)**		**Glucose (mmol/l) (medians)**	**pH (medians)**	**Antimicrobial activity of VF, % of killed *E. coli* cells (medians)**
			Obligatory	**Facultative**			
Norm (absence of BV symptoms)	9	0	6	1	1,9	4,4	78,2
Mild BV	12	8	3	5	1,7	5,0	36,4
Moderate BV	14	11	3	4	1,8	5,0	44,0
Severe BV	18	14	3	5	1,65	5,2	22,4
*r**			**-0,914**	**0,885**	**- 0,830**	**0,975**	**- 0,944**

*r**- correlation coefficient between certain index and severity of BV

Table 4. Some biochemical and microbiological parameters of vaginal secretions in patients with BV and healthy women

severity of BV (*r* = 0,885), but the abundance of obligate microflora - *vice versa* (*r* = -0,914). Low pH of the vagina is reputed as the result of metabolism of lactic acid bacteria. Our data show that the lowest pH value had indeed taken place if there were not symptoms of BV and in the presence of a large number of lactobacilli. At the same time the biggest pH values were corresponded to severe form of BV. In other words, pH values were correlated not only with the severity of BV (r = 0,975), but with the nature of the microflora of this locus: between pH and obligate microflora relationship was characterized by ***r*** = - 0,962, and the pH and facultative microflora of ***r*** = 0,966.

As the concentration of glucose in vaginal secretions may affect the abundance of flora, we assessed the level of glucose in the studied groups of patients. It turned out that the greatest concentrations of glucose were detected in the absence of symptoms of BV, and the smallest - in the group with the severe form of the disease. There has been an inverse relationship of the index with the severity of BV (***r*** = - 0,830), a direct correlation with the abundance of obligate microflora (***r*** = 0,827), but the reverse - with abundance of facultative one (***r*** = - 0,933). Probably, the low level of glucose in the locus leaded to more intense amino acids consumption, which in part resulted in alkalization of the medium (pH increase) due to release of amines.

In determining the total antimicrobial activity resulting from the cumulative action of antimicrobial peptides, the highest index was appeared in a group of women without symptoms of BV, as well as the rise of symptoms was accompanied by decrease of antimicrobial activity (***r*** = - 0,944). Obviously, it is just reducing immune defense of locus resulted the increase of the opportunistic microflora population and the depletion of the normal microflora: the correlation between antimicrobial activity and obligate microflora was ***r*** = 0,926, while between antimicrobial activity and facultative microflora ***r*** = - 0,969. The division of VF proteins in a gradient of polyacrylamide gel demonstrated an association between antimicrobial activity and the presence / intensity of bands corresponded to antimicrobial peptides (**figure 2**). Track number 2 corresponds to the VF, obtained from women with mild BV (severity corresponded to 5 points, antimicrobial activity 100%). On this track, we can distinguish the following polypeptides - calprotectin with mol.mass about 37 kDa, cathelicidine hCAP18 (18 kD), secretory leucoprotease inhibitor – SLI (about 12 kDa), lysozyme (14.5 kDa) and defensin (less than 5 kDa). Track number 1 - the mild BV (severity score was equal to 7 points, antibacterial activity 66,7%) - no calprotectin, distinguishable other proteins, but less intense band. Track number 3 - severe BV (sum of symptoms 14, antimicrobial activity 21,6%) - no calprotectin and subtle defensin. Track number 4 - severe BV (total symptoms -17 points, antimicrobial activity 0%) - no calprotectin and defensin, a weaker band of lysozyme. Reduced concentrations of AMP in the VF of patients with BV have observed previously (Valore et al, 2006), however, the study of this phenomenon on a background of varying severity was carried out for the first time. From these data we can also conclude that the most important AMP in the antibacterial protection of the vagina in these patients are calprotectin, defensin and lysozyme.

Above, based on data from the literature, we compared the concentrations of AMP, which actually can inhibit the growth of *C. albicans*, with the levels of these substances in the vaginal secretions with the purpose of determine, which of these compounds are most important in the protection of this locus. It turned out that the most significant may be calprotectin, defensin HNP-1 and, to some extent, lysozyme. It is known that calprotectin causes depletion of the environment in trace elements, i.e. inhibits the growth of

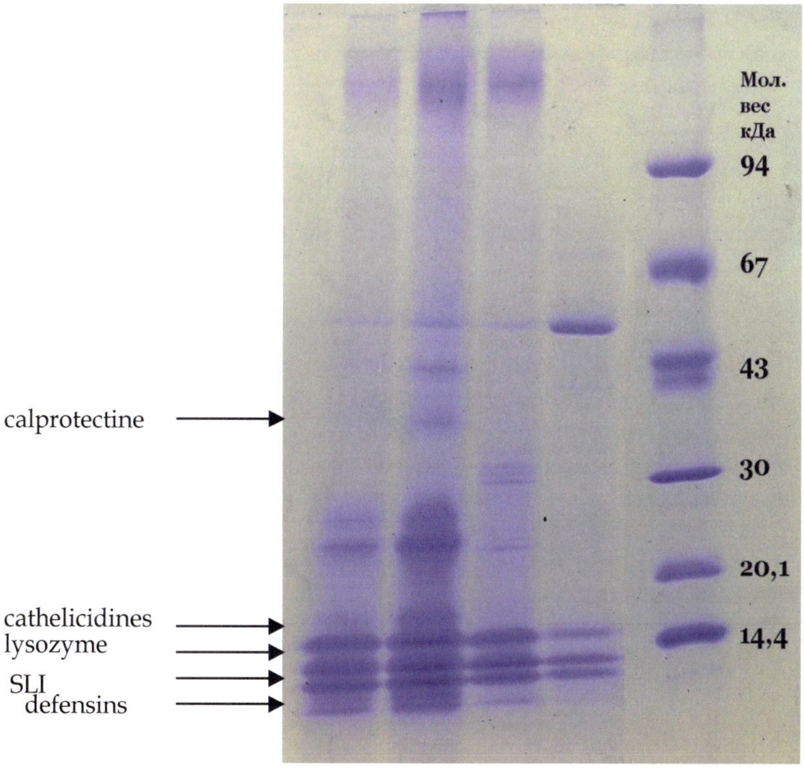

Fig. 2. SDS-PAGE of vaginal fluid samples obtained from patients with BV: track 1 - the mild BV (severity - 7 points, antibacterial activity - 66,7%track); track 2 - mild BV (severity - 5 points, antimicrobial activity - 100%); track 3 - severe BV (severity - 14, antimicrobial activity - 21,6%); track 4 - severe BV (severity -17 points, antimicrobial activity 0%); track 5 – mol.mass markers.

microorganisms (Loomans et al, 1998). Lysozyme destroys the glycoside bonds of polysaccharides of cell walls and damages the cytoplasmic membrane (Ibrahim et al, 2001). Defensin bound to negatively charged cytoplasmic membrane and cause the formation of pores (Schneider et al, 2005). Since these peptides were key in this study, we can conclude that in the mechanism of action they are not specific for microorganisms and act similarly to the pro-and eukaryotes.

Thus, we conclude that in pregnant patients with BV was noted a direct correlation between the severity of the disease and the level of pH of vaginal secretions, and invert correlation with the level of glucose in this locus. At the same time the high degree correlation between the BV severity and microbiological parameters was registered. Important role in the vaginal antimicrobial immune defense belongs to the total activity of AMP : calprotectin, defensins, lysozyme, cathelicidine and secretory leucoprotease inhibitor.

5. Antimicrobial peptides in local defence of skin

Surface epithelium of multicellular organisms is a barrier between the body and environment and works as active immune organ. Skin covered not only by several layers of specialized cells, but antimicrobial substances - fatty acids, chloride ions and antimicrobial peptides. AMP are constitutively synthesized in normal epithelial cells and induced under the influence of several factors - microorganisms, disruption of skin integrity etc.

At last six classes of AMP - dermcidins, cathelicidins, defensins, ribonuclease 7, psoriasin and antileucoprotease - protect the skin from pathogenous and opportunistic microflora, with demonstrating of multiple functions concerned with immune defense.

Biochemical features of defensins, cathelicidins and secretory leucoprotease inhibitor we reviewed above (chapter 2). Below some properties of other peptides are listed.

Dermcidines were created in 2001 and shown to be secreted by merocrine sweat glands (Schittek et al, 2001). This class of peptides have the broad spectrum of action in the large diapason of pH and chloride ions ; molecular mass varied close to 5-8 kDa. They are excreted on the skin surface at 1-10 μg/ml of sweat, at that this concentration is toxic for bacteria and yeasts (Schittek et al, 2001; Flad et al, 2002). In connection with skin have mentioned the following dermcidins: DCD-1,DCD-1L, SSL-46, SSL-45, SSL-29, SSL-25, LEK-45, LEK-44, LEK-43, LEK-42, LEK-41, LEK-26, LEK-24, YDP-42, among which are cationic, anionic and neutral peptides. Interestingly that quantity of the secreted dermcidins depends on the certain body area: zones with intensive sweat excreting (axilla, hands, forehead etc.) had higher levels of dermcidins. Skin produces the peptides with the constant rate and they are stable during 72 hours. However in spite of their stability and broad spectrum of activity, some microorganisms are resistant to dermcidins. Presence of dermcidins is caused the production of specific proteases in *Staphylococcus aureus* and *S. epidermidis* (Lai et al, 2007). Estimation of synthetic dermcidin effect showed the absence of antifungal activity in this peptide (López-García et al, 2006).

Ubiquitous **ribonucleases** (RNKases) play important role in metabolism, angiogenesis, neurotoxicity, and antitumoral activity. Recently was created a new antimicrobial function of ribonucleases (Harder, Schroeder, 2002). The main source of RNKase 7 is keratinocytes. The peptide has molecular mass of 14,5 kDa, and destroys cytoplasmic membrane of different microorganisms even at low temperature (4 °C) and during some minutes (Huang et al, 2007). The lethal dose of RNKase 7 against vancomicin-resistant *Enterococcus faecium* was 30 nM.

Psoriasin (synonym S100A7) was known since the beginning of 90-th years as a peptide participated in inflammatory processes of chemotaxis, oncogenesis, angiogenesis and found in tissues of ear, skin, tongue and amniotic fluid (Madsen et al, 1991). It is the anionic peptide with molecular mass 11,4 kDa. High concentrations of the peptide contain the keratynocytes of patients with psoriasis, for which the infiltration of neutrophyls is typical. Antimicrobial activity of psoriasin was fixed in 2005 when the direct bactericidal action on *E. coli* was demonstrated by neutralization of the peptide by monoclonal antibodies (Glaser et al, 2005). The increase of psoriasin expression was found in presence of bacteria and proinflammatory cytokines , and decrease of the peptide activity took place in presence on zinc ions.

Apparently much in known about skin AMP, but for today no information exists concerning the native immune defense of skin appendages - hair and nails.

5.1 The congenital immune defence of hair

Skin appendages - nails and hair - are known to be affect by microorganisms with keratinase activity - dermatophytes, yeasts , rare staphylococci and propionic bacteria (Mikx, de Jong , 1987). However, not all people of the population are susceptible to these microbial agents, therefore apparently some defense of these loci must exist. Role of fatty acids and chloride ions in nail/hair protection is improbable, and so it remains to propose that AMP may realize the defense function.

Estimation of the presence and activity of AMP in normal hair keratinocytes was the aim of further study.

5.2 Materials and methods

Hair samples were obtained from 5 women 7 to 50 years old, which did not use the chemical coloration and hairdressing. Samples of fresh washed hair cut at a range of 10 cm from background. 360 mg of hair cut of scissors, then grinded by mortar and pestle up to homogeneity, adding drop by drop 8 ml 0,1 M solution of citric acid in 50% water ethanol [Harder, 2001]. The obtained cell-free homogenate was centrifuged during 7 min at the rate 10000 g, supernatant (about 3 ml) was dried at 27^{0}C in Petri dish. Dry extract was washed off by 0,5 ml potassium phosphate buffer ph 8,2 and centrifuged again; the final pH of solution was 6,5 (further we identify this solution as **E**). Control to E was the initial solution of citric acid processed by the same manner (**C**).

As a **test yeast culture** was used strain *Candida albicans* (№ 927 from collection of Mechnikov Institute) grown in glucose-pepton-yeast agar during 2 days at 27^{0}C.

For the study of **antimicrobial activity** we used 3 methods - determination of alive yeast cells by microscopy and inoculation, and estimation of growth inhibition zones. For microscopy 1 loop of culture was suspended in 1 ml of potassium phosphate buffer pH 4,6 , then 100 μl of the suspension was added to 80 μl solution E. After incubation during 2 hours at 32^{0}C, 800 μl of 2 mM bromocrezol purple solution in the same buffer was added (Kurzweilova H, Sigler K., 1993). Mixture was incubated during 1 hour at 32 ^{0} C, centrifuged at the rate 5000 g during 5 min, pellet microscoped at magnification x 1750 and photographed by camera Sony DSC-W7 (**figure 3**).

For the estimation of antimicrobial activity by inoculation method 1 loop of test-culture was suspended in 1 ml of sterile distilled water, and 10 μl of the suspension was added to 3 ml potassium phosphate buffer pH 5,5 (final cells concentration was approximately 10^3 CFU/ml). Then 20 μl of new suspension was added to 80 μl of solution E, mixed and inoculated the Petri dishes with agar immediately after mixing and after the certain time (**table 5**).

Dishes were incubated during 2 days at 27^{0}C, after that number of grown yeast colonies was calculated.

The zones of growth inhibition were estimated in the following way: warm molten agar was inoculated by test-culture cells (approximately 50 CFU/ml), filled in two Petri dishes and leave alone for congelation. After the agar surface drying, 20 μl of solution E and control solution C were put in the center of dishes (**figure 4**). Dishes were incubated during 4 days at 27^{0}C.

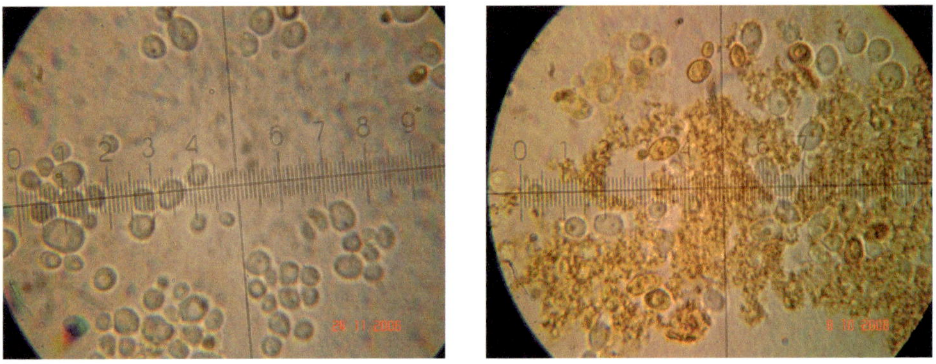

Fig. 3. Cells of *C. albicans* test-culture processed by extract of human hair : white cells – alive, dead and broken cells - yellow, dye – bromocrezol purple.

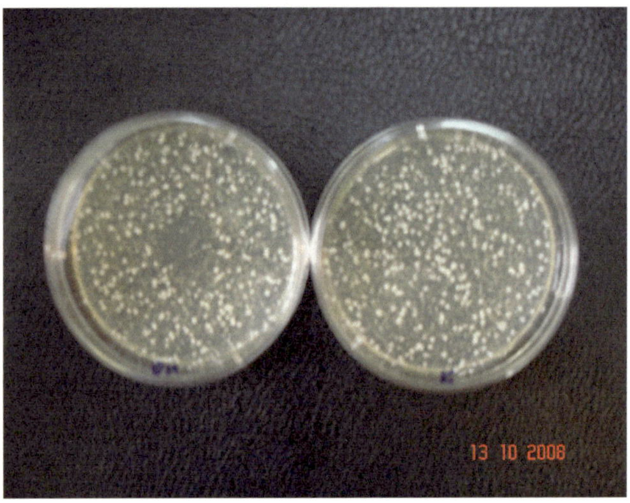

Fig. 4. Antimicrobial activity of hair extract against *C. albicans* test-culture : left dish is experimental, right one is control.

Time of exposition, hours	**% of killed cells, average means from 5 samples (5 experiments)**
0	0
0,5	20,7 ± 8,3
1	32,6 ± 7,3
3	46,7 ± 5,3

Table 5. Estimation of dieing rate of test-culture *C. albicans* cells under the incubation with hair extract.

Separation of proteins was carried out as written in chapter 4.1. Gels were stained by silver nitrate (**figure 5**). As the molecular mass standards LMW protein mixture was used ("Amersham-Pharmacia").

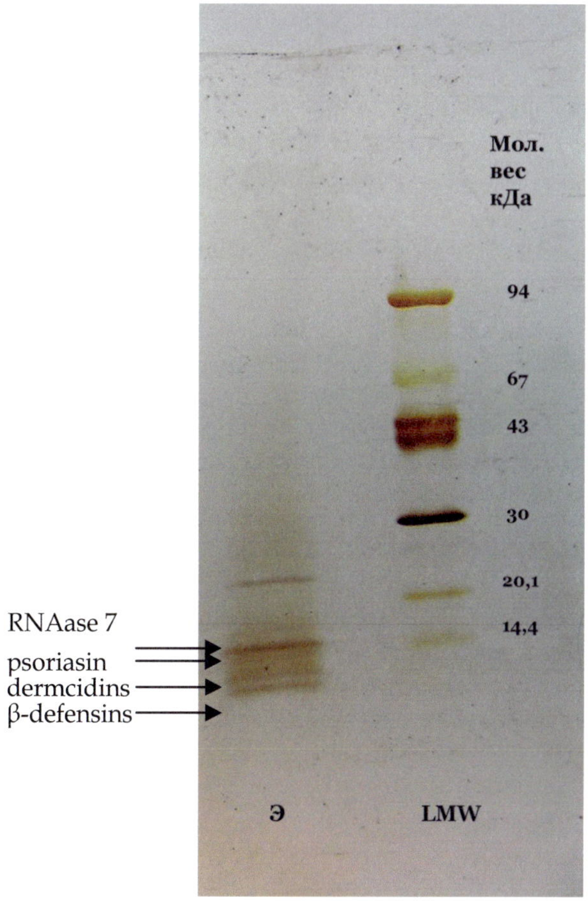

Fig. 5. Separation of proteins in gradient of SDS-PAGE: left track - hair extract, right track - low molecular weight markers.

5.3 Results and discussion

For the extraction of AMP from hair cells we used the solution, which was applied with the same purpose to the skin scales (Harder, Schroeder , 2001). The resulted extract even after centrifugation contained keratin - high molecular protein, the main component of keratinocytes. After neutralization of the extract by basic buffer the second centrifugation was necessary for the removing of keratin residue, which precipitated after the procedure.

Microscopy of test-culture cells processed by the solution E showed that alive cells were almost absent, whereas in the control sample all cells were alive (**figure 3**).

One can see large amount of cell debris, that is evidence of destruction of cell membranes typical for AMP, and disruption of cell walls. Such results did not enable to calculate the percent of killed cells.

The treatment of test-culture by hair extract and the following inoculation of agar showed, that during 3 hours a suspension with initial concentration about 200-300 alive cells in 20 μl, lost about 52% of cells (**table 5**). At that cells, treated with control solution, completely saved alive.

Use of zone inhibition method demonstrated the presence of empty areola in the center of dish (diameter about 15-20 mm) with the absence of such one in the control (**figure 4**).

Thus we have proved the availability of antimicrobial activity in hair by 3 different ways. As the extraction method means the removal of peptides, it would be logically to expect a relationship between their presence and antimicrobial activity. Thereupon we separated the solution E by gradient SDS-PAGE (**figure 5**). It turned out that the extract really contained the low molecular proteins. Most distinct band corresponded to molecular mass about 14,4 kDa. Probably it is RNAase 7, which is usually expressed in skin keratinocytes (Harder, Schroeder , 2002). The second intensity had the the band with moleculad mass about 23 kDa, but among known skin AMP such peptides are absent. Multiple bands located in the diapason from 12 kDa to 3 kDa, they may correspond to psoriasin (11,4 kDa), dermcidins (5-8 kDa) and β-defensins (3,5-4,5 kDa).

Filtration of the extract through the membrane filter with pore diameter 3 kDa showed the absence of antimicrobial activity in the obtained solution.

From the data one can conclude that normal hear extracts displayed the antimicrobial action, which expressed in membrane lysis and disruption of cell walls. This antimicrobial activity is the result of AMP availability. Hair keratinocytes as skin keratinocytes contain the endogenous AMP, which play the important role in the innate defense of human hair from keratinophylic microorganisms.

6. Conclusion

Since the time of AMP creation much information were collected concerning their role in host defense against microbial agents. From our data and some facts from literature it is obvious that just low levels of these "natural antibiotics" in different loci are the reason of transformation of opportunistic microflora to pathogenic. Frequently even humoral and cellular components of immune system are secondary as compared with AMP: good example of this may be observed in vulvovaginal candidosis and bacterial vaginosis. In the innate defense of hair AMP also play the main role what is caused by the specific structure and "dry" consistence of this skin appendages. We may suppose that AMP should participate in the defense of nails.

The development of the investigation line was apparently concerned the working out of different substances similar to AMP as a new pharmaceutical antimicrobial preparations. Now on the base of knowledge about structure and function of AMP new antimicrobial preparations are developed (Bals, 2000). Synthetic and recombinant analogues of the peptides are at the study of pharmaceutical research, including clinical trials of I – III phase.

Lysozyme for example is already traditional preparation used in the cases of local microbial affection. It is possible that the analogues of natural AMP will have not only antimicrobial activity, but could be used as immunomodulators.

The other direction of research may be study of stimulation of AMP synthesis *in vivo* in immune cells - neutrophils, epitheliocytes etc. - by immunomodulating agents. There are many such preparations in pharmaceutical market, which are known to have the stimulating effect on the proliferation of immune cells. However it is still a question if they may increase the AMP synthesis or not.

7. Acknowledgment

Authers express the thankfulness to Olga Serdiuk for the significant help in the carrying out of some analytical procedures.

8. References

Arzumanian, V.G., Bykova, S.A., Serdiyk, O.A. , & Kozlova N.N. (2000). Allergen-Containing Drug from Malassezia SPP. Yeast (Nov 2000). *Bulletin of Experimental Biology and Medicine*, Vol. 130, No.10, pp. 1084-1086

Bals, R. (2000). Epithelial antimicrobial peptides in host defense against infection. *Respiratory Research*, Vol.1, No. 3 (Oct 2000), pp. 141-150

Barousse, M.M. ,Van Der Pol, B.J., Fontenberry, D. (2004).Vaginal yeast colonisation, prevalence of vaginitis, and associated local immunity in adolescents. *Sex Transm Infect* , Vol.80, No.2, (Apr 2004); pp.48-53

Bellamy, W., Wakabayashi, H., Takase, M., Kawase, K., Shimamura, S., & Tomita, M. (1993). Killing of Candida albicans by lactoferricin B, a potent antimicrobial peptide derived from the N-terminal region of bovine lactoferrin. *Medical Microbiology & Immunology*, Vol.182, No. 2, (May 1993), pp. 97-105

Brandtzaeg, P., Gabrielsen, T.O., Dale, I., Müller, F., Steinbakk, M., & Fagerhol, M.K. (1995). The leucocyte protein L1 (calprotectin): a putative nonspecific defence factor at epithelial surfaces. *Advances in Experimental Medicine & Biology*, Vol. 371A, (1995), pp. 201-206

Cohen, M.S., Black, J.R., Proctor, R.A., & Sparling, R.F. (1984). Host defences and the vaginal mucosa: a re-evaluation. *Scand J. Urol. Nephrol.Suppl.*, Vol.86, (1984), pp. 13-22

Cullor, J.S., Mannis, M.J., Murphy, C.J., Smith, W.L., Selsted, M.E., & Reid, T.W. (1990). In vitro antimicrobial activity of defensins against ocular pathogens. *Archives of Ophthalmology*, Vol.108, No. 6, (Jun 1990), pp. 861-864

de Carvalho, R.J.V., Cunha, C.M., Silva, D.A., Sopelete, M.C., Urzedo, J.E., Moreira, T.A., & Moraes, P.S.A. (2003). IgA, IgE and IgG subclasses to Candida albicans in serum and vaginal fluid from patients with vulvovaginal candidiasis. *Revista Da Associacao Medica Brasileira*, Vol. 49, No. 4, (Oct-Dec 2003), pp. 434-438

Demirezen, S., & Beksac M.S. (2004). Relationship between the morphology of Candida cells and vaginal discharge. *New Microbiologica*, Vol. 27, No. 2, (Apr 2004), pp. 173-176

den Hertog, A.L., van Marle, J., van Veen, H.A., Van't Hof ,W., Bolscher, J.G., Veerman, E.C., & Nieuw Amerongen, A.V. (2005). Candidacidal effects of two antimicrobial peptides: histatin 5 causes small membrane defects, but LL-37 causes massive disruption of the cell membrane. *Biochem J.* , Vol. 388, No. Pt 2, (Jun 2005), pp.689-695

Doumas, S., Kolokotronis, A., & Stefanopoulos, P. (2005). Anti-Inflammatory and antimicrobial roles of secretory leukocyte protease inhibitor *Infection and Immunity*, Vol. 73, No. 3, (March 2005), pp. 1271-1274

Feng, Z., Jiang, B., Chandra, J., Ghannoum, M., Nelson, S., & Weinberg, A. (2005). Human Beta-defensins: Differential Activity against Candidal Species and Regulation by Candida albicans. *Journal of Dental Research*, Vol. 84, No. 5, (May 2005), pp. 445-450

Fidel, PL., Jr. (2007).History and Update on Host Defense Against Vaginal Candidiasis. *American Journal of Reproductive Immunology*, Vol. 57, No.1, (Jan 2007), pp. 2 - 12

Flad, T., Bogumil, R., Tolson, J., Schittek, B., Garbe, C., & Deeg, M. (2002). Detection of dermcidin-derived peptides in sweat by ProteinChip technology. *J Immunol Methods*, Vol. 270, No. 1, (Dec 2002), pp. 53–62

Frohm Nilsson, M. , Sandstedt, B., Sørensen, O., Weber, G., Borregaard, N., & Ståhle-Bäckdahl, M. (1999). The Human Cationic Antimicrobial Protein (hCAP18), a Peptide Antibiotic, Is Widely Expressed in Human Squamous Epithelia and Colocalizes with Interleukin-6. *Infect Immun.*, Vol. 67, No. 5, (May 1999), pp. 2561–2566

Glaser, R., Harder, J., Lange, H., Bartels, J., Christophers, E. & Schröder J-M. (2005). Antimicrobial psoriasin (S100A7) protects human skin from Escherichia coli infection. *Nature Immunology*, Vol. 6, No. 1, (Nov 2005), pp. 57-64

Harder, J. & Schroeder, J. (2002). RNase 7, a novel innate immune defense antimicrobial protein of healthy human skin. *Journal of Biological Chemistry*, Vol. 277, No. 48, (Nov 2002), pp. 46779-46784

Huang, Y-C. ,Lin, Y.-M., Chang, T.-W., Chen, C., Wu, S.H., & Liao, Y.D. (2007). The flexible and clustered lysine residues of human ribonuclease 7 are critical for membrane permeability and antimicrobial activity. *Journal of Biological Chemistry*, Vol. 282, No. 7, (Feb 2007), pp. 4626-4633

Ibrahim, H., Thomas, U., & Pellegrini, A. (2001). A helix-loop-helix peptide at the upper lip of the active site cleft of lysozyme confers potent antimicrobial activity with membrane permeabilization action. *J. Biol.Chem.*, Vol. 276, No. 47, (Sep 2001), pp. 43767-43774

Kurnatowska A., & Magnowski J. (2002). Analysis of sIgA concentrations in the contents of the cervical canal of the uterus and of the oral cavity in women with Candida or without fungi in ontocenoses of these organs. *Wiadomosci Parazytologiczne*, Vol. 48, No. 3, pp. 271-276

Kurzweilova, H., & Sigler, K. Fluorescent staining with bromocresol purple: a rapid method for determining yeast cell dead count developed as an assay of killer toxin activity. *Yeast*, Vol. 9, No. 11, (Nov 1993), pp. 1207-1211

Lai, Y., Villaruz, A.E,. Li, M., Cha, D. J., Sturdevant, D., & Otto M. (2007). The human anionic antimicrobial peptide dermcidin induces proteolytic defence mechanisms in staphylococci. *Molecular Microbiology*, Vol. 63, No. 2, (Jan 2007), pp. 497-506

Lambin, P., Rochu, D., & Fine, J.M. (1976). A new method for determination of molecular weights of proteins by electrophoresis across a sodium dodecyl sulfate (SDS)-polyacrylamide gradient gel. *Anal Biochem.*, Vol.74, No. 2, (Aug 1976), pp. 567–575

Loomans, H.J., Hahn, B.L., Li, Q.Q., Phadnis, S.H., & Sohnle P.G. (1998). Histidine-based zinc-binding sequences and the antimicrobial activity of calprotectin. *Journal of Infectious Diseases*, Vol. 177, No.3, (Mar 1998), pp. 812-814

López-García, B., Lee, P., & Gallo, R. (2006). Expression and potential function of cathelicidin antimicrobial peptides in dermatophytosis and tinea versicolor. *Journal of Antimicrobial Chemotherapy*, Vol. 57, No. 5, (Mar 2006), pp. 877-882

Lulloff, S. J., Hahn, B. L., & Sohnle, P.G. (2004). Fungal susceptibility to zinc deprivation. *Journal of Laboratory & Clinical Medicine*, Vol. 144, No. 4, (Oct 2004), pp. 208-214

Lupetti, A., Brouwer, C. P. J. M., Dogterom-Ballering, H., Senesi, S., Campa, M., Van Dissel, J.T., & Nibbering, P.H. (2004). Release of calcium from intracellular stores and subsequent uptake by mitochondria are essential for the candidacidal activity of an N-terminal peptide of human lactoferrin. *Journal of Antimicrobial Chemotherapy*, Vol. 54, No.3, (Sep 2004), pp. 603-608

Madsen, P., Rasmussen, H.H,, Leffers, H., Honoré, B., Dejgaard, K., Olsen, E., Kiil, J., Walbum, E., Andersen, A.H,, & Basse, B. (1991). Molecular cloning, occurrence, and expression of a novel partially secreted protein "psoriasin" that is highly up-regulated in psoriatic skin. *Journal of Investigative Dermatology*, Vol. 97, No. 4, (Oct 1991), pp. 701-712

Marquis, G,, Montplaisir, S., Garzon, S., Strykowski, H., & Auger, P. (1982). Fungi toxicity of muramidase, ultrastructural damage to Candida albicans. *Lab Investig.*, Vol. 46, No. 6, (Jun 1982), pp. 627–636

Mestecky, J., Moldoveanu, Z., & Russell, M.W. (2005). Immunologic uniqueness of the genital tract: challenge for vaccine development. *American Journal of Reproductive Immunology*, Vol. 53, No. 5, (May 2005), pp. 208-214

Mikx, F.H., de Jong, M.H. (1987). Keratinolytic activity of cutaneous and oral bacteria. *Infection & Immunity*, Vol. 55, No. 3, (Mar 1987), pp. 621-625

Moon, J-Y., Henzler-Wildman, K., & Ramamoorthy, A. (2006). Expression and purification of a recombinant LL-37 from Escherichia coli *Biochimica et Biophysica Acta*, Vol.1758, No. 9, (Sep 2006), pp. 1351-1358

Nomanbhoy, F., Steel, C., Yano, J., & Fidel, PL, Jr. (2002).Vaginal and oral epithelial cell anti-Candida activity. *Infection and Immunity*, Vol.70, No.12, (Dec. 2001), pp.7081-7088

Okutomi, T., Tanaka, T., Yui, S., Mikami, M., Yamazaki, M., Abe, S., & Yamaguchi, H. (1998). Anti-Candida activity of calprotectin in combination with neutrophils or lactoferrin. *Microbiology & Immunology.*, Vol. 42, No.11, (Nov 1998), pp. 789-793

Omaetxebarria, M. J. , Moragues, M. D., Elguezabal, N., Rodríguez-Alejandre, A. Brena, S., Schneider, J., Polonelli, L., & Pontón, J. (2005). Antifungal and antitumor activities of a monoclonal antibody directed against a stress mannoprotein of Candida albicans. *Current Molecular Medicine (Hilversum).*, Vol. 5, No.4, (Jun 2005), pp. 393-401

Salmon, V., Legrand, D., Georges, B., Slomianny, M.C., Coddeville, B., & Spik, G. (1997). Characterization of human lactoferrin produced in the baculovirus expression system. *Protein Expression & Purification*, Vol. 9, No. 2, (Mar 1997), pp. 203-210

Samaranayake, Y.H., Samaranayake, L.P., Wu, P.C., & So, M. (1997). The antifungal effect of lactoferrin and lysozyme on Candida krusei and Candida albicans. *APMIS*, Vol.105, No. 11, (Nov 1997), pp. 875-83

Schittek, B., Hipfel, R., Sauer, B., Bauer, J., Kalbacher, H., & Stevanovic, S. (2001). Dermcidin: a novel human antibiotic peptide secreted by sweat glands. *Nat Immunol.*, Vol. 2, No. 12, (Dec 2001), pp. 1133–1137

Schneider, J.J., Unholzer, A., Schaller, M., Schafer-Korting, M., & Korting, H.C. (2005). Human defensins. *J.Molecular Medicine*, .- Vol. 83, No. 8, (2005), pp. 587-595

Shabashova, N.V. , Mirsabalaeva, A.K., Frolova, E.V., Uchevatkina, A.E., Dolgo-Saburova, U.V., Filippova, L.V., & Bobrovskaya M.V. (2006). Some aspects of vaginal mucosa cells immune response in women with chronic recurrent genital candidosis. *Problemy meditsinskoi mycologii (rus)*, Vol.8, No. 2, (June 2006), pp. 97-98.

Sorensen, O. E., Gram, L., Johnsen, A. H., Andersson, E., Bangsbøll, S., Tjabringa, G.S., Hiemstra, P.S., Malm, J., Egesten, A., & Borregaard, N. (2003). Processing of Seminal Plasma hCAP-18 to ALL-38 by Gastricsin: A novel mechanism of generating antimicrobial peptides in vagina. *J. Biol. Chem.*, Vol. 278, No.31, (August 2003), pp. 28540 - 28546

Taggart, C. C., Greene, C. M., Smith, S.G., Levine, R.L., McCray Jr., P.B., O'Neill, S., & Mc Elvaney, N.G. (2003). Inactivation of Human β-Defensins 2 and 3 by Elastolytic Cathepsins . *The Journal of Immunology*, Vol.171, No. 2, (Jul 2003), pp. 931-937

Tomee, J. F. C., Hiemstra, P. S., Heinzel-Wieland, R. & Kauffman, H.F. (1997). Antileukoprotease: an endogenous protein in the innate mucosal defense against fungi. *J. Infect. Dis.*, Vol.176, No. 3, (Sep 1997), pp. 740-747

Tomee, J.F., Koeter, G.H., Hiemstra, P.S., & Kauffman, H F. (1998). Secretory leukoprotease inhibitor: a native antimicrobial protein presenting a new therapeutic option? *Thorax.*, Vol. 53, No. 2, (Feb 1998), pp. 114-116

Valore, E.V. ,Park, C., Igreti, S.L.,& Ganz, T. (2002). Antimicrobial components of vaginal fluid. *J Obstet Gynecol*, Vol.187, No.3, (Sept. 2002), pp. 561-568

Valore, E.V. Wiley, D.J., & Ganz, T. (2006). Reverible deficiency of antimicrobial polipeptides in bacterial vaginosis. *Infection and Immunity*, Vol.74, No.10, (Oct. 2006), pp. 5693-5702

van der Kraan, M. I. A., van Marle, J. , Nazmi, K., Groenink, J., van 't Hof ,W., Veerman, E.C., Bolscher, J.G., & Nieuw Amerongen, A.V. (2005). Ultrastructural effects of antimicrobial peptides from bovine lactoferrin on the membranes of Candida albicans and Escherichia coli. *Peptides*, Vol. 26, No. 9, (Sep 2005), pp. 1537-1542

Viejo-Diaz, M., Andres, M. T., & Fierro, J. F. (2004). Effects of human lactoferrin on the cytoplasmic membrane of Candida albicans cells related with its candidacidal activity. *FEMS Immunology & Medical Microbiology*, Vol. 42, No. 2, (Oct 2004), pp. 181-185

Vonk, A.G., Wieland, C.W., Netea, M.G., & Kullberg, B. J. (2002). Phagocytosis and intracellular killing of Candida albicans blastoconidia by neutrophils and macrophages: a comparison of different microbiological test systems. Journal *of Microbiological Methods*, Vol.49, No.1, (Mar 2002), pp. 55-62

Zasloff, M. (2002). Antimicrobial Peptides in Health and Disease. *The New England Journal of Medicine*, Vol. 347, No. 1, (October 2002), pp. 1199-1200

Novel Anti-Microbial Peptides of *Xenorhabdus* Origin Against Multidrug Resistant Plant Pathogens

András Fodor[1], Ildikó Varga[1], Mária Hevesi[2], Andrea Máthé-Fodor[3], Jozsef Racsko[4,5] and Joseph A. Hogan[5]

[1]*Plant Protection Institute, Georgikon Faculty, University of Pannonia, Keszthely,*
[2]*Department of Pomology, Faculty of Horticultural Science, Corvinus University of Budapest Villányi út Budapest,*
[3]*Molecular and Cellular Imaging Center, Ohio State University (OARDC/OSU), OH,*
[4]*Department of Horticulture and Crop Science, Ohio State University (OARDC/OSU), OH,*
[5]*Valent Biosciences Corporation, 870 Technology Way, Libertyville, IL,*
[6]*Department of Animal Sciences, Ohio State University (OARDC/OSU) OH,*
[1,2]*Hungary*
[3,4,5,6]*USA*

1. Introduction

The discovery and introduction of antibiotics revolutionized the human therapy, the veterinary and plant medicines. Despite the spectacular results, several problems have occurred later on. Emergence of antibiotic resistance is an enormous clinical and public health concern. Spread of methicillin-resistant *Staphylococcus aureus* (MRSA) (Ellington et al., 2010), emergence of extended spectrum beta-lactamase (ESBL) producing *Enterobacteriaceae* (Pitout, 2008), carbapenem resistant *Klebsiella pneumoniae* (Schechner et al., 2009) and poly-resistant *Pseudomonas* (Strateva and Yordanov, 2009) and *Acinetobacter* (Vila et al., 2007) causes serious difficulties in the treatment of severe infections (Vila et al., 2007; Rossolini et al., 2007). A comprehensive strategy, a multidisciplinary effort is required to combat these infections. The new strategy includes compliance with infection control principles: antimicrobial stewardship and the development of new antimicrobial agents effective against multi-resistant gram-negative and gram-positive pathogens (Slama, 2008). During the last few decades, only a few new antibiotic classes reached the market (Fotinos et al., 2008). These facts highlight the need to develop new therapeutic strategies. The increasing incidence of serious infections caused by antibiotics-resistant and multi-resistant microorganisms such as the methicillin-resistant *Staphylococcus aureus* (MRSA) in human; streptomycin-resistant *Erwinia amylovora* (the bacterial pathogen causing fireblight disease) in Rosaceae make it imperative to develop new antimicrobial agents to face the new challenges Kocsis et al., 2009). It has also become obvious that, at least in Europe, antibiotics are not allowed to use for plant protection. Consequently, new antimicrobial compounds of different mode of action are needed which justify research efforts toward new sources. This chapter should be considered as a modest contribution to these efforts.

Microbes often live in polymorphic environments wherein they have to compete for nutrients, space and overcome toxins in order to survive and flourish. Amongst the chemical toolkits, antibiotics play an important role. The vast majority of recently used antibiotics are of eukaryotic origin, and the vast majority of antibiotics researchers focus on antibiotics producing fungi. However, some bacteria also produce compounds of antimicrobial activity. Consequently, these bacteria may be potential sources of novel antimicrobial compounds acting through novel molecular action mechanism. Our aim is to find bacterial sources for novel antibiotics effective against pathogens, which are resistant to antibiotics used in clinical practice and plant medicine.

The entomopathogenic nematode/bacterium (EPN/EPB) symbiotic associations are considered as model systems to address broad biological questions of mutualism, co-evolution and pathogenesis. As an indispensible part of keeping this system competitive in nature, broad spectral antibiotics produced by symbiotic bacteria (EPB) of entomopathogenic nematodes (EPN) and keep monoxenic conditions within insect cadavers in soil conditions such a way. Recently, the genomic sequences of bacterial symbionts, *Photorhabdus luminescens* (Duchaud et al., 2003) and two *Xenorhabdus* species have been completed, and the latter are being analyzed (Ogier et al., 2010). An area with ramifications in plant pathology, veterinary science, and even human health, is the primary and secondary gene products of antimicrobial activity. Webster et al. (2002) and later Brachmann et al. (2006) and Bode (2009) reviewed the results of antibiotics studies in EPB since the work of Akhurst (1982).

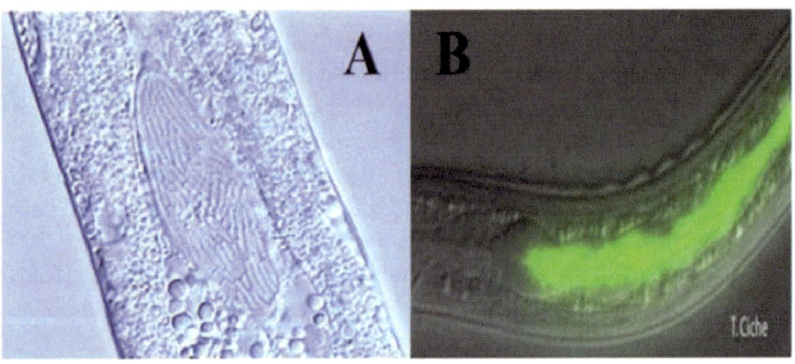

Fig. 1. Localization of *Xenorhabdus* (A) and *Photorhabdus* (B) entomopathogenic nematode symbiotic bacteria (EPB) in the gut of the infective dauer juvenile form of their respective entomopathogenic nematode (EPN) symbiotic partner, *Steinernema* (A) and *Heterorhabditis* (B). After the entering the insect the bacteria kill the host and colonize the cadaver, and serve as food source as well as antimicrobial-producing "safe-guard" for the nematode in soil condition (Courtesy of Dr. Todd Chiche he University of Michigan, USA).

EPB antibiotics have broad target spectra, which provide monoxenic conditions within insect cadavers. Despite descriptions and patenting of antibiotic molecules from *Xenorhabdus* (Webster et al. 1996; Thaler et al. 2001), nothing has been published on their commercial use. Our joint efforts revealed that antimicrobials new profiles are mainly of oligopeptide nature, such as the hexapeptide bicornutin A (Furgani et al., 2008 Böszörményi et al., 2009) which is effective against prokaryotic (*Erwinia amylovora*) and eukaryotic (*Phytophthora nicotianae*) plant pathogens (Böszörményi et al., 2009). In this chapter, we intend to summarize our results since 2009, which mainly include the search of the target specificity of the cell-free conditioned media (CCFM) of two species, *Xenorhabdus szentirmaii* and *X. budapestensis* (Lengyel et al., 2005). Some of the related (sometimes poly-resistant) pathogens belonging respectively to *Pantoea, Klebsiella, Escherichia, Staphylococcus Salmonella, Candida* and *Alternaria* are also of human clinical and/or veterinary importance.

2. Aim, objective and rationale

2.1 Aim

The problems, which should be solved, are related with the increasing poly-resistance of pathogenic microorganisms of clinical, veterinary and plant medicine significance. Strong efforts of the scientific community have been exerted toward finding new antibiotics with novel action mechanisms. Antimicrobials of peptide nature, which induce apoptosis in target cells, have a great potential in control eukaryotic pathogens. Our contribution to the field is the introduction two novel organisms, *X. szentirmaii* and *X. budapestensis* what we have found excellent sources of compounds of strong antimicrobial activity. Their antimicrobial compounds proved antagonistic towards bacteria resistant to other antibiotics (Furgani et al., 2008; Böszörményi et al., 2009) and also toward some eukaryotic pathogens such as oomycetae, fungi and, according to some preliminary data, against pathogenic protozoon (McGwire et al., 2010, personal communication). Thus, importance and possible application of these compounds in agriculture and veterinary medicine was already thoroughly investigated and proved in several studies. Many of them have been patented but none of them is used in the practice. We considered the possible explanations and introduced two novel EPB species of excellent antimicrobial activities. Efficacy of these components against human clinical isolates has never been investigated before. The aim of our study is to investigate the antimicrobial potential of compounds produced by two entomopathogenic bacteria (EPB) as potential tools of controlling poly-resistant pathogens. Within this joint study, identification and purification of the bioactive molecules of the *Xenorhabdus* strains will be performed in different laboratories in Hungary and Ohio (USA).

2.2 Objective

The final goal of this study is to provide a view about the future perspectives of using antimicrobial peptides produced by *Xenorhabdus szentirmaii* and *X. budapestensis* in plant, - veterinary, - and may be in human medicine.

2.2.1 *Xenorhabdus budapestensis* and *X. szentirmaii* as sources of novel pathogen antagonists different from conventional antibiotics concerning their action mechanism

Our previous studies (Furgani et al., 2008; Böszörményi et al., 2009) confirmed that there might be new perspectives concerning the potential of some *Xenorhabdus* antibiotics, and for their use as alternative tools of pathogens of veterinary and plant medicine importance. We found that, *X. budapestensis* and *X. szentirmaii* are the best of theses organisms. We also proved that antibacterial activities could mostly be adsorbed by Amberlite, and could be eluted without significant loss of activity. All of them gave a ninhydrin positive reaction, indicating that the most important compounds are of peptide nature. One of our main objectives is to determine their target specificity and then further purification after that. Our first steps toward this direction resulted in the discovery a hexapeptide (called Bicornutin A) of strong antibacterial activity.

Before going further toward biochemical purification, we wanted to test the target spectrum of the intact CFCM of the two species. This is what we are reporting about in this chapter. We provide new data about their control potential on plant pathogens as well as against multiple resistant human and animal pathogens.

We are also going to give an account on the very strong anti-oomycetal activity, providing a potential tool for plant medicine in nursery and forestry. Finally, they have a selective antifungal effect, which might be exhausted in behalf of plant medicine and human clinical practice.

We have also made the first steps toward a genetic approach of better understanding and improving the antimicrobial potential of by *Xenorhabdus szentirmaii* and *X. budapestensis*.

2.2.2 Potential pathogen targets of EPB antimicrobials in nature

At this point, we included into our Introduction vast majority of the potential target organisms, not only those that have already been analyzed. The reason is to draw the attention of the Reader of the perspectives and potential of using EPB antimicrobial compounds of apoptotic activity in the veterinary and plant medical, and potentially in the clinical practice, in the future.

2.2.2.1 Plant pathogenic bacteria

According to the data presented in the Widely Prevalent Bacteria Lists of Plant Pathology at Bugwood Websites in 2009, Ohio, as far the most infected state of the United States of America, is represented by 62 plant pathogenic bacteria. The climate as well as the plant pathogenic bacterium flora of Hungary and Ohio is rather similar. The most prevalent plant pathogenic bacteria belong to eight different groups. The general aim is to discover efficient antibacterial compounds, which antagonize agriculturally important pathogens in an environmentally friendly way. Our way is to test antibacterial activity of some entomopathogenic (nematode-symbiotic) bacteria on plant pathogenic bacteria. Our target bacteria in this study were as follows: *Agrobacterium, Burkholderia, Clavibacter, Curtobacterium, Dyckeya, Erwinia, Pectobacterium, Ralstonia, Pseudomonas,* and *Xanthomonas* species.

2.2.2.1.1 Agrobacterium species, biovars, strains

Agrobacterium species are Gram-negative, non-spore forming, rod-shaped bacteria attacking 643 dicotyledonous (broad-leaved) plant species. Neither monocotyledons, nor Liliales, nor Arales are *Agrobacterium* targets. The target organs are the roots and the stalks. The most characteristic symptoms are hypertrophies in most of their host plants. Disease symptoms:

formation of tumor-like swellings called galls that can generally be found on the crown of the plant just above the soil. The most important *Agrobacterium* pathogens are listed in **Table 1.** The most widely known species cause crown gall disease. *Agrobacterium* cells have one circular and one rode-shape chromosome. As extra-genomic DNA, they contain plasmids of different biological role (Fig 2, Left).. The most important is the ***Ti*** (tumor inducing) plasmid including a sequence (called TDNA) of eukaryotic nature. The TDNA is excised after infection and is capable of inserting randomly to the chromosomes of the plant. The expression of genes located on the TDNA result in tumor. The plant tumor cells produce compounds that are normally not produced by the plant. These compounds are used as a form of energy by the infecting bacteria. *A. tumefaciens* strains use different carbohydrates. They are classified into three main biotypes called biovars. There are allelic differences of genes located on the circular chromosome resulting in phenotypic differences of biovars concerning host preference and antibiotics sensitivity. Crown gall disease in grapevines (Fig 2, Right), is caused by bacteria belonging to Biovar 3, and (recently renamed as *A. vitis*). As for ecology, *A. tumefaciens* can generally be found on and around root surfaces known as the rhizosphere, where they might be attacked either by chemical or biological antagonists; as well as in the cambria (within the plant) as innocent saprophytes. They seem to use nutrients that leak from the root tissue. They invade the plant if it becomes conditioned (susceptible, wounded).

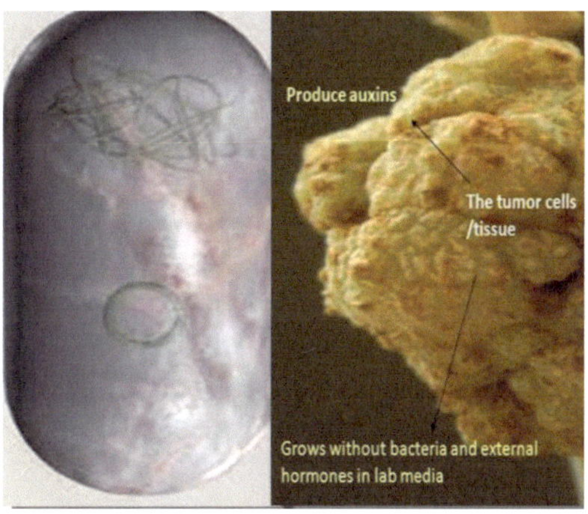

Fig. 2. *A. tumefaciens* cell with visible Ti plasmid ("North") and the genome ("South"); Crown gall disease (on the LEFT) Crown gall disease (on the RIGHT) Courtesy of Prof. George P. Rédei, (University of Missouri-Columbia, MO, USA)

Scientific name	Disease
Agrobacterium tumefaciens	crown gall of dicotyledonous plants
Agrobacterium vitis	crown gall of grape
Agrobacterium rhizogenes	hairy root disease
Agrobacterium rubi	cane gall of *Rubus*

Table 1. ***Agrobacterium*** species and the diseases they cause. (Higher order taxa: full lineage: root; cellular organisms; Bacteria; Proteobacteria; Alphaproteobacteria; Rhizobiales; Rhizobiaceae; Rhizobium/Agrobacterium group). SOURCE: Plant Pathology Bugwood Websites - Widely Prevalent Bacteria Lists

2.2.2.1.2 Clavibacter species, subspecies, and strains

Bacterial canker of is one of the most important and most difficult-to-control tomato diseases. Bacterial canker and wilt disease of tomato caused by ***Clavibacter michiganensis*** subsp. *michiganensis* are spread into different countries all over the world resulting in considerable yield losses (up to 70%), mainly in field-grown tomato production. The bacterium has been present in North America, Canada, Asia, and Africa. In European countries, it is on the list of quarantine pest. Symptom of the disease is wilting and desiccation of the plant both in field. Infected fruits show characteristic "bird's eye" spots; fail to develop and fall or ripen unevenly. Whenever it has been established in vascular tissues of the crop, it and becomes seed-born. The disease has also been recorded in greenhouses (Shoemaker and Echandi, 1976; Agrios, 1997). No sufficient chemical control exists. Biological control agents directly used against bacterial canker such as *Pseudomonas fluorescens, Bacillus* and *Streptomyces* species has been tried (Nishioka et al., 1997). The harmful *Clavibacter* bacteria are listed in **Table 2.**

Scientific name	Disease
Clavibacter michiganensis ssp. michiganensis	bacterial canker and wilt of tomato
Clavibacter michiganensis ssp. nebraskensis	Goss' bacterial wilt on cora
Clavibacter michiganensis ssp. sepedonicus	ring rot of potato
Clavibacter michiganensis ssp. insidiosus	bacterial wilt of alfalfa

Table 2. ***Clavibacter*** species and the plant diseases they cause. Taxonomic position: full lineage: root; cellular organisms Bacteria; Actinobacteria; Actinobacteria (class); Actinobacteridae; Actinomycetales; Micrococcineae; Microbacteriaceae; Clavibacter; *Clavibacter, C. michiganensis.* (SOURCE: Plant Pathology Bugwood Websites Widely Prevalent Bacteria Lists)

2.2.2.1.3 Erwinia species and strains

Fire blight disease caused by *Erwinia amylovora* is one of the most destructive plant diseases that cause severe crop losses in many countries for a long time (Rosen, 1929). At present, the disease threatens commercial fruit industries worldwide. It occurs in many fruit species, especially those belonging to the *Pomaceae* and *Rosaceae* families, such as *Malus domestica, Pyrus communis, Cydonia oblonga* and *Cotoneaster spp*. It appears in the central, southern, and eastern regions of the European continent (Németh, 1998) and present in many states of the USA (Steiner and Zeller, 1996; Paulin, 1997). It appeared in Hungary first in 1995 (Hevesi, 1996). Right after these finding epidemics spread over in many fruit tree orchards (Németh, 1999; Jones et al., 1996). Chemical control and cultural practices did not prove sufficiency to arrest epidemics. Several research programs have been developed continuously to reduce the incidence of this serious disease or overcome the potential of the pathogen (van der Zwet and Beer, 1995). We have provided the first evidence that conditioned cell-free medium of *Xenorhabdus budapestensis* was capable of reducing the spreading of fire blight inflammation on apple trees after artificial infestation in green house conditions. The effects proved unambiguously cytotoxic in laboratory experiments (Böszörményi et al., 2009). **Table 3** provides a list of some plant pathogenic bacterium species belonging to *Erwinia* and *Pantoea* genera.

Scientific name	Disease
Erwinia amylovora	fire blight of Rosaceae
Erwinia tracheiphila	bacterial wilt on corn
Pantoea ananatis	center rot of pineapple
Pantoea stewartii pv. stewartii	Stewart's wilt of maize

Table 3. ***Erwinia*** species and the diseases they cause. SOURCE: Plant Pathology Bugwood Websites Widely Prevalent Bacteria Lists) Taxonomy: Full lineage: cellular organisms; Bacteria; Proteobacteria; Gammaproteobacteria; Enterobacteriales; Enterobacteriaceae; *Erwinia*

2.2.2.1.4 Xanthomonas species, subspecies and strains

Bacteria belonging to genus *Xanthomonas* cause numerous plant diseases with diverse symptoms, including vascular wilts, cankers, soft rots, blights, leaf spots, tumors or galls. ***X. euvesicatoria*** strains are pathogens causing the bacterial spot diseases of *Capsicum annuum* and *Lycopersicon esculentum* a consequence of which is destructive loss in these two economically important crops. Control measures are applied yearly but no complete eradication of the disease has been achieved so far. Biological control provided some promising evidence concerning protection against the disease in small-scale plots. For example, *Pseudomonas fluorescens* gave promising inhibitory effects (Colin et al., 1984; Tzeng et al., 1994). *Xanthomonas* pathogens are listed in **Table 4.**

Scientific name	Disease
Xanthomonas vesicatoria	bacterial spot of tomato and pepper
Xanthomonas campestris pv. papavericola	bacterial blight of poppy
Xanthomonas campestris pv. pelargonii	bacterial blight of geranium
Xanthomonas campestris pv. raphani	bacterial leaf spot of radish and turnip
Xanthomonas campestris pv. vesicatoria	bacterial spot of tomato and pepper
Xanthomonas campestris pv. zinniae	bacterial leaf spot of zinniae
Xanthomonas fragariae	angular leaf spot of strawberry
Xanthomonas gardneri	bacterial spot of tomato & pepper
Xanthomonas hortorum pv. hederae	bacterial leaf spot of ivy
Xanthomonas perforans	bacterial spot of tomato & pepper
Xanthomonas translucens pv. undulosa	wheat leaf streak/black chaff
Xylella fastidiosa ssp. fastidiosa	Pierce's disease of grapevine
Xanthomonas arboricola pv. juglandis	walnut blight
Xanthomonas arboricola pv. pruni	bacterial spot of stone fruits
Xanthomonas axonopodis pv. vitians	bacterial leaf spot of lettuce
Xanthomonas campestris pv. armoraciae	bacterial leaf spot of crucifers
Xanthomonas campestris pv. campestris	black rot of crucifers

Table 4. ***Xanthomonas*** species and the diseases they cause. SOURCE: Plant Pathology Bugwood Websites Widely Prevalent Bacteria Lists Taxonomy, full lineage: root; cellular organisms; Bacteria; Proteobacteria; Gammaproteobacteria; Xanthomonadales; Xanthomonadaceae, *Xanthomonas*.

2.2.2.1.5 Pectobacterium species, subspecies and strains

There is a group of plant pathogens closely related to *Erwinia* called ***Pectobacteria*** In Hungary, ***Pectobacterium chrysanthemi*** and ***P. carotavora*** cause bacterial spot of peach, tobacco wildfire, citrus canker, and some ornamental diseases as well Antibiotics were used to control black rot of cabbage. The list of these bacteria is given in **Table 5.**

Scientific name	Disease
Pectobacterium atrosepticum	black leg of potato
+*Pectobacterium carotovorum ssp. carotovorum*	bacterial soft rot
Pectobacterium chrysanthemi	bacterial wilt and soft rot
Pectobacterium chrysanthemi pv. dieffenbachiae	stem rot
Pectobacterium chrysanthemi pv. zeae	stem rot of maize

Table 5. ***Pectobacteria*** species and the diseases they cause. SOURCE: Plant Pathology Bugwood Websites Widely Prevalent Bacteria Lists. Taxonomy: cellular organisms; Bacteria; Proteobacteria; Gammaproteobacteria; Enterobacteriales; Enterobacteriaceae; Pectobacterium; Pectobacterium carotovorum. +Synonyms: *Erwinia carotovora (ssp.) carotovora.*

2.2.2.1.6 Ralstonia and Burkholderia species, subspecies (plant variants), biovars and races

Ralstonia solanacearum attacks all the solanaceous crop plants. Over 200 hosts are known. Some are listed in **Table 6**. Strains show varying degrees of host specificity. The high economic and social impact of this organism results from its wide geographical distribution in all warm and tropical countries of the globe. Recently, distribution of the pathogen has been extended to more countries that are temperate from Europe and North America as the result of the dissemination of strains adapted to cooler climates. Nevertheless, the diseases are common in Europe but the pathogen is on the list of quarantine pest. Only complex and integrated control strategy can succeed in reducing the disease incidence.

Scientific name	Disease
Ralstonia solanacearum (excl. Race 3 Biovar 2)	southern bacterial wilt
Burkholderia andropogonis	gummosis of grasses
Burkholderia caryophylli	vascular wilt
Burkholderia gladioli pv. gladioli	scab of flower bulb

Table 6. ***Ralstonia*** and ***Burkholderia*** species and the diseases they cause. (SOURCE: Plant Pathology Bugwood Websites Widely Prevalent Bacteria Lists) Taxonomy: **Lineage** (full): root; cellular organisms; Bacteria; Proteobacteria; Betaproteobacteria; Burkholderiales; Burkholderiaceae; *Ralstonia / Burkholderia.*

2.2.2.1.7 Pseudomonas species, subspecies and strains: plant, animal and human pathogens

Pseudomonas aeruginosa is a Gram-negative rod measuring 0.5 to 0.8 μm by 1.5 to 3.0 μm. Almost all strains are motile by means of a single polar flagellum. Its optimum growth temperature is 37 °C, but is also able to grow at 42 °C. *Pseudomonas* may form biofilm or exists in a planktonic form. Genus *Pseudomonas* is cleaved into eight groups. *P. aeruginosa* is the type species and there are another 12 members. Each is well known to plant microbiologists. They cause numerous plant diseases with diverse symptoms, including vascular wilts, cankers, soft rots, blights, leaf spots, tumors or galls. *P. aeruginosa* has become increasingly recognized as an emerging opportunistic pathogen of clinical relevance. Several different epidemiological studies track its occurrence as a nosocomial pathogen and indicate that antibiotic resistance is increasing in clinical isolates. *P. aeruginosa* is an opportunistic pathogen; it exploits some break in the host defenses to initiate an infection. It causes urinary tract infections, respiratory system infections, dermatitis, soft tissue infections, bacteremia, bone and joint infections, gastrointestinal infections and a variety of systemic infections, particularly in patients who are immunosuppressed. The case fatality rate in these patients is nearly 50 percent. *P. aeruginosa* isolates may produce three colony types. Natural isolates from soil or water typically produce a small, rough colony. Clinical samples yield one or another of two smooth colony types. One type has a fried-egg appearance, which is large, smooth, with flat edges and an elevated appearance. Another type, frequently obtained from respiratory and urinary tract secretions, has a mucous appearance, which is attributed to the production of alginate slime. The smooth and mucous colonies are presumed to play a role in colonization and virulence.

Being Gram-negative bacteria, most *Pseudomonas spp.* are naturally resistant to penicillin and to the majority of related beta-lactam antibiotics, but a number of those is sensitive to piperacillin, imipenem, ticarcillin, tobramycin, or ciprofloxacin. This ability to thrive in harsh conditions is a result of their hardy cell wall that contains porins. Their resistance to most antibiotics is attributed to efflux pumps, which pump out some antibiotics before the antibiotics are able to act (Poole, 2004). *Pseudomonas aeruginosa* is a highly relevant opportunistic human pathogen. One of the most worrying characteristics of *P. aeruginosa* is its low antibiotic susceptibility. This low susceptibility is attributable to a concerted action of multidrug efflux pumps with chromosomally encoded antibiotic resistance genes (e.g. *mexAB-oprM*, *mexXY*, etc.) and the low permeability of the bacterial cellular envelopes. Besides intrinsic resistance, *P. aeruginosa* easily develops acquired resistance either by mutation in chromosomally encoded genes, or by the horizontal gene transfer of antibiotic resistance determinants. Development of multidrug resistance by *P. aeruginosa* isolates requires several different genetic events that include acquisition of different mutations and/or horizontal transfer of antibiotic resistance genes. Hypermutation favors the selection of mutation-driven antibiotic resistance in *P. aeruginosa* strains producing chronic infections, whereas the clustering of several different antibiotic resistance genes in integrons favors the concerted acquisition of antibiotic resistance determinants. Some recent studies have shown phenotypic resistance associated to biofilm formation or to the emergence of small-colony-variants may be important in the response of *P. aeruginosa* populations to antibiotic treatment. This justifies testing new natural compounds of antibacterial activity against them. The most important plant pathogenic species are listed in **Table 7**.

Scientific name	Disease
Pseudomonas syringae pv. berberidis	leaf spot/fall of Berberis
Pseudomonas syringae pv. coronafaciens	halo blight of oats
Pseudomonas syringae pv. delphinii	Bacterial leaf spot of delphinium crop
Pseudomonas syringae pv. glycinea	bacterial blight of soybean
Pseudomonas syringae pv. lachrymans	angular leaf spot of cucumber
Pseudomonas syringae pv. maculicola	bacterial leaf spot of cauliflower
Pseudomonas syringae pv. apii	leaf spot of parsley
Pseudomonas syringae pv. atrofaciens	basal glume rot of wheat
Pseudomonas syringae pv. atropurpurea	leaf spot of Italian ryegrass
Pseudomonas syringae pv. mori	bacterial blight of mulberry
Pseudomonas syringae pv. morsprunorum	bacterial canker of stone fruit tees
Pseudomonas syringae pv. papulans	blister spot of apple
Pseudomonas syringae pv. phaseolicola	halo blight of bean
Pseudomonas syringae pv. pisi	bacterial blight of pea
Pseudomonas syringae pv. syringae	bacterial brown spot of bean/canker of stone fruit
Pseudomonas syringae pv. tabaci	wildfire of tobacco
Pseudomonas syringae pv. tomato	bacterial speck
Pseudomonas tolaasii	bacterial blotch on mushrooms
Pseudomonas viridiflava	spots/soft rots on leaves/fruits on many plant species

Table 7. Plant pathogenic ***Pseudomonas*** species and the diseases they cause. (SOURCE: Plant Pathology Bugwood Websites - Widely Prevalent Bacteria Lists) Taxonomy: **Lineage** (full): root; cellular organisms; Bacteria; Proteobacteria; Gammaproteobacteria; Pseudomonadales; Pseudomonadaceae *Pseudomonas*

2.2.2.2 Plant pathogenic Oomycetae

The most harmful plant pathogenic oomycetae are the downy mildew, *Phytophthora* and *Pythium* species, causing planting-off and root-rot (decay) diseases. *Phytophthora* species are well known in agriculture, limiting production of many crops from potatoes to citrus (Erwin and Ribeiro, 1996). Classically, this is a genus of agricultural plant pathogens especially destructive in poorly drained soils or cool wet climates. The name derives from Greek for „plant killer". These are oomycetae, water molds, related to algae and are not true fungi (Mycota). Consequently, they are quite resistant to fungicide, but sensitive to antibiotics (Érsek, 1975). Despite their unique phylogeny, they grow as filamentous hyphae and reproduce by spores, like fungi. They disperse and infect by motile zoospores, and survive unfavorable conditions especially drying, as thick-walled chlamydospores or oospores. Most cause root diseases, but especially on trees, some cause some lethal stem cankers, or infect foliage. *Phytophthora* species are one of the most harmful agricultural pathogens all around the world, often causing serious damages. *Phytophthora infestans* used to cause the Irish potato famine 1845 to about 1860. Although the importance of potato as a staple food to countries lies only with the European countries since then but it extends all the way to the developing countries and especially in Africa. Farmers have tried to stop the disease using synthetic chemicals but it seems to exacerbate from time to time. Even where fungicides are used, continued wet spells often lead to major epidemics, as occurred for example in 2004 in Egypt and 2006 in Northern Peru and Kashmir. Even when epidemic conditions are not widespread, farmers in developing countries may get behind on spraying and lose control of the disease. This increases the chances of developing more resistance to these synthetic chemicals. Hence, the need for us to search for potential biological control for the *Phytophthora* strains which are becoming resistant to fungicides but are sensitive to antibiotics. *P. infestans* is a re-emerging pest ever since its discovery. It still causes major epidemics in potato and tomato crops worldwide (Forbes et al., 1994). For example, in 2000, 15% of the total potato crop of Russia was destroyed due to late blight. Such severe epidemics could trigger a new, potentially catastrophic potato famine (Schiermeier, 2001). Worldwide losses in potato production caused by late blight and measures to control the disease have been estimated at a billion dollar level annually (Duncan, 1999).

There are about 60 species in the genus *Phytophthora* that cause various disease symptoms, including root rot, fruit rot, foliar blight and stem blight on many economically important plants (Erwin and Ribeiro, 1996). A few examples of root rot pathogens are as follows: *Phytophthora sojae* (living on soybean); *P. fragariae* (on strawberries); *P. cryptogea* (on many plant species including tomato and cucumber); and *P. cinnamoni* (on various woody plant species). Other species cause leaf blight symptoms. *Phytophthora* is a good example of such a pathogen along with *P. porri* on leek. There are also many fruit rot pathogens such as *P. capsici* on various plants, and *P. palmivora* and *P. megakarya* on cocoa pods. The classification of *Phytophthora* species based on which part of the plant it predominantly infects is quite arbitrary. Disease symptoms are represented on all plant tissues, above and below ground. The environmental damage caused by *Phytophthora* diseases in natural ecosystems can be tremendous, due to difficulties in controlling the spread of the disease. An example of a severe ecological tragedy is sudden oak death, a disease caused by *P. ramorum*, which has emerged recently along the pacific coast of the United States. *P. ramorum* is destroying oak trees and is probably also spreading to other trees, such as redwoods and to other regions in North America (Knight, 2002). Likewise, *P. cinnamoni*, which has a very wide host range, infecting over 900 species of plants (Zentmyer, 1980), has caused severe epidemics in the jar rah tree forests Western Australia (Podger et al., 1965; Podger, 1972) as well as more recent outbreaks across the world. Hence, one can say the Irish potato famine is therefore not limited to a historic reference. In reality, many *Phytophthora* epidemics are just being kept under control by the use of prophylactic oomycetae (Talbot, 2004). When forest trees are grown in agricultural settings, such as in nurseries, they are vulnerable to agricultural diseases, including *Phytophthora* root rots. The *Phytophthora* species commonly involved in forest nurseries are often the same species affecting agricultural commodities in the area. Douglas-fir seedlings, for example, are affected by *P. megasperma* and other six *Phytophthora* species when raised in poorly drained nursery soils (Hansen et al., 1979). Tree seedlings that had been infected but survived in the nursery are likely to die in the first year after planting into forest sites. These nursery *Phytophthora* species, however, do not survive long in the forest soils and do not spread to surrounding trees (Roth and Kuhlman 1996; Hansen et al., 1980). They are adapted to agricultural soils, and cannot compete in the more complex forest soil microbial community and generally dryer, better-drained forest soils. Other *Phytophthora* species do very well in forests, however. In recent years, it has become clear that, there is a very different community of *Phytophthora* species resident and probably indigenous in more or less undisturbed temperate forests. Many of these are new to science. For example, eight *Phytophthora* species were isolated from oak forest soils in the Forét de Amance in NE France (Hansen and Delatour, 1999). There were no obvious symptoms of *Phytophthora* root rot in this healthy mature stand, yet 12 of 14 soil samples from one site yielded one or more species. Five of the eight *Phytophthora* species were not described or only very recently described. Similar results have been obtained in Germany (Jung et al., 1996, 2002), eastern deciduous forests in the United States (Balci et al., 2007), and the western US (authors' unpublished data). The indigenous forest *Phytophthora* community is numerous and diverse. In most cases, the *Phytophthoras* are confined to the fine roots of the trees, and while they kill fine roots, in normal soil conditions, the trees replace the roots without dramatic growth loss. In Europe *Phytophthora,* species may contribute to the recurrent, chronic disease called oak decline. Oak decline, however, is primarily associated with periods of unusual drought, often coupled with outbreaks of defoliating insects. Under these stressful conditions, loss of additional rootlets to *Phytophthora* contributes to the decline (Hansen and Delatour 1999; Jung et al., 2000). In contrast to the nursery soil *Phytophthora* species that are generally poorly adapted to forest soils, and to the indigenous *Phytophthora* community that persists in a dynamic equilibrium with its host trees, a few species qualify as truly destructive in the forests. These exotic, invasive species can threaten the economic viability and ecological sustainability of the forests they attack. Distinguishing exotic from indigenous organisms is sometimes difficult. The complex processes of coevolution assure, that the host and the pathogen coexist without either consistently affecting the reproductive fitness of the other. It is often presumed that the resulting disease symptoms will be subtle and perhaps difficult to detect, and the ecological impact will be slight. By this thinking, a *Phytophthora* species that kills trees rapidly and over an expanding area must be exotic. This line of reasoning, while compelling in some situations, must be used with caution. *Phellinus weirii* is a pathogenic fungus that causes laminated root rot, a lethal disease of Douglas fir in North American forests (Childs, 1963). The disease is dramatic, altering forest structure and composition and pathways of succession (Hansen and Goheen, 2000), yet the pathogen is indigenous to the forests where it is found. Another presumed characteristic of an alien population is genetic uniformity. An invading population likely started as one or a few individuals making the first beachhead. This would be an evolutionary „bottleneck", and should result in reduced genetic diversity in the new population. *P. cinnamomi* (Fig 3) occurs worldwide and causes severe root rot and dieback on Fraser firs, shortleaf and loblolly pines, azaleas, camellia, boxwood, and many other trees and woody ornamentals (Ferguson & Jeffers, 1999). The disease impacts a range of economic groups including nursery crops managed forests, and Christmas tree farms. Root infected rhododendrons and azaleas and tree saplings develop above ground leaf chlorosis, necrosis, wilt, leaf curl, and death. Stem necrosis may not occur for many weeks after the development of wilting symptoms. Belowground symptoms are most severe in poorly drained soils and include necrosis of young feeder roots and the lower vascular tissues around the crown and just below the soil line. Cankers may become visible at the base of 1-2 year old plant.

Fig. 3. Life cycle of (Left) and a symptom called resin soaked tissue, caused by *P. cinnamomi* Rands. The host is sand pine (*Pinus clausa* (Chapman ex Engelm.) Vasey ex Sarg) Image location: USA. The website of the picture: *Phytophthora* root rot. *Phytophthora cinnamomi* Rands forestryimages.org. Google pictures. Image No: 4823089. From Edward L. Barnard, Florida Department of Agriculture Consumer Services, Bugwood.org.

Phytophthora species reproduce both clonally and sexually. Many sexual species have in-breeding rather than outbreeding mating systems. Genetic diversity can be expected relatively low regardless of origin. As there have been no studies of diversity in indigenous clonal or in-breeding species, there is no diversity baseline against which to compare a suspected invading population (Érsek et al., 1995). *Phytophthora* species and the diseases caused by them are listed in **Table 8.**

2.2.2.3 Plant pathogenic fungi

Plant pathogens provide new challenge because of resistance problems. Necrotrophic plant pathogens have received increasing attention over the past decade. Initially considered to invade their hosts in a rather unsophisticated manner, necrotrophs are now known to use subtle mechanisms to subdue host plants. The gray mould pathogen *Botrytis cinerea* is one of the most comprehensively studied necrotrophic fungal plant pathogens. The genome sequences of two strains have been determined. Targeted mutagenesis studies are unraveling the roles played in the infection process by a variety of *B. cinerea* genes that are required for penetration, host cell killing, and plant tissue decomposition or signaling. Our increasing understanding of the tools used by a necrotrophic fungal pathogen to invade plants will be instrumental to designing rational strategies for disease control (van Kan, 2006). *Alternaria alternata* (Keissl, 1912) has been recorded causing leaf spot and other diseases on over 380 host species. It is opportunistic pathogen on numerous hosts causing leaf spots, rots and blights on many plant parts. It can also cause upper respiratory tract infections and asthma in people with sensitivity (Wiest et al., 1987).

Phytophthora sp.	Disease	Host	Disease Management
P. infestans	Late blight	Potato	Sanitation and fungicides
P. sojae	Root and stem rot	Soybean	Resistant varieties
P. palmivora	Black pod	Cacao	Sanitation
P. alni	Collar rot	Alder	Clean nursery stock
P. cinnamoni	Jar rah dieback	Jar rah eucalyptus	Sanitation
	Little leaf disease	Shortleaf pine	Change species
	Ink disease	Chestnut	Uncontrolled
	Avocado root rot	Avocado	Fungicides, Soil management
P. lateralis	Cedar root disease	Port-Oxford-cedar	Sanitation, avoidance and resistance
P. ramorum	Sudden oak death, Ramorum blight	Fagaceae and Ericaceae	Quarantine and eradication
P. cactorum	Crown rot	Strawberry	Prevention and sanitation
P. parasitica	Phytophthora stem-rot	Snapdragon	Prevention and sanitation
P. capsici	Phytophthora blight	Cucumber, Squash, Melons, Pumpkin,	Practice crop rotation
P. fragariae	Red stele	Strawberry	Proper site selection and preparation
P. megakarya	Black pod disease	Cocoa trees	Chemical control-fungicides
P. syringae	Fruit-rot	Apples	Chemical control
P. primulae	Root and stem rot	Parsley	Quarantine and eradication

Table 8. Some destructive ***Phytophthora*** species and the diseases they cause and available disease management.

2.3 Rationale

The target organisms discussed above are all harmful plant pathogen of economic significance. The options of using chemicals (traditional antibiotics or fungicides) against them are rather limited, because of the realistic danger of environmental pollution. Furthermore, the resistance problems (discussed in the Introduction) also should be overcome by introducing novel antimicrobials of different action mechanisms. EPB could also be used in vivo as an antagonistic biocontrol agent, in some niche like rhizosphere in forest threes.

3. Materials and methods

3.1 Materials

3.1.1 Entomopathogenic Bacteria (EPB)

The colonies of *Xenorhabdus* and *Photorhabdus* on indicator LBTA plates are demonstrated in Fig 4.

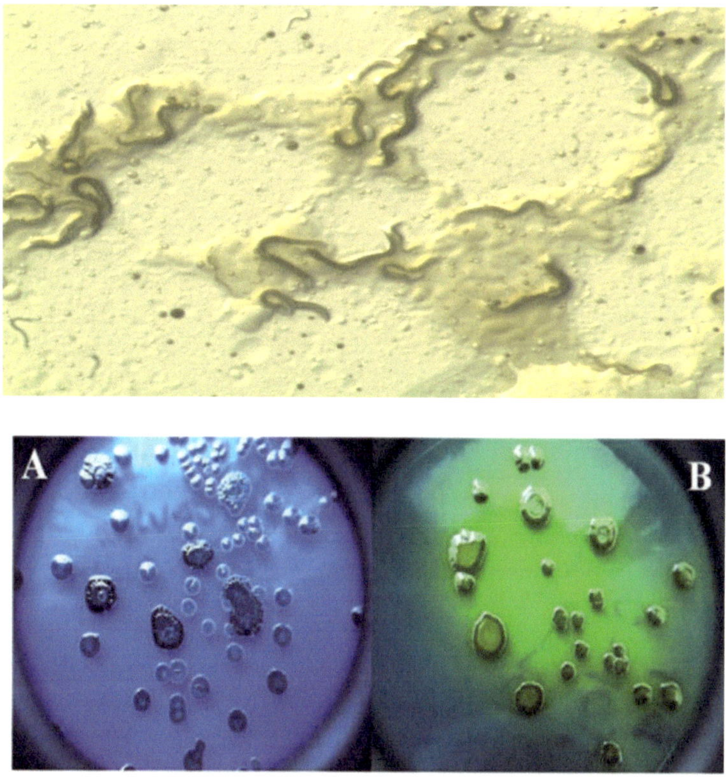

Fig. 4. Colonies of antibiotics-producing primary (1°) cells of a *Xenorhabdus* (A), and a *Photorhabdus* on LBTA indicator plates The bacteria were freshly isolated by Abate Birhan Addisie from the respective nematodes (above). Photo: Andrea Máthé-Fodor (nematodes) and Dr. Csaba Pintér (bacteria)

The list of EPB strains used in our experiments is given in **Table 9**. Some strains were provided by the COST 619 Research Community. Each of them had been deposited at the Georgikon Stock Center of Entomopathogenic Nematodes and Bacteria in Keszthely, Hungary. As for geographic origin, the AZ strains are from St. Miguel Island, Portugal-Azores, HP 88 from the USA, NC19 from North Caroline, USA, Brecon and Q614 from Australia. As for geographic origin, the AZ strains are from St. Miguel Island, Portugal-Azores, HP 88 from the USA, NC19 from North Caroline, USA, Brecon and Q614 from Australia. For further details, see references: Szállás et al., 1997, 2001 Peat et al., 2010; Völgyi et al., 1998, 2000; Lengyel et al., 2005; Brachmann et al., 2006; Furgani et al., 2008; Böszörményi et al., 2009).

Genus, Species	Subspecies	Strain	Reference
Xenorhabdus budapestensis		DSM 16342^{T*}	Lengyel *et al.*., (2005); Furgani *et al.*., (2008);; Böszörményi *et al.*., (2009)
Xenorhabdus .szentirmaii *		DSM 16338^{T}	Lengyel *et al.*., (2005); Brachmann et al.; 2006; Fodor *et al.*., (2007); Furgani *et al.*., (2008); Böszörményi *et al.*., (2009); Bode, (2009)
Xenorhabdus nematophila		ATTC 19061^{T*}	Völgyi et *al.*., .(1998), Völgyi *et al.*., 2000)
Xenorhabdus nematophila		N2-4	Szállás et al.., (1997); Völgyi *et al.*., (1998),
Xenorhabdus innexi		DSM1 6337^{T*}	Lengyel *et al.*., (2005); Furgani *et al.*., (2008);;
Xenorhabdus ehlersii		DSM 16336^{T*}	Lengyel *et al.*., (2005); Furgani *et al.*., (2008);;
Xenorhabdus cabanillassi		RIO-HU	Lengyel *et al.*., (2005); Furgani *et al.*., (2008);;
Photorhabdus luminescens	ssp. *laumondii*	HP88	
Photorhabdus luminescens	ssp. *laumondii*	Brecon	Szállás *et al.*., (1997),; Peat *et al.*., (2010); Marokházi et al; (2003);
Photorhabdus luminescens	ssp. *laumondii*	Az35	Szállás *et al.*., (1997),; Peat *et al.*., (2010);
Photorhabdus luminescens	ssp. *laumondii*	Az36	Szállás *et al.*., (1997),; Peat *et al.*., (2010);
Photorhabdus luminescens	ssp. *laumondii*	Az37	Szállás *et al.*., (1997),; Peat *et al.*., (2010);
Photorhabdus luminescens	ssp. *laumondii*	Az39	Szállás *et al.*., (1997),; Peat *et al.*., (2010);
Photorhabdus luminescens	ssp. *laumondii*	Q614	Szállás *et al.*., (1997),; Peat *et al.*., (2010);
Photorhabdus luminescens	ssp. *laumondii*	Az29	Szállás *et al.*., (1997),; Peat *et al.*., (2010);
Photorhabdus luminescens	ssp. *laumondii*	NC19	Szállás *et al.*., (1997),; Peat *et al.*., (2010);
Photorhabdus luminescens	ssp. *laumondii*	HP88	Szállás *et al.*., (1997),; Peat *et al.*., (2010);

Table 9. The list of the EPB strains used in this study

3.1.2 Plant Pathogenic Bacteria (PPB)

Plant pathogenic (PP) bacterium strains used in this study as test organisms are listed in Table 10 and 11.

SCIENTIFIC NAME	NCAIM * CODE	BIOASSAY		Origin
Erwinia amylovora	NCAIM B 01728	+	+	Hungary, apple; Hungary, M. Hevesi
	Ea 88 *str*R	+	+	USA,
	Ea110 *rif*R	+	+	McGhee & Jones, 2000
	Ea Ca 11 *str*R	+	+	McGhee & Jones, 2000
Klebsiella pneumoniae	HIP32 *chl*R	-	+	Human isolate
Kle. pneumoniae # 696	*carb*R	+	+	Mastitis isolate
Pantoea agglomerans	NPHMOS			Human pathogenic,

Table 10. List of antibiotics resistant bacteria used in overlay and /or agar diffusion bioassays of EPB antimicrobials. Abbreviation: * National Collection of Agricultural and Industrial Microorganisms, Hungary (NCAIM); http://ncaim.uni-corvinus.hu; *Kle* = *Klebsiella*

Altogether twelve *Erwinia amylovora* isolates (isolated by Dr. M. Hevesi, Hungary) were studied, but only *Ea* 1 (NCAIM B 01728) is listed here, since all the others reacted very similarly to all EPB CFCM tested. There are three antibiotic resistant strains: CA11 (rif S, *str* R); Ea88 (*rif* S, *str* R); and Ea110 Ea 110 (*rif* R, *str*S), (McGhee and Jones, 2000) also involved.

Dr. A.L. Jones (Michigan State University) kindly provided them in the list. Plant pathogenic bacteria isolated from different host plants are deposited in National Collection of Agricultural and Industrial Microorganisms, (NCAIM); Hungary. Home page: http://ncaim.uni-corvinus.hu. Other additional bacteria used in the second set of experiments were as follows: *P. agglomerans* 83873, obtained from the National Public Health and Medical Officer Service, Hungary.

SCIENTIFIC NAME	NCAIM * CODE	BIOASSAY		Origin
		Overlay	Agar diff.	Agar diff.
Erwinia amylovora	NCAIM B 01728	+	+	Apple; Hungary, M. Hevesi
Clavibacter michiganense ssp. michiganense	NCAIM B 01531.	+	+	Tomato; Hungary causing wilting and canker
Curtobacterium flaccumfaciens pv. betae	NCAIM B 01612.	+	+	Beans; UK causing leaf spot
Agrobacterium tumefaciens	NCAIM B 01681	+	-	Grapes; Hungary, crown gall disease
Bacillus subtilis	NCAIM B 01623.	+		Soil; Hungary, seed decay
Xanthomonas campestris pv. vesicatoria	NCAIM B 01857.	+	-	Pepper, Hungary, bacterial spot
X. campestris pv. carotae	NCAIM B 01699	+	-	Carrot, Hungary, carrot; bacteriosis
X. arboricola pv. juglandis	BK4	-	+	Hungary, walnut,
X. arboricola pv. juglandis	Szen 10	-	+	Romania, walnut,
X. axonopodis pv. phaseoli	B 01523	-	+	Hungary, beans
Pseudomonas fluorescens	NCAIM B 01670.	+	-	Hungary, soil, saprophytic
Pseudomonas fluorescens	NCAIM B 01667.	+	-	Hungary, soil, saprophytic
Pseudomonas syringae pv. syringae	NCAIM B 01556.	+	+	Hungary, rice, brown spot
Pseudomonas syringae pv. phaseolicola	NCAIM B 01689.	+	-	Hungary, beans, halo blight
Pseudomonas syringae pv. phaseolicola	NCAIM B 01715.	+	-	Hungary, beans, halo blight
Pseudomonas syringae pv. phaseolicola	NCAIM B 01776.	+	-	Hungary, beans, halo blight
Pseudomonas syringae pv. savastanoi	NCAIM B 01823.	+	+	Hungary, oleander, olive knot
Pseudomonas corrugata	NCAIM B 01637.	+	+	Hungary, tomato, pith necrosis
Pseudomonas syringae pv. lachrymans	NCAIM B 01537	-	+	Hungary, Cucumber,
Pseudomonas syringae pv. morsprunorum	NCAIM B 01684	-	+	Hungary, apricot,
Dyckeya chrysanthemi	NCAIM B 01839	-	+	Hungary, tomato
Burkholderia cepacia (syn. Pseudomonas cepacia	NCAIM B 01621	-	+	Italy, onion
Ralstonia solanacearum 879	PD2762	-	+	The Netherlands, potato
Ralstonia solanacearum	1070	-	+	Hungary, potato
Ralstonia solanacearum	1226	-	+	Hungary, potato,

Table 11. List of plant pathogenic bacteria used in overlay and / or agar diffusion bioassay tests. Abbreviation: * National Collection of Agricultural and Industrial Microorganisms, Hungary (NCAIM); http://ncaim.uni-corvinus.hu; *X.* = *Xanthomonas*

3.1.3 Eukaryotic plant pathogens: Phytophthora, Pythium, Botrytis, Alternaria and Fusarium species

Eukaryotic plant pathogens used in this study were oomycetae, such as; Phytophthora nicotianae; P. infestans, P. ramorum, P. cinnamoni, P. citricola; P. citrophthora and P. pelgrandis; and fungi, namely: Botrytis cinerea; Alternaria alternata; Fusarium gramineae. The oomycetal strains used in this study are listed in **Table 12.** All but K-39 had been isolated by András Józsa from Hungary and identified by Józsa & Bakonyi (in preparation) with classical and molecular tools. The fungus isolates Botrytis cinerea; Alternaria alternata and Fusarium gramineae (they are not separately listed) were also isolated in Hungary and kindly provided by Dr. Sándor Kadlicskó (University of Pannonia, Keszthely, Hungary).

SCIENTIFIC NAME	#Strain	Reference
Pythium sp.	JA 317	András Józsa – József Bakonyi
Pythium sp.	JA 319	András Józsa – József Bakonyi
Pythium sp.	JA 301	András Józsa – József Bakonyi
P. citricola	JA 74	András Józsa – József Bakonyi
P. citrophthora	JA 479	András Józsa – József Bakonyi
P. nicotianae	JA 168	András Józsa – József Bakonyi
P. nicotianae	JA H 1/100	András Józsa – József Bakonyi
P. cactorum	JA 163	András Józsa – József Bakonyi
P. pelgrandis	JA 337	András Józsa – József Bakonyi
P. cinnamoni	JA 153	András Józsa – József Bakonyi
P. megaspermium	JA 209	András Józsa – József Bakonyi
P .infestans	K-39	József Bakonyi

Table 12. List of plant pathogenic oomycetae used as test organisms in studying anti-oomycetal activity of Xenorhabdus CFCM agar diffusion tests. #Strains are deposited at the Institute of Plant Protection, Hungarian Academy of Sciences by József Bakonyi. As for Botrytis cinerea, see Rosslenbroich & Stuebler (2000).

3.2 Media and cultures

3.2.1 Bacterium media and cultures

Ingredients of bacterial media used in this study are summarized in Table 13.

Basic components were autoclaved (20', at 121 °C). Some ingredients should be added after autoclave, such as are BTB, TTC, antibiotics and (Leclerc and Boemare, 1991) CFCM. Both EPB and PPB strains grew on LB, and LBA. In overlay bioassay a 5 µl O/EPB culture s dropped to the center of a plate. After 5 days, it was overlaid with text bacteria suspended in a soft (0.6 w/v) agar. The size of the inhibition zone was measured around the colony. When LBA was used as "poisoned agar" 500 ml water was added before, and + 500 ml of diluted CFCM after autoclave.

Name of the media	Final ml	INGREDIENTS						USED FOR
		Before autoclaving				After autoclaving		
	Water	Agar	NaCl	Trypton	Yeast extract	BTB+	TTC++	
LB	1000	-	5 g	10 g	5g	-	-	Culturing EPB and PPB
LBA	1000	15 g	5 g	10 g	5g	-	-	Culturing EPB and PPB; For overlay bioassay
LBTA	1000	15 g	5 g	10 g	5g	1 ml Stock	1 ml Stock	Indicator plate
Mc Conkey	1000	35 g of Commercially available powder						

Table 13. Bacterium media used in this study. Abbreviations: LB = Luria Broth; LBA = Luria Broth Agar; LBTA= Luria Broth Agar supplemented with indicators stains, bromothymol blue (BTC) and triphenyltetrazolium chloride; + and ++: Stock solutions of BTB and TTC, 25 mg/ml and 40 mg/ml chloride (TTC) dissolved in ethanol (Leclerc & Boemare, 1991; Ausubel et al., 1999); EPB and PPB = entomopathogenic (antagonist producing) and plant pathogenic (test) bacteria. On indicator plates antibiotic producing 1° can be distinguish from 2° variants and from contaminants.

3.2.2 Preparation of Cell-Free Conditioned Xenorhabdus Media (CFCM)

Xenorhabdus szentirmaii and *X. budapestensis* had been isolated from their nematode symbiont, *Steinernema rarum* and *S. bicornutum* and identified by us (Lengyel et al., 2005). *Xenorhabdus szentirmaii* and *X. budapestensis* were maintained Luria Broth (LBA) medium and sub-cultured freshly. Because of the instability of the phase I under normal culture condition, glycerinated stocks of the bacteria frozen at -80°C were used as starting material for culture. To ensure the presence of phase I, the glycerinated stocks were incubated in the dark at 28 on LBTA (see above). Phase I is distinguished from phase II by its adsorption of LBTA to produce a blue colony with a clear zone around bacteria culture for antibiotic production was prepared by inoculating a loopful of phase I of *Xenorhabdus szentirmaii* or *X. budapestensis* growing on an LBTA plate into a test tube 5 ml fresh LB medium. At the and of the log phase this volume of cell suspension was added into baker containing 100-300 ml of sterile LB and was cultivated at 28°C on a laboratory shaker at 200 rev min^{-1} for 24-48 h, during which time the optical density respectively. A starter culture was produced first in 5.0 ml of LB media. 5-ml test tubes cultured were used as inoculum for scaling up larger cultures. The liquid cultures were scaled up in to 250, 500 and 1000 ml volume by using the 5 ml overnight cultures as inocula and grown in Erlenmeyer flasks, shaken for a couple of days. Finally, in the mid stationary phase, cultures were centrifuged at 13,000 g and filtered through a 0.22 µm pore size filter. After cooling down, the liquid medium was measured into 5 ml test tubes. A clean growing colony from a Petri-dish that had been grown with the bacterium, a loop-size colony was suspended into 5-ml volume of liquid medium in a test-tube. For each bacterium grown in liquid a non-inoculated control (a test-tube with liquid medium) was left for comparison after the growing period. Aliquot samples were taken regularly to check the antibiotics activity. Not only the antibiotics producing but also the test bacteria were growing on LB or LBA as well.

3.2.3 Cultures for eukaryotic plant pathogens

Ingredients of media used for studying Oomycetae and Fungi are summarized in Table 14.

Name of the media	Final ml	INGREDIENTS				USED FOR
		Before autoclaving				
	Water	Agar	Chopped carrot^{+}	Chopped Potato^{++} or Potato infusion	Ford hook 242 Dried lima bean seed	
CA (Carrot Agar)	1000	15g	200 g			Culturing *Phytophthora, Pythium*
PDA (Potato Dextrose Agar)						Culturing eukaryotic plant pathogens
	1000	20 g		200 g		
LiBA (Lima Bean Agar)	1000	20g			5 g	

Table 14. Ingredients of media used for culturing of and assaying on plant pathogenic oomycetae and fungi in this study. For bioassay, the "poisoned agar" version is used: instead of 1000 ml DW only 500 ml DW is to be added to the media before autoclaving and 500 ml cell free conditioned media (CFCM) of different dilution after autoclaving, before pouring the media into plates.

As for CA, carrots were washed and using a knife peeled and cut into small pieces. 200g of the chopped carrots were put into a grinder. Some water was added in to the chopped carrots during the procedure. The carrots were ground into a fine mixture of water suspension and filtered through a cloth in funnel into a new beaker. 30g agar was added. The mixture was filled up with distilled water (DW) up to a 1000 ml before autoclaved (20'; at 121^{0}C) and dispensed into Petri plates. A. Józsa established this technique in our lab. Lima Bean (LiBA) Agar Media was described) and used for growing stock cultures of *Phytophthora megasperma* var. *sojae* by Hilty & Schmitthenner (1962).

As for PDA, 200 g sliced unpeeled potatoes are boiled in 1 liter DW for 30' and filtered through cheesecloth, or commercially available potato infusion dehydrated form was used. Filtered (sterile) ingredients were added after autoclave (15', 121^{0}C) before dispensed 20-25 ml portions into sterile Petri dishes. Final pH= 5.6 ± 0.2. Medium should not be re-melted more than once. Commercially available media require extra agar supplement with to a final concentration of 20 g/liter. These media could also be used as "poisoned agar" if only 500 ml water is added before autoclave and it is it is supplemented with 500 ml of different dilutions of the CFCM after autoclave.

3.3 Bioassays

3.3.1 Bioassays of antibacterial activity

3.3.1.1 Overlay bioassay

When using overly tests in plates, we incubated plates with a 5 µl of overnight (O/N) liquid culture of the antibiotics producing bacteria for 5 days in a 25-oC incubator. The antibacterial activities of the *Xenorhabdus* and *Photorhabdus* species have routinely been tested against laboratory standard *E. coli* strains. In these tests one colony were suspended in 5 ml of LB and incubated for 24 h (up to nearly stationary phase). The plates were then overlaid with 55 ^{0}C soft (0.8 % w/v) agar containing the suspension of the liquid culture of test bacteria) in 3%. The diameters (given in mm) of the inhibition circles were measured after five days. In later sets of experiment plant, pathogenic bacteria were used as test organisms (various *Pseudomonas spp., Erwinia spp., Clavibacter spp., Agrobacterium tumefaciens, Xanthomonas spp., Curtobacterium flaccumfaciens spp.*). The plates were then incubated at 37 ^{0}C for 24h. The diameter of the inactivation zone around the antibiotics producer *Xenorhabdus* colony was measured.

3.3.1.2 Agar diffusion test

A. Antimicrobial disc assay: In plastic Petri plates of 8.5 cm diameters 15 ml LBA media were poured producing a 7 mm thick solid media. 3 ml of *Erwinia* or other test bacterium cells suspended in soft (0.6 w/v) agar in 50 ^{o}C was poured onto the surface of the media. The density of the bacterial cells was 1:100 of an O/N (stationary phase) *Erwinia* cultures. BBL* blank paper discs had been submerged in the solutions to be tested and then placed on the surface of the agar. The plates were scored on the next day and the diameters of the inactivation zones were measured around each disc. There were three replicates made for each test.

B. Antimicrobial hole test assay: *Erwinia* or other test bacterium cells (1:100 of an O/N (stationary phase cultures) were given to LBA media of 50 ^{o}C temperature (10 V/V %), mixed carefully and poured into plates (18 ml/plate). Right after the agar, solidified holes of 1 cm diameter were made by using a cork borer. 100 µl volume of each solution to be tested was pipette into one hole. The plates were scored on the next day. The diameters of the inactivation zone were measured around each disc. There were at least two replicates made for each test. This technique was standardized in Dr. Hogan's lab this summer (in preparation). The pictures of overlay and of an agar diffusion assays are shown in **Fig 5** and **Fig 6**, respectively.

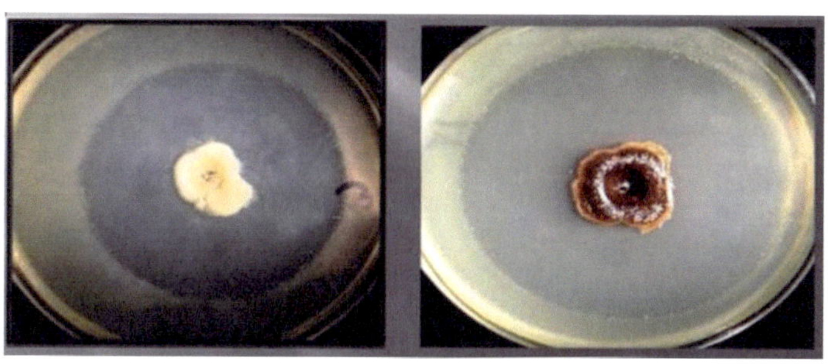

Fig. 5. Antibacterial potential of 5-d old colonies of *Xenorhabdus budapestensis* (Left) and *X. szentirmaii* (Right) on Gram-negative target, *Escherichia coli* OP50. Overlay bioassay on LBA plates. Photo: Dr. Cs. Pintér

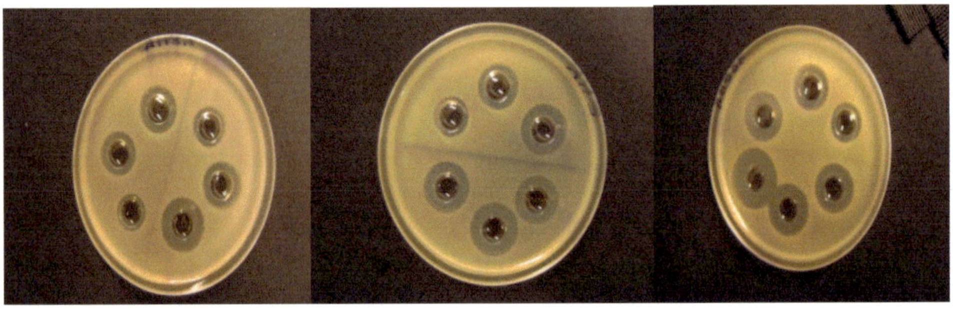

Fig. 6. Agar-diffusion plate diffusion plates Test bacteria: LEFT: Pseudomonas syringae pv. syringae CFCM of Xenorhabdus budapestensis ("West") and X. szentirmaii ("East); MEDIUM: Curtobacterium flaccumfaciens CFCM of X. budapestensis ("North") and X. szentirmaii ("South); RIGHT Xanthomonas axonopodis CFCM of X. budapestensis ("North") and X. szentirmaii ("South). The dilutions of CFCM administered into the holes were 10, 20 and 40 V/V respectively. The sizes of the inhibition zones proved reproducibly dose dependent. Photo: M. Hevesi.

3.3.1.3 Determination of MIC 95 (MID95) values

When exact concentration of bacterial cells needed, it was determined using a spectrophotometer at 600 nm wavelength. Alternatively, for antibiotic production we found that the optimal length of the incubation in liquid is 3-5 days. Considering that the cell-free conditioned *Xenorhabdus* media (CFCM) is a mixture of biologically active compounds of unknown nature, we did not use the term of EC50 and EC95, (concentration of the antibiotics resulting it 0 or 50% growth of the test bacteria). We used the term of "Maximum Inhibiting Dilution" 95 given in V/V % (see Furgani et al., 2008) instead. (100 V/V% is the undiluted CFCM, and 0 V/V% is the LB media). To determine MID 95 we used LB media in 200 µl volume containing 2 µl of test bacterium suspension. It contained of 2 µl test bacterial O/N culture of 1:10 dilution. We worked with 96-hole cell culture or microtiter plates. EMC and EMA cultures were tested in 0, 10, 20, 30, 40, 50, 60, 70, 80, 90 and 100 V/V % dilutions. There were 4 replicates for each dilution. We considered the putative MID 95 (Maximum Inhibitory Dilution, if no bacterial growth was observed by using Sensititre Manual Viewer in 24 h (or by bare eye) at a given concentration. To make determine whether the bacteriostatic affect were complete 100 µl volume of each (visually clean) mixture were plated out. We plated a serial dilution of the culture from - 10^5 dilutions and counted the colonies grown on the plates. The largest dilution (the smallest concentration) of the media was considered as MID95 (practically MID100), if no colony grew in these plates. The untreated controls were diluted and plated similarly. The colonies could be counted only at 10^{-7} dilutions in the control. The LC95 of the commercial antibiotic was determined very similarly using 96-hole cell culture or microtiter plates.

3.3.1.4 In planta bioassays of antagonistic activity on fire blight inflammation

This method was described before (Böszörményi et al., 2009).

3.3.2 Bioassay techniques for quantifying anti-oomycetal and anti-fungal activities

3.3.2.1 Bioassay of Anti-Oomycetal Activities

The antagonistic activities of the CFCM of two *Xenorhabdus* species (*X. budapestensis, X. szentirmaii*) have been tested on four *Phytophthora* and three *Pythium* isolates and evaluated qualitatively and quantitatively. In both experiments, we used "poisoned" carrot agar (PCA, see below) plates of 10 ml volume in 5.5-cm diameter sterile Petri plates or of 20 ml volume in 8.5-cm diameter sterile Petri plates. These media contained double carrot agar and different concentrations (dilutions) of cell-free conditioned media (CFCM) of *Xenorhabdus szentirmaii* (or *X. budapestensis*) in 1:1 ratio. We ran several sets of experiments. In the first run, we compared the dilutions of 0, 25, 50 and 75% v/v dilutions. In the second run, we compared 0, 10, 30 and 40% v/v dilutions of both antibiotics producing bacteria. For confirmation, we ran a third set by using 0, 25, 37.5 and 40% v/v volumes. Each assay carried out on a 1-cm diameter circular-shape carrot-agar disc with *Phytophthora* (or *Pythium*) obtained from culture plate by sterile cork borer, put into the center of each plate. Evaluation: The diameter of the growing mycelia of the *Phytophthora* on the carrot-agar was measured after different days following inoculation

3.3.2.2 Bioassay of Anti-Fungal Activities

The antagonistic activities of the CFCM of two *Xenorhabdus* species (*X. budapestensis, X. szentirmaii*) on one *Botrytis cinerea, Alternaria alternata* and *Fusarium graminearum* isolates were studied in a very similar manner as t Agar (LiBA) instead of CA. The fungi were cultured on PDA plates. The anti-oomycetal studies, with the only difference that media was Potato Dextrose Agar (PDA) or Lima Bean Agar (LiBA).

3.4 Methods of genetic analysis

3.4.1 Transposon tagging of genes of interest

The principle of STM is that a mobile genetic element (Tn10) should be randomly inserted into the genome with the help of the transposon via conjugation. The gene in question is inactivated by insertional mutation. a transposon is used which inserts itself into the gene sequence. When that gene is transcribed and translated into a protein, the insertion of the transposon affects the protein structure and (in theory) prevents it from functioning. In STM, mutants are created by random transposon insertion and each transposon contains a different 'tag' sequence that uniquely identifies it. If an insertional mutant bacterium exhibits a phenotype of interest, such as loosing the ability to produce antibiotics then we will sequence its genome and run a search (on a computer) for the tag used in the experiment. When a tag is located, the gene that it disrupts is also thus located (It will reside somewhere between a start and stop codon which mark the boundaries of the gene).

In our experiments the donor stain was *E. coli* S17 *E. coli* S17-pir pLOF (came from Dr. János Kiss (Agricultural Biotechnology Center, Gödöllő, Hungary), and the recipient was *Xenorhabdus*. The antibacterial activity of population was tested on *E. coli* OP 50 in LBA plates or LB liquid media.

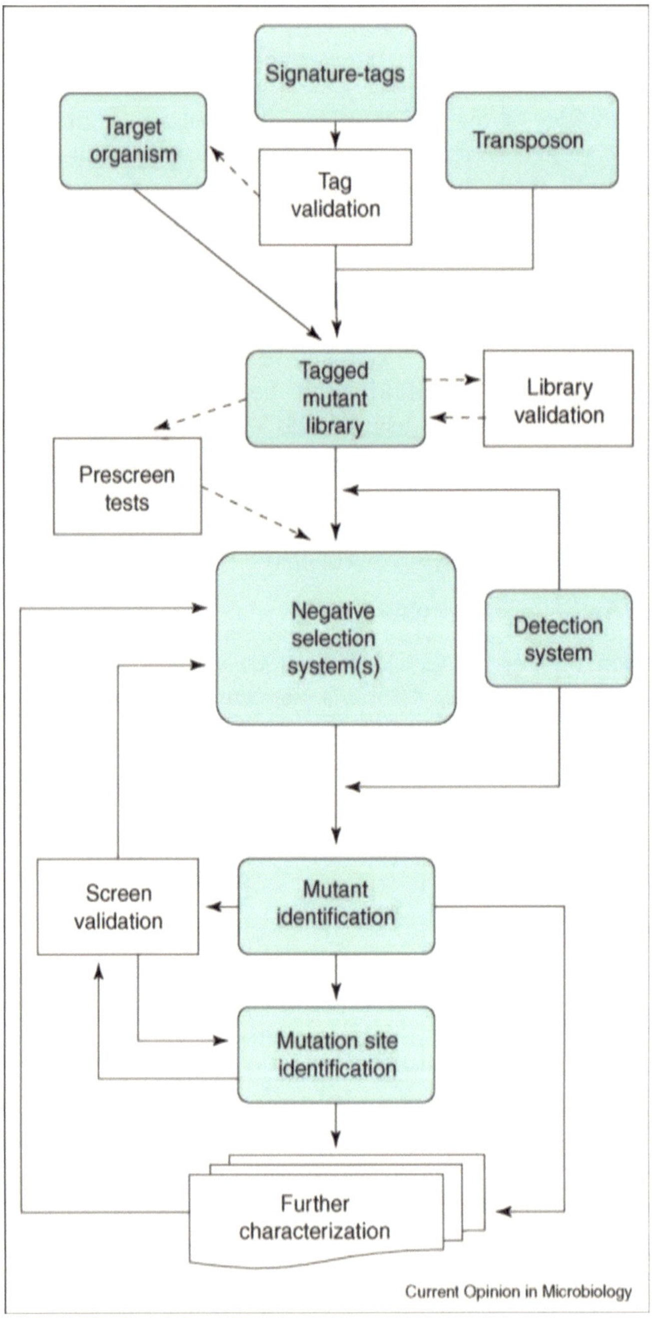

Fig. 7. Flow chart overview depicts Signature Tagged Mutagenesis (STM) of the different modules and their interconnectivity (kindly provided by Prof. Heidi Goodrich-Blair, Univ. Wisconsin. Madison). By using this technique, the genetic background of several important functions related to colonization, pathomechanisms and symbiotic behavior of Xenorhabdus nematophila have been revealed (Heungens et al., 2002; Martens et al., 2003; Richards et al., 2009) in her laboratory.

3.4.1.1 Isolation of spontaneous rifampicin resistant Xenorhabdus mutants

Spontaneous rifampicin resistant (rif^R) mutants of the target organisms are recipients. They were obtained from both *X. budapestensis and X. szentirmaii* by selection on LBA plates containing 50 μg / ml (rif_{50}). The individual colonies were picked up by using sterile loop and transferred to sterile test tube into 5 ml LB containing 50 μg / ml rifampicin. The cultures were grown over night (O/N) on rotary shaker at 200 rpm (at ~ 25 °C for further experiment. Using the same protocol, *E. coli* S17- λ^{pir} pLOF (which carries the transposon gene) was grown in LB containing kanamycin 30μg/ml (kan_{30}).

3.4.1.2 Conjugation

The insert of plasmid was subjected to transposon mutagenesis by the use of a genetically engineered derivative of Tn*10*, which will be referred to as mini-Tn*10*pLOF/Km (Simon et al., 1983; Herrero et al., 1990). This construction behaves as a mobilizable plasmid could be replicated only in the in S17 λ^{pir} background. Consequently, if it could be transferred from a conjugative *E. coli* strain to *Xenorhabdus*, where it is be capable of being inserted into the chromosome but unable for further "jump". The easiest and way is to transfer Tn10 via conjugation from *E. coli* to *Xenorhabdus*. There are several technical alternatives for promoting conjugation, either in a LBA plates or in a filter. One then could select for the respective mutant by double antibiotics selection. The technique what was establishes in our laboratory are given below **(Table 15).** This is a modification of the original protocol of Herrero et al. (1990) by A. Máthé-Fodor.

3.4.1.3 Mutant hunt: Selecting for transconjugants

Each survivor on the selective media was tested for the differential dye uptake on indicator plates. The spontaneous secondary (2°) variants (Völgyi et al., 2000) were discarded. We considered "transconjugants" those cells from the conjugation mixture, which survived and develop colony in the presence of 50μg /ml rifampicin and 30 μg /ml kanamycin. They were tested for antibiotic activity in overlay bioassay (see above) and *E. coli* OP50 was used as test organism. Putative pigment mutants of *X. budapestensis* were identified visually. The putative mutants have been stored in 20% glycerol / LB solution at -80°C as described by Miller (1972). Crystal mutants were screened based on colony color morphology.

BEFORE STARTING THE TRANSPOSON MUTAGENESIS EXPERIMENT	*NOTES*
Choose a donor: **Donor**: S17-1 pLOF (*Km*®) or ChlR®)	
Choose a recipient: *EPB* strain **(Xenorhabdus szentirmaii, X. budapestensis)**	
PREPARATION 1	
ISOLATION OF SPONTANEOUS RIFAMPICIN RESISTANT (rif®) MUTANTS	
Make an overnight (O/N) LB liquid culture of he chosen EPB bacterium	
Seed at least 30 LB plates containing Rif100 with 0.1 ml EPB bacterium suspension	***Keep the plates in dark, Use aluminum folia***
Isolate single colonies, which had been grown for 48-72 hrs. (Not later, because the rifampicin can be decayed)	
Prepare LB rif100 and LB Km30 LB plates	
PREPARATION 2	
Prepare 50 ml of LB rif100 LB liquid media	
Prepare 50 ml Km 30 LB liquid media	
Prepare 30 LB rif100 plates.	***Keep plates at 4 °C***
Prepare 30 Km 30 LB plates	

MUTAGENEZIS PROTOCOL

1st DAY

Pick 2-3 day-old SINGLE colonies from your donor on LB-Km30.
Pick 2-3 days old SINGLE colonies from your recipient on LB-Rif100 plates. — ***Use fresh plates! In case, keep the plates*** at 4 °C.
Inoculate 5-ml liquid LB Km30 with 1 colony of the donor
Inoculate 5-ml liquid LB rif100 with 1 colony of the recipient
Make overnight cultures (O/N) from both.

2nd DAY

Inoculate 50 ml LB Km30 liquid culture with 1ml O/N Donor culture.
Inoculate 50 ml LB rif100 liquid culture with 1 ml O/N Recipient culture
Incubate the donor at 37 °C or 30 °C
Incubate the recipient at 30 °C — Do not shake or roll!
Stop growing the recipient at OD (600nm) 0.2-0.3.
Mixed the parents as follows:
DONOR / RECIPIENT cells is 1:2

Calculation:
[(OD of the donor) x 2] / (OD of the recipient) = (ml of recipient) to be added to each ml of donor

Take a sterile glass Petri plate
Pipette 1 ml of the donor onto the plate.
Pipette 4 ml of LB liquid media to dilute it.
Take a sterile syringe, to which bacterium filter could be adjusted
Take out the 5 ml diluted donor cell suspension with the sterile syringe
Adjust sterile 0.22 µm filter to it.
Push through the filter and discard supernatant
Disconnect the filter and the syringe
Wash donor cells. Take 5 ml of sterile LB media into the syringe (use a new sterile one). — Repeat this step twice or 3X
Adjust the sterile 0.22 µm filter containing the donor cells to it.
Do the same with the recipients:
Put the calculated volume of recipient suspension onto another sterile plate.
Dilute it up to 5 ml by adding sterile LB.
Take up the 5 ml diluted recipient suspension with ANOTHER sterile syringe
Adjust the SAME FILTER you had used with the donors to the syringe with the recipient cell suspension
Push the suspension through the filter.
Disconnect the filter and the syringe.
Take 5 ml of sterile LB media into the syringe (you should use a new sterile one)
Take out the 0.22 µm filter paper of which the donor and recipient cells are with a forceps onto the surface of a sterile LB agar plate for a couple of hours
After a couple of hours transfer the filter paper onto the surface of an LB plate containing IPTG35
Incubate overnight, let the cells make love (conjugate)

3rd DAY

Wash the cells off filter with 4 ml of LB liquid media
Each of 100 µL aliquots should be spread on separate Km30 + Rif100 LB agar plates
Incubate them O/N at 30 °C — ***Only trans-conjugants will grow***

4th DAY

Prepare about 60 fresh LB Km30 plates what you need to use on the fifth day

5th DAY

Transfer individual colonies grown up on the surface of the Km30Rif100 LB plates onto Km30 plates in a replicable order by using sterile toothpicks. We usually put 25 colonies onto one plate

Table 15. Transposon tagging (mutagenesis) protocol used in our laboratory by A. Máthé-Fodor.

4. Results and discussion

4.1 Results on plant pathogen bacteria

4.1.1.1 Effects on Xanthomonas strains

The CFCM of both *X. budapestensis* and *X. szentirmaii* exerted strong growth-inhibiting effects against the *Xanthomonas* strains studied. Part of the results came from overlay bioassays others from Agar Diffusion tests.

Data are presented in **Table 16**

SCIENTIFIC NAME	NCAIM * CODE	Overlay bioassay, Diam. inactivation zone (mm)		Agar diffusion assay – diam. the inactivation zones in mm	
		EMA	EMC	EMA	EMC
Xanthomonas campestris pv. carotae	NCAIM B 01669.	51.00±1.22	48.50±11.67	NT	NT
Xanthomonas campestris pv. vesicatoria	NCAIM B 01857.	58.50±0.50	19.00±2.60	NT	NT
Xanthomonas arboricola pv. Juglandis (Szen 10)	-	NT	NT	17.94±1.83	24.77±1.40
Xanthomonas axonopodis pv. phaseoli	-	NT	NT	20.52±3.18	22.04±2.76
Xanthomonas axonopodis pv. phaseoli	-	NT	NT	20.52±3.18	22.04±2.76

Table 16. Growth inhibiting effects of *Xenorhabdus* antimicrobials on *Xanthomonas* strains in overlay bioassay and agar-diffusion tests. In different degree, but each strain proved sensitive. Abbreviations: EMA = *X. budapestensis*; EMC = *X. szentirmaii*.; NCAIM: National Collection of Agricultural and Industrial Microorganisms, Hungary.

4.1.1.2 Effects on Erwinia, Pectobacteria and Pantoea strains

Each *E. amylovora* strain was equally sensitive to antimicrobials of different EPB strains, especially to those of *X. budapestensis* and *X. szentirmaii* regardless their resistance to any other conventional antibiotics (**Table 17**).

SCIENTIFIC NAME	NCAIM * CODE	Streptomycin, 200 ug/ml	*X. nematophila*	*Photorhabdus luminescens* ssp.		
			N2-4	*laumondii* ARG	*akhurstii* IS5	*akhurstii* EG2
Erwinia amylovora	NCAIM B 01728	40±0.3	36±0.5	23±0.5	21±0-5	13±05
	Ea 88 *str*R	40±0.5	35±0.5	24±0.5	22±0.5	13±0.5
	Ea110 *rif*R	13±0.2	35±0.5	24±0.5	23±0.5	14±0.5
	Ea Ca 11 *str*R	26±02	36±0.5	30±0.5	22±0.5	13±5

Table 17. Growth inhibiting effects of different EPB antimicrobials on *E. amylovora* strains resistant to different conventional antibiotics - results of overlay bioassays

EPB antimicrobials were active against chloramphenicol resistant human pathogenic *Pantoea, Klebsiella pneumoniae* and mastitis isolation *Kle. pneumoniae* #696 (see Table 18). This facts support the idea of considering *Xenorhabdus*, antimicrobials as potential tools against poly-resistant harmful plant pathogens, belonging to *Agrobacterium tumefaciens, Clavibacter michiganense ssp michiganense, Curtobacterium flaccumfaciens pv. betae* and *Dyckeya chrysanthemi* species (see Fig. 8, Table 18).

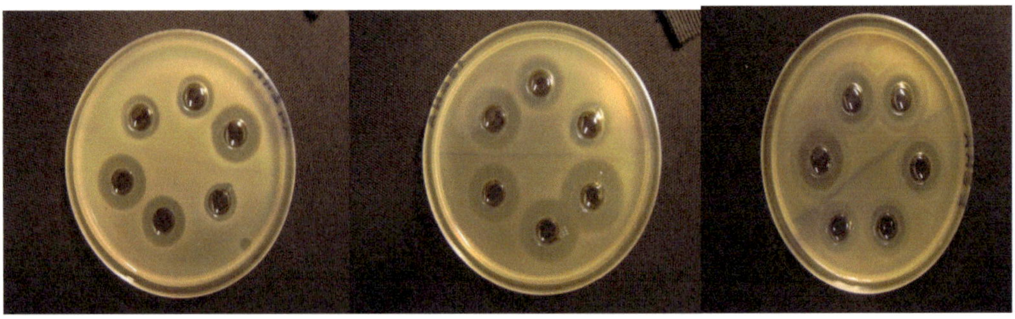

Fig. 8. Sensitivity of antibiotics resistant Gram-negative bacteria to the antibacterial activity of *Xenorhabdus budapestensis* ("North", "North" and "South-East", respectively, and *X. szentirmaii*. Test organism from left to right: Human pathogen *Klebsiella pneumoniae (HIP32)* (chloramphenicol resistant); *Erwinia amylovora* Ca 11(rif S, str R); *Pantoea agglomerans* (human pathogen, closely related to *Erwinia* species). See data in Tables 17 and 18.

SCIENTIFIC NAME	**NCAIM * CODE**	Overlay bioassay, Diam. inactivation zone (mm)		Agar diffusion assay – diam. the inactivation zones in mm	
		EMA	EMC	EMA	EMC
Erwinia amylovora Ea1	Ea1	44.00±0.50	43.33±1.44	20.79±0.69	24.82±1.13
Erwinia carotavora ssp. atroseptica	NCAIM B 01611	51.33±1.22	49.00±3.28	NT	NT
Pantoea agglomerans (related human pathogen)		NT	NT	17.43±4.81	17.84±0.83
Klebsiella pneumoniae (HIP32) chloramphenicol resistant		NT	NT	18.59±2.96	20.05±0.43
Dyckeya chrysanthemi		NT	NT	19.24±0.84	24.64±0.83
Agrobacterium tumefaciens	NCAIM B 01681	36.67 ±1.15	24.33 ±4.04	NT	NT
Clavibacter michiganense ssp michiganense	NCAIM B 01531.	47.33±2.33	NT	NT	NT
Curtobacterium flaccumfaciens pv. betae	NCAIM B 01612	34.16±0.76	47.67±2.08	19.18 ±0.88	23.10 ±0.09

Table 18. Growth inhibiting effects of *Xenorhabdus* CFCM on sensitive plant pathogenic and related bacteria. Abbreviations: EMA = *X. budapestensis;* EMC = *X. szentirmaii.;* NCAIM: National Collection of Agricultural and Industrial Microorganisms, Hungary. Home page: http://ncaim.uni-corvinus.hu Number of replicates: 3

4.1.1.3 Effects on Pseudomonas strains

The *Pseudomonas* strains reacted rather differently. Two independent strains *P. fluorescens* proved sensitive to *X. budapestensis* but hardly reacted to *X. szentirmaii* in overlay bioassay. Off the *Pseudomonas syringae* variants *pv. glycineae* proved far the most reactive and the reaction was similar toward the two CFCM. There were quantitative differences between the reaction of three *P. syringae pv. phaseolicola* isolates to the two CFCM. *P. syringae pv. syringae* double more sensitive to *X. budapestensis* than to *X. szentirmaii* in overlay bioassay but the results of agar diffusion assays did not differed significantly. Data are presented in **Table 19.** *P. syringae pv. savastanoi* proved completely inactive in repeated agar diffusion tests, but showed significant activity in overlay bioassays. *Pseudomonas corrugata, Pc 12, Burkholderia cepacia (syn. Pseudomonas cepacia),* as well as *Ralstonia solanacearum 1070, Ralstonia solanacearum 1240P. corrugata Pc 12* proved completely resistant to the antibacterial potential of both *X. szentirmaii* and *X. budapestensis* in agar diffusion tests.

4.2 Results on eukaryotic plant pathogens

4.2.1 Results on Oomycetales: Phytophthora and Pythium strains

In order to see the potential EPB antibiotics on eukaryotic plant pathogens, the anti-microbial activities of *X. szentirmaii* DSM 16638 (EMC) were assayed on some different *Phytophthora* and *Pythium* isolates qualitatively.

SCIENTIFIC NAME	NCAIM * CODE	Overlay bioassay, Diam. inactivation zone (mm)		Agar diffusion assay – diam. the inactivation zones in mm	
		EMA	EMC	EMA	EMC
Pseudomonas fluorescens	NCAIM B 01670	38.33 ±4.00	26.00 ±5.29	NT	NT
Pseudomonas fluorescens	NCAIM B 01667	41.00±2.00	19.33±3.05	NT	NT
Pseudomonas syringae pv. syringae	NCAIM B 01556	45.50±0.50	23.83±2.02	16.88±0.88	20.13±4.3
Pseudomonas syringae pv. phaseolicola	NCAIM B 01689	49.00±4.00	42.33±3.21	NT	NT
Pseudomonas syringae pv. phaseolicola	NCAIM B 01776	27.50±4.00	24.00±5.20	NT	NT
Pseudomonas syringae pv. phaseolicola	NCAIM B 01715	NT	45.17±4.25	NT	NT
Pseudomonas syringae pv. glycineae	NCAIM B 01574	48.75±10.04	45.33±3.05	NT	NT
Ps. syringae pv. lachrymans		NT	NT	22.54±1.58	23.97±1.75
Ps. syr. pv. morsprunorum		NT	NT	14.60±0.78	14.88±1.27

Table 19. Effects of *Xenorhabdus* CFCM on sensitive plant pathogenic *Pseudomonas*. Abbreviations: see Table 16.

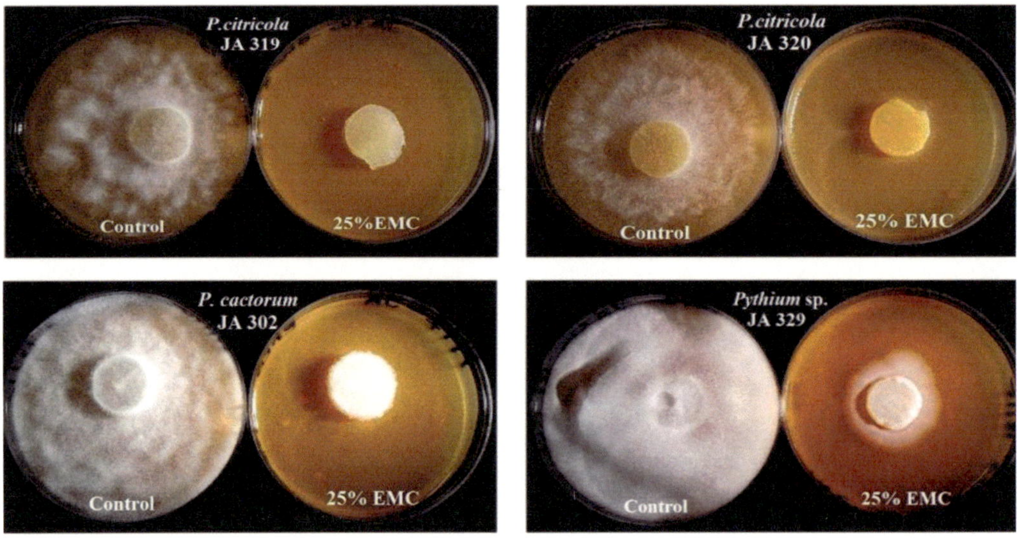

Fig. 9. Three different "Petri-plate phenotypes" showed of the extremely sensitive plant pathogen oomycetes treated with a low dose of *X. szentirmaii* CFCM on "poisoned carrot agar (PCA) plates.

As demonstrated on **Fig 9** each strain proved extremely sensitive to the presence of EPB antimicrobials in the "poisoned agar", even at low doses as 25 V/V%. There were three "phenotypes" of the plates. A). There were no mycelial growth (shown by in two independent isolate of *P. citricola;* B) There were some vertical growth from the disc, but no horizontal growth, (*P. cactorum*) indicating, that the antimicrobial compound was so cytotoxic, that it killed each cell which got contact to the agar, but the active compounds could not penetrate through the disc; C). Some *Pythium* was able for a horizontal growth. These data may provide information about reversibility or irreversibility of the treatment concerning a given pathogen species. In these studies we used 25 V/V % of CFCM in Petri plates of 5.5cm and 8.5 mm diameter, respectively. A CA-disc of 1-cm diameter with *Phytophthora* (*Pythium*) mycelia was put in the center of the plates and the growth of the colonies was evaluated 5 days thereafter. The radial mycelial growth was completely inhibited. Once a mycelium got in contact with the agar containing the antibiotics it died. The vertical growth of the mycelia was not inhibited, indicating, that agar disc on which the colonies were grown acted as a barrier (**Table 20**). The different doses were also compared.

We found that 75 V/V %- of the conditioned media resulted in complete inhibition.

50 V/V %- of the conditioned media did not result in complete inhibition of isolates JA 320 (*P. citricola*) and JA 301-es (*Pythium* sp.). The growth retardation were ~ 35% in comparison to the control (100 %). 25 V/V %- of the conditioned media did result in complete inhibition of three isolates: JA 320 (*P. citricola*) and JA 301-es (*Pythium* sp.) growth retardation were ~ 54%, while that of JA 317 Pythium 34 % in comparison to the control (100 %).

	Treatment					
Phytophthora Pythium strains	**Control**			**25% EMC Carrot-Agar**		
P. citricola JA 320	+++	+++	+++	++	+	-
P. citricola JA 319	+++	+++	+++	-	-	-
P. plurivorum JA 309	+++	+++	+++	-*	-*	-*
P. cactorum, JA 302	++++	++++	+++	-**	-**	-**
Pythium sp. JA 317	++++	++++	++?	-*	+*	-*
Pythium sp. JA 329	++++	+++	+++	-*	+*	-*
Pythium sp. JA 301	++++	++++	+++	++	+	-

Table 20. Qualitative analysis of the antimicrobial activity of EMC on the isolates on the 5th day

The quantitative data (diameters of the growing colony) determined on the 5th day of the experiment concerning *X. budapestensis* are summarized in **Table 21.** The original diameter of the mycelia disc that had been initially placed at the centre of the Petri-dish was 10.0 mm on the start (0) day. We found that 75 V/V %- of the conditioned media resulted in complete inhibition. 50 V/V %- of the conditioned media did not result in complete inhibition of two isolates: In case of JA 320 (*P. citricola*) the retardation rate was ~39 %, while that of and JA 301 (*Pythium* sp.) growth retardation were ~ 28% in comparison to the control (100 %).

Isolate	Scientific name	EMA experiment: Diameter of the CA discs with mycelia			
		0 (Control)	25 V/V%	50 V/V %	75 V/V%
JA319	*Phytophthora citricola*	37.0	10.0	10.0	10.0
JA320	*Phytophthora citricola*	44.5	21.0	18.0	10.0
JA302	*Phytophthora cactorum*	52.0	10.0	10.0	10.0
JA309	*Phytophthora plurivora*	49.5	16.0	10.0	10.0
JA317	*Pythium sp*	52.5	10.0	10.0	10.0
JA301	*Pythium sp.*	42.0	10.0	10.0	10.0
JA301	*Pythium sp.*	51.0	24.0	14.0	10.0

Table 21. Effects of CFCM of *X. budapestensis* on the growth of *Phytophthora* and *Pythium* isolates (on the size of colonies of the test organisms) on the 5th day.

4.2.1.1 The kinetics of mycelial growth of Phytophthora and Pythium strains

Altogether 8 *Phytophthora* species (see **Tables 21** and **22**) were involved in the experiments aimed at determining the time and dose dependence of the anti-oomycetal activities of the intact CFCM of *X. szentirmaii* and *X. budapestensis.*

As for *Xenorhabdus szentirmaii* (EMC), on the third day, its CFCM exhibited complete inhibition in all dilutions except for *P. nicotianae* JA 168, *P. pelgrandis* JA 337 and *P. megasperma* JA 209 which showed some growth at 10% v/v. *P. pelgrandis* JA 337 appeared to depict much growth when compared to the control. On the 6th day, there were some little growths with *P. nicotianae* JA 168 and *P. megasperma* JA 209 at 20% v/v but *P. citrophthora* JA 479, *P. nicotianae* H-1/00, *P. cactorum* P163 and *P. cinnamoni* JA 153 could not show any growth even at 10 V/V %. On the 9th day, some growth could be seen with the *P. nicotianae* JA 168, *P. nicotianae* H-1/00, *P. cinnamoni,* and JA 153 and *P. megasperma* JA 209 at 20% v/v. *P. citrophthora* JA 479 and *P. pelgrandis* JA 337 could not show any growth even at 10% v/v. On the 14th day, only *P. cinnamoni* JA 153 and *P. megasperma* JA 209 could show some growth at 20% v/v and above but no strain exhibited any growth at 40% v/v.

As for *Xenorhabdus* budapestensis (EMA), on the third day its CFCM allowed some growth for *P. citrophthora* JA 479, *P. pelgrandis* JA 337, *P. cinnamoni* JA 153 and *P. megasperma* JA 209 could at 10 and 20 % v/v but no strain exhibited any growth at 30% v/v and above. On the 6th day only *P. citrophthora* JA 479, *P. cinnamoni* JA 153 and *P. megasperma* JA 209 had showed some growth at 20% v/v with *P. nicotianae* JA 168 and *P. cactorum* P163 not depicting any growth at all even at 10% v/v. On the 9th day, only *P. nicotianae* JA 168 and *P. nicotianae* H-1/00 did not show any growth at 20% v/v but there was no growth at all at 30 and 40% v/v. On the 14th day, only *P. citrophthora* JA 479, *P. nicotianae* JA 168 and *P. megasperma* JA 209 showed growths at 30 and 40% v/v.

4.2.1.2 Anti-oomycetal activity quantitatively be adsorbed by Amberlite XAD 1148R

We found that autoclaved sample of Amberlite XAD 1148R could quantitatively adsorb the anti-oomycetal activities from the CFCM of *X. szentirmaii* when used according to the manufacturer's suggestion (**Fig 10**).

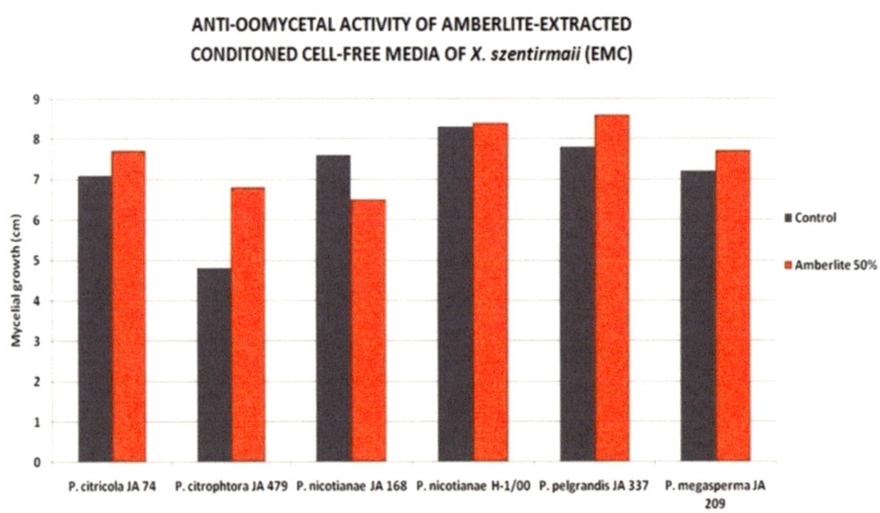

Fig. 10. Growth of *Phytophthora* citricola (JA74, JA 479), *P. nicotianae* (JA168 and H-1/100), *P. pelgrandis* (JA337) and *P. megasperma* (JA 209) on untreated CA plates (control, black) and CA media containing 50 V/V% CFCM from *Xenorhabdus szentirmaii* which had been incubated with sterile autoclaved Amberlite-XAD 1148R O/N.

	Treatments: *X. budapestensis* CFCM in 2X LB liquid media (V / V %)				
3rd day					
	0	10	20	30	40
P. citricola JA 74	48.0	10.0	10.0	10.0	10.0
P. citrophthora JA 479	54.0	14.0*	15.0*	10.0	10.0
P. nicotianae JA 168	37.0	10.0	10.0	10.0	10.0
P. nicotianae H-1/00	52.0	10.0	10.0	10.0	10.0
P. cactorum P163	48.0	10.0	10.0	10.0	10.0
P. pelgrandis JA 337	46.0	14.0*	10.0	10.0	10.0
P. cinnamoni JA 153	55.0	18.0*	14.0*	10.0	10.0
P. megasperma JA 209	42.0	18.0*	10.0	10.0	10.0
6th day					
P. citrophthora JA 479	48.0	10.0	10.0	10.0	10.0
P. nicotianae JA 168	54.0	14.0*	15.0*	10.0	10.0
P. nicotianae H-1/00	37.0	10.0	10.0	10.0	10.0
P. cactorum P163	52.0	10.0	10.0	10.0	10.0
P. pelgrandis JA 337	48.0	10.0	10.0	10.0	10.0
P. cinnamoni JA 153	46.0	14.0*	10.0	10.0	10.0
P. megasperma JA 209	55.0	18.0*	14.0*	10.0	10.0
9thth day					
P. citricola JA 74	80.0	29.0*	26.0*	10.0	10.0
P. citrophthora JA 479	82.0	55.0*	35.0*	10.0	10.0
P. nicotianae JA 168	88.0	18.0*	10.0	10.0	10.0
P. nicotianae H-1/00	88.0	37.0*	10.0	10.0	10.0
P. cactorum P163	78.0	23.0*	15.0*	10.0	10.0
P. pelgrandis JA 337	82.0	43.0*	19.0*	10.0	10.0
P. cinnamoni JA 153	88.0	48.0*	42.0*	10.0	10.0
P. megasperma JA 209	88.0	32.0*	10.0	10.0	10.0
14th day					
P. citricola JA 74	84.0	38.0*	30.0*	10.0	10.0
P. citrophthora JA 479	88.0	64.0*	44.0*	38.0*	29.0*
P. nicotianae JA 168	88.0	25.0*	35.0*	15.0*	15.0*
P. nicotianae H-1/00	88.0	49.0*	47.0*	10.0	10.0
P. cactorum P163	80.0	35.0*	19.0*	10.0	10.0
P. pelgrandis JA 337	88.0	55.0*	28.0*	10.0	10.0
P. cinnamoni JA 153	88.0	62.0*	59.0*	10.0	10.0
P. megasperma JA 209	88.0	54.0*	52.0*	49.0*	50.0*

Table 22. Diameter of the *Phytophthora* colonies (mm) at different *X. budapestensis* CFCM doses at different days of the experiment

	Treatments: ***X. szentirmaii*** CFCM in 2X LB liquid media (V / V %)				
	3rd da				
	0	10	20	30	40
P. citricola JA 74	48.0	10.0	10.0	10.0	10.0
P. citrophthora JA 479	60.0	10.0	10.0	10.0	10.0
P. nicotianae JA 168	36.0	19.0*	10.0	10.0	10.0
P. nicotianae H-1/00	60.0	10.0	10.0	10.0	10.0
P. cactorum P163	50.0	10.0	10.0	10.0	10.0
P. pelgrandis JA 337	45.0	22.0*	10.0	10.0	10.0
P. cinnamoni JA 153	50.0	10.0	10.0	10.0	10.0
P. megasperma JA 209	40.0	15.0*	10.0	10.0	10.0
	6th day				
P. citrophthora JA 479	84.0	10.0	10.0	10.0	10.0
P. nicotianae JA 168	58.0	28.0*	13.0*	10.0	10.0
P. nicotianae H-1/00	88.0	10.0	10.0	10.0	10.0
P. cactorum P163	82.0	10.0	10.0	10.0	10.0
P. pelgrandis JA 337	70.0	22.0*	10.0	10.0	10.0
P. cinnamoni JA 153	80.0	10.0	10.0	10.0	10.0
P. megasperma JA 209	86.0	22.0*	16.0*	10.0	10.0
	9thth day				
P. citricola JA 74	62.0	30.0*	10.0	10.0	10.0
P. citrophthora JA 479	88.0	10.0	10.0	10.0	10.0
P. nicotianae JA 168	62.0	36.0*	26.0*	10.0	10.0
P. nicotianae H-1/00	88.0	24.0*	11.0*	10.0	10.0
P. cactorum P163	88.0	10.0	10.0	10.0	10.0
P. pelgrandis JA 337	76.0	35.0*	10.0	10.0	10.0
P. cinnamoni JA 153	88.0	19.0*	14.0*	10.0	10.0
P. megasperma JA 209	88.0	28.0*	21.0*	10.0	10.0
	14th day				
P. citricola JA 74	72.0	45.0*	10.0	10.0	10.0
P. citrophthora JA 479	88.0	21.0*	10.0	10.0	10.0
P. nicotianae JA 168	66.0	43.0*	10.0	10.0	10.0
P. nicotianae H-1/00	88.0	34.0*	10.0	10.0	10.0
P. cactorum P163	82.0	13.0*	10.0	10.0	10.0
P. pelgrandis JA 337	82.0	44.0*	10.0	10.0	10.0
P. cinnamoni JA 153	88.0	33.0*	29.0*	20.0*	10.0
P. megasperma JA 209	88.0	41.0*	36.0*	30.0*	10.0

Table 23. Diameter of the *Phytophthora* colonies (mm) at different *X. szentirmaii* CFCM doses at different days of the experiment

4.2.2 Results on fungi: Botrytis, Alternaria and Fusarium strains

The *in vitro* antifungal activities of the cell-free filtrates of *X. szentirmaii* and *X. budapestensis* were assayed against different species of phytopathogenic fungi that are common in plant fungal infections and are known as major causes of disease in agriculture crops.

These include *Botrytis cinerea, Alternaria alternata,* and *Fusarium graminearum*. The three fungi reacted rather differently.

Depending on the concentration, the mycelial growth ceased. *Botrytis cinerea* ceased growing at 25 V/V % of the media, while the controls grew over the plates (of 9.5 mm diameter).

The reaction of *A. altera* to the antifungal compounds of *X. budapestensis* was different (**Fig 11**). While the untreated controls, similarly to those of *Fusarium* overgrew the plates, the *A. altera* colonies on the "poisoned" agar plates were less then a third in size.

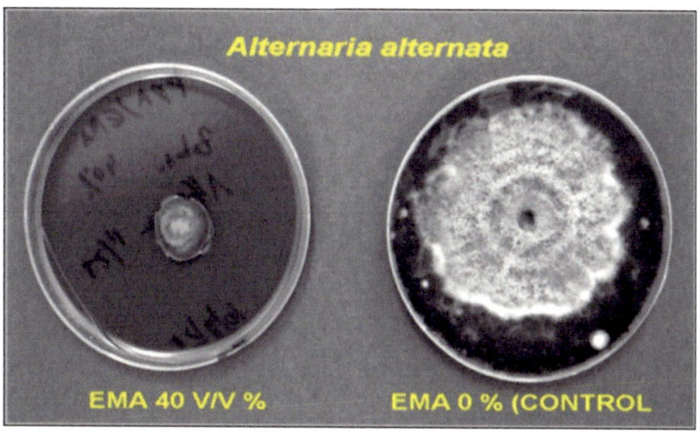

Fig. 11. 3-day-old poisoned (left) and control PDA plates containing 40 V/V % of *X. budaestensis* (EMA) CFCM and an *Alternaria aternata* inoculum (Photo: Dr. Csaba Pintér).

The mycelial growth was very poor, but the number of conidia was significantly higher in the treated plates than in the control plates in one week. Both the growth retardation and the conidial production was dose dependent.

F. gramineae controls overgrew the Petri plates within 3 days. During this time there is only a retarded growth could be observed at higher CFCM concentrations (**Fig 12**)

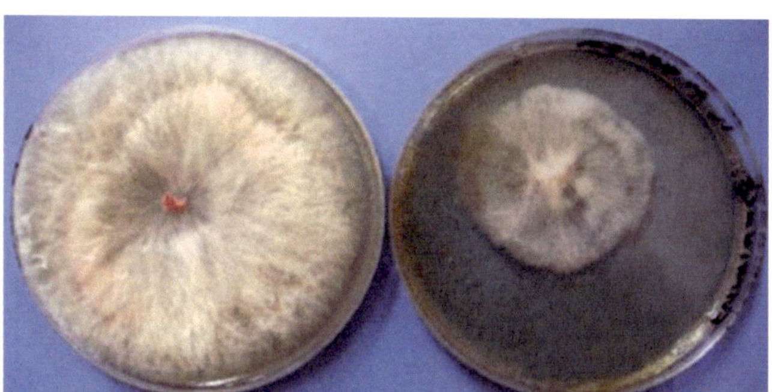

Fig. 12. 3-day-old poisoned (left) and control PDA plates containing 40 V/V % of *X. budaestensis* (EMA) CFCM and an *Fusarium graminarum* inoculum (Photo: Dr. Csaba Pintér).

Finally, the growth was complete at each concentration, but speed of the growth depended on the concentration, as demonstrated in **Fig 13**.

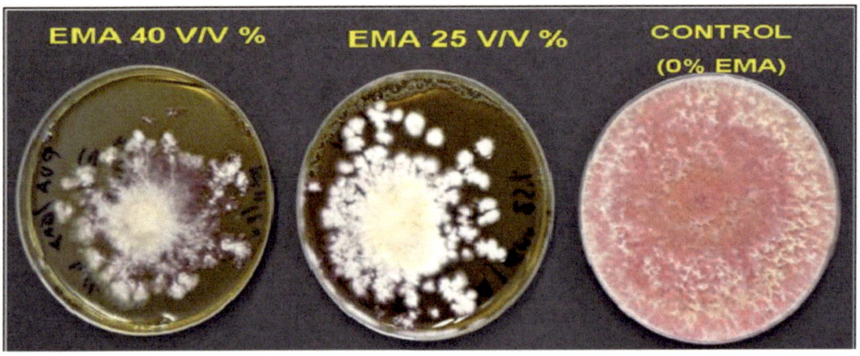

Fig. 13. Demonstration the reversible antifungal potential of the CFCM of *Xenorhabdus budaestensis* (EMA) on *Fusarium gramineae* on "poisoned" Potato Dextrose Agar Plate. (Photo: Dr. Csaba Pintér).

4.2.2.1 Reversible and irreversible antifungal activity

Our data suggest that either the active compounds in the CFCM, which exert antibacterial, anti-oomycetal or anti-fungal activities are probably different. An argument for this hypothesis is that the antibacterial activity of *X. budapestensis* was stronger but the anti-oomycetal effects were somewhat weaker then that of *X. szentirmaii*. Both compounds exerted an irreversible and strong fungicide effects on *Botrytis* while the effect on the growth of *Fusarium* proved temporary and reversible. In general, the antifungal effects of *X. szentirmaii* were more convincing. The effects concerning *X. szentirmaii* are shown on **Figs 14, 15, and 16.** An the basis of data below, we concluded that *Xenorhabdus szentirmaii* produces compound(s) which might be useful in controlling both *Botrytis* and *Alternaria* in a proper concentration, in proper way. *Alternaria*, however, may have a potential of adapting a survival strategy through sporulation.

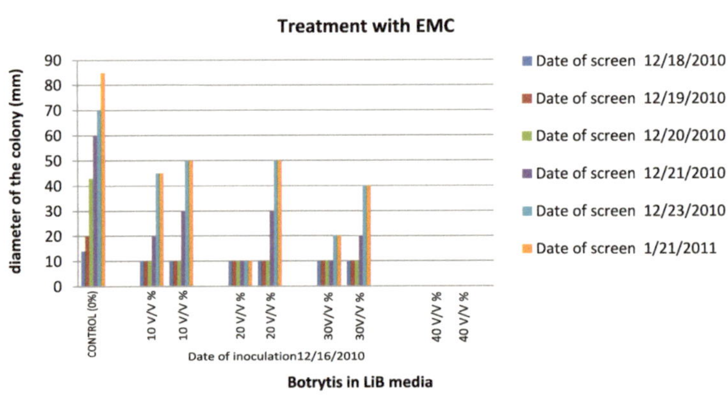

Fig. 14. Intact CFCM of *X. szentirmaii* inhibited completely the growth of *B. cinerea* at least for two weeks up to 30 V/V%. The largest dose (40 V/V %) quantitatively destroyed the pathogen.

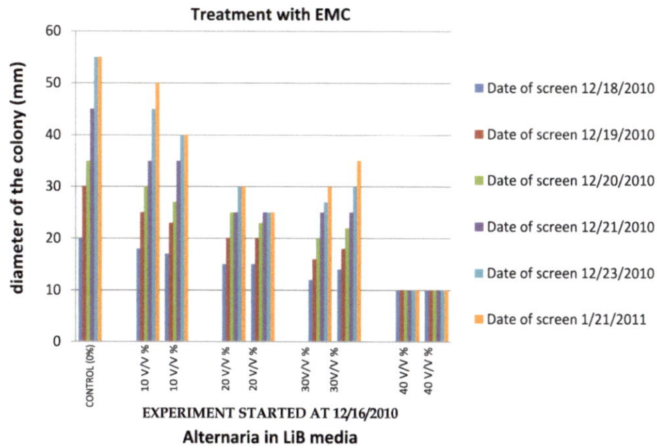

Fig. 15. Intact CFCM of *X. szentirmaii* significantly slowed down the growth of *Alternaria alternata* in Lima Bean agar media.

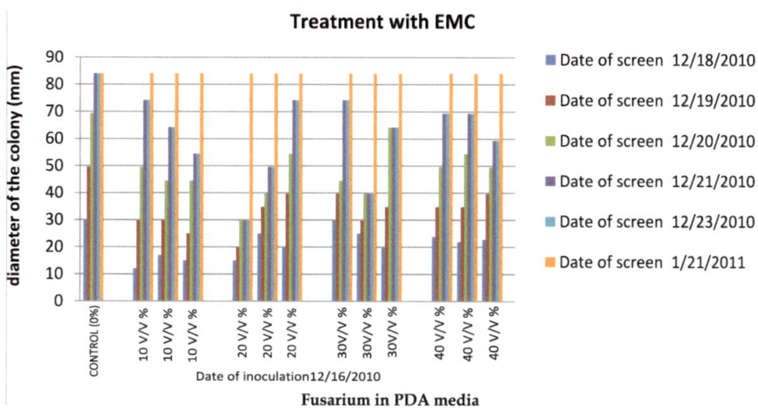

Fig. 16. Intact CFCM of *X. szentirmaii* exerted only temporary slowing effects on the growth of *F. gramineae* in As for *Alternaria alternata*, the speed of the growth as well as the final size of the colony was more or less dose-dependent up to 30 V/V %. The largest dose (40 V/V %) quantitatively destroyed the pathogen. The results are slightly influenced by the media. The antifungal effects are unambiguously (significantly) stronger in PDA than in LiBA media (data not shown).

As for *Fusarium graminearum*, only the speed but not the final size of the colonies proved dose-dependent. Even the largest dose (40 V/V %) could not destroy the pathogen. The results are slightly influenced by the media. The antifungal effects are unambiguously (significantly) stronger in PDA than in LiBA media (data not shown).

4.4 Preliminary results of genetic analysis

4.4.1 A search for antibiotics non-producer mutants in *X. budapestensis*

As for our mutant hunting in *X. budapestensis* we altogether tested 38 of the 177 *rif*R *kan*R transconjugants for antibiotics their antibiotics hypo, - or hyper-production **(Fig 17).**There were 26, which could not show any antibiotics production in our experimental conditions.

We considered them as antibiotics non-producers with inactivate structural gene(s). There were also 13 (# 15, 69, 70, 73, 74, 75, 76, 88, 90, 91, 92, 93 and 135) overproducers. They should be carefully retested. Amongst them, we are searching for regulatory mutants. Those which antibiotics production was uncertain discarded.

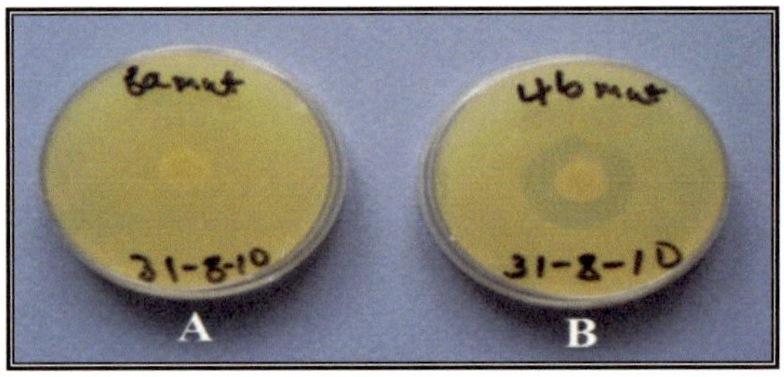

Fig. 17. Overlay bioassay mutant antbiotics non-producing mutant (left, A), and a low producing mutant (B)

4.4.2 A search for exocrystal mutants in *X. szentirmaii*

This unique natural product of *X. szentirmaii* **(Fig 18 and 19)** was made accessible from the recently described *X. szentirmaii* (Fodor et al., 2007). The surface of the colonies of *X. szentirmaii* has a striking phenotype by their purple metallic color.

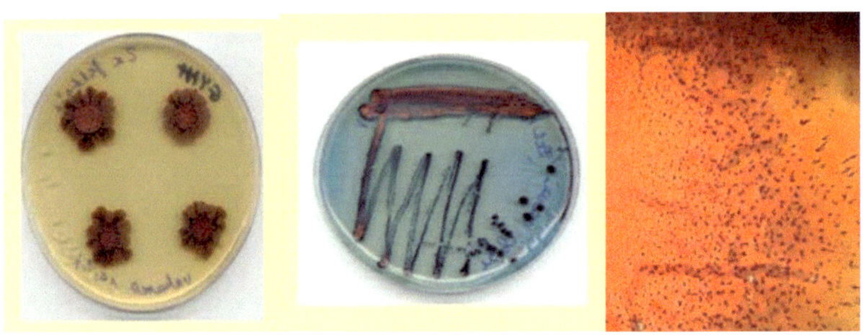

Fig. 18. *Xenorhabdus szentirmaii* colonies on LBA (Left)) and LBTA (Medium) plates; Crystals on the surface of a colony, 40X magnification in through Leica stereomicroscope (Right) Photo: A. Máthé-Fodor

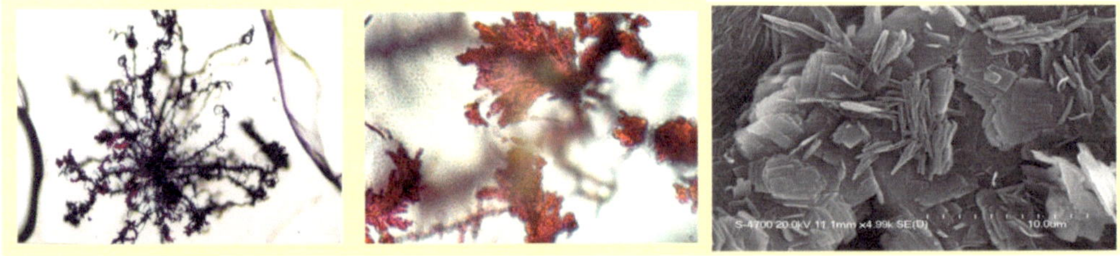

Fig. 19. Isolated antibiotic poly-iodinin crystal from the agar light microscopy, 125 X (D) and 1,000X (E) and with SEM (S-4700 20.0 kV 11.1 mmX4.99 SE. Photo: A. Máthé-Fodor. The crystal was isolated by using a double layer of sterile cellophane covering an LB plate and over-layered with bacterium suspension. This pigment crystal was identified by Haynes and Zeller as iodinin, (Fodor et al., 2008.

We altogether isolated 22 crystal antimicrobial crystal mutants from *X. szentirmaii* **(Fig 20).**

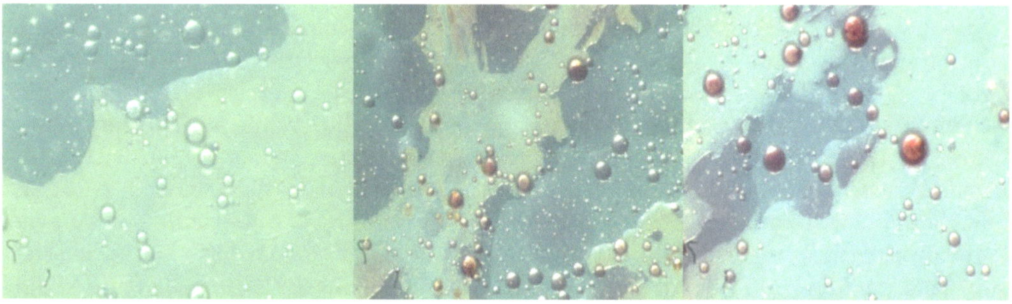

Fig. 20. Light mutant produce colorless (Left); WT produces colored (Medium) and dark mutant (Right) produce dark oligomer soluble in the oil of ENGM plate (Fodor et al., 2010b). The antibiotic pigment crystal was discovered and isolated by A. Máthé-Fodor in 2003). The structure was identified by Haynes and Zeller as iodinin, (see: Fodor et al., 2007).

5. Conclusions and perspectives

We have to think about new perspectives of antimicrobials produced by entomopathogenic bacteria. It should be a rational explanation why the intact cell-free media of some *Xenorhabdus* strains is so powerful against different microorganisms, while the isolated (and patented, Webster et al., 1996) compounds are useless from practical aspect (Fodor et al., 2008). The cell-free conditioned media (CFCM) of both *Xenorhabdus szentirmaii* and *X. budapestensis* is a mixture of natural compounds, which are successfully used in soil condition against the most different pathogens. Interestingly, none of the single isolated, identified, patented and re-synthesized compound reached the market. The list of them is given by Dr. Bode (2009, **Fig. 21**). The authors admitted (Brachmann et al., 2006; Susa et al., 2008) that the isolated "antibiotics" proved rather weak in different bioassays.

On the other hand, the intact media, at least those of *X. szentirmaii* and *X. budapestensis* are quite effective against different bacteria, including multi-resistant plant, human pathogens (including MRSA, data are not given), mastitis pathogens (Furgani et al., 2008), plant parasitic oomycetae and fungi. They are effective against the plant pathogenic protozoon *Leichmania donovani* (McGwire et al., unpublished data). We also found that the different Xenorhabdus species uses different chemical "weapons" when competing with each other (Fodor et al., 2010a) indicating that the "toolkit" needed for survival in the nature is more complex and probably need synergistic interactions. The data presented here seen, that majority of the PPB was rather sensitive, they proved some resistant. All the tested *Xanthomonas, Erwinia* and *Pectovora* strains were extreme sensitive, while *Ralstonia* proved resistant and so some *Pseudomonas* strains. The availability of both sensitive and resistant *Pseudomonas* strains pride an option of genetic analysis of action mechanism of EPB antimicrobial in plant pathogen bacteria.

Badosa et al (2007) developed and screened a 125-member library of synthetic linear undecapeptides based on a previously described peptide H-K(1)KLFKKILKF(10)L-NH(2) (BP76) that inhibited *in vitro* growth of the plant pathogenic bacteria *Erwinia amylovora, Xanthomonas axonopodis pv. vesicatoria*, and *Pseudomonas syringae* pv. *syringae* at low micromolar concentrations. Peptides were designed using a combinatorial chemistry approach by incorporating amino acids possessing various degrees of hydrophobicity and hydrophilicity at positions 1 and 10 and by varying the N-terminus. Library screening for *in vitro* growth inhibition identified 27, 40 and 113 sequences with MIC values below 7.5 microM against *E. amylovora, P. syringae* and *X. axonopodis*, respectively. Cytotoxicity, bactericidal activity and stability towards protease degradation of the most active peptides were also determined. Seven peptides with a good balance between antibacterial and hemolytic activities were identified. Several analogues displayed a bactericidal effect and low susceptibility to protease degradation. The most promising peptides were tested in vivo by evaluating their preventive effect of inhibition of E. amylovora infection in detached apple and pear flowers. The peptide H-KKLFKKILKYL-NH(2) (BP100) showed efficacies in flowers of 63-76% at 100 microM, being more potent than BP76 and only less effective than streptomycin, currently used for fire blight control. We suppose that our natural compounds are chemical relations with this peptide family. The only sequenced oligopeptide (bicornutin A, Böszörményi et al., 2009) what we found in *X. budapestensis* is a hexapeptide but quite different from those described by Badosa et al (2007). Linear and cyclic peptides have recently been discovered in *Xenorhabdus* species by Lang et al (2008).

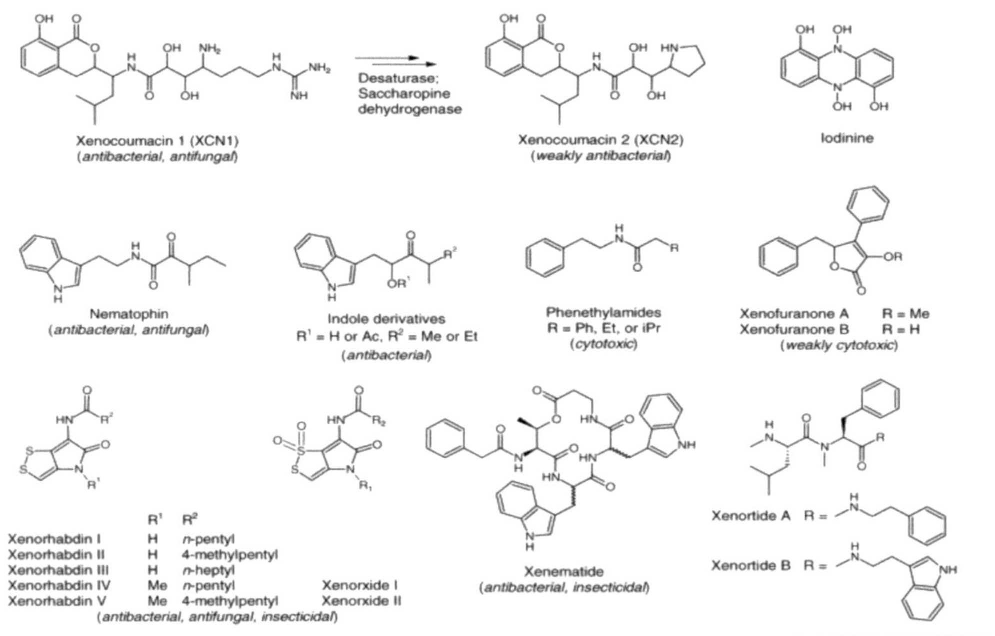

Fig. 21. EPB antibiotics, which had so far been identified (Bode, 2009)

Oomycetes from the genus *Phytophthora* are fungus-like plant pathogens that are devastating for agriculture and natural ecosystems. Due to their particular physiological characteristics, no efficient treatments against diseases caused by these microorganisms are presently available. To develop such treatments, it appears essential to dissect the molecular mechanisms that determine the interaction between *Phytophthora* species and host plants.

Available data are scarce, and genomic approaches were mainly developed for the two species, *Phytophthora infestans* and *Phytophthora sojae*.

However, these two species are exceptions from, rather than representative species for, the genus. *P. infestans* is a foliar pathogen, and *P. sojae* infects a narrow range of host plants, while the majority of *Phytophthora* species are quite unselective, root-infecting pathogens Attard et al., 2008). *Phytophthora ramorum*, causal agent of sudden oak death, is an emerging plant pathogen first observed in North America (Knight, 2002); associated with mortality of tanoak (*Lithocarpus densiflorus*) and coast live oak (*Quercus agrifolia*) in coastal forests of California during the mid-1990s. (Rizzo et al., 2005). The pathogen is now known to occur in North America and Europe and have a host range of over 40 plant genera. (The anti-oomycetal of the CFCM of *X. szentirmaii* and *X. budapestensis* make these two bacteria as potential tools in fighting Phytophthora, especially in forests, in the rhizosphere of the trees.

In the last half century, antibiotics revolutionized the human medicine, veterinary science, animal breeding and agricultural plant medicine. The use of antibiotics without any control, however, resulted in serious problems, mainly due to the massive selection for antibiotics resistant and multi-resistant pathogens. Gram-positive organisms are the most common bacterial pathogens that cause diseases in humans, with streptococci and staphylococci occurring most frequently. Immunization has been extremely successful in eradicating some Gram-positive infections, such as diphtheria and tetanus, and relatively successful for pneumococci. *Staphylococcus aureus* vaccines are under investigation. In terms of antimicrobial susceptibility, some Gram-positive organisms have remained sensitive to most antimicrobials, whereas others, including staphylococci, pneumococci and enterococci, have developed clinically relevant resistance. Extensive exposure to antimicrobials in the hospital setting has caused the spread of clones mainly in the hospital environment, yet multiresistance is now also found in community-acquired diseases. Community-acquired methicillin-resistant S. aureus (CA-MRSA) and resistant pneumococci are the most important examples, but even viridans streptococci are becoming resistant to some antibiotics. Moreover, MRSA and vancomycin-resistant enterococci (VRE) are found in pets and farm animals. Because of these concerns, new antimicrobials have been developed during the past decade, including quinupristin/dalfopristin, linezolid, trigecycline, daptomycin and dalbavancin. Also under investigation are beta-lactams, streptogramins and quinolones with activity against MRSA, penicillin-resistant pneumococci and VRE. Finally, infection-control measures, including the identification of carriers of multiresistant organisms and appropriate isolation, must continue to be implemented (Metzger et al., 2009). Increasing incidence of infections caused by poly-resistant organisms is associated with prolonged hospital stays, increased health care costs, and increased morbidity and mortality (Dötsch et al., 2009). When using a molecule antibiotics, it is only the matter of time when the first resistant pathogen arises and then the use of the antibiotics will be as a selective factor. Biofilm formation of some pathogens also act against the efficacy of traditional antibiotics (Kustos et al., 2005).Furthermore, in the past antibiotics were also used in an uncontrolled manner in animal husbandry. For instance, antibiotics were used in poultry for many years for their growth promoting effects.

Antibiotics growth promoters mostly have their effects by modifying the intestinal microbiota, targeting mainly Gram-positive bacteria, which are associated with poor health and performance of the animals (Bredford, 2000). Taking into account this typical spectrum of activity of the antibiotics used as growth promoters, and taking into account that the concentrations used in feed are below the minimum inhibitory concentrations as tested *in vitro*, one may expect any beneficial effect of growth promoter against intestinal carriage of zoonotic agents. Most of zoonotic agents found, for instance, in the chicken gastro-intestinal track are Gram-negatives.

Antibiotics have been used since the 1950s to control certain bacterial diseases of high-value fruit, vegetable, and ornamental plants. Today, the antibiotics most commonly used on plants are oxytetracycline and streptomycin. In the USA, antibiotics applied to plants account for less than 0.5% of total antibiotic use. Resistance of plant pathogens to oxytetracycline is rare, but the emergence of streptomycin-resistant strains of Erwinia amylovora, *Pseudomonas* spp., and *Xanthomonas campestris* has impeded the control of several important diseases. A fraction of streptomycin-resistance genes in plant-associated bacteria are similar to those found in bacteria isolated from humans, animals, and soil, and are associated with transfer-proficient elements. However, the most common vehicles of streptomycin-resistance genes in human and plant pathogens are genetically distinct. Nonetheless, the role of antibiotic use on plants in the antibiotic-resistance crisis in human medicine is the subject of debate (MacManus et al., 2002)

The multi-resistance in Gram-negative bacteria is more frequently efflux-mediated (Poole, 2004). The integral inner membrane resistance-nodulation-division (RND) components of three-component RND-membrane fusion protein-outer membrane factor multidrug efflux systems define the substrate selectivity of these efflux systems. To gain a better understanding of what regions of these proteins are important for substrate recognition, a plasmid-borne mexB gene encoding the RND component of the MexAB-OprM multidrug efflux system of Pseudomonas aeruginosa was mutagenized *in vitro* by using hydroxylamine and mutations compromising the MexB contribution to antibiotic resistance identified in a ΔmexB strain (Sobel et al., 2003). Of 100 mutants that expressed wild-type levels of MexB and showed increased susceptibility to one or more of carbenicillin, chloramphenicol, nalidixic acid, and novobiocin, the mexB genes of a representative 46 were sequenced, and 19 unique single mutations were identified. While the majority of mutations occurred within the large periplasmic loops between transmembrane segment 1 (TMS-1) and TMS-2 and between TMS-7 and TMS-8 of MexB, mutations were seen in the TMSs and in other periplasmic as well as cytoplasmic loops. By threading the MexB amino acid sequence through the crystal structure of the homologous RND transporter from Escherichia coli, AcrB, a three-dimensional model of a MexB trimer was obtained and the mutations were mapped to it.

Unexpectedly, most mutations mapped to regions of MexB predicted to be involved in trimerization or interaction with MexA rather than to regions expected to contribute to substrate recognition. Intragenic second-site suppressor mutations that restored the activity of the G220S mutant version of MexB, which was compromised for resistance to all tested MexAB-OprM antimicrobial substrates, were recovered and mapped to the apparently distal portion of MexB that is implicated in OprM interaction. As the G220S mutation likely impacted trimerization, it appears that either proper assembly of the MexB trimer is necessary for OprM interaction or OprM association with an unstable MexB trimer might stabilize it, thereby restoring activity (Middlemiss & Poole (2004).

Due to the fatal consequences the prophylactic or curative use of antibiotics more and more prohibited by EU regulations, such as N° 2160/2003, amended by EU N° 1091/2005 regulation in the agriculture since January, 2006. The use of "ionophore" class of antibiotics (stereo isomers of sacred water) is permitted, but their efficacy is rather questionable (Van Immerseel et al., 2002). The non-digestible feed ingredients, mainly saccharids, called prebiotics are used to make the host less vulnerable from some pathogens. Probably cannot replace antibiotics either. The tendency is to use complex materials with more than one mode of action. The permission of using probiotics, defined as living microbial feed supplements, which beneficially affect the host by improving its microbial balance (Van Immerseel et al., 2010) however, may provide an option to replace antibiotics by some natural factors with similar efficacy and reducing the risk of selecting for poly-resistant pathogens. Herein, we provided an example of a natural product of potential use against different plant, veterinary and even human pathogens. The perspective is to develop a "probiotic" microorganism, which carries (but cannot transmit) those genetic units of *Xenorhabdus budapestensis* and /or *X. szentirmaii*, which were a natural component, such as *L. salivations* (Pascual et al., 1999) of the niche where should be applied, and the genetic construction would be cloned downstream of an inducible bacteria promoter. As a first step toward this direction, we made the following: We have altogether isolated 22 antimicrobial crystal mutants from *X. szentirmaii* and a few antibiotics non-producers from *X. budapestensis*. The antibiotics mutants should be retested. As for the crystal mutants, we followed the protocol of Prof. Heidi Goodrich-Blair (University of Wisconsin-Madison). We have started the following experiment:

We have isolated genomic DNA from each and digested with restriction enzymes *Xba*I, *Knp*I, *Spe*I, *and SalI* respectively. We also digested pBluescript SK vector with the same enzyme. We ligated the fragments into pBluescript SK vector, which is AmpR. The fragments digested with different enzyme from a given mutants were united. The salts were removed by dialysis. The DNA was electroporated into *E. coli* DH5 alpha. We were selecting on ampicillin (150) kanamycin (50) LB plates. The-double resistant clones were grown in 3 ml LB + Km and plasmids were isolated, digested with *Eco*RI. The lengths of the fragments were determined and then sequenced.

6. Acknowledgement

This study has generously been supported by TÁMOP-4.2.2-08/1/2008-0018 - entitled as ***"Livable environment and healthier people – Bioinnovation and Green Technology research at the University of Pannonia, Hungary"***. The project is being co-financed by the European Social Fund with the support of the European Union. We intend to express our thanks and appreciations to Prof. Dr. ***Éva* Lehoczky, *D. Sc*,** the Head of this program for financing my microbiological experiments. We would like to thank for the wonderful pictures to Dr. Csaba **Pintér**. We acknowledge Prof. Dr. Bradford S. **McGwire** for his thoughtful help in preparing the manuscript and providing an invaluable option to work in his Laboratory with my antimicrobial antagonists with him and with Dr. Manju **Kulkarni**, for a week, needed for our conclusions. In the manual work, the high quality lab work of our FAO students (Josephat **Muvevi** and, Eliud Magu **Mutitu** (Kenya); Hundessa Wakuma **Bayissa** and Abate Birhan **Addise** (Ethiopia); Zakria **Faizi** and Mohammad Iqbal **Karimi**

(Afghanistan) was indispensible. We need to thank Mrs. Janet **McCormick** for her laboratory help in the Hogan's Laboratory. Prof. Dr. Károly **Dublecz,** Dean of the Georgikon Faculty, also kindly supported this work in more way than one. The *Phytophthora* strains were kindly provided by graduate student András **Józsa** and his supervisor Dr. József **Bakonyi**. The fungus isolates were kindly provided by Dr. Sándor **Kadlicskó**.

We would also like to thank **Valent Biosciences Corporation** (870 Technology Way, Libertyville, IL 60048, USA) for heir generous contribution covering our publication costs.

7. References

Agrios G. (1997) Plant Pathology, Fourth Edition. Academic Press, pp. 1-635.

Akhurst, R. J. (1982) Antibiotic activity of *Xenorhabdus ssp.* bacteria symbiotically associated with insect pathogenic nematodes of the families *Heterorhabditidae* and *Steinernematidae J. Gen. Microbiol.* 128, 3061–3065.

Attard, A., Gourgues, M., Galiana, E., F., Ponchet, M., & Keller, H. (2008) Strategies of attack and defense in plant-oomycete interactions, accentuated for *Phytophthora parasitica* Dastur (syn. P. Nicotianae Breda de Haan) *J Plant Physiol* 165, 83-94.

Ausubel, F.M., Brent, R., Kingston, R.E., Moore, D.D., Seidman, J.G., Smith, J.A. & Struhl, K (Eds.) (1999) Short Protocols in Molecular Biology (Fourth Edition).John Wiley & Sons, Inc. ISBN 0-471-32938-X. USA, pp. 1-1 - 1-6; 5-1 - 6-30. John Wiley & Sons, New York;

Badosa, E., Ferre, R., Planas, M., Feliu, L., Besalú, E., Cabrefiga, J., Bardají, E., & Montesinos, E. (2007) A library of linear undecapeptides with bactericidal activity against phytopathogenic bacteria *Peptides* 28, 2276-2285

Balci, Y., Balci, S., Eggers, J., Juzwik, J., Long, R., MacDonald, W. & Gottschalk, K. (2007): *Phytophthora* species associated with forest soils in eastern and north central U.S. oak ecosystems. *Plant Disease* 91, 705-710

Bode, H. B. (2009) Entomopathogenic bacteria as a source of secondary metabolites *Curr Opin Chem Biol.*13, 224-230

Bedford, M. (2000) Removal of antibiotic growth promoters from poultry diets: implications and strategies to minimize subsequent problems *World's Poultry Science* Journal 56, 347-354

Böszörményi, E., Érsek, T., Fodor, A., Fodor, A. M., Földes, S., Hevesi, M., Hogan, J. S., Katona, Z., Klein, M. G., Kormány, A., Pekár, S., Szentirmai, A., Sztaricskai, F., & Taylor, R. A. (2009) Isolation and activity of *Xenorhabdus* antimicrobial compounds against the plant pathogens *Erwinia amylovora* and *Phytophthora nicotianae J. Appl. Microbiol* 107, 746-759

Brachmann A. O., Forst, S., Furgani, G.M., Fodor, A., & Bode H.B. (2006) Xenofuranones A and B: phenylpyruvate dimers from *Xenorhabdus szentirmaii*. *J Nat Prod* 69, 1830-1832.

Childs, T. W. (1963). *Poria weirii* root rot. *Phytopathology*. 53, 1124-1127

Colin, J., Pusse L. & Diouf A. M. (1984 Demonstration and application of the antagonism of fluorescent *Pseudomonas* spp. towards phytopathogenic bacteria *Mededelingen van de Faculteit Landbouwwetenschappen Rijksuniversiteit Gent*. 49, 587-596

Dötsch A., Becker T., Pommerenke C., Magnowska Z., Jansch L., & Haussler S. (2009) Genome-wide Identification of Genetic Determinants of Antimicrobial Drug Resistance in *Pseudomonas aeruginosa. Antimicrob Agents Chemother*, 53, 2522–2531

Duchaud, E., Rusniok, C., Frangeul, L., Buchrieser, C., Givaudan A., Taourit, S., Bocs, S., Boursaux-Eude, C., Chandler M., Charles J.F., Dassa, E., Derose, R., Derzelle, S., Freyssinet, G., Gaudriault, S., Médigue, C., Lanois, A., Powell, K., Siguier, P., Vincent, R., Wingate, V., Zouine, M., Glaser, P., Boemare, N. E., Danchin, A., & Kunst, F. (2003). The genome sequence of the entomopathogenic bacterium *Photorhabdus luminescens Nat Biotechnol* 21, 1307-1313

Duncan, J. (1999): *Phytophthora-* an abiding threat to our crops. *Microbial Today* Vol. 11, 114-116

Ellington M.J., Ganner M., Warner M., Cookson B.D., & Kearns A.M.: (2010) Polyclonal multiply antibiotic-resistant methicillin-resistant *Staphylococcus aureus* with Panton-Valentine leucocidin in England. *J Antimicrob Chemother.*, 65, 46-50

Érsek, T. (1975): The sensitivity of *Phytophthora infestans* to several antibiotics. *Z Pflanzensch.* 82, 614-617

Érsek, T., English, J. T. & Schoelz, J. E. (1995) Creation of species hybrids of *Phytophthora* with modified host ranges by zoospore fusion. *Phytopathology,* 85, 1343-1347

Erwin, D.C. & Ribeiro, O. K. (1996): Introduction to the genus *Phytophthora* In: *Phytophthora* Diseases world wide, 1-7 American Phytopathological Society St. Paul, MN

Ferguson, A. J. & Jeffers, S. N. (1999) Detecting multiple species of *Phytophthora* in container mixes from ornamental crop nurseries *Plant Disease* 83, 1129-1136

Fodor, A., Érsek, T., Fodor, A. M., Forst, S., Hogan, J., Hevesi, M., Klein, M.G., Stackebrandt, E., Szentirmai, A., Sztaricskai, F., & Zeller, M (2008). New aspects on *Xenorhabdus* antibiotics research. *Insect Pathogens and Insect Parasitic Nematodes IOBC/wars Bulletin* Vol. 31, pp. 157-164

Fodor, A., Fodor, A.M., Forst, S., Hogan, J. S., Klein, M.G., Lengyel, K., Sáringer, Gy., Stackebrandt, E., Taylor, R. A. Y. & Lehoczky, É. (2010a) Comparative analysis of antibacterial activities of *Xenorhabdus* species on related and non-related bacteria *in vivo. J. Microbiol. Antimicrobials* 2, 30-35.

Fodor, A., Fodor, A.M., Lehoczky, É, Jagdale, G., Grewal P.S. Klein, M. G. (2010b) ENGM: an NGM-like solid media suitable for doing genetics on the entomopathogenic nematode *Heterorhabditis bacteriophora The Worm Breeders Gazette* 18, (Number 2). p. 10

Fotinos, N. Convert M., Piffaretti J.C., Gurny R., & Lange N. (2008). Effects on Gram-Negative and Gram-Positive Bacteria Mediated by 5-Aminolevulinic Acid and 5-Aminolevulinic Acid Derivatives. *Antimicrob Agents Chemother*, 52, 1366–1373

Forbes, G. A. & Jarvis, M.C. (1994): Host resistance for management of potato late blight In: (Ed) Zehnder G, Jansson R, & Raman K. V. (1994) Advances in Potato Pest Biology and Management St. Paul, Minnesota: American Phytopathological Society, 439-457

Furgani, G., Böszörményi, E., Fodor, A., Fodor, A.M., Forst, S., Hogan, J., Katona, Z., Klein. M.G., Stackebrandt, E., Szentirmai, A., Sztaricskai, F., & Wolf, S. (2008) *Xenorhabdus antibiotics:* a comparative analysis and potential utility for controlling mastitis caused by bacteria *Journal of Applied Microbiology* 104,745-758

Hansen, E.M. & Delatour, C. (1999) *Phytophthora* species in oak forests of North West France *Ann. For. Sci.* 56, 539-547.

Hansen, E. M. & Goheen, E. (2000) *Phellinus weirii* and other native root pathogens as determinants of forest structures and process in western North America. *Ann. Rev. Phytopath.*, 38, 515-53.

Hansen, E.M., Hamm, P.B., Julis, A.J. & Roth, L.F. (1979) Isolation, incidence, and management of *Phytophthora* in forest tree nurseries in the Pacific North west. *Plant Disease Reporter* 63, 607-611

Hansen, E. M., Hamm, P. B., Julis, A. J. & Roth, L. F. (1980) Survival spread and pathogenicity of *Phytophthora* spp. on Douglas-fir seedlings planted on forest sites. *Phytopathology*, 70, 422-425

Herrero, M., de Lorenzo, V., & Timmis, K.N. (1990) Transposon vectors containing non-antibiotic resistance selection markers for cloning and stable chromosomal insertion of foreign genes in gram-negative bacteria. *Journal of Bacteriology.* 172, 6557-6567.

Hevesi M. (1996): Appearance of fire blight on apple in Hungary *Növényvédelem* 32, 225-228

Hilty, J. W. & Schmitthenner, A. F. (1962). Pathogenic and cultural variability of single zoospore isolates of *Phytophthora megasperma* var. *sojae*. *Phytopath.* 52, 859-862

Heungens, K., Cowles, C.E., & Goodrich-Blair, H. (2002) Identification of *Xenorhabdus nematophila* genes required for mutualistic colonization of *Steinernema carpocapsae* nematodes *Mol Microbiol.*45, 1337-1353.

Jones A. L., McManus P. S., & Chiu C. S. (1996) Epidemiology and genetic diversity of streptomycin resistance in *Erwinia amylovora* in Michigan *Acta Hort.* 338, 333-340

Jung, T., Blaschke, H., & Osswald, W. (2000) Involvement of *Phytophthora* species in central European oak decline and the effect of site factors on the disease. *Plant Pathology* 49,706-718

Jung, T., Blaschke, H. & Neumann, P. (1996) Isolation, identification and pathogenicity of *Phytophthora* species from declining oak stands. *European J. Forest Pathology.* 26, 253-272

Jung, T., Hansen, E. M., Winton, L., Osswald, W. & Delatour, C. (2002) Three new species of *Phytophthora* from European oak forests. *Mycological Research*, 106, 397-411

Kan, J. A., van, (2006a) Licensed to kill: the lifestyle of a necrotrophic plant pathogen. *Trends Plant Sci.* 11, 247-253. Epub 2006 Apr 17. Review

Kan J, A. van, (2006b): A magyarországi *Fusarium graminearum* populáció trichotecén kemotípusai Növényvédelem, 31/3., pp. 103-108;

Knight, J. (2002) Fears mount as oak blight infects redwoods. *Nature* 415, 251 (17 January 2002) | doi:10.1038/415251a

Kocsis, B., Kustos, I., Kilár F., Nyúl, A., Jakus, P.B., Kerekes, Sz., Villarreal, V., Prókai, L., & Lóránd T. (2009). Antifungal unsaturated cyclic Mannich ketones and amino alcohols: Study of mechanism of action *Eur. J. of Medical Chemistry,* 44, 1823-1829

Kustos I., Kustos T., Kilár F., Rappai G., & Kocsis B. (2005) Effect of Antibiotic Treatment on Bacterial Attachment to DePuy Enduron™ Orthopedic Implant. *Chemotherapy,* 51, 286-290

Lang, G., Kalvelage T., Peters, A., Wiese, J. & Imhoff, J.F. (2008) Linear and cyclic peptides from the entomopathogenic bacterium *Xenorhabdus nematophilus, J Nat Prod* 71, 1074–1077.

Leclerc, M.C., & Boemare, N.E. (1991) Plasmids and Phase Variation in *Xenorhabdus* spp. *Applied and Environmental Microbiology,* 57, 2598-2601

Lengyel, K., Lang, E., Fodor, A., Szállás, E., Schumann, P., & Stackebrandt, E. (2005) Description of four novel species of *Xenorhabdus*, family Enterobacteriaceae: *Xenorhabdus budapestensis* sp. nov., *Xenorhabdus ehlersii* sp. nov., *Xenorhabdus innexi* sp. nov., and *Xenorhabdus szentirmaii* sp. nov. *Syst Appl Microbiol.* 28,115-122

Marokházi, J., Waterfield N, LeGoff, G, Feil, E, Stabler, R, Hinds J, Fodor A, & ffrench-Constant, R.H. (2003). Using a DNA microarray to investigate the distribution of insect virulence factors in strains of *Photorhabdus* bacteria *J Bacteriol.*185, 4648-4656

Martens E. C., Gawronski-Salerno, J., Vokal, D.L., Pellitteri, M.C., Menard, M. L., & Goodrich-Blair, H. (2003) *Xenorhabdus nematophila* requires an intact iscRSUA-hscBA-fdx operon to colonize *Steinernema carpocapsae* nematodes. *J Bacteriol.* 185, 3678-3682.

McGhee, G. C. & Jones, A. L., (2000) Complete nucleotide sequence of ubiquitous plasmid pEA 29 from *Erwin amylovora* strain Ea 88: Gene organization and intraspecies variation. *Applied and Environmental Microbiology,* 66, 4897-4907

McManus, P.S., Stockwell, V.O., Sundin, G.W., & Jones, A.L. (2002) Antibiotic use in plant agriculture. *Ann Rev Phytopathol* 40, 443-465

Metzger R., Bonatti H., & Sawyer R. 2009. Future trends in the treatment of serious Gram-positive infections. *Drugs Today (Barc.),* 45, 33-45

Middlemiss, J.K., & Poole K. (2004). Differential impact of MexB mutations on substrate selectivity of the MexAB-OprM multidrug efflux pump of *Pseudomonas aeruginosa. J Bacteriol.* 186, 1258-1269

Miller, J. H. (Ed). (1972). Experiments in molecular genetics [Cold Spring Harbor, N.Y.] Cold Spring Harbor Laboratory, ISBN:0879691069

Muvevi, J. W. (2011) The antimicrobial potential of *Xenorhabdus szentirmaii* and *X. budapestensis* nematode symbiotic bacteria on oomycetal agricultural pathogens, (*Phytophthora, Pythium*) MSc Thesis, University of Pannonia, Keszthely, Hungary.

Németh, J. 1998: Occurrence and spread of fire blight (*Erwinia amylovora*) in Hungary (1996-1998) management of the disease. *Acta Hort.*489, 177-183.

Németh, J. (1999) Fire blight (*Erwinia amylovora*) of pome fruits in Hungary. National Phytosanitary measures and management of the disease. EPPO Bulletin (29) 1-2: 135-144.

Nishioka M., Furuya N., Nakashima N. & Matsuyama N. 1997: Antibacterial activities of metabolites produced by *Erwinia spp.* Against various phytopathogenic bacteria *Annals of the Phytopathological Society of Japan.* 63, 99-102.

Ogier, J.-C., Calteau, A., Forst, S., Goodrich-Blair, H., Roche, D., Rouyn, Z., Suen, G., Zumbihl, R., Givaudan, A., Tailliez, P., Médigue, C., & Gaudriault,, S. (2010). Units of plasticity in bacterial genomes: new insight from the comparative genomics of two bacteria interacting with invertebrates, *Photorhabdus* and *Xenorhabdus BMC Genomics* 11, 568-589 The electronic version of this article is the complete one and can be found online at:http://www.biomedcentral.com/1471-2164/11/568

Pascual, M., Hugas, M., Badiola, J. I., Monfort, J.M., & Garriga, M. (1999). Lactobacillus salivarius CTC2197 prevents *Salmonella* Entertains colonialization in chicken. *Applied and Environmental Microbiology* 65, 4981-4986.

Paulin J. P., 1997: Fire blight: epidemiology and control (1921-1996) Nachrichtenblatt des DeutschenPflanzen schutzdienstes 49, 116-125

Peat, S.M., ffrench-Constant, R.H., Waterfield, N.R., Marokházi, J, Fodor, A, & Adams, B.J. (2010) A robust phylogenetic framework for the bacterial genus *Photorhabdus* and its use in studying the evolution and maintenance of bioluminescence: a case for 16S, gyrB, and glnA. *Mol Phylogenet Evol. 57*:728-740. Epub

Pitout, D.: Multiresistant Enterobacteriaceae: new threat of an old problem *Expert Rev Anti Infect Ther* 2008, 6, 657-669

Podger, F. D. (1972) *Phytophthora cinnamomi*, a cause of lethal disease of indigenous plant communities in Western Australia. *Phytopathology*. 62, 972-981.

Podger, F. D., Doepel, R. F. & Zentmyer, G. A. (1965) Association of *Phytophthora cinnamomi* with a disease of *Eucalyptus marginata* forest in Western Australia. *Plant Dis. Rep.* 49, 943-9472010 Aug 21

Poole K. (2004) Efflux-mediated multiresistance in Gram-negative bacteria *Clin Microbiol Infect.* 10, 12-26. A Review.

Richards, G.R., Vivas, E.I., Andersen, A.W., Rivera-Santos, D., Gilmore, S., Suen, G., & Goodrich-Blair, H. (2009) Isolation and characterization of *Xenorhabdus nematophila* transposon insertion mutants defective in lipase activity against Tween. *J Bacteriol.* 191, 5325-5331.

Rizzo D. M., Garbelotto, M., & Hansen, E.M. (2005). *Phytophthora ramorum*: integrative research and management of an emerging pathogen in California and Oregon forests. *Annu Rev Phytopathol.* 43, 309-335

Rosen, H. R. (1929). A study of the fire blight patogén, *Bacillus amylovorus* within living tissues *Science* 70, 329-330

Rosslenbroich H. J., & Stuebler D. (2000). *Botrytis cinerea* history of chemical control and novel fungicides for its management *Crop Protection* 19, I 557-561

Rossolini G.M., Mantengoli E., Docquier J.D., Musmanno R.A., & Coratza G. (2007) Epidemiology of infections caused by multiresistant gram-negatives: ESBLs, MBLs, panresistant strains. *New Microbiol.*, 30, 332-339

Roth, L. F. and Kuhlman, E. G. (1996) *Phytophthora cinnamomi*, an unlikely threat to Douglas-fir forestry. *Forest Science* 12, 147-159

Slama, T.G. (2008) Gram-negative antibiotic resistance. There is a price to pay. *Critical Care, 12(Suppl 4)*:S4

Schechner V., Straus-Robinson K., Schwartz D., Pfeffer I., Tarabeia J., Moskovich R., Chmelnitsky I., Schwaber M.J., Carmeli Y., & Navon-Venezia S. (2009) Evaluation of PCR-based testing for surveillance of KPC-producing carbapenem-resistant members of the Enterobacteriaceae family. *J Clin Microbiol.*, 47, 3261-3265

Schiermeier, Q. (2001) Russia needs help to fend off Potato famine, researches warn. *Nature,* 410, 1011

Shoemaker P. B. & Echandi E. 1976: Seed and plant bed treatments for bacterial cancer of tomato. *Plant Disease Reporter* 60: 163-166.

Sobel ML, McKay G, A., & Poole, K. (2003) Contribution of the MexXY multidrug transporter to amino glycoside resistance in *Pseudomonas* aeruginosa clinical isolates. *Antimicrob Agents Chemother.* 47, 3202-3207

Steiner P.W. and Zeller W. (1996) Fire blight in Hungary. Report to the Hungarian 1-16 Maryland, Darmstadt

Strateva T., & Yordanov D. (2009) *Pseudomonas aeruginosa* - a phenomenon of bacterial resistance. *J Med Microbiol*, 58, 1133-1148

Susa, J. A., Alexander, O. B., Glazer, I., Lango, L., Schwär, G., David, J. C., & Bode, H. B. (2008) Bacterial biosynthesis of a multipotent stilbene. *Angew Chem Int Ed Engl.*, 47, 1942-1945.

Szállás, E., Koch, C. Fodor, A., Burghart, J., Buss, O., Szentirmai, A., Nealson, K. H. & Stackebrandt, E. (1997): Phylogenetic evidence for the taxonomic heterogeneity of *Photorhabdus luminescens Int. J. Syst. Bacteriol 47,* 402-407

Szállás, E., Pukall, R., Pamjav, H., Kovács, G., Buzás, Z., Fodor, A. & Stackebrandt, E. (2001). Passengers who missed the train: comparative sequence analysis, PhastSystem PAGE-PCR-RFLP and automated RiboPrint Phenotypes of *Photorhabdus* strains In: (Ed.) Griffin, C. T., Burnell, A. M., Downes, M. J. & Mulder, R. (2001) Development in Entomopathogenic Nematode/Bacterial Research Luxemburg: European Commission Publications. 36-53

Talbot, N. J. (2004a) Let there be blight: functional analysis of virulence in *Phytophthora infestans Mol Microbiol.* 51, 913-915

Talbot, J. N. J. (2004b): Plant-Pathogen Interactions. *Annual Plant Reviews*, 11: 220-221.

Thaler, J-O., Givaudan, A., Dubals, M., & Latorse, P. (2001) Co-depot INRA-Rhobio Brevet no 0100415. Patent EP1351977;

Tzeng, K. C., Lin Y.C. & Hsu S.T. 1(994) Foliar fluorescent pseudomonads from crops in Taiwan and their antagonism to phytopathogenic bacteria. *Plant Pathology Bulletin,* 3, 24-33.

Van Immerseel, F., Eeckhaut, V., Teirlynck, E., Pasmans, F., Haesebrouck, F & Ducatelle, R. (2010) 16th European *Symposium on Poultry Nutrition* 16, 243-249.

Vila J., Martí S., & Sanchez-Céspedes J. (2007) Porins, efflux pumps and multidrug resistance in *Acinetobacter baumanii J Antimicrob Chemother*, 59, 1210-1215.

Völgyi, A., Fodor, A. Szentirmai, & A., Forst, S. (1998). Phase Variation in *Xenorhabdus nematophilus. Appl Environ Microbiol.* 64, 1188-1193. PMID: 16349534 [PubMed - as supplied by publisher]

Völgyi, A., Fodor, A., & Forst, S. (2000). Inactivation of a novel gene produces a phenotypic variant cell and affects the symbiotic behavior of *Xenorhabdus nematophilus. Appl. Environ. Microbiol.* 66,1622-1628 PMID: 10742251 [PubMed - indexed for MEDLINE.

Webster, J.M., Li, J, & Chen, G. (1996) Indole derivatives with antibacterial and antimycotic properties United States Patent 5569668 http://www.freepatentsonline.com/5569668.html

Webster, J. M., Chen, G., Hu, K. & Li, J. (2002): Bacterial metabolites In: In: *Entomopathogenic Nematology* Ed. Gaugler R. pp. 99-114, CABI International, London

Wiest P., Wiese K., Jacobs M. R, Morrissey, Anne B., Abelson T. I., Witt W., & Lederman M. M. (1987). "*Alternaria* Infection in a Patient with Acquired Immunodeficiency

Syndrome: Case Report and Review of Invasive Alternaria Infections". Reviews of Infectious Diseases (The University of Chicago Press), (4):799-803.

Zentmyer, G. A. (1980): *Phytophthora cinnamomi* and the diseases it causes. Monograph. American Phytopathological Society, St Paul Minnesota. No. 10, 1- 96

Zwet T., van der, & Beer S. V. (1995) Fire blight - Its Nature. Prevention and Control. A Practical guide to Integrated Disease Management *Agr. Inf. Bull.* 631. pp. 91

10

Metal Complexes as Prospective Antibacterial Agents

Joshua A. Obaleye*, Adedibu C. Tella and Mercy O. Bamigboye
Laboratory of Synthetic Inorganic Chemistry
Department of Chemistry, University of Ilorin, Kwara State
Nigeria

1. Introduction

Ninety elements occur naturally on earth. Out of these, nine are radioactive and among the remaining eighty one that could support life, sixty one are metals. Our bodies are 3% metal. Thus, it is surprising that some of the most serious challenges to human life, externally, the pollutants cadmium, mercury and lead are attracting more attention, whereas internally, there is a constant battle against sodium and calcium that are rejected by cells and accumulated elsewhere in the body during the ageing process. Furthermore, some diseases release metals into the blood stream. Their use in the fight against diseases was first described by Schubert in 1965. Man just like other vertebrates requires cations of the metals to facilitate a great many essential life processes. Moreover, many of the metals are essential for all other forms of life process. Around 5000 years ago the Egyptians used copper metal to sterilize water and gold was used in a variety of medicines in Arabian and China, but the practice emanated from the value of pure metal rather than from therapeutic effects.

Metals have played an important role in medicine for years, ever since human being started to walk on the planet. Many are essential to our diets in varying quantities, although people have only recently realized their significance.

This could probably be attributed to our increased awareness of personal and family health. Most of the major classes of pharmaceutical agents contain examples of metal compounds which are in current clinical use. Inorganic compounds (metal complexes) have been used to treat various diseases and ailments for many centuries.

Introducing metal ions into a biological system may be carried out for therapeutic or diagnostic purposes, although these purposes overlap in many cases. Yet despite the obvious success of metal complexes as diagnostic and chemotherapeutic agents, few pharmaceutical or chemical companies have serious in-house research programs that address these important bioinorganic aspects of medicine. Metals not only provide templates for synthesis, but they also introduce functionalities that enhance drug delivery vectors. It should be recognized that traditional studies of inorganic drugs at a fundamental level are not complete without a program on metal pharmacology. Many organic drugs require interaction with metals for activity. They interact with metals at their target site or

* Corresponding Author

during their metabolism or disturb the balance of metal ion uptake and distribution in cells and tissue. The unique properties of metal complexes tend to offer advantages in the discovery and development of new drugs. The metal complexes are amenable to combinatorial synthetic methods, and an immense diversity of structural scaffolds can be achieved. Metal centers are capable of organizing surrounding atoms to achieve pharmacophore geometries that are not readily achieved by other means. Additionally, the effects of metals can be highly specific and can be modulated by recruiting cellular processes that recognize specific types of metal-macromolecules interactions.

Understanding these interactions can lead the way towards rational design of metallo-pharmaceuticals and implementation of new co-therapies. Metal complexes appear to provide a rich platform for the design of novel chemotherapeutic drugs. We can choose the metal itself and its oxidation state, the numbers and types of coordinated ligands and the coordination geometry of the complexes. The ligands can not only control the reactivity of the metal, but also play critical role in determining the nature of secondary coordination-sphere interactions involves in the recognition of biological target sites such as DNA, enzymes and protein receptors. Also the ligands themselves can sometimes undergo biologically-important redox reactions or other modifications(e.g hydrolysis) *in vivo* mediated by the metal. These variables provide enormous potential diversity for the design of metallodrugs. Metal-ligand (coordination) bonds are usually much-weaker than covalent bonds and so ligand substitution reactions will be common in biological media. Most metallodrugs are therefore, pro-drugs. They can undergo ligand substitution and redox reactions before reaching the target site. A displaced ligand may itself attack a target site and controlled ligand release can play a role in the mechanism of action. The ongoing battle against resistance of organism towards drugs is far over. The importance of developing new drug chemotherapeutic drugs cannot be over emphasized. This trend was in the past intiated by the successes of metal-containing antitumor drugs such as cisplatin. Over the last four decades thousand of Inorganic drugs have been screened for their medicinal activity in a wide range of diseases, but only a handful have made it into the clinic.

Huge success was achieved in the area of anticancer drugs and antimicrobial drugs. Some of the metal-based drugs already in market are cisplatin (anticancer drug), cardolite (myocardial imaging agent drug), silverderma (skin burn drugs marketed in spain by Aldo Union), flammazine (Skin diseases drug marketed by Durpha) and Matrix metalloproteinase inhibitors (cancer and inflammatory disease marketed by British Biotech). The main objective of this paper is to review some of the previous work done by researchers with the hope of shielding light on the need to develop new drugs. We shall also draw together some of the work carried out by our group within our laboratories in the field of metal-containing Antibacterial agents.

2. Resistance of drugs

One way of restoring the activity of organic drugs for which resistance has emerged is to modify the structure to contain a metal, and some of these compounds are metal complexes. As early as 1975, it was reported by Edwards *et al*, (1975) that substituting the aromatic groups in the antibiotics penicillin and Cephalosporine with ferrocenyl moieties produced compounds with altered antibacterial activity compared to the starting materials Against various strains of *Staphyloccus aureus*, ferrocenyl penicillin showed comparable activity to benzyl-penicillin and also *β- lactamase,* which is one of the enzymes responsible for bacterial resistance to penicillin-type antibiotics. Many synthetic drugs have been discovered over the years for the treatment of malarial disease like chloroquine, sulphadoxine and pyrimethamine being among the most effective. However, malarial parasites resistant to these drugs are now widespread in America, Asia and Africa. Resistance to antimalarial drugs first to chloroquine and then to others was first noticed in the 1950s and since then, it has spread all over the world. Resistance of *Plasmodium falciparum* to chloroquine has become a major health concern of developing world. Therefore, it becomes highly necessary to come up with alternative antimalarial drugs with different structures and mode of action to deal with the development of resistance to the drugs in current use. Many researchers worked extensively on discovery of new therapeutic drugs to combat this problem of resistance. A number of papers on modification of the structure of the existing antimalarial drugs by incorporation of metals into their molecular structure appeared in literature. Notable among them are those of (Spacu *et al,*1968; Wasi *et al,*1987 ; Hubel *et al,* 2000; Biot *et al,*1999; Navarro *et al,* 2001; Sanchez-Delgado *et al,* 1993 and Tsangaris *et al,* 1974, Biot *et al.*2000)

Majority of these complexes were found to possess higher antimalarial activities than their parent drugs. The most recent and remarkable is the work carried out by Biot and co-workers. They inserted ferrocene(organometallic compound) into molecular structure of some antimalarial drugs.There is strong evidence that significant structural change to the side chain either through altering its length or through the introduction of more structural motifs such as ferrocene circumvents chloroquine resistance. Since the parasite needs iron for its development inside the red blood cells. Attempt was made by Biot *et al,* (1999) to insert a ferrocenyl group into the side chain of chloroquine, thus, producing a hybrid compound called Ferroquine (Figure 1). They combined poison (chloroquine) and bait (ferrocene) in the same molecule.

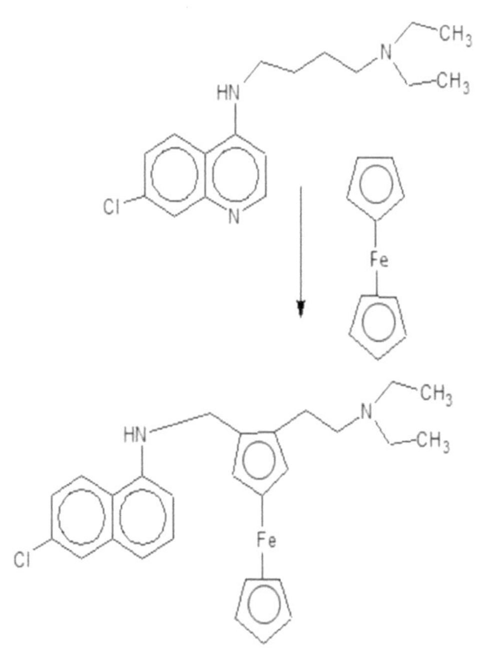

Fig. 1. Ferroquine

Ferroquine synthesized by them was found to be much more potent in mice than chloroquine. Tests have shown that ferroquine is active against both chloroquine sensitive and chloroquine resistant strains of *Plasmodium* and that it is safe and effective in mice, as well as being non-mutagenic. It was discovered that even when resistance to the drug builds up in mice, it can be reversed. The complex is a good candidate for further development. It is a promising organometallic analog of chloroquine..

They also used the same strategy to incorporate ferrocene into mefloquine. The pathway used for the synthesis of the ferrocenic mefloquine(Figure 2) includes coupling of an aminomethyl substituted ferrocene carboxaldehyde with a lithioquinoline compound. Mefloquine is covalently linked to a substituted ferrocenyl unit. The complexes exhibited a broad strong hydroxyl absorption band (3000cm^{-1}) characteristic of a hydroxyl group coordinated to an iron atom.(Biot *et al.* 2000).

The ^{1}H-NMR and mass spectra of the two diastereoisomers of the complex were similar except for the resonance of the Fe-CH-OH proton. The resonance of diastereomer (A) appeared at 8-6.5ppm (singlet, 1H) and resonance of diastereomer (B), appeared at 8 - 6.2 ppm due to a different aniosotropic zone of ferrocenic skeleton.

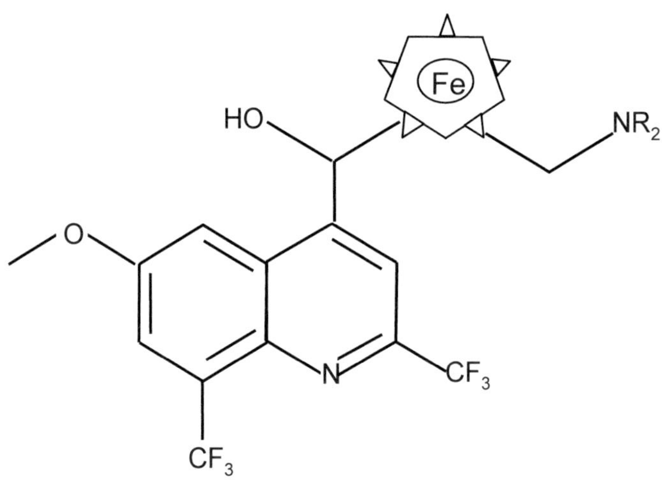

Fig. 2. Ferrocenic mefloquine

Artemisinin ferrocenic complexes were synthesized by Delhaes *et al* (2000). Novel ferrocenic artemisinin derivatives were found potent as artemisinin (QHS). Their antimalaria activity and affinity to bind with Ferritoporphyrin (IX) were studied. All the compounds showed capacity to bind with ferritoporphyrin (IX) resulting from the addition of different drugs concentrations. The association stoichiometry of compounds to Ferritoporphyrin (IX) was found to be 1:2.

Our research group also contributed to efforts being made to search for novel chemotherapeutic drugs against the resistant strains of *Plasmodium falciparum*. The synthetic strategy involves enhancing the activity of antimalarial drugs through the incorporation of transition metals into their molecular structures. In 1997, Obaleye and Nde-aga reported the preparation and characterization of amodiaquine hydrochloride and chloroquine complexes. They reported that Amodiaquine HCI coordinated through O-H and N-H to the metal ion and chloroquine coordinated to the metal ion through (N-H), (C-N) and (C=N) functional groups acting as either a bidentate or tridentate ligand. All the complexes possessed antimalarial activity as confirmed by studies on mice infected with *Plasmodium yoelli*. Obaleye *et al* (2009) carried out *in vivo* antimalarial activities and toxicological studies of some quinoline methanol metal complexes. Antimalarial activities of these complexes were investigated using mice infected with *Plasmodium berghei*. The results showed that four of the metal complexes $[(MefH^+)_2Fe(SO_4)_2]^{2-}$, $(MeFH+)CuCl_4.4H_2O$, $[Fe(QUIN)Cl_2.H_2O]SO_4.3H_2O$ (Figure 3) and $[Zn(QUIN)ClSO_4\ ]\infty$ exhibited significant higher antimalarial activity ($P<0.05$) than chloroquine and their parent ligands respectively.

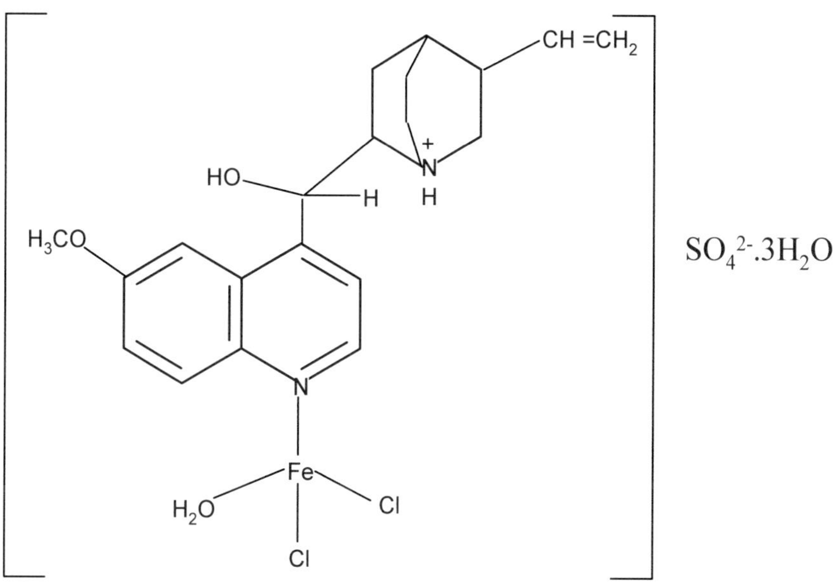

Fig. 3. Structure of $[Fe(QUIN)Cl_2.H_2O]SO_4.3H_2O$

The effects of these complexes on alkaline phosphatase (ALP) activity of kidney, liver and serum of Albino rats were investigated. Based on the results obtained, the complexes were found to be non-toxic and possess better antimalarial activity than the conventional antimalarial chloroquine. Tella and Obaleye (2010) synthesized two metal complexes of Co(II) and Cd(II) Trimethoprim. A distorted tetrahedral geometry is suggested for their structures (figure 4) and Trimethoprim behaves as a monodentate ligand.

The metal binds through the pyrimidine N(1) of the ligand. The complexes have been screened for antiplasmodial activity against *plasmodium berghei* and the results show that they are less active than the parent ligand. Toxicological study was carried out by investigating the effect of administration of the complexes on Alkaline phosphatase activity of kidney, liver and serum of Albino rats and they were found to be nontoxic. It can be seen from these studies by our group that incorporation of metal into molecular structure can either enhance the antimalarial activity or less or even make it the same.

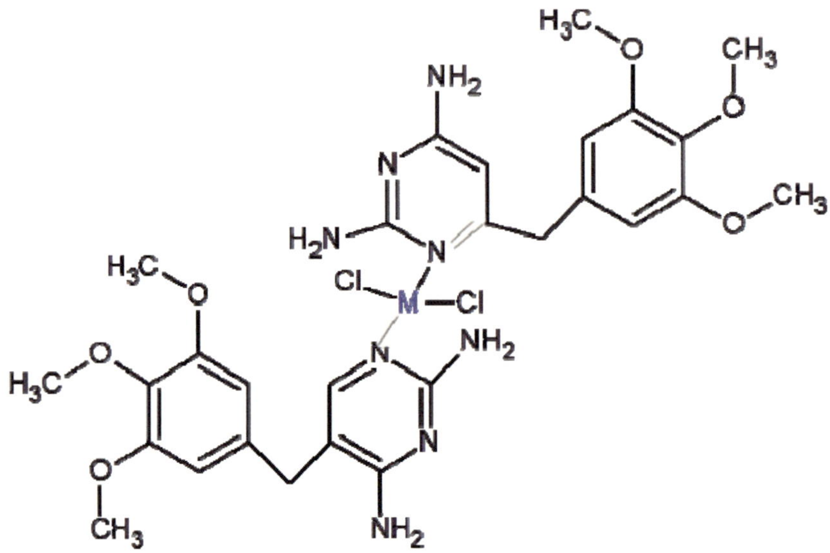

M= Co(II) or Cd(II)

Fig. 4. Structure of Trimethoprim metal complexes

Similarly, many organic drugs used to treat the parasitic diseases *leismaniasis* and *chagas* disease, as well as those used to treat *helminth* worm infections are becoming increasingly ineffective due to drug resistance. *Leishmania donovani,* which causes Leismaniasis, transferred via bite of a sandfly, infects approximately 10-15 million people worldwide. The disease may be fatal if not treated and effectivity of traditional organic drugs such as Pentamidine, Amphotericin B, Aminosidine and Antimonials is declining due to drug resistance (Ashford *et al,*1992; Quellette *et al,*1993). Pentamidine one of the organic antiparasitic drugs, has been complexed to several different metal centres and its activity has been evaluated against different parasite species. Of these an Organo-osmium derivative was found to have a 7.5 fold higher therapeutic index than pentamidine alone in treating leishmaniasis curing infected mice in a single dose(Loiseau *et al,*1992; Zinsstag *et al,*1991; Mesavalle *et al,*1993).

3. Anticancer metal complexes

Metal-based drugs are the most widely used drug in chemotherapy. The gallium, titanium salts have been shown to have anti-cancer activity. In the mid-1960s, Bernett Rosenberg and his co-workers(1965) serendipitously discovered that cis-dichlorodiammineplatium(II) (cis-$[Pt(NH_3)_2Cl_2]$, cisDDP, cisplatin) exhibited antitumor activity but trans isomer, trans-$[Pt(NH_3)_2Cl_2]$, did not. Cisplatin, (cis-$[PtCl_2[(NH_3)_2]$ also known as cis-DDP (Figure 5) is perhaps the best known example of a small molecule metal-containing drug. The history of the discovery and development of Cisplatin remains a remarkable scientific story. Its use and effectiveness in cancer chemotherapy since entry into the clinic in the late 1970s has been thoroughly documented (Lipport, 1999; Kelland *et al,* 2000; Wong and Giandomenico,1999). Cisplatin is cited for treatment of germ-cell cancers, gestational trophoblastic tumors, epithelial ovarian cancer and small cell lung cancer as well as for palliation of bladder, cervical, nasopharyngeal, esophageal, head and neck cancers. Cisplatin is a truly remarkable drug, in that for the last 30 years, it has been used to treat more than 70% of all cancer patients.

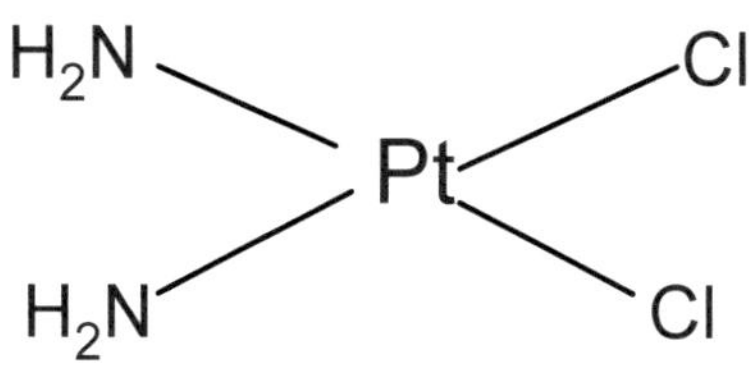

Fig. 5. Structure of cisplatin

Despite this success, there is still a limited range of tumors sensitive to cisplatin intervention. Some cancers are inherently resistant (O' Dwyer *et al*, 1999; Highley *et al*,2000). Due to this shortcomings, the second- generation compounds based on the cisplatin structure were developed in an attempt to improve toxicity and/or expand the range of useful anticancer activity. Carboplatin entered the clinic in 1998, principally in response to the necessity to reduce the toxic side effects of the parent drug(Christian, 1992). Despite the lower toxicity,carboplatin is essentially active in the same set of tumors as cisplatin and a broader spectrum of activity is not indicated. The problem associated with the use of cisplatin have driven the development of new inorganic anticancer therapies such as medaplatin and oxaliplatin(Van Rijt and Sadler, 2009). Progress to develop many chemotherapeutic transition metals drugs has been quite slow. Metallocenes and metallocene dichloride are potential candidates e.g. Titanocene compound ($TiCl_2Cp_2$). Titanocene dichloride was first recognized as an anticancer agent by Koepf and Koepf-Maier, (1979) and until recently being evaluated for activity against cisplatin resistant ovarian and metastatic renal-cell carcinomas (Clarke *et al*, 1999).The drug seems to bind to the protein in a similar way to iron, resulting in the cylopentadienyl ligands being released (Guo *et al*, 2000) and allowing the titanium metal to be delivered to the cancer cells. Titanocene dichloride demonstrates general anti-proliferation activity and has been shown to be effective against five types of cancer cells(Koepf-Maier and Koepf, 1987). Other metallocenes have more specific activity for example ferrocifens. Ferrocifens (Figure 6) was synthesized by Jaouen and Co-workers in 1994. They inserted ferrocene into molecular structure of Tamoxifen. Tamoxifen in form of its active metabolite hydroxyltamoxifen, is widely used in cancer hormone therapy and belongs to a class known as the Selective oestrogen receptor modulators (SERMS). It was discovered 20 years ago. It is used at all stages of the disease and recently, it has been shown to have a role in the prevention of cancer. It is however, only effective in 60% of cases and can cause problem of resistance when used for long time. Tamoxifen also increases the risks of endometrial (uterine) cancer and blood clotting in the lungs. Due to these shortcomings, Jaouen and co-workers in 1994 developed new SERMS that are as active as tamoxifen against as many types of breast cancer as possible. In order to find alternative SERMS, they decided to attach metallocenes to tamoxifens with the hope of improving its effect(Jaouen *et al*, 2000)

One series of Ferrocifens was made by replacing the aromatic phenyl in the β position of tamoxifen with an aromatic ferrocenyl of slightly greater bulk and lipophilicity, ferrocifen was synthesized via Mc-Murry coupling reaction. Ferrocifens has higher antiproliferative effect against (ER_{α} - and $ER\beta$ +) breast cancers than tamoxifen. The process of conversion is shown below(figure 6).

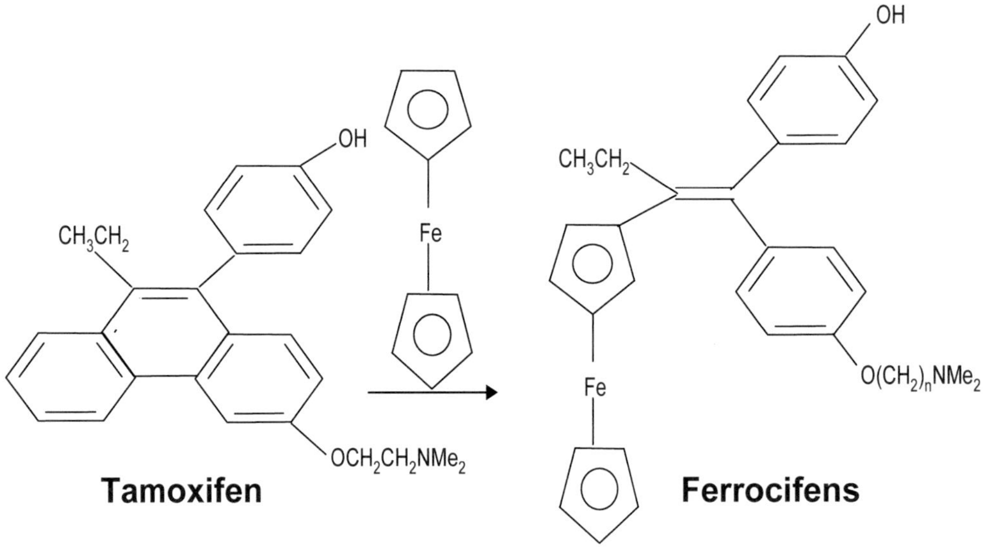

Fig. 6. Structure of Ferrocifens

It has been widely reported in the literature that the cytoxic effect of ferrocene complexes (Fe^{2+}) is associated with their oxidation to ferricinium type ($Fe^{\cdot 3+}$) radical and $O_2^{\cdot -}$ radical. Thus,

$$Fe^{2+} + O_2 \rightarrow Fe^{\cdot 3+} + O_2^{\cdot -}$$
$$Fe^{2+} + O_2^{\cdot -} + 2H^+ \rightarrow Fe^{\cdot 3+} + H_2O_2$$
$$Fe^{2+} + H_2O_2 \rightarrow Fe^{\cdot\cdot 3+} + OH^- + OH^{\cdot}$$

$O_2^{\cdot -}$ radicals are only slightly reactive towards DNA, but $OH\cdot$ radicals are very reactive and provoke various types of lesions, making these radicals highly genotoxic.

By introducing an organometallic function that liberates OH radicals within the cell via an oxidant/antioxidant cascade, we have the SERMS (and possibly for other systems) a potential solution to a difficult and essentially paradoxical problem. Combining metallocenes with other organometallic centres such as Gold provides another means of modifying drug activity and specificity. In this case the complex shows promising activity against bladder and colon cancers(Violette *et al,* 1995). Complexes of group VIIIB metals especially Rhodium, Indium and Platinum have been reported to have considerable antibacterial activity and induce lysis in *lysogenic bacteria*. Platinum II and IV amino halide complexes were found to have antitumor activity (Adrien, 1985). Dinuclear Platinum(II) complex[Pt_2_N,N^1 – bis(2-dimethylamino oxamide) Cl_4] was prepared by reaction of K_2PtCl_4 with the ligand N, N^1-bis(2-dimethylamino oxamide) in aqueous solution (Messori, *et al*,2003). X-ray diffraction analysis revealed that the platinum ions simultaneously bound to the ligand, on opposite sides. The coordination environment of both platinum centers is square planar with identical $NOCl_2$ donor sets. Preliminary *in vitro* studies point out that the dinuclear Platinum complex exhibits significant growth -inhibiting properties on a panel of cultured human tumor cell lines, although less pronounced than those of cisplatin.

Several recent studies(Adrien,1985; David, 1972; Messori, *et al* 2003) investigated twenty six inorganic compounds for antiviral and anticancer activity. These included metal complexes of already established anticancer drugs such as six mercaptopurine. It was suggested that the transfer of metal ion from the ligand to the viruses associated with cancer was a mechanism for releasing the anticancer drug in the locality of the tumor. Metal chelates of dl- methonine and ethionine were prepared, antibacterial activity were attributed to the ligands that render the metal ion fat soluble and thus made it capable of acting within the cell(Adrien, 1985).

Komeda *et al* (2000) reported another four azole bridged dinuclear platinum(II) complexes. All the four complexes were found to show much higher cytotoxicity than those of cisplatin.

4. Anti-inflammatory agents

Metal complexes of organic drugs have also been used as anti-inflammatory and anti-arthritic agents. Extensive research is being conducted into Au, Cu and Zn anti-inflammatory drugs that have fewer side effects with similar or higher efficacy than the parent organic drugs commonly in use. Gold compounds were first used in 1929 by French doctors to treat rheumatold arthritis(Shaw, 1999; Snyder *et al*,1987) and remain important in the treatment of rheumatic diseases. Several injectable transition gold complexes like sodium aurothiomalate, aurothioglucose and sodium aurothiopropanol are used clinically in the treatment of severe cases of rheumatold arthritis. (Rafique *et al*, 2010) The developments of orally Auranofin(also known as Ridaura was a major improvement over the early "injectable gold" preparation which were polymeric). Zhuo *et al*, (2000) and Sorenson,(1976) reported the synthesis and characterization of anti inflammatory Zn(II) and Cu(II) complexes of indomethacin. The studies were undertaken in order to reduce the side effects associated with the clinical use of indomethacin and related carboxylate-containing nonsteroidal anti-inflammatory drug (NSAID). NSAID drug exhibits favorable anti-inflammatory, analgesic and antipyretic properties, but it has undesirable side effects of inducing gastro-intestinal ulceration and haemorrhages. The complexes of indomethacin are superior to uncomplexed indomethacin for treatment of a range of conditions and most importantly reduce considerably lower incidences of gastro-intestinal damage.

Investigation of anti-inflammatory activity of complexes of diclofenac was carried out by Kovala-Demertzi *et al*,2000, 2001). Diclofenac is one of the widely used non-steroidal anti-inflammatory drugs (NSAIDs.) The Binuclear copper(II) complex of diclofenac $[Cu(L)_2(H_2O)]_2 . 2H_2O$ was found to have an anti-inflammatory profile superior to diclofenac when inhibiting inflammations due mainly to the activation of lipooxygenase and or to the complement systems. Other metal complexes of diclofenac synthesized by this group are Co(II), Ni(II) and Pd(II). These complexes exhibit a superior anti -inflammatory profile, inhibiting inflammations and phagocytosis and act as antioxidant compounds, properties that are absent in diclofenac.

The group also synthesized and characterized organotin complexes of Diclofenac (Kourkoumelis *et al.*, 2003).The complexes were found to be dimeric. The anti-inflammatory activity of complexes were not investigated. Extensive work was also carried out on the complexation of piroxicam with organometallic compound. Di Leo *et al,* 1998 synthesized Platinum(II)-piroxicam (Figure 7) complex by reaction of ziese salt $K(PtCl_2)(\eta^2\text{-}C_2H_4).H_2O$ with piroxicam in ethanol. The complex $PtCl_2)(\eta^2\text{-}C_2H_4).(Hpir)].0.5C_2H_5OH$- was synthesized. Platinum was linked to two chloride ions trans to each other, N(1) atom from the pyridyl ring of piroxicam and to an $\eta^2\text{-}C_2H_4$ molecule.

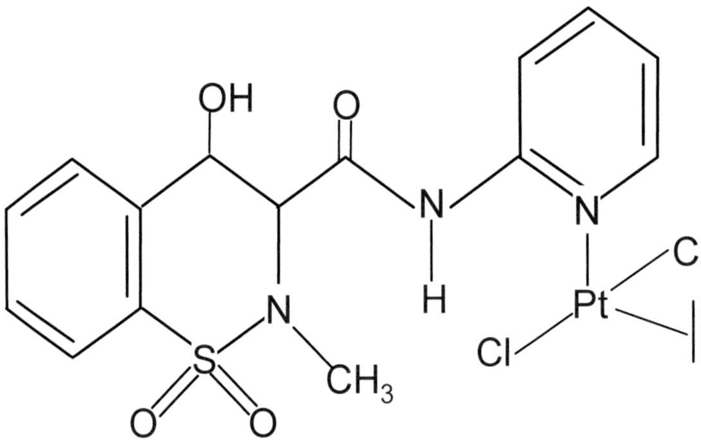

Fig. 7. Structure of Platinum(II)-Piroxicam

5. Antifungal agents

Many metal complexes have powerful antifungal activities and and are already in common day to day use such as silverderma (Silver complex of sulfadiazine) and Flammazine (Zinc complex of silverdiazine.)

Navarro *et al,* (2001) synthesized and characterized complexes of Copper (II) and Gold (I) with Clotrimazole and ketoconazole. It was found out that the ligands coordinated to Gold through imidazole N(1) atom of each ligand with linear structure. The clotrimazole and ketonazole coordinated to copper (II) through N (3) of the ligand atom with square planar structure. The new compounds were tested for *in vitro* activity against cultures of *epimastigotes* of *trypanosoma cruzi.* At concentration equivalent to 10.6M of total clotrimazole and ketonazole in dimethylsulfoxide, all the complexes exhibited higher inhibitory activity than their respective parental compound. Sanchez-Delgado *et al*(1993) reported enhancement of efficacy of complex of Ruthenium clotrimazole against *trypanosome Cruzi* as compared to clotrimazole ligand. Bankole *et al*(1979) reported the synthesis of organosilicon derivatives of p-aminosalicylic, salicylic and benzoic acids. It was discovered that the presence of silicon in the p-aminosalicyclic acid- silicon complex- prolonged and increased the antitubercular activity of p-amino salicyclic acid in the body. The reaction scheme for the synthesis of p-aminosalicyclic acid silicon complex is shown below (Figure 8).

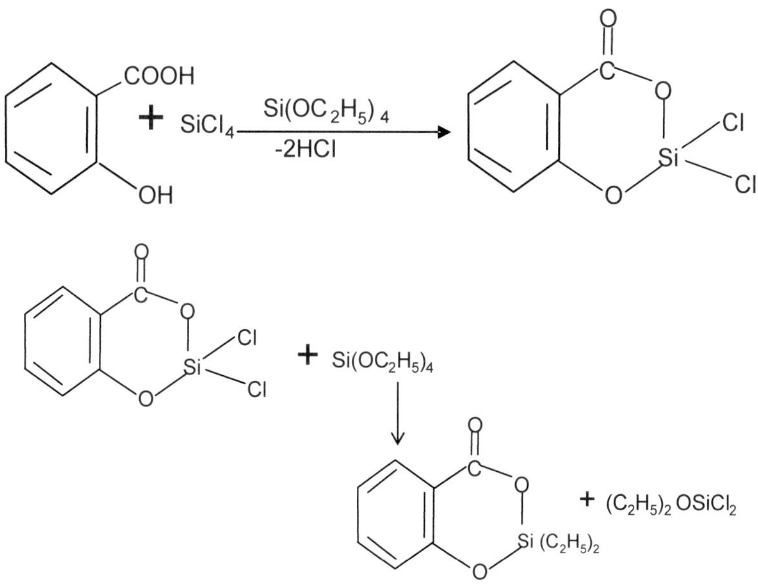

Fig. 8. Structure of Para-aminosalicyclic acid silicon complex

6. Antihypertensive agents

Hypertension can be a long term illness and it is becoming increasingly common, severely limiting the quality of life of the sufferer. Because of the long term nature of the condition and therefore the long term nature of the medication, there is a strong incentive to develop drugs that can regulate blood pressure without causing side effects or becoming resistant over the course of the treatment. Many of the current leading therapies are based on organic drugs, although some inorganic compounds also exhibit excellent activity.

Anion of Sodium nitroprusside anion can be used to release NO in biological system and has been investigated as a potential hypotensive (Tuzel,1974). Sodium nitroprusside can be administered by infusion and reduces blood pressure within two months, the effect depending on the rate of NO release. Other similar compounds have been studied for potential application as vasodilators including vanadium, cobalt and molybdenium analogues(Hayton,2002).

Essien and Coker, (1987) reported the complexation of antihypertensive drug with calcium. Calcium nifedipine(Figure 9) was synthesized by reaction of calcium salt with nifedipine. The infrared spectrum revealed a strong evidence of possible complexation occurring at the carbonyl group (C=O) of the nifedipine. Two atoms of calcium complexed each to one pair of C=O groups of 2 molecules of nifedipine.

Recently, Golcu *et al*,(2005) carried out the synthesis of binuclear copper (II) complex of Antihypertensive drug Pindolol. The biological activity of the parent drug pindolol was compared with the complex. The binuclear Cu(II) complex of pindolol was found to be highly active against *Bacillus megaterium, Aeromonas hydrophilia, Escherichia Coli, Candida albicans bacteries* and *Saccoramyces cerevisia, Rodotorula rubra, Kluyveromyces fragilis yeasts.* However, the free ligand was found not to be against these bacteria and yeasts.

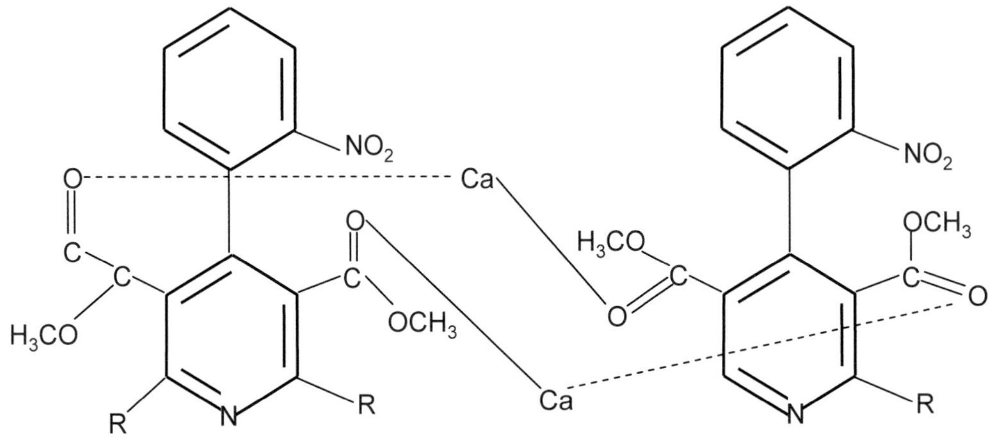

Fig. 9. Structure of Nifedipine - Ca.

7. Antibiotics agents

Many metal complexes have powerful antimicrobial activities and some of them are already in market.Silver bandages for treatment of burns.

A lot of antimicrobial metal complexes synthesized by researchers appeared in literature. We shall review some of these work. Emphasis will also be laid on the work carried out. Behrens *et al.* (1986) synthesized the transition metal complexes of Nalidixic acid.

Nalidixic acid is used in the clinical treatment of urinary tract infections caused by *gram-negative bacteria.* The mode of coordination of the drug was investigated by spectroscopic studies. From the spectra data, nalidixic acid anion binds through the carboxylate group either as a chelate or as bridge to give polymeric structure. Zupanicic *et al.*(2001) synthesized $[CfH_2]_2[ZnCl_4].2H_2O$ from Ciprofloxacin and $ZnCl_2$ in dilute HCl. The compound was shown to be ionic consisting of tetrachlorozincate(II)dianion and two protonated monoatomic ciprofloxacin molecules. The second one which is a complex $[Cu(Cf)(H_2O)_3]SO_4$. $2H_2O$ was synthesized by Turel *et al.*(1999). The complex was prepared by direct reaction of copper sulphate pentahydrate with ciprofloxacin in distilled water. X-ray crystallographic studies showed that the ciprofloxacin atom is bonded to the metal through carbonyl oxygen and carboxylic oxygen atom. Water molecules also coordinated to the copper.

Obaleye *et al.* (2001) synthesized and characterized metal (II) complexes of tetracycline HCl (Figure 10). For Mn, Fe, Zn, Co and Cd metals, the coordination of the metal to tetracycline is through one of the hydroxyl bands of tetracycline and oxygen of the carboxamide group, the proposed structures of the complexes were tetrahedral.

For Ni and Cu, the proposed structure of the complexes (Figure 11) were still tetrahedral, but the coordination is via oxygen of the v(C=O) and hydroxyl band of the drug. By using well - known antibiotics. Ogunniran *et al* (2007) complexed Ampicillin Trihydrate,Chloramphenicol and Oxytetracycline with Ni(II), Fe(III) and Co(II) chloride salts. Thus, the three ligands acted as terdentate. The values of Zone of inhibition for *E.Coli, S. Aureus and K.Pneumonas* revealed enhanced antimicrobial activities upon complexation with metal salts.

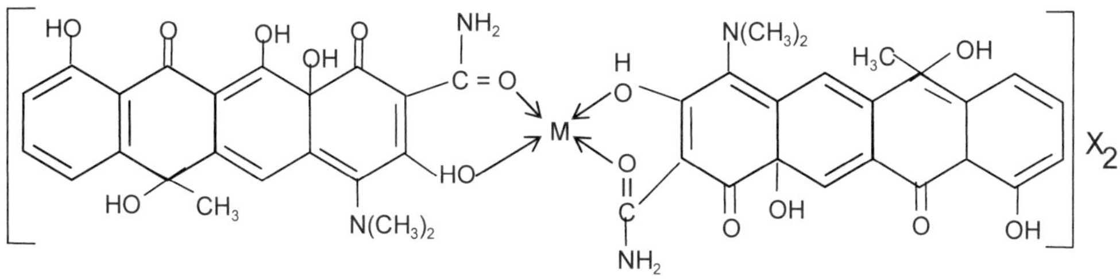

Fig. 10. Metal complexes Tetracycline [M = Mn, Fe, Co, Zn or Cd (II)].

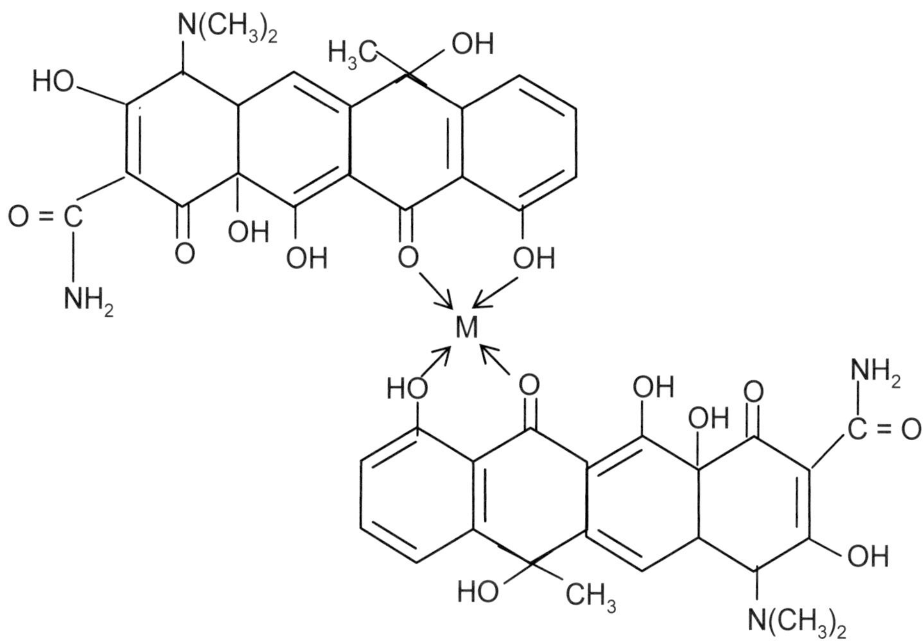

Fig. 11. Metal complexes of Tetracycline [Cu(II) or Ni(II)].

In 2007, Obaleye and co- workers synthesized two iron(II) complexes of ciprofloxacin by reaction of the ligand with iron(III) Chloride hexahydrate in different solutions. [Fe(Cip) $_2Cl_2$]Cl. $6H_2O$ and $(H_3Cip)FeCl_4$]Cl. H_2O were prepared. The antibacterial activities of the products against microorganisms were tested and it was established that their activities were comparable with those of their parent drug. Toxicological studies were carried out in which therapeutic dose of the ciprofloxacin drug and the metal complexes were administered to albino rats and the results showed that the metal complexes are not toxic.

Attempt was also made to incorporate metal salts into mixed ligands. Cu and Co complexes of mixed sulphadiazine-cloxacillin were synthesized and characterized(Tella *et al*, 2010). Infra-red spectra revealed that coordination of the metal to the sulphadiazine is through nitrogen of the pyrimidine and sulphonyl groups while in cloxacillin, coordination with the metal occurred through the oxygen of the carbonyl group of β- lactam ring. Octahedral structure was proposed for the complexes. Antimicrobial screening was also evaluated which showed that the complexes exhibit higher activities than their corresponding ligands.

Recently, Tella *et al*,(2011) investigated the possibility of transition metals coupling of antibiotics into cellulose. Chelates of Co(II), Zn(II) and Mn(II) cellulose- Antibiotics(Figure 12) were synthesized to form insoluble immobilized matrix with antibiotics. It can be established from this study that it is possible to form active immobilized antibiotics by simple chelation with metal salts.

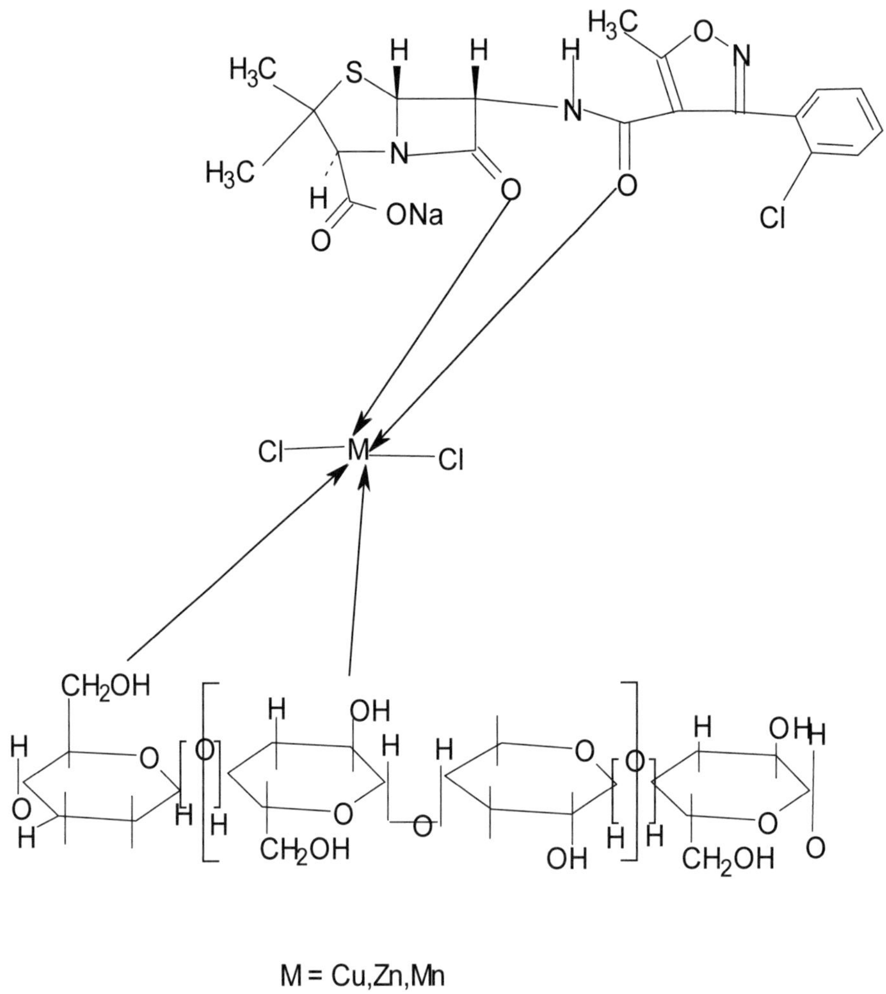

Fig. 12. Metal cellulose-Antibiotics Chelates.

From the antibacterial studies carried out, the products might be of greater applicability as food packaging material, antibacterial surface(water storage tanks, industrial membranes and chromatographic columns)..

8. Antibacterial agents

Casanova *et al.* (1993) synthesized single-crystal complex of [Zn (sulfathiazole)$_2$] H_2O. The sulfathiazole was found to act as a bidentate ligand, chelates to two Zn ion as a bridge through the N thiazole and N amino atoms. Rudzinski *et al* (1982) prepared several Sulfonamide -schiff base complexes of selenium(IV)(Figure 13) and tellurium(IV) (Figure14). Selenium coordinated through azomethine nitrogen, hydroxyl oxygen of the sulphonamide schiff base with chloride ion of the selenium salt to complete octahedral structure, while tellurium coordination site is only azomethine nitrogen, with chloride ion of the tellurium salt to complete the octahedral structure.

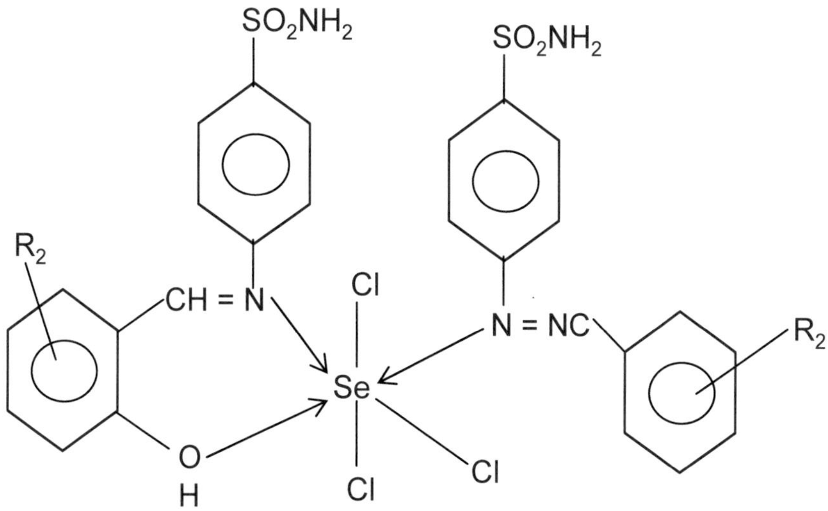

Fig. 13. Selenium (IV) sulphonamide Schiff base complex

The complexes were found to be active against bacterial even better than the parent ligand.

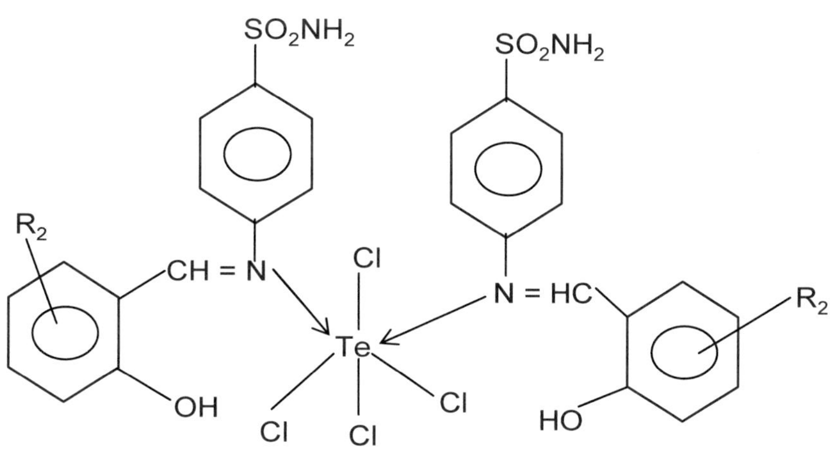

Fig. 14. Tellurium (IV) sulphonamide Schiff base complex

These two complexes proved to be biologically active as evidenced by pharmacological tests. Garcia-Raso *et al* (2000) synthesized single crystal of zinc-sulfamethoxazole complex [Zn(sulfamethoxazolato)$_2$ (pyridine)$_2$ (H_2O)$_2$](Figure15).

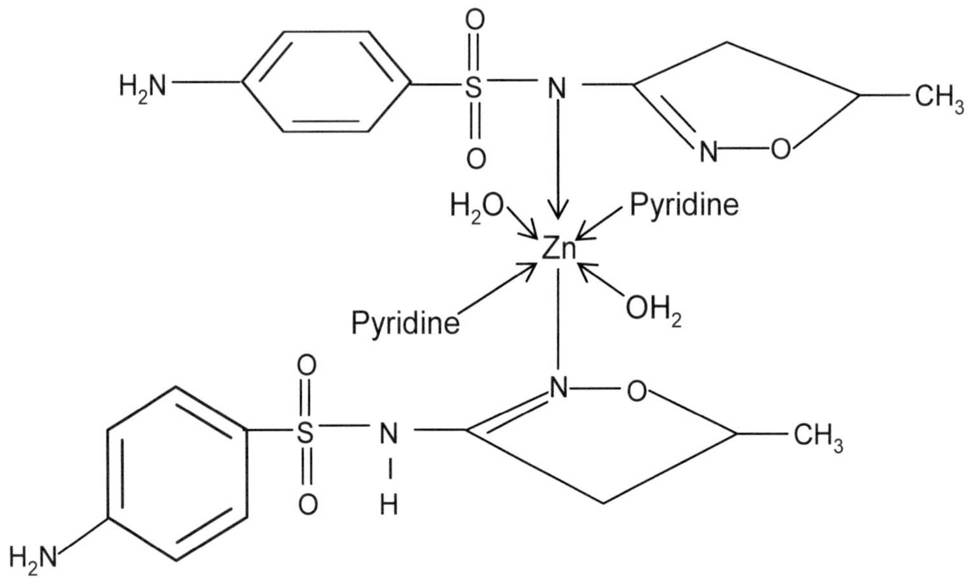

Fig. 15. Zinc sulfamethoxazole complex

The geometry around Zn(II) ion can be described as a slightly distorted compressed octahedron. Two pyridine and two isoxazole N atoms are located in the equatorial plane and two oxygen atoms of two water molecules are placed in the apical positions.

Ajibade and Kolawole, (2008) reported trivalent complexes of sulphadiazine. The complexes were tested for in-vitro activity against cultures of the resistant strains of *Plasmodium falciparium, Tripamastigotes T.B. rhodesiense* and *Amastigotes L. donovani* to determine their antiprotozoa activities. The Fe(III) complex is more active than the other complexes against the parasitic protozoa.

Recently, our group studied mode of coordination and Antimicrobial activities of complexes of some sulphonamides. Tella and Obaleye(2009) synthesized five complexes of copper (II) 4,4-diaminodiphenylsulphone (Dapsone) using copper salts of counter ion(sulphate, nitrate and chloride) in different reaction media(Solvents). The structure of the compounds were elucidated by spectroscopic techniques. Dapsone coordinated to the metal in monodentate and bidentate manner. with all the complexes having tetrahedral structures (Figures 16 and 17)

The biological activities data showed that the complexes are more active against *Escherichia coli, Klebbsiella pneumonia* and *staphylococcus aureus* than the free ligand.

Antimalarial activities of the complexes and the ligand were investigated using mice infected with *Plasmodium berghei.* All the complexes exhibited lower activity than the ligand and chloroquine. Toxicological studies carried out showed that the complexes are not toxic, as indicated from the effect of administration of the complexes on alkaline phosphatase activities of kidney, liver and serum of Albino rats. The serum ALP activity showed no significant change(P>0.05), suggesting non-damaging effect on the plasma membrane of liver and kidney cells.

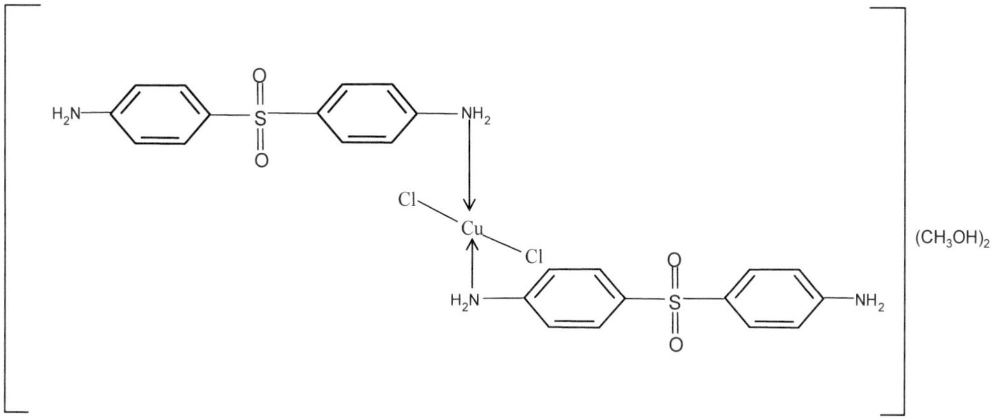

Fig. 16. Structure of $[Cu(DAP)_2Cl_2](CH_3OH)_2$

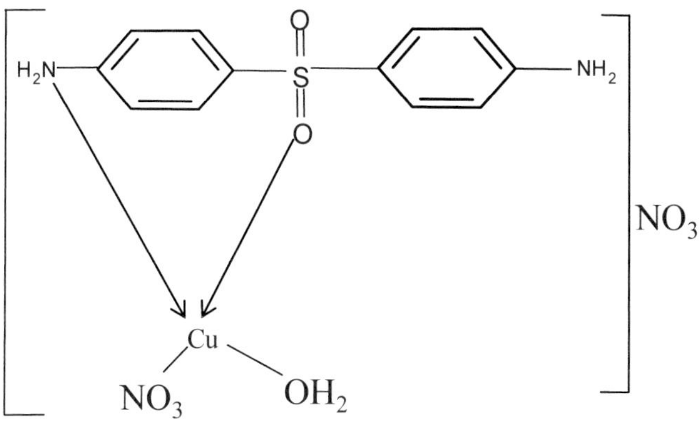

Fig. 17. Structure of $[Cu(DAP)(NO_3)(H_2O)]NO_3$

Metal complexes of sulphadimidine synthesized by Tella and Obaleye (2009) were established to possess higher antibacterial activities than the ligand. The complexes showed greater activities against the *Escherichia coli, Klebbsiella pneumonia* and *staphylococcus aureus.* This is in agreement with the findings of other researchers.

9. Vitamin metal complexes

Vitamins are essential for the normal growth and developments of a muticellular organisms. Once growth and development are completed, vitamins remain essential nutrients for the healthy maintenance of cells, tissues and organ that make up a multicelluar organisms. It has been established that complexation of metal with vitamins enhances the activity of the vitamins. Many workers made attempt to synthesis metal complexes of thiamine. From early complexation studies, it was evident that thiamine and its derivatives do not readily form true complexes with direct metal-thiamine bonds but instead, they give ionic salts mainly of the type $(HT)^{2+}(MX_4)^{2-}$ due to the net positive charge on the thiazolium ring and the easy protonation of pyrimidine N(1) atom. Such ionic salts synthesized by them are $[ThH]^{2+}$ $[HgCl_3]_2^-$, $[ThH]^{2+}[CdCl_4]^{2-}$ and $[ThH]^{2+}$ $[HgCl_4]^{2-}$. (Garcia-Raso *et al*, 2000,

Richardson,1975). However, the successful preparation and structure determination of $Cd(Th)Cl_3.O.6H_2O$,$Cd(Th)\ Cl_3]_2.2H_2O$, $Cu(Th)\ Cl_2$, $Zn(Th)\ Cl_3.\ O.4H_2O$ (Hadjilias, 1983; Gramer,1984; Bencini,1987) proved that metal-thiamine complexes exist. Another study on vitamin is the work carried out by Mosset *et al*(1978). They synthesized $Cd(Py)Cl_2$ from reaction of aqueous solution (pH=6.5) of equimolar amount of pyridoxine and cadmium chloride. From NMR and X-ray structural studies, the structure consists of infinite chains of Cd-bridge-Cd. The Cd atom, in an octahedral environment is bound to three chlorine atoms and three oxygen atoms of the ligand. Two chloride ions make a double bridge between two equivalent cadmium atoms. The pyridoxine molecule acts as a bidentate ligand through two oxygen and as a bridge between two equivalent atoms. Zinc complex of pyridoxine was also reported by Thompson *et al*(1980). $[Zn(Py)_2(H_2O)_2](NO_3)_2$ was synthesized from reaction of zinc nitrate and pyridoxine. The X- ray crystallographic studies revealed that Zinc atom lie on centers of symmetry and are chelated to the 4-amino methyl and phenolate groups of pyridoxine zwitterions. Octahedral coordination is completed by water molecules. Dakovic *et al* (2008) reported the synthesis of nicotinamide metal complexes of Zn(II) and Hg(II). The nicotinamide with carboxamide group in the meta-position to the pyridine nitrogen atom acts as a monodentate –N ligand for the zinc(II) and mercury (II) ion coordinating to metal ions through the pyridine nitrogen atom and leaving both carboxamide moieties available. The biological activities of all these complexes were not investigated. It can be seen from literature review that there is little or no work carried out on investigation of biological activities by many researchers. Attempt was made by our research group to investigate antimicrobial activities of vitamin-metal complexes. Investigation of antimicrobial activities of transition metal complexes of vitamin C was carried out. Obaleye *et al.* (1994) ; Obaleye and Ojiekwe, (1983) synthesized and screening Co(II), Zn(II), Mn(II),Fe(III),Hg(II),Cu(II) and Cd(II) complexes of vitamin C against four strains of bacterial species- *Escherichia coli, Klebbsiella pneumonia, staphylococcus aureus and Bacillus substilis* and two fungal species- *Aspergillus flavus and Aspergillus niger.* The complexes have little or no activity on the bacterial species studied. Generally, percentage inhibition of Ascorbic acid on fungal species was the greatest among all compounds tested.

10. Antidote

Metal complexes have been used as antidote since 1945, for chronic intoxication arising from therapy or household contamination or hasten excretion of radioactive element. These antidotes circulate in the blood stream without causing much depletion of the body 's essential metals. A lot of ligands have been used as antidote to combat metal poisoning. Dimercaptol is used to counter poisoning by compound of gold and mercury. The use of ascorbic acid as a possible antidote was demonstrated by Key pour *et al.* (1986).

In an attempt to discover possible antidote for drug or metal poisoning. We investigated the interaction of pyrimidine and sulphonamide drugs with some transition metals by determining their stability constants in order to assess their potentiality as antidote for metal-overload poisoning(Tella and Obaleye, 2010). The stability constant(β) were found to be log 10.68, 5.5 and 4.8 for Trimethoprim, sulphadiazine and sulphadimidine with respect to metal salts. The order of stability constants(β) was found to be Cu(II) > Fe(III).>Ni(II) >Co(II) > Zn(II) in accordance with Irvin-williams series. The stability constant data revealed that this ligand may be used as antidote or chelating agent for medical treatment of metals overload or poisoning.

11. Conclusion

The structure of known biologically active molecules is modified to result in new molecules known as metal coordinated complexes. The goal of such modification is to get a molecule that is improved in some way, such as potency, stability, reduced side effect or targeted delivery. The improvement is achieved without sacrificing the molecules' desirable properties. In this paper, we have been able to shield light on effort made so far by our group and other researchers in discovery of new antibacterial agents by modification of existing known biological agents through the incorporation of metal into their molecular structure. We have been able to establish that some of these complexes are biologically active than their parent ligands, making them promising candidates to join league of metal-based drugs already in market. It should be known that traditional studies of organic drugs at a fundamental level are not complete without a parallel program on metal pharmacology. In general, because metal can undergo ligand exchanges, metal complexes are pro-drugs, ligand substitution can activate the metal complex toward binding to target molecules.

It should be recognized that a metal is not just a metal : it is a metal ion plus its ligand. The metal ion plus the ligand determine the biological activity.

We should know that despite the obvious success of metal complexes (Cisplatin, Silverderma, Flammazine and others) as chemotherapeutic agents, few pharmaceutical companies have serious in-house research programs that address these important bioinorganic aspects of medicine. Research programs in organic and metallo-drugs should not be seen as mutually exclusive. They overlap extensively and the combination is likely to be a powerful force for the future.

12. Acknowledgement

The authors are grateful to World Bank Supported Science and Technology Education Post - Basic Project (STEP- B), University of Ilorin Senate Research Grant Committee and University of Cape Town, South Africa for financial and material support. JAO appreciates TWAS,Italy and Institute of Chemistry,CAS,Beijing,China for Visiting Scholar award. ACT is also grateful to Queen's University, School of Chemistry and Chemical Engineering, Belfast, United Kingdom for fellowship. JAO also thank Prof.(Mrs.) E.A. Balogun and Dr.O.Awotunde from UNILORIN for their support.

13. References

[1] Adrien, A. (1985). *Selective Toxicity*, 7th Ed. Chapman and Hall, New York, 432- 468.

[2] David, R.W. (1972). Anticancer drug design involving complexes of amino acid and metal ions. *Coord. Chem. Review*, 5, 123-133

[3] Messori,L., Shaw,J., Camalli, M., Mura, P., Marcon, G. (2003). Structural features of a new dinuclear platinum (II) complex with significant antiproliferative activity. *Inorg. Chem.* 42, 6166-6168.

[4] Komeda,S., Lutz, M., Spek, A.L., Chikuma, M., Peedijk, J.(2000). New antitumor - active Azole - bridged dinuclear platinum (II) complexes synthesis, characterization, Crystal structures and Cytotoxic studies. *Inorg. Chem.* 39, 4230 - 4236.

[5] Zhou,Q., Hambley, T.W., Kennedy, B. J., Ley, P. A Turner, P., Warwick, B., Biffin, J. R., Regtop, H. L.(2000). Synthesis and characterization of Anti-Inflammatory Dinuclear

and Mononuclear. Zinc Indomethacin complexes. Crystal structures (Zn_2 (Indomethacin)$_4$ (L)$_2$] and (Zn (Indomethacin)$_2$ (L)$_2$] *Inorg. Chem.* 39, 3742-3748

[6] Sorenso,J.R.J(1976). Copper Chelates as possible active forms of the antiarthritic agents. *J.Med.Chem,19,135-148.*

[7] Navarro, M., Fajardo, E.J., Lechmann, T; Delgado, R.A., Atencio, R., Silva, P., Liva, R. , Urbina, J.A. (2001). Toward a novel metal-based chemotherapy against tropical diseases; synthesis and characterization of new copper (II) and Gold (I) Clotrimazole and ketoconazole complexes and evaluation of their activity against trypanosome cruzi.*Inorg. Chem.* 40, 27, 6879 - 6884.

[8] Sanchez - Delgado, R.A., Lazardi, K., Rincon, L., Urbina, J., Hubert, A.J., Noels, A.N. (1993). Toward a novel metal based chemotherapy against tropical diseases II. Enhancement of efficacy of clotrimazole against *trypanosoma cruzi* by complexation to Ruthenium in $RuCl_2$ (clotrimazole)$_2$. *J. Med. Chem.* 36, 2041-2043.

[9] Bankole, F.O. (1979). Organosilicon derivatives of salicylic acid. *J. Pharm. and Med. Sci.* 4(5), 249-250..

[10] Essien, E.E., Coker, H.A.B. (1987) Interaction of nifedipine with calcium. *The Nig. J. Pharm.* 18(4)21-22.

[11] Obaleye, J.A., Adeyemi, O.G. , Balogun, E.A. (2001). Some metal tetracycline Complexes: Synthesis, characterization and their effects against malaria parasites. *Int. J. Chem.* 11(2), 101-106.

[12] Casanova J., Alzuet, G.,Ferrer, S., Borras, J., Garcia - Granda, S., Perez - Carreno, E. (1993). Metal Complexes of sulfanilamide Derivatives Crystal structure of [Zn (sulfathiazole)$_2$]. H_2O]. *J. Inorg. Biochem.*,51, 689-699.

[13] Rudzinski, W.E., Aminabhavi, T.M., Biradar, N.S., and Patil, C.S. (1982). Biologically active sulfonamide Schiff base complexes of Selenium (IV) and Tellurium (IV). *Inorg. Chim. Acta.* 67, 177-182.

[14] Richardson M.F., Franklin, K., Thompson, D.M. (1975).Reaction of metals with vitamins I. Crystal & molecular structure of thiamium tetrachlorocadmate monohydrate. *J.Am. Chem. Soc.* 97, 3204-3209.

[15] Hadjiliadis, N. , Yannopoulos, A. (1983). Complexes of Mercury (II) with Thiamine. *Inorg. Chim. Acta*, 69, 109-115.

[16] Gramer, R.E., Maynard, R.B., and Evangalista, R.S. (1984). Synthesis, crystal and molecular structure of a Cu (II) complex of vitamin B1, Cu (thiam) Cl_2. *J.Am. chem. Soc.* 106, 111-116.

[17] Bencini, A., Borghi, E. (1987). Complexes of vitamin B1 with Transition metal ions Crystal and molecular structure of Zn (thiamine) Cl_3. $0.4H_2O$. *Inorg. Chim. Acta.* 135, 85-91.

[18] Mosset, A., Nepveu-juras, F., Haran, R., Bonnet, S.S. (1978). Complexation of the vitamin B_6 with Cd^{2+} cation: NMR and X-ray structural study. *J. Inorg. Nucl. Chem.* 40, 1259-1263.

[19] Thompson D.M., Balenovich, W., Hornich, L.H. M., Richardson, M.F. (1980). Reactions of metal ions with vitamins. The crystal structure of a Zinc complex of pyridoxamine (vitamin B_6). *Inorg. Chim. Acta.* 46, 199-203.

[20] Obaleye, J.A., Orjiekwe, C.L. (1983). Synthesis and characterization of copper (II) and Zinc (II) complexes of Ascorbic acid. *Int. J. Chem.* 4, 37.

[21] Zupanicic, M., Turel, I., Bukorec, P., White, A.J.P., Williams, D.J. (2001). Synthesis and characterization of two novel Zinc (II) complexes with ciprofloxacin. Crystal structure of $(C_{17}H_{19}N_3O_3F)_2$. $(ZnCl_4)$. $2H_2O$. *Croatica Chemica Acta.* 74(1),61-74

[22] Turel, I., Golic, L., Ramirez, O.L.R. (1999). Crystal structure and characterization of a new Copper (II) cirpofloxacin (cf) complex [Cu (cf) $H_2O)_3$] $SO_4.2H_2O$. *Acta. Chim. Slov.* 46 (2), 203-211.

[23] Spacu, P., Gheorghiu, K. , Nicolaescu, A. (1968). Coordination Compounds of Ni (II) and Ni (III) with Paludrine. *Coord. Chem. Reviews,* 12, 413-415.

[24] Wasi, N., Singh, H.B., Gajanana, A., Raichowdhary, A.N. (1987). Synthesis of metal complexes of antimalarial drugs and *in-vitro* evaluation of their activity against *Plasmodium falciparum. Inorg. Chim. Acta.* 135(2), 133-137.

[25] Obaleye, J.A. , Nde-aga, B.J. (1997) Some antimalarial drug metal Complexes. Afr. J. Sc. 1, 10-12.

[26] Tsangaris, J.M., Baxevanidis, G.T., ZeitsChrift Naturforschung, Teil Bi (1974). Complexes of Quinine with Copper (II), nickel (II), Cobalt (II) and Chromium (III) Chlorides. *Zeitschrift fuer Naturforschung,* 29 (7/8) 532-537.

[27] Ogunniran, K.O., Tella, A.C., Alensela, M., Yakubu, M.T. (2007) Synthesis, physical properties, antimicrobial potentials of some antibiotics complexed with transition metals and their effects on alkaline phosphatase activities of selected rat tissue. *Afr. J. biotech.* 6(10), 1202 -1208.

[28] Keypour, H., Silver, J., Wilson, M.T. , Hamed, M.T. (1986). Studies on the reactions of ferric ion with Ascorbic acid. A study of solution chemistry using Mossbauer Spectroscopy and stopped-Flow techniques. *Inorg. Chim. Acta,* 125, 97-106.

[29] Biot, C., Delhaes, L.A., N'Daiye,C.M., Maciejewski,L., Camus, D., Dive, D and Brocard, J.S. (2000). Synthesis and potential metabolites of ferrochloroquine and related compounds. *Biorg. Med. Chem.* 7, 2843-2846.

[30] Biot, C., Delhaes, L., Abessolo, H. (1999). Novel Metallocenic compounds as antimalarial agents: Study of the position of ferrocene in chloroquine. *J. Org. chem.* 589, 1, 57-65.

[31] Delhaes, L., Biot, C., Berry, L., Maciejewski, L., Camus, D.,Brocard, J.S., Dive, D. (2000). Novel ferrocenic Artemisinin derivatives: Synthesis, *in-vitro* antimalarial activity and affinity of binding with ferrophotoporphyrin (IX*) Bioorg. Med. Chem.* 8, 2739.

[32] Hubel, R.; Polborn, K.; Knizek, J.; Heinrich,; N and Wolfgang B. (2000). Metal complexes of Biologically important ligands CXXVI Palladium (II) and Platinum (II) complexes with the antimalarial drug mefloquine as ligand. *Anorg. Allg. Chem.* 629, 1701-1708.

[33] Di Leo,D., Berrettin, F and Renzo, C.(1998) Synthesis of Platinum(II)-Piroxicam Compounds.Crystal structure of trans-dichloro(η^2-ethene(Piroxicam)Platinum(II). *J.Chem.Soc. Dalton Trans,*1993-2000.

[34] Garcia-Raso, A., Fiol, J.J., Rigo, S., Lopez-Lopez, A., Molins, E., Espinosa, E., Borras, E., Alzuet, G., Borras, J., Castineiras, A. (2000). Coordination behaviour of sulfanilamide derivatives, Crystal structure of [Hg (Sulfamethoxypyridazinato)$_2$], (Cd (Sulfadimidinato)$_2$ (H_2O)]. $2H_2O$ and [Zn (Sulfamethoxazolato)$_2$ (pyridine)$_2$ ($H_2O)_2$]. *Polyhedron,* 19, 991 - 1004.

[35] Obaleye J.A, Tella, A.C and Arise, R.O(2009). Advances in Natural and Applied Sciences,3(1),43-38.

[36] Ashfordm, R, Desjeux, P, de raadt,P (1992). Parasitol. Today, 8, 104

[37] Quellette, M, Papadopoulou, B(1993). Parasitol. Today, 9.150

[38] Loiseau, P M, craciunescu, DG, Doadriovillarejo J.C, Certedfombona, G, Gayral, P.(1992). Tropical medicine and Parasitology, 34.110.

[39] Insstag, J; Brun, R; Craciunescu, D.G; lglesias, E.P(1991). Tropical medicine and Parasitology, 42 : 41

[40] Mesa valle, C.M; Mortaleda Lindez, V; Craciunescu, D; Alonso, M; Osuna, A(1993).Arzneimittel Forsch Drug Res. 43: 1010.

[41] Rosenberg, B; Van camp, L ; Krigas, T.(1965). Nature, 205, 698.
[42] VanRujt, S.H and Sadler, P.J.(2009) Drug Discovery Today. 14, 23/29, 1089-1097
[43] Lippert B(1999). Ed, Cisplatin : Chemistry and Biochemistry of a leading Anticancer Drug, Wiley-VCH, New York.
[44] Kelland, L.R; Farell, N.P.(2000). Platinum-based drugs in Cancer Therapy: Human Press, Totowa, New Jersey.
[45] Wong, E;Giandomenico, C.M.(1999), Chem. Rev. 99, 2451-2466.
[46] Christian, M.C.(1992). Semin. Oncol. 19,1970.
[47] O'Dwyer, P.J; Stevenson, J.P; Johnson, S.W.(1999) Clinical status of cisplatin, carboplatin and other platinum-based Antitumor Drugs. in cisplatin: Chemistry and Biochemistry of a leading Anticancer Drug ; Lippert, B; Ed. Wiley-VCH, New- York, 31-72.
[48] Rafique, S; Idress, M; Naim, A; Alcbar, H and Athar, A.(2010). Biotechnology and Molecular Biology Reviews, 5(2), 38-45.
[49] Highley, M.S; Calvert, A.H. (2000) Clinical Experience with Cisplatin and Carboplatin ; Kelland, L.R; Farell, N.P; Eds: Human Press, Totowa N.J. 89-113.
[50] Shaw, C.(1999). Gold Complexes with anti-artritic, Antitumor and Anti-HIV Activity; In uses of inorganic chemistry in medicine, Farell, N; Ed. Cambridge, 26-57.
[51] Jaouen, G; Top, S; Vessieres, A; Leclercq, G; Quivy, J; Jin, L; Groisy, A.(2000). CR. Acad. Sci. IC, 3,89.
[52] Tella, A.C and Obaleye, J.A.(2009). E-journal of Chemistry,6(S1), s311-S323.
[53] Violette, M; Gautheron, B; Kubicki, M.M; nifantev, I.E; Fricker, S.P(1995). Metal-based Drugs, 2, 311.
[54] Koepf-Maier, P; Koepf, H.(1987). Chem. Rev. 87, 1137.
[55] Snyder, R.M; Mirabelli, C.K; Crooke, S.T.(1987), Semin. Arthritis Rheum, 17, 71-80.
[56]Guo, M.L; Sun, H.Z; Mc Ardle, H; Gambling, L; sadler, P.J.(2000). Biochemistry, 29, 10023.
[57] Koepf, H. Koepf-Maier, P.(1979). Angew Chemie, 91, 509.
[58] Clarke, MJ; Zhu, F.C; Frasca, D.R.(1999). Chem. Rev. 99.2511.
[59] Edwards, E.I; Epton, R; Marr, G.(1975). J. Orgnometal. Chem. 85, c23.
[60] Tella, A.C and Obaleye, J.A.(2010). Int.Chem. Sci., 8(3), 1675-1683.
[61] Tuzel, H.I(1974).Clin. Pharmacol.14, 494.
[62] Hayton, T.W; legzdins, P; sharp, W.B(2002). Chem.. Rev. 102, 935.
[63] Tella, A.C and Obaleye, J.A(2010). Orbital-The Electronic Journal of Chemistry, 2(1), 11-26.
[64] Tella, A.C and Obaleye, J.A.(2010). Int. J. Biol. Chem. Sci.4(6), 2181-2191..
[65] Ajibade,P.A and Kolawole, G.A.(2008). Bulletin Chem. Soc. Ethiopia, 22(2), 261-268.
[66] Tella, A.C; Obaleye, M.O and Obaleye, J.A(2010).Integrated Journal of Science and Engineering, 9, 1,58-63.
[67] Kourkoumelis, N; Demertzis, M.A; Kovala- Demertzi, D; Koutsodimou, A; Moukarika,A.(2004). Spectrochimica Acta Part A 60, 2253-2259
[68] Golcu, A; Dolaz, M; Dager, E.K.(2005). KSU Journal of Science and Engineering, 8(1),4-9.
[69] Behrens, N.B and Diaz, G.A(1986). Inorganica Chimica Acta. 125,21-26.
[70] Dakovic, M; Popovic, Z; Giester, G; Rajic-Linaric,M.(2008). Polyhedron,27, 465-472.
[71] Obaleye, J.A and Famurewa, O.(1989). Biosci. Res. Commun. 1, 87.
[72] Obaleye, J.A; Akinremi, C.A; Balogun, E.A and Adebayo, J.O.(2007). African Journal of Biotechnology,6(24), 2826-2832.
[73] Tella, A.C ; Obaleye, M. O and Akolade, E.O.(2011). Middle- East Journal of scientific Research,7(3),260-265.
[74] Obaleye J.A.,Orjiekwe C.L. and Famurewa O.(1994).Effects of Some Novel Ascorbic Acid-Metal Complexes on SelectedBacterial and Fungal Species. *J.Sci.I.R. of Iran* 5(4) 154-157.

11

Bacteriostatic Agents

Marzieh Rezaei[1], Majid Komijani[1] and Seyed Morteza Javadirad[2]

[1]Department of Biology, Faculty of Science, Nour Danesh Institute of Higher Education, Hafez St., Meymeh, Isfahan,

[2]Genetic Division, Biology Department, Faculty of Sciences, University of Isfahan, Isfahan,

[1,2]Iran

1. Introduction

In this chapter we begin to study the effect of the antibacterial agents used for control of microbial growth

There are some essential related terms for studying the antibacterial agents that are mentioned as in following:

a. **Biocide**: A widespread chemical or physical agent which inactivates microorganisms.
b. **Bacteriostatic**: property of a specific biocide agent which is able to bacterial multiplication.
c. **Bactericidal**: A specific term referring to the property by which a biocide is able to kill bacteria.
d. **Disinfectants**: Products or biocides used to reduce only the number of viable microorganisms on the inanimate objects
e. **Septic**: Characterized by the presence of pathogenic microbes in living tissue.
f. **Antiseptic**: A biocide or product that inhibits the growth of microorganisms in or on living tissue.
g. **Aseptic**: Free of or using methods to keep free of, microorganisms.
h. **Antibiotics**: Naturally occurring or synthetic organic compounds which inhibit or destroy selective bacteria, generally at low concentrations.
i. **Sterilization**: is defined as the process where all the living microorganisms, including bacterial spores are killed. Sterilization can be achieved by physical, chemical and physiochemical means.
j. **Asepsis** is the employment of techniques (such as usage of gloves, air filters, uv rays etc) to achieve microbe-free environment.

Large numbers of antibacterial agents are of clinical interest. The mechanisms by which compounds with antibacterial activity inhibit growth or cause bacterial death are varied and depend on the affected targets. Some strategies of antibacterial agents are introduced as following:

Damage to DNA

Ultraviolet light, ionizing radiations and DNA-reactive chemicals are example of physical and chemical agents that act by damaging DNA. Among the DNA-reactive chemicals, alkylating agents react covalently with purine and pyrimidine bases to form DNA interstrand cross-links. Ultraviolet light, for example, induces cross-linking between adjacent pyrimidines on one or the other of the two polynucleotide strands, forming pyrimidine dimers; ionizing radiations produce breaks in single and double strands.

Protein denaturation

The tertiary structure of the protein is readily disrupted by a number of physical or chemical agents, causing the protein to become nonfunctional. The disruption of the tertiary structure of a protein is called protein denaturation. A range of antibacterial agents inhibit the translation of the messenger RNA (mRNA) chain into its corresponding peptide chain.

Disruption of cell membrane

The cell membrane is known as a selective barrier, allowing some solutes to pass through and excluding others. Some bactericidal agents may alter the physical and chemical properties of the membrane, preventing its normal functions and therefore killing or inhibiting the cell. The structure of the cytoplasmic membranes in bacterial cells can be readily disrupted by certain agents. Polymyxins are the most important antibiotics which act on the membranes of Gram-negative bacteria. Amphotericin B and Nystatin are other toxic molecules named as polyene antifungal agents which have inhibitory action on membrane function

Disruption of cell wall

Destroying or preventing the synthesis of cell wall occurred after exposure to agents such as Lysozyme and Penecillin, respectively. The disruption of the cell wall may cause the cell lysis.

Synthesis of peptidoglycan precursors starts in the cytoplasm; wall subunits are then transported across the cytoplasmic membrane and finally inserted into the growing peptidoglycan molecule. Several different stages are therefore potential targets for inhibition. β-lactams , Bacitracin and Cycloserines are inhibitors of synthesis of peptidoglycan . β-lactams are the most important and the glycopeptides which are active only against Gram-positive bacteria. Cycloserines mainly used as a 'second-line' medication for treatment of tuberculosis, discussed later in this chapter have many fewer clinical applications.

Removal of free sulfhydryl groups

A large number of the antibiotics have demonstrated chemical reactivity toward compounds containing sulfhydryl groups. There have also been observed marked differences in reactivity of individual antibiotics toward various types of sulfhydryl-containing compounds.

Enzyme proteins containing cysteine have side chains terminating in sulfhydryl groups. In addition to these, coenzymes such as coenzyme A and dihydrolipoate contain free sulfhydryl groups. Such enzymes and coenzymes cannot function unless the sulfhydryl groups remain free and reduced. Oxidizing agents thus interfere with metabolism by forming disulfide linkages between neighboring sulfhydryl groups.

The most widespread methods used for controlling microorganism are the application of chemical and physical agents.

The most widely used physical methods include heat, radiation, and filtration which can destroy or remove undesirable microorganisms. Here we discuss how these methods work and discuss some practical examples.

Heat

One of the simplest means of sterilization is heat. Heat acts by oxidative effects as well as denaturation and coagulation of proteins. Those articles that cannot withstand high temperatures can still be sterilized at lower temperature by prolonging the duration of exposure. Dry heat acts by protein denaturation, oxidative damage and toxic effects of elevated levels of electrolytes. The moist heat acts by coagulation and denaturation of proteins. Moist heat is superior to dry heat in action. Temperature required to kill microbe by dry heat is more than the moist heat. The minimum time required to kill a suspension of organisms at a predetermined temperature in a specified environment is known as Thermal death time.

A temperature of 100°C will kill all bacteria, but in laboratory-scale cultures, within 2-3 minutes; a temperature of 121°C for 15 minutes whit 15 pound per inch is utilized to kill spores.

Radiation

Two types of radiation are used, ionizing and non-ionizing. Non-ionizing rays are low energy rays with poor penetrative power while ionizing rays are high-energy rays with good penetrative power.

Non-ionizing rays: Rays of wavelength longer than the visible light are non-ionizing. Microbicidal wavelength of UV rays lie in the range of 200-280 nm, with 260 nm being the most effective. UV rays are generated using a high-pressure mercury vapor lamp. It is at this wavelength that the absorption by the microorganisms is at its maximum, which results in the germicidal effect. UV rays induce formation of thymine-thymine dimers, which ultimately inhibit DNA replication. UV readily induces mutations in cells irradiated with a non-lethal dose. Microorganisms such as bacteria, viruses, yeast that are exposed to the effective UV radiation are inactivated within seconds. Since UV rays don't kill spores, they are considered to be of use in surface disinfection. UV rays are employed to disinfect hospital wards, operation theatres, virus laboratories, corridors, etc.

Ionizing rays: Ionizing rays are of two types, particulate and electromagnetic rays.

Electron beams are particulate in nature while gamma rays are electromagnetic in nature. High speed electrons are produced by a linear accelerator from a heated cathode. Electron beams are employed to sterilize articles like syringes, gloves, dressing packs, foods and pharmaceuticals.

Sterilization is accomplished in few seconds. Unlike electromagnetic rays, the instruments can be switched off.

Filtration

In the filtration method microbes do not kill, it just separates them out. Membrane filters with pore sizes between 0.2-0.45 µm are commonly used to remove particles from solutions that can't be autoclaved.

Various applications of filtration include removing bacteria from ingredients of culture media, preparing suspensions of viruses and phages free of bacteria, measuring sizes of viruses, separating toxins from culture filtrates, counting bacteria, clarifying fluids and purifying hydrated fluid. Different types of filters are Earthenware filters, Asbestos filters, Sintered glass filters, Membrane filters and Air Filters.

The other antimicrobial agents are those chemicals which destroy pathogenic bacteria from inanimate surfaces. They are listed in the table (1).

Chemical	Mode of action	Uses
Alcohols	Denaturing proteins and Solublizing lipids	Antiseptic used on skin
Formaldehyde (8%)	Reacting with NH_2, SH and COOH groups	Disinfectant, kills endospores
Tincture of Iodine	Inactivating the proteins	Antiseptic used on skin Disinfection of drinking water
Chlorine (Cl_2) gas	Formation of hypochlorous acid (HClO), a strong oxidizing agent	Disinfect drinking water; general disinfectant
Heavy metals	Inactivating the proteins	Disinfection of skin and laboratories
Mercuric chloride	Inactivation of proteins by reacting with sulfide groups	Disinfectant, although occasionally used as an antiseptic on skin
Detergents	Disruption of cell membranes	Skin antiseptics and disinfectants
Ethylene oxide gas	Alkylating agent	Disinfectant used to sterilize heat-sensitive objects such as rubber and plastics
Ozone	Produces lethal oxygen radicals	Purification of water, sewage
Phenols	decreasing the surface tension	Disinfection of laboratory devices, toilet and recycle bin

Table 1. Chemical antibacterial agents

Antibiotics fight against bacteria by inhibiting certain vital processes of bacterial cells or metabolism. Based on these processes, we can divide antibiotics into five major classes:

1. Cell wall inhibitors, such as Penicillin and Vancomycin.
2. Inhibitors of cell membrane function, such as Polymyxin B and Daptomycin.
3. Protein synthesis inhibitors, such as Aminoglycoside.

4. Inhibitors of nucleic acid synthesis, such as Fluoroquinolones, which inhibits DNA synthesis, and Rifampin, which inhibits RNA synthesis.

Inhibition of cell wall synthesis

The cell wall contains chemically distinct polysaccharides. The polysaccharides contain the amino sugars *N*-acetyl glucosamine (GlcNAc) and acetylmuramic acid (MurNAc). All β-lactam drugs are selective inhibitors of bacterial cell wall synthesis and therefore active against growing bacteria.

The bacterial cell wall-a unique structure in most bacteria can be affected in several ways: at different stages of synthesis (Fosfomycin, Cycloserine) or transport (Bacitracin, Mureidomycins) of its metabolic precursors, or by a direct action on its structural organization (β-lactams, Glycopeptides). The initial step in drug action is binding of the drug to cell receptors (Penicillin-binding proteins; PBPs). β-lactam drugs act as a false substrate for D-alanyl-D-alanyl transpeptidases , so they inhibit the transpeptidation reaction and peptidoglycan synthesis. In the next step, inhibitor of autolytic enzymes in the cell wall is inactivated. This activates the lytic enzyme and results in lysis if the environment is isotonic. So, β-lactam drugs are only active against rapidly dividing bacteria and growth lag phase ones are more stable to cell wall synthesis inhibitors.

Penicillins, Cephalosporins, Vancomycin, and Cycloserine inhibit the cell wall synthesis. Several other drugs, including Bacitracin, Teicoplanin, Vancomycin, Ristocetin, and Novobiocin, inhibit early steps in the biosynthesis of the peptidoglycan. In an effective inhibitory mechanism these drugs must be penetrated in the early stages of the cell wall synthesis took place inside the cytoplasmic membrane. In the case of Glycopeptides such as Vancomycin and Teicoplanin, attachments to D-ALA-D-ALA terminal end of peptidoglycan precursors occur. This inhibits the action of transglycosidase and transpeptidases, resulting in cell wall impairment.

The difference in susceptibility of gram-positive and gram-negative bacteria to various Penicillins or Cephalosporins would be attributed to the structural differences in their cell walls. Transpeptidases are located in periplasmic space that is directly accessible in gram-positive bacteria but not in Gram-negatives; so, theses drugs need to cross the outer bacterial cell membrane of Gram-negatives (passive diffusion) or pass through porin channels.

Some factors (eg, amount of peptidoglycan, presence of receptors and lipids, nature of cross-linking, activity of autolytic enzymes) affect the penetration, binding and activity of the drugs

Inhibition of cell membrane function

The cytoplasmic membrane is a selective permeability barrier, carries out active transport functions, and thus controls the internal composition of the cell. Macromolecules and ions can escape from the membrane as a result of cytoplasmic membrane disruption or cell damage. The cytoplasmic membrane of bacteria and fungi is more rigid than animal or plant cells and can be disrupted by certain agents. Consequently, selective chemotherapy is suggested.

Amphotericin B, Colistin, Ionospheres, Daptomycin , the Imidazoles and Triazoles are other examples of agents which inhibit the function of cell membrane. The detail mechanisms of action of other cell membrane inhibitors are shown in the table (2).

Inhibition of protein synthesis

Protein synthesis can be blocked by a large variety of compounds that affect any of the phases of this process, including activation (Mupirocin), initiation (Oxazolidinones, Aminoglycosides), binding of the tRNA amino acid complex to ribosomes (Tetracyclines, Glycylcyclines) and elongation (Amphenicols, Lincosamides, Macrolides, Ketolides, Streptogramins, Fusidic acid). In details, Tetracycline, Minocycline and Doxycycline, reversibly bind to the 30S subunit of ribosome and inhibit binding of aminoacyl-t-RNA to the acceptor site (A-site) on the 70S ribosome. Aminoglycosides also, bind to the A-site of the 30S subunit (the equivalent of mammalian 40S subunit) in an energy dependent process. In contrast to Tetracycline, the binding of Aminoglycosides to the A-site of the 30S subunit is irreversible. This mode of action means that Aminoglycosides act as bactericidal agents while Tetracycline belong to bacteriostatic agent group. This frustrating binding, freeze the 30S initiation complex (30S-mRNA-tRNA), disturbs elongation of the peptide chain. At the second step, aminoglycosides impair translational accuracy that finally lead to misreading of the mRNA sequence and/or premature termination of protein synthesis.

On the other hand, the large subunit of bacterial ribosomes (the equivalent of mammalian 60S subunit) occupied by Macrolides and some non-macrolides such as Chloramphenicol and Lincosamides. Premature dissociation of peptidyl tRNA from ribosome during elongation process occurred base on the attachment to 23S rRNA of the 50S ribosomal subunit. Consequently, peptidyl tRNA translocation from A to P site inhibited and peptide bond formation would be blocked; so, truncated peptide would be released after that. Similar mechanism is used by Lincosamides (Lincomycin and Clindamycin) that bind 50S subunit of ribosomes to inhibit protein synthesis.

Protein synthesis is also inhibited by another Macrolide (erythromycin) with a completely different way. Erythromycin prevents assembly of 50S subunit and as a result, no functional ribosome emerged that could trigger protein synthesis. This mechanism is also used by a new class of synthetic antibacterials (Linezolid) that inhibit the formation of the initiation complex.

Fusidic acid binds to elongation factor G (EF-G) and inhibits release of EF-G from the EF-G/GDP complex.

Rifampin, Rifamycin, Rifampicin bind to DNA-dependent RNA polymerase and inhibit initiation of RNA synthesis.

Inhibition of nucleic acid synthesis

Examples of drugs acting by inhibition of nucleic acid synthesis are the Quinolones, Pyrimethamine, Rifampin, Sulfonamides, Trimethoprim, and Trimetrexate.

Nitroimidazoles, Nitrofurans affect DNA directly. Trimethoprim and Sulfamides block bacterial metabolic pathways. Some compounds are unable to kill bacteria but can block bacterial mechanisms of resistance, enhancing the activity of other antimicrobials administered in combination. Among this group of agents, only certain β-lactamase inhibitors are currently in clinical use.

All Quinolones and Fluoroquinolones inhibit microbial DNA synthesis by blocking DNA gyrase.

In the following table are grouped characteristics of each class of antibiotics and mode of action.

Class	Example	Mode of action
Aminoglycoside	Gentamicin,Tobramycin,Amikacin	Bactericidal; inhibit protein synthesis
β-lactam/β-lactamase inhibitors	Ampcillin-sulbacam, Ticaracillin-clvulnate, Piperaciin-Tazobactam	Bactericidal; inhibit cell wall synthesis
Cephalosporin	Cefotaxime,Ceftriaxone, Ceftazidime, Cefepime	Bactericidal; inhibit cell wall synthesis
Fluoroquinolone	Levofloxacin, Ciprofloxacin, Moxifloxacin	Bactericidal; block DNA replication
Glycopeptide	Vancomycin	Bactericidal; inhibition of cell wall synthesis
Glycylcycline	Tigecycline	Bacteriostatic; inhibit protein synthesis
Macrolide	Erythromycin,Clarithromucin, Azithromycin	Bacteriostatic; inhibit protein synthesis
Oxazolidinone	Linezolid	Bacteriostatic; inhibit protein synthesis
Polymyxins	Polymyxin B, Colistin	Bactericidal; disrupt cell membrane
Tetracycline	Doxycycline, Tetracycline, Minocycline	Bacteriostatic; inhibit protein synthesis

Table 2. Antimicrobial agent classification and mode of action

Resistance to antimicrobial drugs

Infectious microorganisms can develop ways to exhibit resistance to drugs. This antibiotic resistance is due to the increasing use of antibiotics. There are many different mechanisms by which microorganisms can survive. Acquired resistance is often caused by mutations in chromosomal genes, or by the acquisition of mobile genetic elements, such as plasmids or transposons, which carry the antibiotic resistance genes.

1. Production of destroying enzyme

Organism may acquire genes encoding enzymes, such as β-lactamases, that destroy the antibacterial agent before it can have an effect. This enzyme destroys and inactivates the penicillin G drug. An important strategy of organisms for resistance to penicillins is due to penicillin-destroying enzymes (β -lactamases).β -Lactamases disrupts the antimicrobial activity of penicillins and cephalosporins by opening the β -lactam ring. Some inhibitors that have a high affinity for β -lactamase are Clavulanic acid, sulbactam, and tazobactam. Gram-negative bacteria produce some adenylylating, phosphorylating, or acetylating enzymes for resistant to aminoglycosides

2. Altering the permeability of the drugs

Bacteria may acquire efflux pumps that extrude the antibacterial agent from the cell before it can reach its target site and exert its effect. For example, changing the permeability of the drug (e.g Polymyxins) is one of the strategies of organism to exhibit resistance.

3. Altering the structural target for the drug

Bacteria may acquire several genes for a metabolic pathway which ultimately produces altered bacterial cell walls that no longer contain the binding site of the antimicrobial agent. Organisms which are resistant to erythromycin alter the receptor on the 50S subunit of the ribosome through methylation of a 23S ribosomal RNA. The loss or alteration of PBPs is another resistance mechanism of some drugs (e.g. Penicillins and Cephalosporins).

4. Altering the metabolic pathway

Some Sulfonamide-resistant bacteria do not require extracellular PABA but can utilize preformed folic acid.

5. Altering the function of enzyme

that can still perform its metabolic function but is much less affected by the drug. In Trimethoprim-resistant bacteria, the dihydrofolic acid reductase is inhibited far less efficiently than in trimethoprim-susceptible bacteria.

Factors affecting antimicrobial activity

Some antimicrobial agents are microbicidal under one set of conditions and microbistatic under others. Factors that influence the activity of antimicrobial agents are (1) the susceptibility of the microorganism, (2) the concentration or dose of the agent, (3) the length of exposure, (4) the number of microorganisms, and (5) environmental conditions.

Microbial susceptibility

Microbes vary in their response to different disinfectants. For example, vegetative cells of the Mycobacteria that cause tuberculosis and leprosy, however, are covered by a waxy coating that protects them from many antimicrobial chemicals. In addition, the hepatitis B virus and some fungal spores are resistant to most disinfectants and are persistent problems in hospitals. *Bacillus* and *Clostridium* are especially difficult to eliminate.

Concentration or dose of the agent

Diluting microbicidal chemicals usually weakens their antimicrobial activity. At lower concentrations they become microbistatic or lose antimicrobial activity altogether. The antimicrobial effects of temperature or radiation also depend on intensity of the exposure.

Low dose may inhibit growth, whereas high doses may result in sterilization. With a few important exceptions, the more concentrated, the more target organisms will be destroyed.

Length of exposure

Microorganisms die when physical or chemical conditions irreversibly damage essential cell components. All organisms present, however, do not die rapidly and simultaneously when a critical exposure is achieved, because microbial death is a function of time-the longer microbes are exposed to potentially lethal conditions, the more microbes will be killed. For many germicides, if exposure time is long enough, the probability of even a single cell surviving becomes so low that sterilization is practically assured. In contrast, microbistatic agents are effective only as long as they are present and must be used during the entire time that inhibition is to be maintained.

Number of microorganisms

Antimicrobial effectiveness also depends on the initial concentration of the microbial population. As the number of microbial contaminants increases, either the exposure period to or concentration of the agent must increase to achieve acceptable levels of decontamination.

Environmental conditions

Temperature, pH, and moisture affect the efficiency of most antimicrobial agents. In addition, some chemical agents are absorbed by organic materials (blood, mucus, feces, and tissue) that severely reduce antimicrobial effectiveness. These agents therefore cannot be used on the skin. Some antimicrobial agents are impeded by soaps and detergents that remain as thin films on skin or object surfaces. This difficulty can be avoided by through rinsing prior to disinfection or antisepsis.

Enzybiotics

The heavy use of antibiotics during the last century has resulted in widespread bacterial resistance. Overcoming resistance requires the development of antibiotics aimed at new targets in microorganisms. Preferably, such targets should be highly conserved in bacteria and required for pathogenesis, but not found in humans.

One of the most recently delivered classes of antibiotics are enzybiotics, the lytic enzymes named because of their enzymatic mode of action in degrading bacteria. Originally, enzybiotics named bacteriophage lytic enzymes that destroy the cell wall of the host bacterium quickly. According to the nature of bacteriophages that select their host specifically, the bacteriophage lytic enzymes were so specific. As the first attempt, Nelson and co-workers designate an enzybiotics for fighting with group A *Sreptococcus pyogenes* (*S. pyogenes*), the primary etiologic agent of bacterial pharyngitis (an inflammation of the throat or pharynx). But we must mention that, enzybiotics refer to all kind of enzymes with any kind of sources that have the ability to overcome bacterial infection.

The nick name of enzybiotics is peptidoglycan hydrolyses which induced the enzymatic cleavage of peptidoglycan covalent bonds of bacterial cells. As we know, the backbone of peptidoglycan consists of alternating residues of GlcNAc and MurNAc. The tetrapeptide side chains branching off from MurNAc are cross-linked by the pentaglycine bridges. This is the major mode of enzybiotics action that leads to the hypotonic lysis of poor bacteria by degrading different parts of their protective cell wall. Here, we would discuss the well-known enzybiotics according to, their site of invasion to the peptidoglycan backbone of bacteria.

1. Amidase enzybiotics

Lysins are the major class of Amidase enzybiotics that work on covalent bonds of bacterial peptidoglycan. They are the products of bacteriophages double-stranded DNA and named endolysin as a result of their non-bacterial sources. Another class of Amidase enzybiotics is autolysins that emerge from host bacterium. Both classes could fall in to three sub-classes according to their location of bond cleavage in bacterial peptidoglycan backbone. i) *N* - acetylmuramoyl- L- alanine amidases that break up the covalent bond between MurNAc and the first amino acid (L- alanine) of tetrapeptide side chain. ii) endopeptidase that break up the covalent bond between internal amino acids of tetrapeptide side chain especially between the first L- alaninen and the second D- Glu. Another endopeptidase act on pentaglycine bridges and dissociate the internal links of the backbone. iii) the third brother enjoy cleavage of covalent bond between *N* - acetylglucosamine and *N* - acetylmuramic acid by muramidases, transglycosylases and glucosaminidases activities (Figure1).

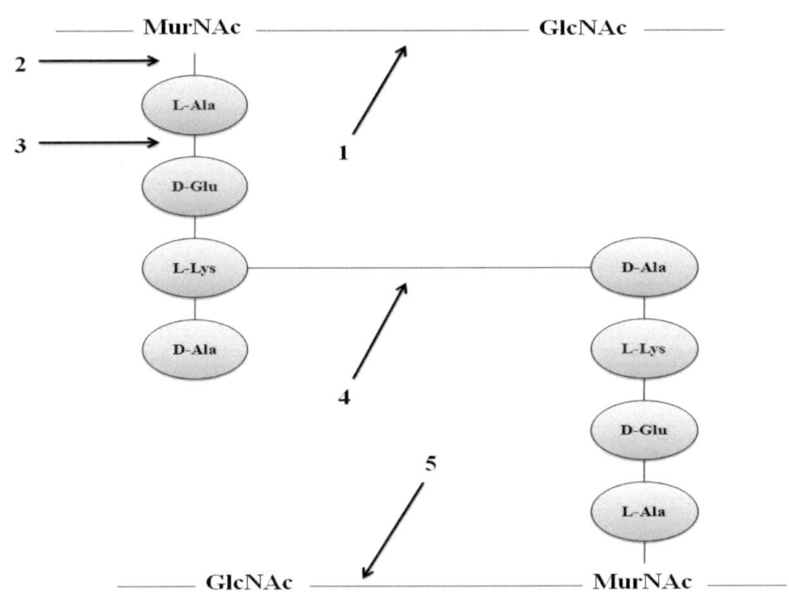

Fig. 1. The sites of cleavage by main classes of enzybiotics are shown by the numbered arrows (1) muramidases and transglycosylases; (2) amidases; (3 and 4) endopeptidases; (5) glucosaminidases

One surprise point in applying lysins as new antibiotic is its strain specific function. For example Pal only lyses pneumococcal strains while Ply3626 of kills *Clostridium perfringens*

(*C. perfringens*) strains. The most potent lysine discovered in group C streptococci C1 phage named PlyC Amidase first reported in 1957. PlyC Amidase is the only known multimeric lysine that did not act on streptococci groups B, D, F, G, L, and N so, it is limited to pathogenic streptococci groups A and C. Pal1, encoded by pneumococcal bacteriophage Dp-1, lyses Penicillin-resistant *Streptococcus pneumoniae* strains as well as penicillin- sensitive ones. Cpl-1 muramidase ia another anti-pneumococcal enzyme that came from Cp–1 phage. A broad range lysin (PlyGBS lysin) was also reported in *Streptococcus agalactiae* bacteriophage NCTC 1126. PlyGBS lysin contain two endopeptidase and muramidase domains that enable it efficient against groups A, C, G, and L streptococci. The other broad range lysin that kills groups A, B, C, E, G streptococci and also *Enterococcus faecalis* is B30 lysin. B30 lysin also includes two previous domains and extracted from *S. agalactiae* B30 phage. Other antibacterial lysins are found in *Staphylococcus aureus*, coagulase-negative staphylococci, *Bacillus anthracis, Bacillus cereus, Listeria monocytogenes, C. perfringens* phages and one of them (PlyG lysine) is used for the identifi cation of *B. anthracis* by the U.S. Centers for Disease Control and Prevention.

2. Endopeptidase enzybiotics

Endopeptidase or endoproteinase are proteolytic peptidases that in contrast to exopeptidases, break peptide bonds of nonterminal amino acids. Lysostaphin and zoocin A are two endopeptidase encoded by *Staphylococcus simulans* biovar *staphylolyticus* and *Streptococcus equi* ssp.

3. *N*-acetylmuramidases enzybiotics

Lysozymes, or *N* - acetylmuramidases, are produced by cells of many different animal species, plants, insects, bacteria, and viruses. Lysozymes are unique antibiotics, so that they contain enzematic and also non-enzematic mechanisms for fighting with foreign bacterias. There are some facts about lysosymes take make them so different from other enzybiotics. First of all, Lysozymes are the only peptidoglycan hydrolases that have been used on a larger scale in humans for the past several decades. Second, antibacterial action of lysozyme is based on the enzymatic cleavage of peptidoglycan, and nonenzymatic mechanisms based on activation of autolysins and also cytoplasmic membrane destabilization resulting from the removal of divalent ions from the membrane surface.

Plants antibiotics

Traditionally, people get used to herbal treatment as a natural way to fight diseases almost in all continents. Some of them take it easier to go to herbalist than a doctor, and also prefer natural drugs than chemically synthesized ones. Especially, in ancient civilizations such as Iran, Iraq, India and China it has been so common to use plants as an effective tool for diseases therapy. Hippocrates (in the late fifth century B.C.) mentioned 300 to 400 medicinal plants. Avicenna (Ibn Sīnā) in his famous book "*The Canon of Medicine*" lists 800 tested drugs, including plant and mineral substances, and describe their specific properties according to the known diseases of that time.

Plants antibiotics come in to view after the arrival of antibiotics in the 1950s, when scientists realize that new antibiotics might be obligatory because of the antibiotic resistance. Major classes of plants antimicrobial components are phenolics, terpenoids, essential oils, alkaloids, lectins and polypeptide and polyacetylenes. For example, the roots of *Glycyrrhiza glabra* contain glabridin and hispaglabridin B. The former is active element against both *Mycobacterium tuberculosis* H37Ra and H37Rv strains at 29.16 g/mL concentration. It was also clear that, glabridin was more active against Gram-positive strains than Gram-negative. Antimycobacterial activity of glabridin like component (hispaglabridin B) did not find because of their structural differences. Glabridin have two free phenolic hydroxyls which might be crucial in antibacterial activity while hispaglabridin might be inactivated after the corporation of one hydroxyl group in protected benzopyrene ring. Methicillin resistant *Staphylococcus aureus* (MRSA), and, *Helicobacter pylori* (*H. pylori*) are also sensitive to Glabridin. Anti-*H. pylori* activities are also shown in *rachyspermum copticum* and *Xanthium brasilicum* with minimum inhibitory concentrations within the range of 31.25-250 micro g/ml.

Grabidin strongly inhibits adenosine 3', 5'-cyclic monophosphate (cAMP) phosphodiesterase. Glabridin is associated with reduction in protein kinase C (PKC) activity and since PKC is required for low density lipoprotein (LDL) oxidation; so, grabidin induces reduction of LDL oxidation. This phenomenon is considered to be of major importance in atherosclerosis attenuation because LDL is associated in early atherogenesis. Glabridin can inhibit both mono- and diphenolase tyrosinase activities and because of the involvement of tyrosinase in melanin biosynthesis, glabridin may serve as candidates for skin-lightening agents. On the other hands, grabidin could serve as whitening agents for treatment of various dermatological disorders (melasama, age spots, and sites of actinic damage) that arise from the excessive accumulation of epidermal pigments. Cytochrome P450 3A4 enzymes is the major human drug metabolizing enzyme, that inactivated by glabridin antioxidant irreversible. The effect on P450 enzymes inactivation may play a role in the reported antiatherosclerotic activity of glabridin.

One of the most exciting plants with antibiotic activity is genus *Allium with* common garlic (*Allium sativum* L.) that known as Russian penicillin. In a study, two Allioideae alkaloids, canthin-6-one and 8-hydroxy-canthin-6-one, are extracted from *Allium neapolitanum*. They displayed minimum inhibitory concentrations (MICs) in the range 8-32 microg/mL against a panel of fast-growing Mycobacterium species and 8-64 microg/mL against multidrug-resistant (MDR) and MRSA. Antibacterial activities of *Eucomis autumnalis* and *Cyathula uncinulata* against ampicillin-resistant and kanamycin-resistant strains of E.*coli revealed a low MIC range of 0.27 mg/ml and 0.39 mg/ml respectively.*

New candidate antibiotics for MRSA are from Leguminosae family and the most active ones are the flowers of *A. auriculiformis* and *B. kockiana*. Some of these medicinal plants, such as *B. kockiana*, *B. purpurea*, *C. pulcherrima*, and *C. surattensis* have been used traditionally to treat various diseases in Malaysia. In the case of *A. auriculiformis* two acylated bisglycoside saponins, acaciaside A and B isolated, were found to exhibit antibacterial and antifungal activity.

In a study of resistant and standard strains of *Escherichia coli (E. coli)*, it has been reported that *Anagyris foetida* (Leguminosae) and *Lepidium sativum* (Umbelliferae) enhanced the activity of amoxicillin against resistant *E. coli* strain. Clarithromycin in combinations with *Gundelia tournefortii* L. (Compositae), *Eruca sativa* Mill. (Cruciferae), and *Origanum syriacum* L. (Labiateae), shows enhanced activity against the resistant *E. coli* strain. This strategy, the use of herbals and drugs in a multi targeted approach, named "herbal shotgun" or

"Synergistic multi-target effects". As a result, herbal-drug combinations affect not one but several targets, cooperating in an agonistic-synergistic way.

Umbelliferae family is known as an herbal antibiotic family after its comparison with standard antibiotics. The results show that cefixime and chloramphenicol resistant *Enterococcus faecalis* and *Pseudomonas aeruginosa* are sensitive to *Anethum graveolens, Foeniculum vulgare* and *Trachyspermum ammi.*

In a study of folk medicine, the effects of traditional therapeutic plants used by Haudenosaunee peoples of New York State have been administered. Four predicted plants (*Achillea millefolium, Ipomoea pandurata, Hieracium pilosella,* and *Solidago canadensis*), showed antimicrobial properties strongly against *Salmonella typhimurium.*

Strong attempts have been performed for treatment of brucellosis, a high morbidity zoonosis caused by brucella, with natural antibiotics in Iran. Among four effective herbs tasted as antibacterial agents against *Brucella melitensis (B. melitensis), Oliveria decumbens* was chosen as the most effective plant for further studies. A *B. melitensis* strain that show resistance to tetracycline, nafcillin, oxacillin, methicillin, and colistin, inhibited by methanolic extract of *Oliveria decumbens* after 7 hours. Synergistic effect between *Oliveria decumbens* extracts and two other anticiotics (doxycycline and tetracycline) indicated as well.

Cuminum cyminum L. from Apiaceae family has been used since many years in Iranian traditional medicine. This aromatic herb is an astringent that has been used in the treatment of mild digestive and bronchopulmonary disorders as well as a cough remedy or analgesic. It has been shown that cumin seed essential oil significantly enhance antibacterial efficacy of ciprofloxacin against *Klebsiella pneumoniae* (*K. pneumoniae*). The authors theorize that essential oil damage cell wall or it modify the outer membrane proteins which lead to enhancement of ciprofloxacin activity against *K. pneumoniae.* As a result, they believe that herbal shotgun approach may be used in the case of cumin seed essential oil in future as a semi-natural way for fighting with *K. pneumoniae related disorders.*

Another medicinal plant that used extremely among Iranian population is *Zataria multiflora* (*Z. multiflora*) that belongs to the Lamiaceae family. Iranian traditional folk remedies, mainly used *Z. multiflora* as an antiseptic, analgesic, and carminative. In a wide local study of both gram-positive and gram-negative bacteria with important clinical impacts, some exciting results have been emerged. In the case of gram-positive cocci, and in presence of *Z. multiflora* essential oil (0.44 to 1.41 μL/mL) growth inhibition of both MRSA and MSSA has been observed. Same results with higher essential oil MICs have been reported for vancomycin-resistant E. faecalis (VREf) and vancomycin-sensitive E. faecalis (VSEf).

The growth of *E. coli* O157:H7, the cause of many food-borne outbreaks in different countries, inhibited at essential oil concentrations of 0.12 μL/mL for one ecotype of *Z. multiflora.* Two other gram-negative bacteria, *Salmonella entrica* and *Shigella flexneri,* inhibited and also killed at concentrations ranging from >0.12 to 2 μL/mL.

A phosphorylated structure, similar to the adenine, was isolated from the berries of *Solanum incanum (S. incanum)* Linnaeus (Thorn Apple or Bitter Apple). It was astonishing that crystals of this compound inhibit the growth of gram-positive and gram-negative bacteria, yeasts, dermatophytes, and some agricultural pathogens effectively. The zone of inhibition for 6.5mm diffusion disks was between 15-26mm with the highest inhibition for *S. pyogenes*

(26mm), *C. perfringens* (25mm) and *Clostridium septicum* (25mm) in bacterial group. It has been shown that, *S. incanum* crystals contain steroidal glycoalkaloids, solanine, which may act as a saprogenic surface active agent at high concentration. It must mensioned that solanine is found mainly in any part of solanum family plants (solanacea), including the leaves, fruit, and tubers and is rather high in the green peel and the sprouts.

Phage therapy, a candidate for antibiotic replacement

Phage therapy means the use of lytic bacteriophages as an alternative to antibiotics especially against the infection of resistant bacteria. Bacteriophages are bacterial viruses that invade bacterial cells and abberated as "phages". Phages have a developmental cycle within the host bacteria which can be lytic or lysogenic. The former involves a series of events that lead to the lysis of bacterial cell, but the lysogenic cycle comprises replication of phage nucleic acid together with the host genes for several generations.

There are some important benefits for phage therapy such as host-specific that did not observed in the case of routine antibiotics. Bacteriophages specificity could show important impacts in clinical use. Another advantages for phage therapy is its lower side effects and lesser therapeutic dose according to phages self-replicating in its target bacterial cell. Therefore, phage therapy is harmless to the eukaryotic host undergoing therapy theoretically.

Some bacteriophages synthesize degrading enzymes that breakdown the biofilms of bacteria that facilitate the bacterial cell lysis. Bacterial resistance to phages, if emerged, could be overcome according to the fact that mutation of phages occur with the same rate as bacteria. Cheap production of fighting bacteriophages is another advantage of

It must be mentioned that, phage therapy suffer from serious problems that make it unrealable.

Selection of appropriate mixture of high virulence phages against the target bacteria, poor understanding of heterogeneity and ecology of both the phages and the bacteria, are the most important problems. Resistance of bacteria to lysis by phages was another important challenge especially in the case of Pseudomonas plecoglossicida.

2. References

Ajami M, Eghtesadi S, Pazoki-Toroudi H, Habibey R, Ebrahimi SA. Effect of crocus sativus on gentamicin induced nephrotoxicity. Biol Res. 2010; 43(1): 83-90.

Barbosa-Filho JM, Agra MF, Oliveira RA, Paulo MQ, Trolin G, Cunha EV, Ataide JR, Bhattacharyya J. Chemical and pharmacological investigation of Solanum species of Brazil--a search for solasodine and other potentially useful therapeutic agents. Mem Inst Oswaldo Cruz. 1991; 86 Suppl 2: 189-91.

Barrett JF, Dolinger DL, Schramm VL, Shockman GD. The mechanism of soluble peptidoglycan hydrolysis by an autolytic muramidase. A processive exodisaccharidase. J Biol Chem. 1984 Oct; 259 (19): 11818-27.

Beaman-Mbaya V, Muhammed SI. Antibiotic action of Solanum incanum Linnaeus. Antimicrob Agents Chemother. 1976 Jun; 9(6): 920-4.

Bisi-Johnson MA, Obi CL, Hattori T, Oshima Y, Li S, Kambizi L, Eloff JN, Vasaikar SD. Evaluation of the antibacterial and anticancer activities of some South African medicinal plants. BMC Complement Altern Med. 2011 Feb;11: 14-8.

Clark, J.R. and March, J.B. "Bacteriophages and biotechnology: vaccines, gene therapy and antibacterials". *TRENDS in Biotechnology*. 2006. Volume 24, Number 5. p. 212-218.

Danielle J. "Islamic Pharmacology in the Middle Ages: Theories and Substances", European Review. 2008; 16 (2): 219–227.

Darralyn McCall., David Stock and Phillip Achey. Introduction to Microbiology. Chapter 8. Control of Microorganisms. 11th edition. Blackwell Science.

Darwish RM, Aburjai TA. Effect of ethnomedicinal plants used in folklore medicine in Jordan as antibiotic resistant inhibitors on Escherichia coli. BMC Complement Altern Med. 2010 Feb;10: 9-21.

Fred C. Tenover. Mechanisms of Antimicrobial Resistance in Bacteria. The American Journal of Medicine 2006. 3-10.

Frey FM, Meyers R. Antibacterial activity of traditional medicinal plants used by Haudenosaunee peoples of New York State. BMC Complement Altern Med. 2010 Nov;10: 64-73.

Gao SY, Wang QJ, Ji YB. Effect of solanine on the membrane potential of mitochondria in HepG2 cells and [Ca2+]i in the cells. World J Gastroenterol. 2006 Jun; 12(21): 3359-67.

Gupta VK, Fatima A, Faridi U, Negi AS, Shanker K, Kumar JK, Rahuja N, Luqman S, Sisodia BS, Saikia D, Darokar MP, Khanuja SP. Antimicrobial potential of Glycyrrhiza glabra roots. J Ethnopharmacol. 2008 Mar, 116(2): 377-80.

Hajimahmoodi M, Shams-Ardakani M, Saniee P, Siavoshi F, Mehrabani M, Hosseinzadeh H, Foroumadi P, Safavi M, Khanavi M, Akbarzadeh T, Shafiee A, Foroumadi A. In vitro antibacterial activity of some Iranian medicinal plant extracts against Helicobacter pylori. Nat Prod Res. 2011 Jul; 25(11): 1059-66.

Ian Chopra and Marilyn Roberts. Tetracycline Antibiotics: Mode of Action, Applications, Molecular Biology, and Epidemiology of Bacterial Resistance. MICROBIOLOGY AND MOLECULAR BIOLOGY REVIEWS 2001; 232–260.

Jawetz., Melnick & Adelberg's. Medical Microbiology. Chapter 28. Antimicrobial Chemotherapy. 25th edition. The McGraw-Hill Companies.

K. Gupta, S. Kaushal, S. C. Chopra. Tigecycline: A novel glycylcycline antibiotic. Indian J Pharmacol 2006. 217-19.

Karen L. Bowlware, MD, Terrence Stull, MD. Antibacterial agents in pediatrics. Infect Dis Clin N Am 2004; 513–531.

Kaur GJ, Arora DS. Antibacterial and phytochemical screening of Anethum graveolens, Foeniculum vulgare and Trachyspermum ammi. BMC Complement Altern Med. 2009 Aug; 9: 30-9.

Kent UM, Aviram M, Rosenblat M, Hollenberg PF.The licorice root derived isoflavan glabridin inhibits the activities of human cytochrome P450S 3A4, 2B6, and 2C9. Drug Metab Dispos. 2002 Jun;30(6):709-15.

Kusano A, Nikaido T, Kuge T, Ohmoto T, Delle Monache G, Botta B, Botta M, Saitoh T. Inhibition of adenosine 3',5'-cyclic monophosphate phosphodiesterase by flavonoids from licorice roots and 4-arylcoumarins. Chem Pharm Bull (Tokyo). 1991 Apr; 39(4): 930-3.

Lance R. Peterson. A review of tigecycline the first glycylcycline. International Journal of Antimicrobial Agents 2008; S215- S222.

Lee KR, Kozukue N, Han JS, Park JH, Chang EY, Baek EJ, Chang JS, Friedman M.Glycoalkaloids and metabolites inhibit the growth of human colon (HT29) and liver (HepG2) cancer cells. J Agric Food Chem. 2004 May; 52(10): 2832-9.

Marne Gaynor and Alexander S. Mankin. Macrolide Antibiotics: Binding Site, Mechanism of Action, Resistance. Current Topics in Medicinal Chemistry 2003; 3, 949-961.

MD Mathur, S Vidhani, PL Mehndiratta. Bacteriophage Therapy : An Alternative to Conventional Antibiotics. JAPI 2003. 593-596.

Michael T. Madigan and John M. Martinko. Biology of Microorganisms. Chapter 20. Microbial Growth Control. 11th edition. Pearson Prentice Hall.

Motamedi H, Darabpour E, Gholipour M, Seyyed Nejad SM. In vitro assay for the anti-Brucella activity of medicinal plants against tetracycline-resistant Brucella melitensis. J Zhejiang Univ Sci B. 2010 Jul; 11(7): 506-11.

Nariman F, Eftekhar F, Habibi Z, Falsafi T. Anti-Helicobacter pylori activities of six Iranian plants. Helicobacter. 2004 Apr; 9(2):146-51.

Nelson D, Loomis L, Fischetti VA. Prevention and elimination of upper respiratory colonization of mice by group A streptococci by using a bacteriophage lytic enzyme. Proceedings of the National Academy of Sciences, USA. 2001; 98: 4107-12.

Nerya O, Vaya J, Musa R, Izrael S, Ben-Arie R, Tamir S. Glabrene and isoliquiritigenin as tyrosinase inhibitors from licorice roots. J Agric Food Chem. 2003 Feb; 51(5): 1201-7.

O'Donnell G, Poeschl R, Zimhony O, Gunaratnam M, Moreira JB, Neidle S, Evangelopoulos D, Bhakta S, Malkinson JP, Boshoff HI, Lenaerts A, Gibbons S. Bioactive pyridine-N-oxide disulfides from Allium stipitatum. J Nat Prod. 2009 Mar; 72(3): 360-5.

Rosenblat M, Belinky P, Vaya J, Levy R, Hayek T, Coleman R, Merchav S, Aviram M. Macrophage enrichment with the isoflavan glabridin inhibits NADPH oxidase-induced cell-mediated oxidation of low density lipoprotein. A possible role for protein kinase C. J Biol Chem. 1999 May; 274(20): 13790-9.

Shaffiee A, Javidnia K. Composition of essential oil of *Zataria multiflora. Planta Med.* 1997; 63: 371–2.

www.aic.cuhk.edu.hk/web8/index.htm

www.pathmicro.med.sc.edu/book/welcome.htm

www.textbookofbacteriology.net/control_3.html

12

New Improved Quinlone Derivatives Against Infection

Urooj Haroon[1], M. Hashim Zuberi[1], M. Saeed Arayne[2] and Najma Sultana[2]
[1]*Federal Urdu University for Arts, Science and Technology,*
[2]*Department of Chemistry, University of Karachi,*
Pakistan

1. Introduction

There had been a tremendous advancement in the quality and quantity of worldwide research production in the field of microbiology. Among various biomedical fields, microbiology has been the subject of extensive study due to the problems imposed on human health by countless infectious diseases known so far. Identification of infectious agents and to adapt measures for its eradication is considered a major tool for alleviating human from the burden of infections. Progressively it is important to modulate the structure of earlier marketed antibacterial agents to produce newer agents with superior antimicrobial profile.

Fluoroquinolones, useful in the treatment of many bacterial infections, attacks DNA gyrase and topoisomerase IV on bacterial chromosomal DNA (Ball et al., 1998, Blandeau, 1999). However, widespread use of fluoroquinolones has caused the resistant rates of various Gram-negative bacilli (e.g., *Pseudomonas aeruginosa, Escherichia coli,* and *Salmonella*), to approach the critical points.

To solve the problem of increasing antimicrobial resistance, it is crucial to design and produce new quinolones that could provide effective therapy for infections caused by organisms resistant to older agents. Most of the quinolones currently on the market or those which are under development, displays only moderate activity against many Gram-positive cocci, including Staphylococci and Streptococci. This insufficient activity has not only limited their use in infections caused by these organisms, such as respiratory tract infections, but has also been believed to be one of the reasons for the rapidly developing quinolone resistance. Therefore, recent efforts have been directed toward the synthesis of new quinolone antibacterials that can provide improved Gram-positive antibacterial activity, while retaining good Gram-negative activity

The extensive research efforts have enabled a better definition of the structural moieties or elements around the basic pharmacophore of quinolones that offer the best combination of clinical efficacy and reduced resistance selection in Gram-positive and Gram negative bacteria.

The general structure of quinolone antibacterial agents consist of a 1-substituteted-1,4 dihydro-4-oxopyridine-3-carboxylic moiety A combined with an aromatic or hetero-aromatic ring B.

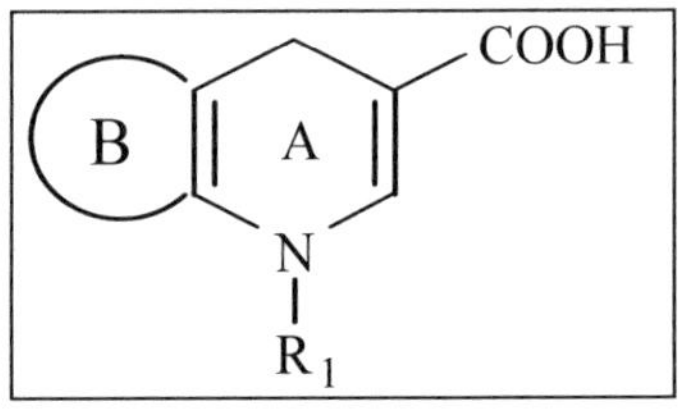

Fig. 1.

Most of the current agents have a carboxyl group at position 3, a keto group at position 4, a fluorine atom at position 6, and a nitrogen heterocycle moiety at the C-7 position (Sarro & Sarro, 2001). In addition at the C-7 position, substitution of bulky functional groups is permitted (Emami et al., 2006). The substituent at N-1 and C-8 positions of the quinolones should be relatively small and lipophilic to enhance self-association and also control potency and pharmacokinetics. Groups at C-5 and C-6 position controls potency for Gram-positive activity. SAR studies on quinolones explain that C-3 and C-7 positions are vital for antibacterial activities.

Most of the quinolone antibacterial research has been focused on the functionality at C-7 position as it is the most adaptable site for chemical change (Anderson & Osheroff, 2001, Foroumadi et al., 2005) **416**. The 3-carboxylic group is considered important for DNA gyrase binding (Chu et al., 1985, Domagala et al., 1988, Sarro & Sarro, 2001). Modifications on this position of quinolone are normally not accepted (Mitscher, 2005). However, there are some exceptions to this rule; of which, one was ester pro-drug analogues that were converted *in-vivo* back to the acid (Kondo et al., 1988). In one case, replacement of 3-carboxylic acid group of ciprofloxacin with bioisostere- fused isothiazolo ring was found more potent (4 to 10 times) than ciprofloxacin and possessed enhanced activity against DNA gyrase (Chu & Fernandes, 1989). A series of the C3/C3 bis-fluoroquinolones tethered with an 1, 3, 4-oxadiazole ring were reported showing antitumor activity (Hu et al., 2012) while 1,8 Imidazo fused quinolones exhibited moderate antibacterial activity (Venkat Reddy et al., 2009). Recently two naphthyl ester quinolone derivatives have been reported to demonstrate photo-oxidant properties (Vargas et al., 2008).

Advances in quinolone field are likely to provide better compounds capable of dealing with the resistant strains. These research efforts have been rewarded by very significant improvements in antibacterial potency as well as *in vivo* efficacy. It is evident from literature that most of the research on quinolone is directed towards group modification of the C-7 basic group of the quinolone and the effect of substitution at C-3 position of quinolones has not been studied to a large extent to produce new agents with better antimicrobial profile. Accordingly to explore the potential of 3-carboxylic quinolone derivatives as anti Gram positive and Gram negative agents, we have recently reported some novel enoxacin (Saeed Arayne et al., 2009) , levofloxacin (Sultana et al., 2011) and ofloxacin (Arayne et al., 2012) carboxamide analogues by introducing new functionality at C-3 position. In continuation of our research program to establish the structure-activity relationships of 3-carboxylic quinolone derivatives, we here in report some novel enoxacin analogues which have been prepared by introducing new functionality at C-3. These compounds were prepared by two series of reactions. In reaction **1**, the carboxylic group at C-3 position of enoxacin was esterified in methanol followed by reaction **2**, whereby the ester group was subjected to nucleophilic attack at the carbonyl carbon by various aliphatic (urea, thiourea, acetamide, thioacetamide) and aromatic amines (phthalimide, benzamide) yielding amides and regenerating alcohols.

2. Experimantal

2.1 Physical measurements

Melting points were obtained manually by capillary method. Infrared spectra were recorded on FT-IR spectrophotometer (shimadzu prestige-21 200 VCE coupled to a P IV- PC and loaded with IR solution version 1.2 software). The potassium bromide disks were placed in the holder directly in the IR laser beam. Spectra were recorded at a resolution of 2 cm^{-1} and 50 scans were accumulated. NMR spectra were recorded on Bruker FT - NMR 500 MHz with the compounds dissolved in DMSO. Chemical shifts are reported in parts per million (δ) relative to tetra methyl silane as an internal standard. Significant ^1H NMR data are tabulated in the following order: number of proton(s) and multiplicity (s, singlet; d, doublet; t, triplet; q, quartet; and m, multiplet). The mass spectra were recorded on Finnign-MAT212 under electron impact (EI) ionization condition. Thin layer chromatography (TLC) was performed on HSF-254 TLC plate and compounds visualized under UV lamp.

2.2 Antibacterial studies

Disk Diffusion technique developed by Bauer et al (Baur et al., 1966) was adopted to determine the antibacterial activity of enoxacin and its derivatives against 10 different clinical isolates of Gram positive (*Staphylococcus aureus, Bacillus subtilis, Streptococcus pneumoniae*) and Gram negative organisms (*Klebsiella Pneumoniae, Proteus mirabilis, Shigella flexneri, Escherichia coli, Pseudomonas areuginosa, Citrobacter species,* and *Salmonella typhi*).

50 µg mL^{-1} stock solutions of the drug and their derivatives were prepared in hot methanol. The stock solution was diluted to 3 different concentrations i.e. 5, 10 and 20 µg mL^{-1}. Commercially available filter paper discs were impregnated with the prepared solutions of the drug and the derivatives, dried and applied on the surface of the agar plate over which a culture of micro organism was already streaked. After 24 hours of incubation the clear zone of inhibition around the disc was determined, this is proportional to the bacterial susceptibility for the antimicrobial agent present in the disk. Three replicas were made for each treatment to minimize error.

2.3 Antifungal studies

Enoxacin and its derivatives were tested for antifungal activity against the fungi; *Aspergillus purasiticus, Saccharomyces cervics, Candida albicans* and *Fusraium solani.*

100 µg mL^{-1} stock solutions of the compounds were prepared in hot methanol. The stock solutions were diluted to 3 different concentrations i.e. 30, 40 and 50 µg mL^{-1}. Commercially available filter paper discs were impregnated with the prepared solutions of the drugs and the compounds under study, dried and applied on the surface of the agar plate over which a culture of micro organism was already streaked. After 24 hours of incubation the clear zone of inhibition around the disc was determined, this is proportional to the fungal susceptibility for the antimicrobial agent present in the disk. Three replicas were made for each treatment to minimize error.

2.4 Phagocyte chemiluminescence assay

The assay was performed as described by Helfand et al. protocol (Helfand et al., 1982). Briefly enoxacin and its derivatives were diluted in three concentrations, (0.5, 5 and 25 μg mL^{-1}) in Hanks Balanced Salt Solution containing calcium and magnesium ($HBSS^{++}$). 25 μL diluted blood (1:50 dilution in sterile PBS, pH 7.4) or polymorphoneutrophils or mouse peritoneal macrophages (1x 10^6 mL^{-1}) were added to the culture reaction. After 15 minutes of incubation for whole blood and 30 minutes for isolated cells, 25 μL (7 x 10^5 M) luminol (G-9382 Sigma Chemical Co. USA) or 25 μL (0.5 mM) lucigenin were added, followed by 25 μL (20 mg mL^{-1}) SOZ (Sigma Chemical Co. USA) or 25 μL (40 nM) PMA.

$HBSS^{++}$ alone was run as a control. The level of ROS for the compounds was monitored for 50 minutes by the Luminometer (Luminoskan RS, Labsystems, Finland). The total ROS level was recorded as total light produced and recorded during 50 minutes scan.

2.5 T-Cell proliferation assay

Cell proliferation was evaluated by standard thymidine incorporation assay following a method reported by Nielsen et al. (Nielsen et al., 2000). 10 mL blood was collected by vein puncture in heparin containing tube from a healthy volunteer. 50 mL sterile falcon tube was placed in safety cabinet (at sterile conditions) to this was added blood, ficoll paque and RPMI incomplete media (without fetal bovine serum) in equal volume. 5 mL ficoll paque was added in three 15 mL sterile empty centrifuge tubes and centrifuged at 400xg for 20 minutes at room temperature with no break. Peripheral blood mononuclear cells (PBMCs) appeared at the junction of two layers, the upper layer was discarded and the lower containing the PBMCs was transferred in another sterile tube, washed with RPMI incomplete media and centrifuged again at 300xg for 10 minutes at 4°C. The supernatant was discarded and re-suspended pellet in 1mL RPMI incomplete media and stored on ice. The viability of the cells was checked by trypan blue and cells were counted by hemocytometer. 50 μL of 5% complete RPMI was added into each well of a sterile 96 well plate in sterile environment using safety cabinet followed by samples (drug and its derivatives) having the concentration between 6.25 and 100 μg mL^{-1} and adjusting the final volume to 0.3 mL. While well 'A' contained only 5% complete RPMI to be used as control. 50 μL PBMC cells (1x10^6/mL) were added in suspension of 5% complete RPMI to each well except blank followed by the addition of 50 μL PHA except negative control and blank and volume of well was made up to 0.2 mL by 5% complete RPMI .The mixture was incubated for 72 hours in CO_2 incubator at 37°C. After incubation 25 μL of thymidine was added in each well except blank and the plate again incubated in CO_2 incubator at 37°C for 18 hours. After incubation cells were harvested using glass fiber filter and the cell harvester. In cell harvesting the plates were washed with 70% ethanol 5 times, filtered and left until they were completely dry, then dissolved in separate vials using the scintillation solution and scanned by scintillation counter.

2.6 Cytokinne production from mononuclear cells

The monocytes were grown in 75 mm^2 flask in RPMI-1640 supplemented with 10% FBS until they attained 70% confluency and passaged twice a week. On reaching confluency the cells were plated in 24-well tissue culture plates at a concentration of 2.5×10^{-5} cells/mL. The cells were differentiated into macrophage like cells by using phorbol myristate acetate (PMA) at a final concentration of 20 ng mL^{-1} and incubated at 37°C in 5% CO_2 for 24 hours.

After 24 hr incubation with PMA, cells were stimulated with bacterial lipopolysaccharide (LPS) 50 ng mL^{-1} and treated with compounds (drug and its derivatives) at a concentration of 25 μg mL^{-1} and incubated for 4 hours at 37°C in 5% CO_2. The supernatants were collected after 4 hrs which was then analyzed for the level of TNF-α. The plate was incubated for further 18 hrs and then supernatants were collected for the analysis of IL-1β level. Human TNF-α and IL-1β Duo set Kits (R&D systems, Minneapolis, USA) were used for cytokine quantification according to manufacturer's instructions (Singh et al., 2005).

2.7 General procedure for preparation of derivatives (A6-A12)

Enoxacin and all the chemicals used were of analytical grade. Amines used for the synthesis were urea, thiourea, acetamide, thioacetamide, hydroxylamine, benzamide and phthalimide). The compounds were prepared as summarized in figure 2 and 3.

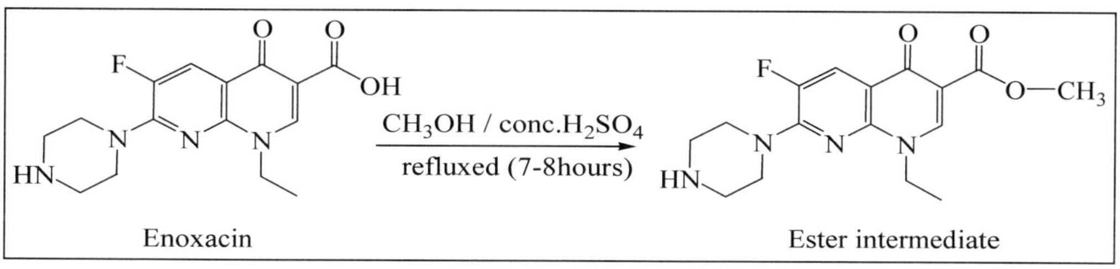

Fig. 2. Synthesis of ester intermediate

0.01 mole of enoxacin (3.21 g) was dissolved in hot methanol (80 mL) to which 1.2 mL sulphuric acid was added and the solution was refluxed for about 7-8 hours till the consumption of drug in ester formation (monitered by TLC). The solution was cooled down to room temperature and the precipitate obtained was washed with chloroform and dried for 1 hour at 80°C to give methyl ester of the drugs. The corresponding esters were subjected to nucleophilic attack by adding 0.01 molar methanol solutions of amines respectively with continuous stirring to generate carboxamides. While for compound **A11** and **A12,** NaOH was added in the methanolic mixture of benzamide and phthalimide and warmed for 30 mins prior to their addition in the reaction flask containing enoxacin ester intermediate. The reaction was refluxed for about 2-3 hours till completion, indicated by TLC. The volume of the reaction mixture was then reduced by rotary-evaporation. The precipitates were filtrated off, washed with chloroform and re-crystallized from methanol-chloroform (2:8) mixture till pure compounds were achieved (checked on TLC and constant melting point).

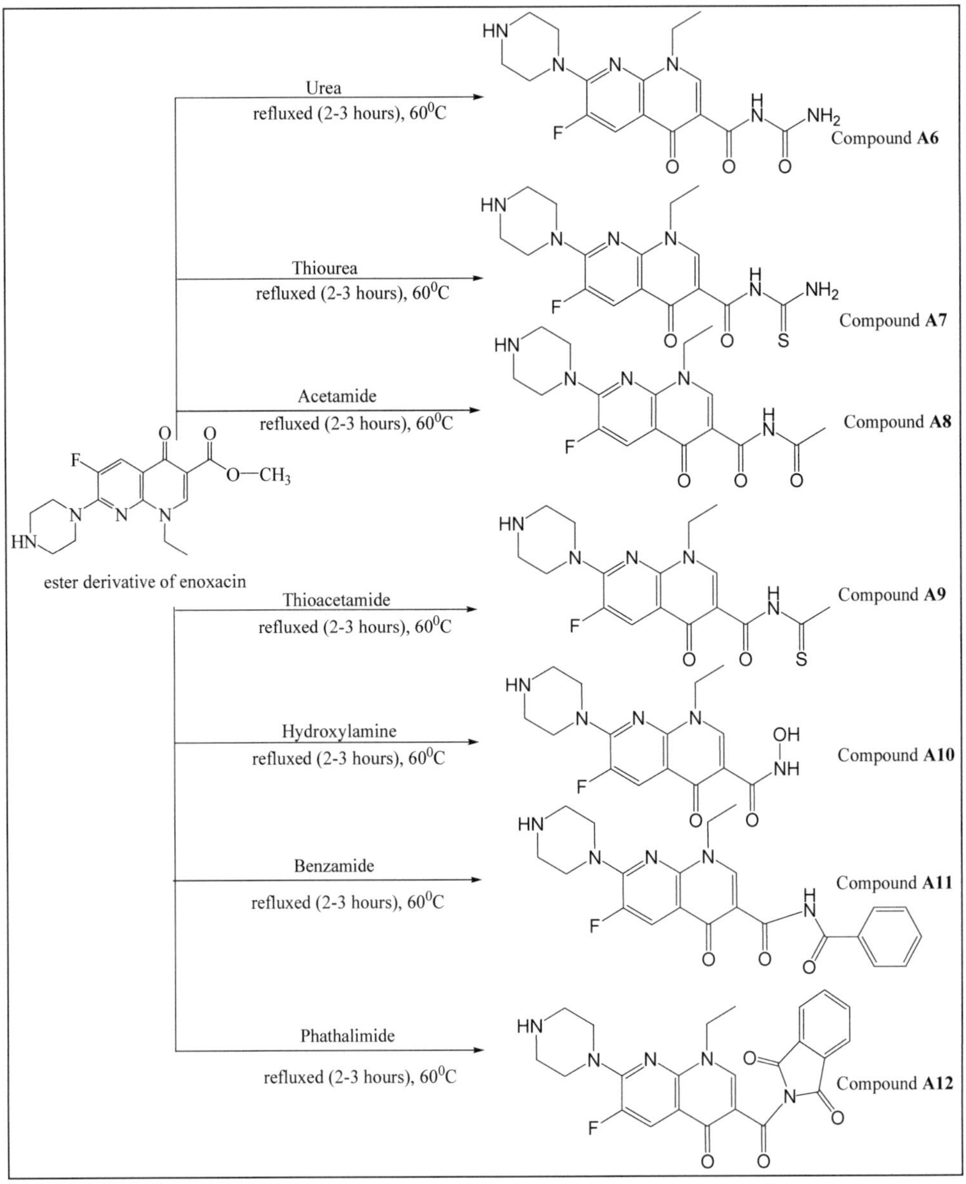

Fig. 3. Synthesis of compound **A6-A12**

3. Result and disscussion

3.1 Spectral analysis

3.1.1 IR studies

In the IR spectrum of enoxacin, the OH stretching vibration of carboxylic acid appeared as a broad band in the range 3500–3100 cm^{-1} and interfered with the NH stretching vibration of the secondary amino group of the piperyzinyl ring which absorbed in the same region at

3100 cm^{-1}. Two strong carbonyl absorptions at 1690 and 1640 cm^{-1} were observed due to the presence of keto carboxylic acid.

The spectra of all the compounds showed absence of acidic OH peak of enoxacin and a distinct, strong and un-obscured NH stretch was observed near 3200-3100 cm^{-1} indicating that carboxylic site reacted with the selected amines forming amides. However compound **A10** showed a sharp band at 3300 cm^{-1} due to the presence of hydroxyl group in their structure. While compounds **A7** and **A9** showed a medium intensity band near 2100 cm^{-1} due to the presence of C=S group, additionally, IR spectra of the compounds revealed that carbonyl absorption of carboxylic group shifted towards right near 1665-1640 cm^{-1} in the spectra of all the compounds which suggests the utilization of carboxylic moiety in amide formation.

3.1.2 ^{1}HNMR studies

Important ^{1}H NMR signals of enoxacin were observed at chemical shifts of 1.40 (t, 3H, J = 7.0 Hz, $-CH_3$ methyl), 2.0 (s, 1H, amine), 2.62 (s, 4H, piperazine), 3.85 (s, 4H, piperazine), 4.48 (q, 2H, J = 7.0 Hz, $-CH_2-$ ethyl), 8.10, 8.95 (s, 2H, naphthyridine) and 11.0 (s, 1H, COOH). On comparing main peaks of enoxacin with its derivatives, all the signals of enoxacin were present in the ^{1}HNMR spectra of the compounds except the acidic proton resonance at 9.8 ppm. Moreover, all these derivatives displayed resonance of naphthyridine protons down-field by (0.4-0.7ppm). Unlike enoxacin all the derivatives showed a singlet in the region 7.9-8.0 ppm which corresponds to the absorption of sec-amide. Further, the spectra of the compounds also exhibited other signals corresponding to their respective chemical structure (mentioned in the spectral data). The signals for the aliphatic and piperazine protons were practically unchanged since they lie far from the reaction site of the drug.

3.1.3 Mass spectrometric studies

1. The electron impact mass spectrum (EIMS) of enoxacin showed molecular ion (M^+) peak at (m/z, %): (320, 35). However compounds (**A6-A12**) showed a very low percentage of M^+ peaks with respect to their proposed structures.

3.1.4 Spectral data of compounds A6-A12

A6 1-(1-ethyl-6-fluoro-4-oxo-7-(piperazin-1-yl)-1,4-dihydro-1,8-naphthyridine-3-carbonyl)urea

M.p. 257°C IR (KBr) ν_{max}: 1655, 1625 (C=O) and 3150 (N-H) cm^{-1}, ^{1}H NMR (DMSO, 500 MHz) 1.41 (t, 3H, J = 7.2 Hz, $-CH_3$ ethyl), 2.59 (s, 4H, piperazine), 3.84 (s, 4H, piperazine), 5.6 (s, 4H, amine), 4.45 (q, 2H, J = 7.0 Hz, $-CH_2-$ ethyl), 7.98 (s, 1H, sec. amide), 8.20 (d, 1H, H_5-naphthyridine, $J_{H,F}$ = 12.49 Hz), 9.18 (s, 1H, H_2-naphthyridine), MS (m/z, %): (362, 10.5) M^+.

A7 1-(1-ethyl-6-fluoro-4-oxo-7-(piperazin-1-yl)-1,4-dihydro-1,8-naphthyridine-3-carbonyl)thiourea

M.p. 287°C IR (KBr) ν_{max}: 1655, 1630 (C=O), 2100 (C=S), 3300 (OH), 3200 (N-H) and 2100 (C=S) cm^{-1}, ^{1}H NMR (DMSO, 500 MHz) 1.43 (t, 3H, J = 7.1 Hz, $-CH_3$ ethyl), 2.62 (s, 4H, piperazine), 3.83 (s, 4H, piperazine), 4.2 (s, 4H, amine), 4.45 (q, 2H, J = 7.2 Hz, $-CH_2-$ ethyl), 7.95 (s, 1H, sec. amide), 8.21 (d, 1H, H_5-naphthyridine, $J_{H,F}$ = 12.51 Hz), 9.19 (s, 1H, H_2-naphthyridine), MS (m/z, %): (378, 9.5) M^+.

A8 N-acetyl-1-ethyl-6-fluoro-4-oxo-7-(piperazin-1-yl)-1,4-dihydro-1,8-naphthyridine-3-carboxamide

M.p. 295°C IR (KBr) ν_{max}: 1655, 1630 (C=O) and 3140 (N-H) cm^{-1}, ^1H NMR (DMSO, 500 MHz) 1.42 (t, 3H, J = 7.0 Hz, $-CH_3$ ethyl), 2.02 (s, 3H, methyl), 2.61 (s, 4H, piperazine), 3.84 (s, 4H, piperazine), 4.46 (q, 2H, J = 7.1 Hz, $-CH_2-$ ethyl), 7.85 (s, 1H, sec. amide), 8.21 (d, 1H, H_5-naphthyridine, $J_{H,F}$ = 12.56 Hz), 9.19 (s, 1H, H_2-naphthyridine), MS (*m/z*, %): (361, 89.5) M^+.

A9 N-ethanethioyl-1-ethyl-6-fluoro-4-oxo-7-(piperazin-1-yl)-1,4-dihydro-1,8-naphthyridine-3-carboxamide

M.p. 237°C IR (KBr) ν_{max}: 1645, 1628 (C=O), 2100 (C=S), 3300 (OH), 3150 (N-H) and 2120 (C=S) cm^{-1}, ^1H NMR (DMSO, 500 MHz) 1.41 (t, 3H, J = 7.1 Hz, $-CH_3$ ethyl), 2.529 (s, 3H, methyl), 2.62 (s, 4H, piperazine), 3.81 (s, 4H, piperazine), 4.44 (q, 2H, J = 6.9 Hz, $-CH_2-$ ethyl), 7.98 (s, 1H, sec. amide), 8.21 (d, 1H, H_5-naphthyridine, $J_{H,F}$ = 12.54 Hz), 9.19 (s, 1H, H_2-naphthyridine), MS (*m/z*, %): (377, 5.5) M^+.

A10 1-ethyl-6-fluoro-N-hydroxy-4-oxo-7-(piperazin-1-yl)-1,4-dihydro-1,8-naphthyridine-3-carboxamide

M.p. 267°C IR (KBr) ν_{max}: 1655, 1634 (C=O), 3300 (OH) and 3180 (N-H) cm^{-1}, ^1H NMR (DMSO, 500 MHz) 1.43 (t, 3H, J = 7.0 Hz, $-CH_3$ ethyl), 2.62 (s, 4H, piperazine), 3.84 (s, 4H, piperazine), 4.45 (q, 2H, J = 7.0 Hz, $-CH_2-$ ethyl), 6.58 (s, 1H, OH), 7.97 (s, 1H, sec. amide), 8.18 (d, 1H, H_5-naphthyridine, $J_{H,F}$ = 12.56 Hz), 9.09 (s, 1H, H_2-naphthyridine), MS (*m/z*, %): (335, 11) M^+.

A11 N-benzoyl-1-ethyl-6-fluoro-4-oxo-7-(piperazin-1-yl)-1,4-dihydro-1,8-naphthyridine-3-carboxamide

M.p. 255°C IR (KBr) ν_{max}: 1645, 1635 (C=O) and 3155 (N-H) cm^{-1}, ^1H NMR (DMSO, 500 MHz) 1.42 (t, 3H, J = 7.0 Hz, $-CH_3$ ethyl), 2.62 (s, 4H, piperazine), 3.86 (s, 4H, piperazine), 4.43 (q, 2H, J = 7.0 Hz, $-CH_2-$ ethyl), 7.21 (m, 5H, phenylic H), 7.99 (s, 1H, sec. amide), 8.18 (d, 1H, H_5-naphthyridine, $J_{H,F}$ = 12.56 Hz), 9.1 (s, 1H, H_2-naphthyridine), MS (*m/z*, %): (423, 7) M^+.

A12 2-(1-ethyl-6-fluoro-4-oxo-7-(piperazin-1-yl)-1,4-dihydro-1,8-naphthyridine-3-carbonyl)isoindoline-1,3-dione

M.p. 278°C IR (KBr) ν_{max}: 1655, 1630 (C=O) and 3150 (N-H) cm^{-1}, ^1H NMR (DMSO, 500 MHz) 1.43 (t, 3H, J = 7.0 Hz, $-CH_3$ ethyl), 2.60 (s, 4H, piperazine), 3.85 (s, 4H, piperazine), 4.45 (q, 2H, J = 7.0 Hz, $-CH_2-$ ethyl), 7.69 (s, 2H, phenylic H), 8.3 (s, 2H, phenylic H), 7.94 (s, 1H, sec. amide), 8.15 (d, 1H, H_5-naphthyridine, $J_{H,F}$ = 12.56 Hz), 9.09 (s, 1H, H_2-naphthyridine), MS (*m/z*, %): (449, 10) M^+.

3.2 Antimicrobial studies

3.2.1 Antibacterial activity

All compounds showed activities nearly similar to enoxacin against all the *Gram-positive* test strains. Generally, the 7-piperazinyl enoxacin have better *Gram-negative* than *Gram-positive* antimicrobial potency (Chin & Neu, 1983, Paton & Reeves, 1988, Wood, 1989). Studies revealed that the mode of action of quinolones involves inhibition of essential type II bacterial topoisomerases such as DNA gyrase and topoisomerase IV (Sissi & Palumbo, 2003). Commonly DNA gyrase is more sensitive to *Gram-negative* bacteria and topoisomerase IV more sensitive to *Gram-positive* bacteria. As topoisomerase IV is the primary target of the quinolones with bulky functional group at N-4 position of piperazine ring, therefore the activities of all the compounds were found to be similar to enoxacin.

However, against all the *Gram-negative* organisms, strong improvement in the antibacterial activity of the derivatives **A11, A12** and **A6** was observed in comparision to enoxacin (Tables 1 and 2). The increase in activity was greatest against *Salmonella typhi, Pseudomonas aeruginosa* and *Escherichia coli.* The enhancement in the antibacterial activity of these compounds might be a result of better interaction with DNA gyrase. It was observed that the activity of compounds **A7-A10** decreased against all the *Gram-negative* strains. It is proposed that in terms of structure–activity relationship, the antibacterial activity profile of enoxacin derivatives against *Gram-negative* bacteria was enhanced by the phenyl attachment via amide linkage at the 3-position of the enoxacin molecule. While reaction with aliphatic amines caused a diminution in antibacterial activity of enoxacin, exception was compound **A6**.

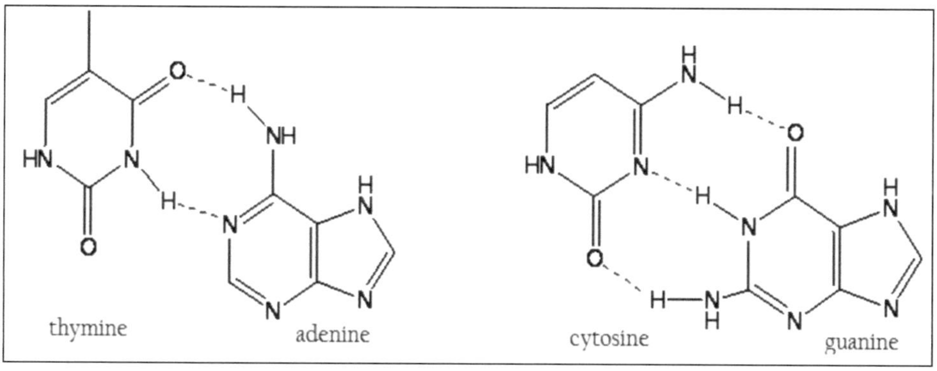

Fig. 4. Base pairing between adenine/thymine & cytosine/guanine

Previous studies showed that carboxylic group of quinolones was necessary to bind to bacterial DNA gyrase through hydrogen bonding. Looking at the structure of DNA, it is apparent that the hydrogen bonding between the two DNA strands is also through the amide group of heterocyclic bases.

Since the inhibitory action of quinolones is not simply accomplished by inhibiting the bacterial enzyme function but they also actively poison cells by trapping the two topoisomerases on DNA as drug/enzyme/DNA complexes in which double-strand DNA breaks are held together by protein. It is proposed that converting carboxylic group into amide group would help quinolone to better undergo binding with bacterial DNA due to structural complementarities.

Microorganisms	Enoxacin µgmL-1			A6 µgmL-1			A7 µgmL-1			A8 µgmL-1		
	5	10	20	5	10	20	5	10	20	5	10	20
S. aureus	14	16	16	12	16	18	11	15	15	10	12	13
Citrobacter	13	14	19	12	14	18	10	11	12	9	10	10
S. pneumoniae	11	13	17	11	12	16	11	12	13	12	14	15
S. flexneri	10	12	15	12	14	18	11	11	12	10	12	13
E. coli	12	19	20	13	19	23	10	11	12	11	12	14
S. typhi	10	15	20	12	15	28	9	10	11	11	14	18
P. aeruginosa	13	14	14	14	16	18	11	12	13	9	10	11
B. substilis	10	12	14	12	14	16	9	10	11	11	12	14
K. pneumoniae	12	13	16	13	15	17	10	12	14	10	11	12
P. mirabilis	11	12	14	11	13	16	9	11	12	10	10	12
C. hofmannii	10	10	12	11	13	14	9	10	11	9	10	11

Table 1.

Microorganisms	A9 µgmL-1			A10 µgmL-1			A11 µgmL-1			A12 µgmL-1		
	5	10	20	5	10	20	5	10	20	5	10	20
S. aureus	11	12	14	11	14	14	11	12	15	10	14	16
Citrobacter	12	12	13	14	15	17	10	12	21	10	15	22
S. pneumoniae	12	14	16	13	14	15	11	14	16	13	14	17
S. flexneri	11	12	14	11	13	14	11	12	16	11	13	17
E. coli	12	14	16	12	14	18	13	20	26	12	18	25
S. typhi	11	13	16	12	15	16	14	17	26	11	15	27
P. aeruginosa	10	12	13	10	11	12	10	14	20	10	11	19
B. substilis	11	13	13	11	14	14	11	13	13	11	14	14
K. pneumoniae	11	12	13	14	15	16	11	12	18	10	15	16
P. mirabilis	11	12	13	10	12	12	9	12	17	10	12	16
C. hofmannii	9	10	11	8	9	10	9	11	11	9	9	12

Table 2.

3.2.1 Antifungal activity

Quinolone antibiotics have been shown to be potent inhibitors of DNA gyrase and have been useful antibacterial drugs in clinical practice (Marklein, 1996). Laboratory work has indicated that inhibition of DNA gyrase is not specific and activity against nonbacterial targets has been shown (Dykstra et al., 1994, Fostel et al., 1996). Nakajima et al., (Nakajima et al., 1995) have reported that quinolone augmented the activity of amphotericin B and fluconazole against a variety of fungi in both *in vitro* and *in vivo* assays of combined antifungal activity. Based on theses studies, we aimed to evaluate the antifungal activities of enoxacin and its derivatives against common fungal pathogens; *Aspergillus purasiticus, Saccharomyces cervics, Candida albicans* and *Fusraium solani*. Results show that except compound **A6**, none of the tested compounds possessed clinically useful antifungal activity. Compound **A6** showed moderate inhibitory activity against *Candida albicans* (zone of inhibition = 17mm). The increased activity is probably due to some intracellular interaction of the derivative within the fungal cell; however the mechanism of the observed antifungal effect is still warranted (Sugar et al., 1997).

3.3 Immunomodulatory activities

Fluoroquinolones have been studied for their modulatory activity on the immune response. The best investigated agent seems to be moxifloxacin which showed immunomodulatory actions both *in vitro* and *in vivo*. However, the molecular mechanism causing the immunomodulatory effects of fluoroquinolones are still under investigation (Tauber & Nau, 2008). Recently we have reported few carboxamide analogues of enoxacin showing potentials to mediate immune response.

In order to test the immunomodulatory effect of enoxacin and its derivatives, we investigated their effect on phagocytosis, T-cell proliferation and cytokine release (particularly IL-1β and TNF-α) by macrophages.

3.3.1 Effect on phagocytes oxidative burst

Phagocytic cells on activation induce release of reactive oxygen species (oxidative burst) which is then quantified by a luminol- lucigenin enhanced chemiluminescence assay. With the luminol probe which detects the intracellular ROS using serum opsonize zymosan as activator, compounds **A11** and **A12** showed moderate inhibition on whole blood (IC_{50} 10.5 and 14.4μgmL^{-1} respectively). Similar inhibitory effects were observed on the isolated neutrophils (IC_{50} = 11.2 and 13.6μgmL^{-1} respectively) and on macrophages (IC_{50}, 13.2 and 11.4μgmL^{-1}). In another set of experiment lucigenin probe and PMA as activator were used as a replacement for luminol and SOZ, only compounds **A12** again demonstrated a moderate inhibition on the isolated neutrophils (12.6 μgmL^{-1}) respectively). Enoxacin and other compounds did not exert any significant inhibition up to the highest concentration (25μgmL^{-1}) in the system tested.

3.3.2 Effect on T-cell proliferation

The anti-proliferation of the test compounds was determined by measuring the inhibition of (PHA)-induced T-cell proliferation by determining radioactive thymidine incorporation. Results show (Figure 7) that compound **A11** significantly suppressed T-cell proliferation in dose dependent manner. A dose as low as 3.15μgmL^{-1} of compound **A11** caused approximately 50% reduction in T cell proliferation compared to control. While enoxacin and other compounds did not have any effect on T-cell proliferation up to the highest concentration (50μgmL^{-1}) in the system tested.

Figure 5 shows the immunomodulatory effects of compound **A11** exhibiting prominent effect on PHA-induced T-cell proliferation. Cells were incubated with different concentrations of the test compound in RMPI along with PHA for 72 hrs at 37°C in CO_2 environment. Cells were further incubated for 18 hrs after the addition of thymidine $[H]^3$ and the radioactivity count as CPM reading was recorded using scintillation counter. The effect of compound on the T-cell proliferation is compared with the control. Each plot and error bar represents readings ± SD of three repeats.

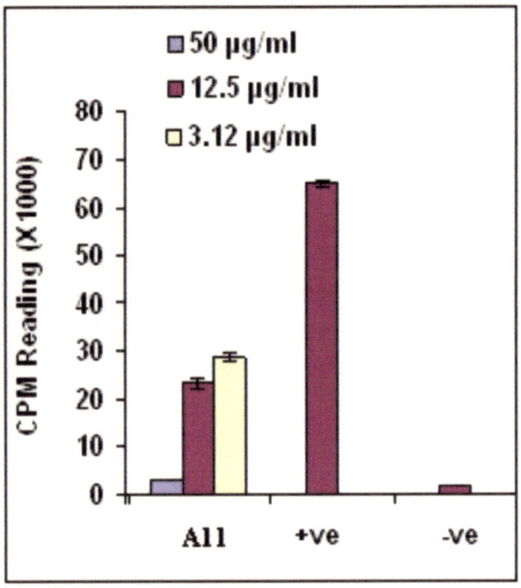

Fig. 5. Immunomodulatory effects of compound A-11

3.3.3 Effect on cytokine release by macrophages

The immuno-suppressive effect of enoxacin and its derivatives were tested on the release of selected cytokines including IL-1β and TNF-α by PHA-induced macrophages. Only compound **A11** at concentration of 25µgmL^{-1} fairly suppressed the production of TNF-α showing 70% inhibition. Enoxacin and other compounds demonstrated no effect on any of the tested cytokines released form activated macrophages.

4. Conclusion

In conclusion we have synthesized some novel enoxacin derivatives bearing various aliphatic and aromatic substituents at C3 position via amide linkage. It was found that introduction of amide linkge with phenyl substituent at C3 position, produced noticeable enhancement in the *in vitro* activity of enoxacin particularly against Gram negative organism. In addition, unlike enoxacin, Compound **A6** also exhibited effective anti-fungal activity against *Candida albicans*. Moreover, the phagocytic function, T-cell proliferation and cytokine release was moderately affected in presence of compounds **A11** and **A12**, thus showing potentials to be anti-inflammatory. Conclusively, the test compounds showing diverse beneficial biological activities could serve as new lead molecular entities for treating various conditions such as infections, organ transplantation, cancer and for the treatment of rheumatoid arthritis and related autoimmune disorders. However, appropriate clinical trials are necessary before using the antimicrobial and immunomodulatory property of the test compounds in clinical practice.

5. References

Anderson, V. E. & Osheroff, N. (2001). Type II Topoisomerases as Targets for Quinolone Antibacterials Turning Dr. Jekyll into Mr. Hyde. *Current pharmaceutical design*, Vol.7, No.5, pp. 337-353, 1381-6128

Arayne, M. S.; Sultana, N.; Haroon, U.; Zuberi, M. H. & Rizvi, S. B. S. (2012). Synthesis, characterization and biological activity of a series of carboxamide derivatives of ofloxacin. *Archives of pharmacal research*, Vol.33, No.12, pp. 1901-1909, 0253-6269

Ball, P.; Fernald, A. & Tillotson, G. (1998). Therapeutic advances of new fluoroquinolones. *Expert opinion on investigational drugs*, Vol.7, No.5, pp. 761-783, 1354-3784

Baur, A. W.; Kirby, W. M.; Sherris, J. C. & Turck, M. (1966). Antibiotic susceptibility testing by a standard single disc method. *Am J Clin Pathol*, Vol.45, pp. 493-496,

Blandeau, J. M. (1999). Expanded activity and utility of the new fluoroquinolones: a review. *Clinical therapeutics*, Vol.21, No.1, pp. 3-40, 0149-2918

Chin, N. X. & Neu, H. C. (1983). In vitro activity of enoxacin, a quinolone carboxylic acid, compared with those of norfloxacin, new beta-lactams, aminoglycosides, and trimethoprim. *Antimicrobial agents and chemotherapy*, Vol.24, No.5, pp. 754, 0066-4804

Chu, D. T. & Fernandes, P. B. (1989). Structure-activity relationships of the fluoroquinolones. *Antimicrobial agents and chemotherapy*, Vol.33, No.2, pp. 131, 0066-4804

Chu, D. T. W.; Fernandes, P. B.; Claiborne, A. K.; Pihuleac, E.; Nordeen, C. W.; Maleczka Jr, R. E. & Pernet, A. G. (1985). Synthesis and structure-activity relationships of novel arylfluoroquinolone antibacterial agents. *Journal of medicinal chemistry*, Vol.28, No.11, pp. 1558-1564, 0022-2623

Domagala, J. M.; Heifetz, C. L.; Hutt, M. P.; Mich, T. F.; Nichols, J. B.; Solomon, M. & Worth, D. F. (1988). 1-Substituted 7-[3-[(ethylamino) methyl]-1-pyrrolidinyl]-6, 8-difluoro-1, 4-dihydro-4-oxo-3-quinolinecarboxylic acids. New quantitative structure activity relationships at N1 for the quinolone antibacterials. *Journal of medicinal chemistry*, Vol.31, No.5, pp. 991-1001, 0022-2623

Dykstra, C. C.; McClernon, D. R.; Elwell, L. P. & Tidwell, R. R. (1994). Selective inhibition of topoisomerases from Pneumocystis carinii compared with that of topoisomerases from mammalian cells. *Antimicrobial agents and chemotherapy*, Vol.38, No.9, pp. 1890, 0066-4804

Emami, S.; Shafiee, A. & Foroumadi, A. (2006). Structural features of new quinolones and relationship to antibacterial activity against Gram-positive bacteria. *Mini reviews in medicinal chemistry*, Vol.6, No.4, pp. 375-386, 1389-5575

Foroumadi, A.; Emami, S.; Mehni, M.; Moshafi, M. H. & Shafiee, A. (2005). Synthesis and antibacterial activity of N-[2-(5-bromothiophen-2-yl)-2-oxoethyl] and N-[(2-5-bromothiophen-2-yl)-2-oximinoethyl] derivatives of piperazinyl quinolones. *Bioorganic & medicinal chemistry letters*, Vol.15, No.20, pp. 4536-4539, 0960-894X

Fostel, J.; Montgomery, D. & Lartey, P. (1996). Comparison of responses of DNA topoisomerase I from Candida albicans and human cells to four new agents which stimulate topoisomerase dependent DNA nicking. *FEMS microbiology letters*, Vol.138, No.2 3, pp. 105-111, 1574-6968

Helfand, S. L.; Werkmeister, J. & Roder, J. C. (1982). Chemiluminescence response of human natural killer cells. I. The relationship between target cell binding, chemiluminescence, and cytolysis. *The Journal of experimental medicine*, Vol.156, No.2, pp. 492, 0022-1007

Hu, G.; Yang, Y.; Yi, L.; Wang, X.; Zhang, Z.; Xie, S. & Huang, W. (2012). Part II: Design, synthesis and antitumor action of C3/C3 bis-fluoroquinolones linked-cross 2, 5-[1, 3, 4] oxadiazole. pp.

Kondo, H.; Sakamoto, F.; Kawakami, K. & Tsukamoto, G. (1988). Studies on prodrugs. 7. Synthesis and antimicrobial activity of 3-formylquinolone derivatives. *Journal of medicinal chemistry*, Vol.31, No.1, pp. 221-225, 0022-2623

Marklein, G. (1996). Quinolones in everyday clinical practice: respiratory tract infections and nosocomial pneumonia. *Chemotherapy*, Vol.42, No.1, pp. 33-42, 0009-3157

Mitscher, L. A. (2005). Bacterial topoisomerase inhibitors: quinolone and pyridone antibacterial agents. *Chemical reviews*, Vol.105, No.2, pp. 559-592, 0009-2665

Nakajima, R.; Kitamura, A.; Someya, K.; Tanaka, M. & Sato, K. (1995). In vitro and in vivo antifungal activities of DU-6859a, a fluoroquinolone, in combination with amphotericin B and fluconazole against pathogenic fungi. *Antimicrobial agents and chemotherapy*, Vol.39, No.7, pp. 1517, 0066-4804

Nielsen, M. B.; Gerwien, J.; Nielsen, M.; Geisler, C.; Röpke, C.; Svejgaard, A. & Ødum, N. (2000). MHC class II ligation induces CD58 (LFA-3)-mediated adhesion in human T cells. *Experimental and clinical immunogenetics*, Vol.15, No.2, pp. 61-68, 0254-9670

Paton, J. H. & Reeves, D. S. (1988). Fluoroquinolone antibiotics. Microbiology, pharmacokinetics and clinical use. *Drugs*, Vol.36, No.2, pp. 193, 0012-6667

Saeed Arayne, M.; Sultana, N.; Haroon, U.; Ahmed Mesaik, M. & Asif, M. (2009). Synthesis and biological evaluations of enoxacin carboxamide derivatives. *Archives of pharmacal research*, Vol.32, No.7, pp. 967-974, 0253-6269

Sarro, A. D. & Sarro, G. D. (2001). Adverse reactions to fluoroquinolones. An overview on mechanistic aspects. *Current medicinal chemistry*, Vol.8, No.4, pp. 371-384, 0929-8673

Singh, U.; Tabibian, J.; Venugopal, S. K.; Devaraj, S. & Jialal, I. (2005). Development of an in vitro screening assay to test the antiinflammatory properties of dietary supplements and pharmacologic agents. *Clinical chemistry*, Vol.51, No.12, pp. 2252,

Sissi, C. & Palumbo, M. (2003). The quinolone family: from antibacterial to anticancer agents. *Current Medicinal Chemistry-Anti-Cancer Agents*, Vol.3, No.6, pp. 439-450, 1568-0118

Sugar, A. M.; Liu, X. P. & Chen, R. J. (1997). Effectiveness of quinolone antibiotics in modulating the effects of antifungal drugs. *Antimicrobial agents and chemotherapy*, Vol.41, No.11, pp. 2518, 0066-4804

Sultana, N.; Arayne, M. S.; Rizvi, S. B. S. & Haroon, U. (2011). Synthesis, Characterization and Biological Evaluations of Ciprofloxacin Carboxamide Analogues. *Bulletin of Korean Chemical Society*, Vol.32, No.2, pp. 6,

Tauber, S. C. & Nau, R. (2008). Immunomodulatory properties of antibiotics. *Curr Mol Pharmacol*, Vol.1, No.1, pp. 68-79,

Vargas, F.; Zoltan, T.; Rivas, C.; Ramirez, A.; Cordero, T.; Diaz, Y.; Izzo, C.; Cárdenas, Y. M.; López, V. & Gómez, L. (2008). Synthesis, primary photophysical and antibacterial properties of naphthyl ester cinoxacin and nalidixic acid derivatives. *Journal of Photochemistry and Photobiology B: Biology*, Vol.92, No.2, pp. 83-90, 1011-1344

Venkat Reddy, G.; Ravi Kanth, S.; Maitraie, D.; Narsaiah, B.; Shanthan Rao, P.; Hara Kishore, K.; Murthy, U. S. N.; Ravi, B.; Ashok Kumar, B. & Parthasarathy, T. (2009). Design, synthesis, structure-activity relationship and antibacterial activity series of novel imidazo fused quinolone carboxamides. *European journal of medicinal chemistry*, Vol.44, No.4, pp. 1570-1578, 0223-5234

Wood, M. J. (1989). Tissue penetration and clinical efficacy of enoxacin in respiratory tract infections. *Clinical pharmacokinetics*, Vol.16, pp. 38, 0312-5963

13

Biocompatibility and Antimicrobial Activity of Some Quaternized Polysulfones

Silvia Ioan and Anca Filimon
*"Petru Poni" Institute of Macromolecular Chemistry Iasi
Romania*

1. Introduction

In recent decades, considerable attention has been devoted to the investigation of new applications of polysulfones, which are mainly because of their specific properties. The literature shows that polysulfones and their derivatives are widely used as new functional materials in biochemical, industrial, and medical fields because of their structure and physical properties, such as their good optical properties, high thermal and chemical stability, mechanical strength, resistance to extreme pH values, and low creep (Barikani & Mehdipour-Ataei, 2000; Higuchi et al., 1988; Johnson, 1969; Mann & West, 2001). The chain rigidity is derived from the relatively inflexible and immobile phenyl and SO_2 groups, whereas their toughness is derived from the connecting ether oxygen. Although these materials have excellent overall properties, their intrinsic hydrophobic nature precludes their use in membrane applications, which require a hydrophilic character. Therefore, such polymers should be modified to improve their performance for specific applications (Johnson, 1969; Khang et al., 1995). The chemical modification of polysulfones, especially the chloromethylation reaction, is a subject of considerable interest from both theoretical and practical points of view; there is interest in obtaining the precursors for functional membranes, coatings, ion exchange resins, ion exchange fibers, and selectively permeable membranes (Higuchi et al., 2002; Tomaszewska et al., 2002). Functionalized polymers, chloromethylated and quaternized polysulfones, have evidenced several interesting properties, which recommend them for a wide range of industrial and environmental applications. Quaternization with ammonium groups is an efficient method for increasing their hydrophilicity. Accordingly, these polymers can be used for multiple applications, *e.g.* as biomaterials and semipermeable membranes. Also, the different components of a block or graft copolymer may segregate in bulk to yield nanometer-sized patterns or mesophasic structures. Numerous applications involve nanodomained solids. By matching the periodicity of the patterns with the wavelength of visible light, literature studies have demonstrated that block copolymers, including polysulfones, act as photonic crystals. Segregated block copolymers, polysulfones included, have been also used as precursors for the preparation of various nanostructures, including nanospheres, nanofibers, annotates, and thin membranes-containing nanochannels. Thin membranes containing nanochannels have been used as membranes, pH sensors, or templates for the preparation of metallic nanorods. Furthermore, in the last decades, blends of polysulfone or modified polysulfones and other synthetic polymers have continued to be a subject of intense both academic and industrial investigation, because of their simplicity and effectiveness in the mixture of two different polymers for producing new materials.

In such applications, the addition of functional groups to the polysulfone enhance some properties of the material, such as hydrophilicity (which is of special interest for biomedical applications) (Guan et al., 2005), antimicrobial action (Filimon et al., 2009; Yu et al., 2007), and solubility characteristics (Filimon et al., 2007; Ioan et al., 2006), to allow higher water permeability and better separation (Idris et al, 2007; Kochkodan et al., 2008). In addition, functional groups are an intrinsic requirement for affinity, ion exchange, and other specialty membranes (Guiver et al, 1993).

In this context, the paper presents the synthesis and some properties of new polysulfones for biomedical applications. Studies are carried out on the quaternization reaction of chloromethylated polysulfones with N,N-dimethylethanolamine (Filimon et al., 2010; Ioan et al., 2006a, 2006b, 2007), N,N-dimethylethylamine and N,N-dimethyloctylamine (Ioan et al., 2011a), for obtaining water soluble polymers with various amounts of ionic chlorine, or on the new polysulfones with bulky phosphorus pendant groups, obtained by chemical modification of the chloromethylated polysulfones by reacting the chloromethyl group with the P-H bond of 9,10-dihydro-oxa-10-phosphophenanthrene-10-oxide (Ioan et. al., 2011b).

The different properties, such as morphological aspect (by atomic force microscopy), where the type of nonsolvents in casting solutions of polymer (identified by viscometry and rheology investigations (Ioan et al., 2006b; Filimon et al., 2010)) significantly influenced membrane morphology and where ordered domains depend on the charge density of polyelectrolytes, the hydrophilic/hydrophobic characteristics (by contact angle method), as well as the history of the formed films, correlated with a good adhesion of the red blood cells and with a good cohesions of the platelets on the surface of the quaternized polysulfone films (determined from surface properties), are investigate for specific biomedical applications. Furthermore, bacterial adhesions to the surfaces (by analyzing the inhibition zones) are studied for applications of modified polysulfones as semipermeable membranes. Thus, the analysis of antibacterial activity, using *Escherichia coli* ATCC 10536 and *Staphylococcus aureus* ATCCC 6538 microorganisms, contribute to extending the possible applications of quaternized polysulfones membranes in biomedical domains.

2. Synthesis of some new functionalized polysulfones

The quaternized polysulfones tested for biomedical applications, such as PSFQ, PSF-DMEA and PSF-DMOA, are obtained by quaternized reaction (Luca et al., 1988) of chloromethylated polysulfones (CMPSF) with N,N-dimethylethanolamine (Filimon et al., 2010; Ioan et al., 2006a, 2006b, 2007;), N,N-dimethylethylamine, and N,N-dimethyloctylamine (Ioan et al., 2011a), respectively. Also, polysulfones with bulky phosphorus pendant groups (PSF-DOPO) are obtained by chemical modification of the chloromethylated polysulfone, performed by reacting the chloromethyl group with the P-H bond of 9,10-dihydro-oxa-10-phosphophenanthrene-10-oxide (DOPO) (Petreus et al., 2010). The general chemical structures of the studied polysulfones, PSF, CMPSF and quaternized polysulfones PSFQ, PSF-DMEA, PSF-DMOA and PSF-DOPO are presented in Scheme 1.

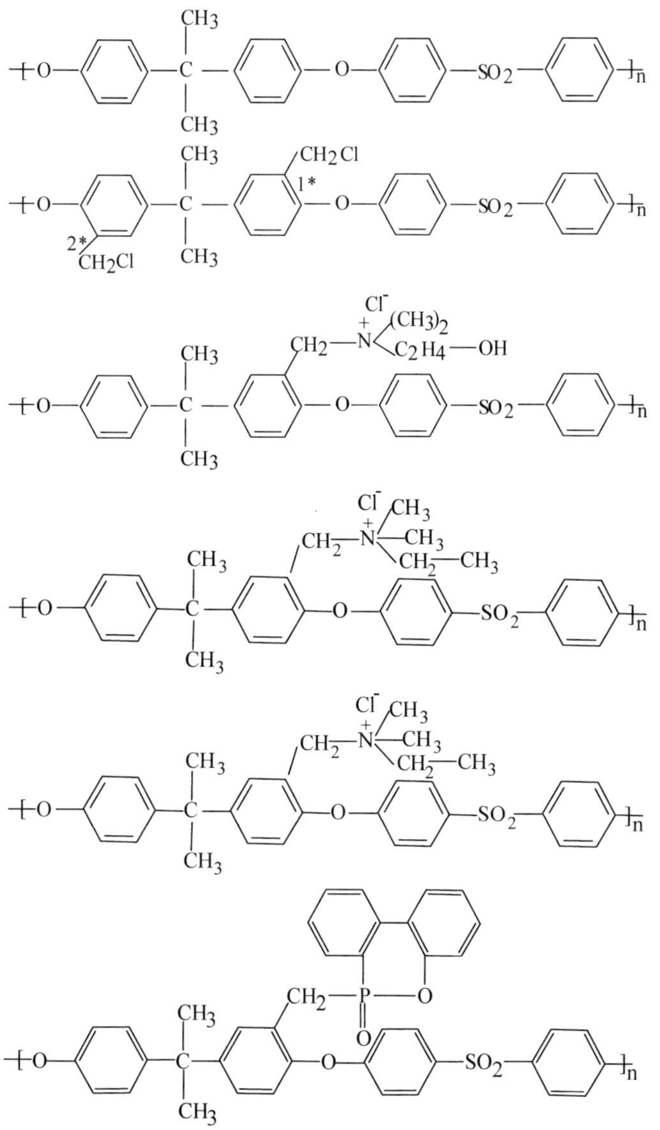

PSF

CMPSF

PSFQ

PSF-DMEA

PSF-DMOA

PSF-DOPO

Scheme 1. Chemical structures of polysulfone (PSF), chloromethylated polysulfone (CMPSF), quaternized polysulfones (PSFQ, PSF-DMEA, PSF-DMOA) and polysulfone with bulky phosphorus pendant groups (PSF-DOPO)

Table 1 lists the chlorine content, substitution degree, molecular weights of the structural units, m_0, number-average molecular weights, M_n, and intrinsic viscosities determined in N,N-dimethylformamide (DMF) at 25°C of polysulfone and chloromethylated polysulfones (Ioan et al., 2006a). The characteristics of quaternized polysulfones PSFQ1 and PSFQ2 (Ioan et al., 2006b), and also PSF-DMEA and PSF-DMOA (Filimon et al., 2010) are presented in Table 2.

Samples	Cl , %	DS	m_0	M_n	$[\eta]$, dL/g
PSF	0	0	442.51	39000	0.3627
CMPSF1	3.34	0.437	463.68	41000	0.3929
CMPSF2	10.53	1.541	517.17	46000	0.4703
CMPSF3	12.13	1.828	530.83	47000	0.6970

Table 1. Chlorine content, substitution degree, DS, molecular weights of the structural units, m_0, number-average molecular weights, M_n, and intrinsic viscosities in DMF at 25°C, $[\eta]$, of polysulfone and chloromethylated polysulfones

Samples	Obtained from:	Cli, %	m_0	M_n
PSFQ1	CMPSF1	2.15	479.47	42000
PSFQ2	CMPSF2	5.71	647.31	57000
PSFQ3	CMPSF3	6.21	691.29	61000
PSF-DMEA	CMPSF2	2.89	551.70	49000
PSF-DMOA	CMPSF2	3.23	627.30	56000

Table 2. Ionic chlorine content, Cl_i, molecular weights of the structural units, m_0, and number-average molecular weights, M_n, of quaternized polysulfones

On the other hand, the substitution of chlorine with the bulky cyclic phosphorus compound is carried out at elevated temperature, using a large excess of phosphorus reactant. The reactive P-H group interacts with CH_2Cl group of chloromethylated polysulfone. The occurrence of HCl evolved from the reaction proved the substitution.

Samples	Elemental composition, %					DS	m_0^*	M_n
	C	H	S	O	Cl			
PSF	73.07	5.12	7.47	14.16	0.18	-	443	39000
CMPSF1	68.81	4.81	6.96	15.22	4.20	0.56	470	41000
CMPSF2	66.58	4.68	6.72	15.44	6.58	0.90	486	43000
CMPSF3	61.27	4.25	6.12	17.86	10.50	1.53	528	47000

Table 3. Carbon, hydrogen, sulfur, oxygen and chlorine content, substitution degree, DS, molecular weight of the structural units, m_0, and number-average molecular weights, M_n, of polysulfone and chloromethylated polysulfone

Also, the chloromethylation reaction of polysulfone may occur in position 1* for CMPSF1 and CMPSF2, when DS < 1. Thus, according to Table 1, the difference between these two samples lies in the different values of the chlorine content. For sample CMPSF3 with DS > 1, the chloromethylation reaction occurs in positions 1* and 2*. Table 3 lists the carbon, hydrogen, sulfur, oxygen and chlorine content, degree of substitution, molecular weights of the structural units, m_0, and number-average molecular weight, M_n, of the chloromethylated polysulfones, determined from the polymerization degree of the polysulfone (DP $\cong$ 88) and molecular weights of the structural units of chloromethylated polysulfones (Petreus et al. 2010).

The characteristics of phosphorus-modified polysulfones are presented in Table 4 (Ioan *et al.*, 2011b).

Samples	Obtained from:	Elemental composition, %						DS	m_0	M_n
		C	H	S	O	Cl	P			
PS-DOPO-1	CMPSF1 (4.2 Cl %)	72.5	4.9	5.7	13.6	0.4	2.9	0.59	491.5	43000
PS-DOPO-2	CMPSF2 (6.6 Cl %)	71.3	4.7	5.2	14.4	0.4_5	3.4	0.72	569.4	50000
PS-DOPO-3	CMPSF3 (10.5 Cl %)	71.1	4.3	4.5	13.7	0.3_5	6.0	1.85	975.0	90000

Table 4. Elemental composition, substitution degree, DS, molecular weight of the structural units, m_0, and number-average molecular weights, M_n, of phosphorus modified polysulfones

3. Role of casting solutions on the functionalized polysulfones surface morphology

Some studies have reported that the chain shape of a polymer in solution could affect the morphology of the polymer in bulk (Hopkins et al., 2006; Huang et al., 2000; Qian et al., 2005). The PSFQ membranes used in atomic force microscopy (AFM) are prepared with different solvent mixtures, including DMF/MeOH, DMF/water, MeOH/DMF and MeOH/water. The solvent systems are selected as a function of the ionic chlorine content of PSFQ. Thus, DMF solvates PSFQ1 with an ionic chlorine content of 2.15% more intensely than the mixed DMF/MeOH and DMF/water solvents with high contents of MeOH and water, respectively. On the other hand, MeOH solvates PSFQ2 with an ionic chlorine content of 5.71%, more strongly than the mixed MeOH/DMF and MeOH/water solvents with high contents of DMF and water, respectively (Filimon et al., 2007); Ioan et al., 2006b). AFM is used to examine film surface and to measure their surface topography. All of the images presented in Figures 1-3 are recorded under ambient conditions, at different size scales to provide morphological details. Each micrograph shows that the membrane surface is not smooth, existing in ordered domains, in which are distributed pores and nodules of different size and intensities. Increasing the nonsolvent content in the casting solutions favores modification of the ordered domains, more visibly in images with a 500 nm x 500 nm or 3 μm x 3 μm scanning area. Thus, the average surface roughness in the 3 μm x 3 μm scanning area decreases from 0.354 nm for the PSFQ1 membranes obtained in 60 DMF/40 MeOH (Figure 1 (1b, b′), in two- dimensional (2D) and three-dimensional (3D) AFM images, respectively), to 0.193 nm for the PSFQ1 membranes realized in 20 DMF/80 MeOH (Figure 1 (2b, b′), in 2D and 3D AFM images, respectively). Also, the AFM images from Figure 1, for a more extended scanning area, show that increasing the nonsolvent content favores the increase of the pores number and of their characteristics, such as area, volume, diameter, and also root-mean-square roughness of the surface (rms), whereas their depth decreases. This changing trend in morphology is probably due to the modification of the chain conformation of quaternized polysulfones, which is influenced by the quality of the mixed solvents (Qian et al., 2005; Kesting, 1990).

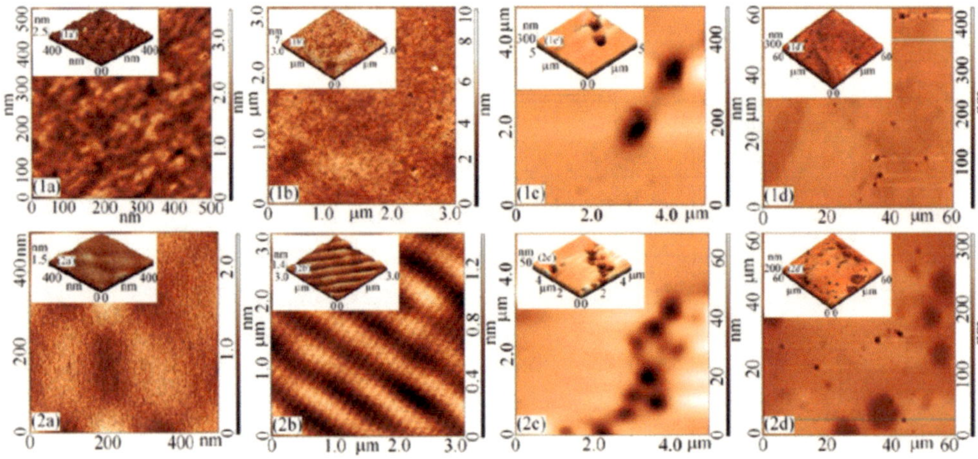

Fig. 1. AFM images of PSFQ1 membranes obtained from different DMF/MeOH solvent mixtures. (1) 60/40 DMF/MeOH: (a, a') 2D and 3D images - scanned area 500 nm x 500 nm , (b, b') 2D and 3D images - scanned area 3 μm x 3 μm , (c, c') 2D and 3D images - scanned area 5 μm x 5 μm , (d, d') 2D and 3D images - scanned area 60 μm x 60 μm ; (2) 20/80 DMF/MeOH: (a, a') 2D and 3D images - scanned area 500 nm x 500 nm , (b, b') 2D and 3D images - scanned area 3 μm x 3 μm , (c, c') 2D and 3D images - scanned area 5 μm x 5 μm , (d, d') 2D and 3D images - scanned area 60 μm x 60 μm

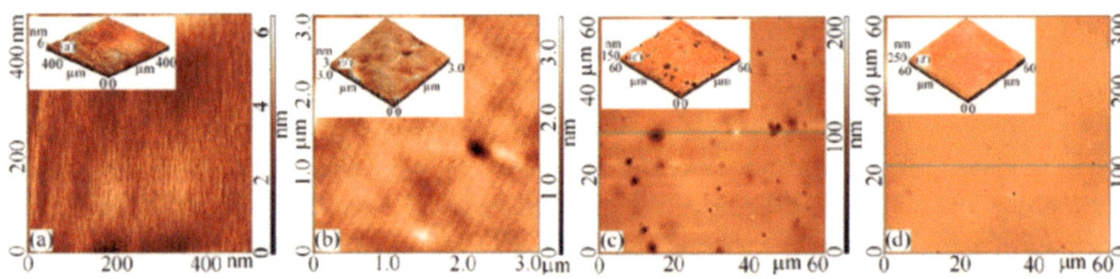

Fig. 2. AFM images of PSFQ1 membranes obtained from: (a, a') 60/40 DMF/water, 2D and 3D images - scanned area 500 nm x 500 nm ; (b, b') 60/40 DMF/water, 2D and 3D images - scanned area 3 μm x 3 μm ; (c, c') 60/40 DMF/water, 2D and 3D images - scanned area 60 μm x 60 μm ; (d, d') 40/60 DMF/water, 2D and 3D images - scanned area 60 μm x 60 μm

A higher charge density in PSFQ2 determines the appearance of nodules, as illustrated in Figure 3 (a, a' and b,b') . The presence of water as a nonsolvent in solutions used for casting membranes influences the AFM images; increasing the water content leads to lower area pores for both polymer membranes under study (Figure 2 (c, c' and d,d') for PSFQ1 membranes, and Figure 3 (c, c' and d,d') for PSFQ2 membranes) and to a lower number of nodules in the PSFQ2 membranes (Figure 3 (c, c' and d, d')). In addition, it may be assumed that the association phenomena of MeOH and water over different composition domains of their mixtures might be one of the factors affecting the morphology of the PSFQ2 membrane surface. Thus, mixed solvents are formed from water and MeOH associated with water, at high water contents; in contrast, at high MeOH compositions, the mixtures consist largely of MeOH and water associate with MeOH. These phenomena change the PSFQ2 solubility, and determine the modification of the solution properties (Filimon et al., 2007), as well as morphology - Figure 3 (c, c′ and d, d′).

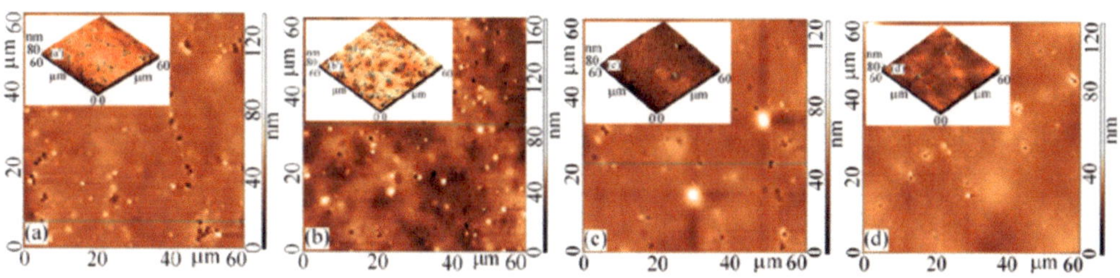

Fig. 3. AFM images of PSFQ2 membranes at scanned area 60 μm x 60 μm , obtained from: (a, a′) 2D and 3D images, 60/40 MeOH/DMF; (b, b′) 2D and 3D images, 20/80 MeOH/DMF; (c, c′) 2D and 3D images, 60/40 MeOH/water; (d, d′) 2D and 3D images, 20/80 MeOH/water

Table 5 gives the average values of the pore characteristics and surface roughness parameters of membranes prepared from different solvent/nonsolvent mixtures. On the other hand, obviously, the number of nodules from the AFM images of the PSFQ2 membranes increases as the nonsolvent content increases.

Samples	Cast solvents	Pore characteristics				Surface roughness		
		Area	Vol.	Depth	Diameter	rms	nhp	nhh
PSFQ1	60/40 DMF/MeOH	0.499	74.72	237.62	0.793	12.36	279	253
	20/80 DMF/MeOH	1.441	83.47	88.54	1.350	18.36	205	180
	60/40 DMF/water	0.342	8.85	46.21	0.665	8.83	170	135
	40/60 DMF/water	0.165	4.60	45.74	0.460	6.53	200	190
PSFQ2	60/40 MeOH/DMF	0.186	1.00	6.82	0.489	5.45	66	50
	20/80 MeOH/DMF	0.216	0.88	6.99	0.528	5.62	64	42
	60/40 MeOH/water	0.485	5.84	20.07	0.783	9.72	104	50
	20/80 MeOH/water	0.145	0.25	3.01	0.431	2.72	45	30

Table 5. Pore characteristics including the area (μm x μm), volume (μm x μm x nm), depth (nm), and diameter (μm), and surface roughness parameters including the root-mean-square roughness (rms, nm), nodule height from the height profile (nhp, nm), and nodule average height from the histogram (nhh, nm) of membranes prepared from the PSFQs with different solvent mixtures

On the other hand, Figures 4 and 5 exemplify the bi- and three-dimensional structures evidence by AFM investigations of PSF-DMEA films prepared with 100/0, 75/25, 50/50 and 25/75, and also with 75/25, 50/50 and 40/60 of DMF/MeOH and DMF/water compositions solvent mixtures, respectively. According to the AFM images, increasing the nonsolvent content in the casting solutions favores modification of surface morphology.

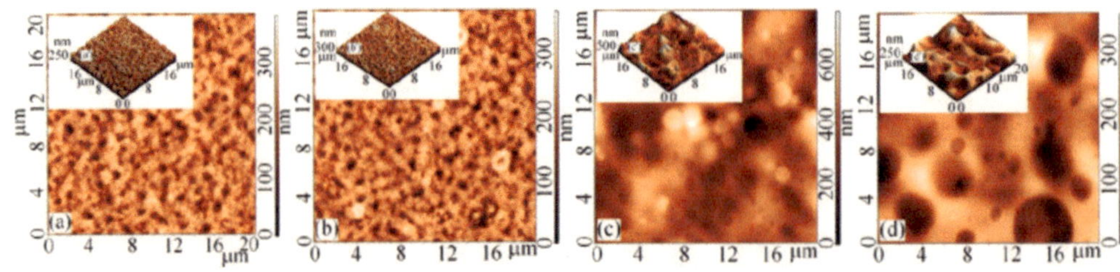

Fig. 4. 2D and 3D AFM images with 20x20 μm^2 scanned areas of the PSF-DMEA films obtained from DMF/MeOH solutions: (a, a') - 100/0; (b, b') - 75/25; (c, c') – 50/50; (d, d') – 25/75

Thus, Figure 4 and Table 6 show that average surface roughness attains a maximum value at 50/50 DMF/MeOH, and favores the appearance of the smallest number of pores with highest depth values. Also, the area, diameter, length and mean width increase with increasing the nonsolvent content. It should be noted that the thermodynamic quality of the solvent mixtures over the studied domain increases with the addition of nonsolvent, at approximately 50/50 DMF/MeOH becoming constant, while the preferential adsorption of nonsolvent takes a maximum value, according to literature data (Filimon et al., 2010). In addition, the presence of water as a nonsolvent in the solutions used for casting films influenced the AFM images presented in Figure 5; a higher water content decreases the thermodynamic quality of the DMF/water solvent mixtures so that, at 50/50 DMF/water, a minimum value of surface roughness and a maximum number of pores with minimum values of area, depth, diameter, length and mean width, are observed.

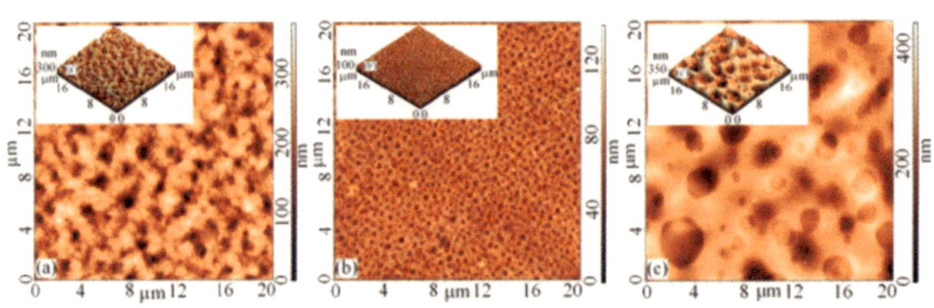

Fig. 5. 2D and 3D AFM images with 20x20 μm^2 scanned areas of the PSF-DMEA films obtained from DMF/water solutions: (a, a') - 75/25; (b, b') – 50/50; (c, c') – 40/60

Figure 6 plot also two- and three-dimensional structures of PSF–DMOA films prepared with DMF/MeOH solvent mixtures of various compositions (100/0, 75/25, 50/50, and 45/55). According to these images, increasing the nonsolvent content in the casting solutions favores modification of the surface morphology. Thus, the average surface roughness decreased with increasing MeOH content and favores increases in the pore number and pore characteristics (area, depth, diameter, and mean width) according to Table 6. The thermodynamic quality of solvent mixtures over the studied domain increases with the addition of the nonsolvent (Filimon et al. 2010; Ioan et al., 2011a).

Solvent mixtures	Pore characteristics						Surface roughness		
	Number of pores	Area	Depth	Diameter	Length	Mean width	Sa	rms	nhh
PSF-DMEA, DMF/MeOH									
100/0	234	0.24	272.78	0.57	0.86	0.31	33.97	42.87	231.14
75/25	268	0.27	259.59	0.62	0.94	0.32	31.89	41.47	217.26
50/50	52	0.47	348.25	0.78	1.09	0.47	74.09	95.12	402.61
25/75	37	1.10	242.07	1.18	1.73	0.62	48.36	59.80	193.33
PSF-DMEA, DMF/water									
75/25	147	0.51	245.19	0.86	1.25	0.48	37.41	47.76	227.79
50/50	1080	0.06	89.57	0.23	0.47	0.15	6.87	9.27	77.43
40/60	42	2.19	230.10	1.64	2.43	0.86	44.98	57.61	281.80
PSF-DMOA, DMF/MeOH									
100/0	9	10.43	25.64	3.33	6.24	1.45	14.11	16.82	58
75/25	-	-	-	-	-	-	14.43	19.90	82
50/50	34	0.70	11.37	0.89	1.56	0.42	3.09	4.41	17
45/55	44	0.80	16.02	0.98	1.53	0.49	2.77	4.24	28
PSF-DMOA, DMF/water									
75/25	5	3.12	13.65	1.97	3.26	0.96	1.59	2.34	15
60/40	27	0.04	19.54	0.22	0.35	0.11	9.19	11.83	70
50/50	18	2.09	5.55	1.46	2.63	0.64	1.52	2.17	8

Table 6. Pore characteristics, including number of pores, area (μm^2), depth (nm), diameter (μm), length (μm), and mean width (μm), and surface roughness paramaters, including average roughness (Sa, nm), root mean square roughness (rms, nm), and nodule average height from the histogram (nhh, nm) of PSF-DMEA and PSF-DMOA films prepared from solutions in DMF/MeOH and DMF/water, with $20 \times 20\ \mu m^2$ scanned areas, corresponding to the 2D AFM images (Ioan et al., 2011)

On the other hand, the presence of water as a nonsolvent in the solutions used for casting films influenced the AFM images presented in Figure 7; a higher water content decreased the thermodynamic quality of the DMF/water solvent mixtures.

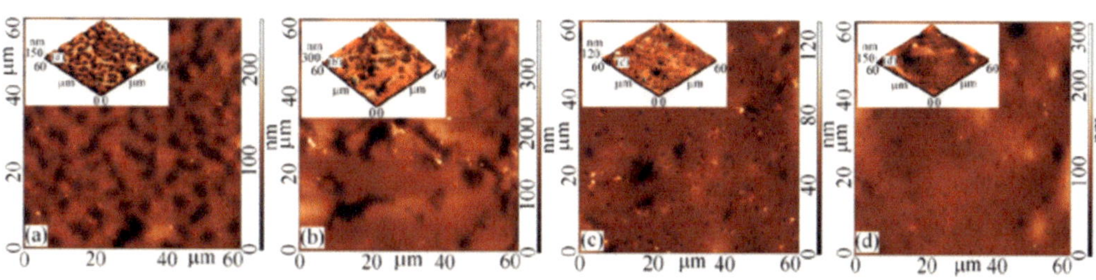

Fig. 6. 2D and 3D AFM images with $60 \times 60\ \mu m^2$ scanned areas of the PSF-DMOA films obtained from DMF/MeOH solution: (a, a') 100/0; (b, b') 75/25; (c, c') 50/50; (d, d') 45/55

Therefore, for a 60/40 DMF/water mixture, the average surface roughness, number of pores, and pore depth are maximum with a minimum area (Table 6). It can be assumed that the specific interactions with the mixed solvents employed in this study modify the PSF-DMOA solubility and determine the modification of the solution properties.

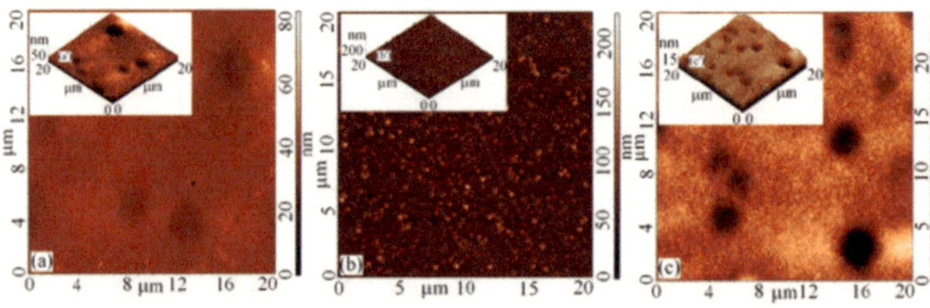

Fig. 7. 2D and 3D AFM images with 20 x 20 μm^2 scanned areas of the PSF-DMOA films obtained from DMF/water solution: (a, a') 75/25; (b, b') 60/40; (c, c') 50/50

In the same context, from AFM images (Figure 8) one can see that the bulky phosphorus pendant groups in PSF-DOPO determine the formation of domains with pores in an approximately continuous matrix (Ioan et al., 2011b).

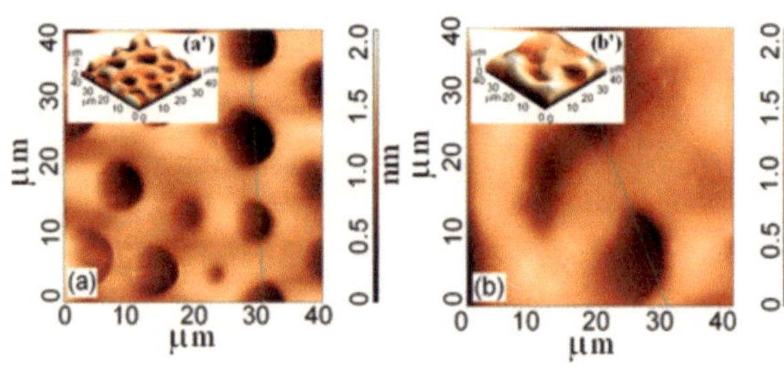

Fig. 8. 2D and 3D-AFM images for PS-DOPO-1 (a, a') and PS-DOPO-2 (b, b')

Samples	Pore characteristics				Surface roughness		
	Area	Volume	Depth	Diameter	rms	nhp	nhh
PS-DOPO-1	23.25	14.39	1.51	5.02	363.16	0.45	1.40
PS-DOPO-2	28.96	15.19	0.95	7.61 6.00*	343.11	0.70	1.30
PS-DOPO-3				25.00*			

* From scanning electron microscopy, according to previous data (Petreus *et al.*, 2010)

Table 7. Pore characteristics including area (μm x μm), volume (μm x μm x nm), depth (μm), and diameter (μm), and surface roughness parameters including root-mean-square roughness (rms, nm), and nodules height from the height profile (nhp, μm) and nodules average height from the histogram (nhh, μm) of polysulfone with bulky phosphorus pendant groups membranes obtained from AFM investigations

Pore characteristics (including area, volume, depth, and diameter) and surface roughness parameters (including root-mean-square roughness (rms), and nodules height from the height profile (nhp) from profile analysis, and nodules average height, from the histogram (nhh)), are presented in Table 7. The dimensions of pores increase and their depth decreases with increasing the substitution degrees, so that, for the PS-DOPO-3 sample, the dimensions exceed the limit of the AFM apparatus.

On the other hand, AFM studies support the conclusion that the increasing density of bulky phosphorus pendant groups from the side chain and decreases roughness and, implicitly, confirm poor adhesion of the modified polysulfone films. Also, the bulky phosphorus pendant groups, found in position 1* for samples PS-DOPO-1 and PS-DOPO-2, and in positions 1* and 2* for sample PS-DOPO-3, modify the rigidity and hydrophobicity, and determine different forms of entanglement in polymer solutions, influencing membrane morphology and AFM images.

4. Surface tension parameters

Surface tension parameters of quaternized polysulfones are calculated by the geometric mean method (GM) (equation (1)) (Kälble et al., 1969; Owens et al., 1969) and the acid/base method (LW/AB) (equation (2)) (van Oss et al, 1988a; van Oss et al, 1988b). In this context, the surface tension data of the doubly-distilled water (W), ethylene glycol (EG), glycerol (G) and formamide (FA) used in the calculations of the surface tension parameters of PSFQ are presented in Table 8.

$$\frac{1+\cos\theta}{2}\cdot\frac{\gamma_{lv}}{\sqrt{\gamma_{lv}^{d}}}=\sqrt{\gamma_{sv}^{p}}\cdot\sqrt{\frac{\gamma_{lv}^{p}}{\gamma_{lv}^{d}}}+\sqrt{\gamma_{sv}^{d}}\ ;\ \gamma_{sv}=\gamma_{sv}^{d}+\gamma_{sv}^{p} \qquad (1)$$

where θ is the contact angle determined for test liquids, subscripts "lv" and "sv" denote the interfacial tension between liquid-vapor and surface-vapor, respectively, while superscripts "p" and "d" denote the polar and disperse components, respectively, of total surface tension, γ_{sv}.

$$1+\cos\theta=\frac{2}{\gamma_{lv}}\cdot\left(\sqrt{\gamma_{sv}^{LW}\cdot\gamma_{lv}^{LW}}+\sqrt{\gamma_{sv}^{+}\cdot\gamma_{lv}^{-}}+\sqrt{\gamma_{sv}^{-}\cdot\gamma_{lv}^{+}}\right);\ \gamma_{sv}^{LW/AB}=\gamma_{sv}^{LW}+\gamma_{sv}^{AB} \qquad (2)$$

where $\gamma_{sv}^{AB}=2\cdot\sqrt{\gamma_{sv}^{+}\cdot\gamma_{sv}^{-}}$, superscript "LW/AB" indicates the total surface tension, and also, superscript "AB" and "LW" represent the polar component obtained from the electron-donor, γ_{sv}^{-}, and the electron-acceptor, γ_{sv}^{+}, interactions, and the disperse component, respectively.

Literature (Della Volpe et al., 2004) shows that an improper utilization of three liquids without dispersive liquids, or with liquids prevalently basic or prevalently acidic, strongly increases the ill-conditioning of the system. In addition, the contact angles should be measured with a liquid whose surface tension is higher than the anticipated solid surface tension (Kwok et al., 2000).

Table 9 shows the contact angle values between water, ethylene glycol, glycerol or formamide and PSF, CMPSF and PSFQ membranes. The surface tension of PSF evidences the lowest hydrophilicity, induced by the aromatic rings connected by one carbon and two methyl groups, oxygen elements, and sulfonic groups, while chloromethylation of PSF with the functional group -CH_2Cl increases hydrophilicity (see the values of surface tension for PSF and CMPSF in Table 10). Moreover, the results indicate that the PSFQ membranes are the most hydrophilic ones from the studied samples (lowest water contact angle), due to the N,N-dimethylethanolamine hydrophilic side groups. Hence, it is observed that total surface tension, γ_{sv} or $\gamma_{sv}^{LW/AB}$, and the polar component, γ_{sv}^{p} or γ_{sv}^{AB}, increase with the degree of substitution of CMPSF and with the quaternization degree of the ammonium groups for PSFQ samples.

Test liquids	γ_{lv}	γ_{lv}^{d}	γ_{lv}^{p}	γ_{lv}^{-}	γ_{lv}^{+}
Water (W) (Ström et al., 1987; (Rankl et al., 2003))	72.8	21.8	51.0	25.50	25.50
Ethylene glycol (EG) (Yildirim et al., 1997)	48.0	29.0	19.0	47.00	1.92
Glycerol (G) (van Oss et al., 1989)	64.0	34.0	30.0	57.40	3.92
Formamide (FA) (van Oss et al., 1989)	58.0	39.0	19.0	39.60	2.28
Methylene iodide (MI) (Rankl et al., 2003)	50.80	50.80	0	0.72	0
1-Brom-naphtalin (1-Bn) (Rankl et al., 2003)	44.40	44.40	0	0	0
Red blood cell (Vijayanand et al., 2005)	36.56	35.20	1.36	0.01	46.2
Platelet (Vijayanand et al., 2005)	118.24	99.14	19.10	12.26	7.44

Table 8. Surface tension parameters (mN/m) of the liquids used for contact angle measurements: total surface tension, γ_{lv}, disperse component of surface tension, γ_{lv}^{d}, polar component of surface tension, γ_{lv}^{p}, electron-donor contribution on polar component, γ_{lv}^{-}, and electron-acceptor contribution on polar component, γ_{lv}^{+}

Polymers \ Solvents	W	G	FA	EG
PSF	79	70	65	60
CMPSF1	78	72	66	57
CMPSF2	73	69	65	54
CMPSF3	71	68	64	52
PSFQ1	64	63	64	51
PSFQ2	64	56	69	48
PSFQ3	32	28	32	29

Table 9. Contact angle of different probe liquids (in °)

Therewith, the relative ratio of the polar component to the total surface tension ranges from approx. 39% for PSF and 43% for CMPSF1, with the substitution degree DS<1, to 57-61% for CMPSF, with the substitution degree DS>1, and to 79% for PSFQ. The apolar component, γ_{sv}^{d}, decreases from PSF to CMPSF and PSFQ. The total surface tensions of PSF and CMPSF1, where the substitution degree DS<1, are dominated by the apolar component, while the total surface tension of CMPSF2, CMPSF3, (DS>1) and PSFQ are dominated by the polar term, with the electron donor interactions, γ_{sv}^{-}, smaller than the electron acceptor ones, γ_{sv}^{+}. Thus, the functional groups -CH_2Cl attached by chloromethylation process increase the polarity for chloromethylated polysulfones with a substitution degree DS>1; also, the N, N-dimethylethanolamine side groups introduced by the quaternization process increase the polarity. The change is moderate for chloromethylated polysulfone with DS>1, with different chloride content.

Samples	GM method			LW/AB method				
	γ_{sv}^{d}	γ_{sv}^{p}	γ_{sv}	γ_{sv}^{LW}	γ_{sv}^{AB}	γ_{sv}^{-}	γ_{sv}^{+}	$\gamma_{sv}^{LW/AB}$
PSF	17.81	11.21	28.49	18.77	9.34	8.74	28.10	33.23
CMPSF1	16.13	12.32	28.45	21.27	7.46	11.43	28.73	25.97
CMPSF2	13.23	17.56	30.79	16.85	12.38	17.02	29.23	42.35
CMPSF3	12.66	19.43	32.09	15.79	14.14	17.56	29.93	47.23
PSFQ1	8.12	29.88	38.00	12.70	19.29	27.24	31.98	60.31
PSFQ2	8.04	31.11	39.15	8.15	42.56	29.98	43.71	97.38
PSFQ3	8.08	31.20	39.18	7.82	56.82	49.34	64.74	82.70

Table 10. Surface tension parameters (mN/m) for polysulfone (PSF) film , chloromethylated polysulfone films, (CMPSF1, CMPSF2 and CMPSF3) prepared from solutions in chloroform, and quaternized polysulfone films (PSFQ1, PSFQ2 and PSFQ3) prepared from solutions in methanol

Moreower, the surface tension parameters of quaternized polysulfones PSF-DMEA, and PSF-DMOA, with surface properties of water, methylene iodide, 1-brom-naphtalin test liquids from Table 8, and the contact angles measured between these solvents and quaternized polysulfone films from Table 11, are presented in Table 12.

Solvent mixtures	Contact angle		
	W	MI	1-Bn
PSF-DMEA			
100/0 DMF/MeOH	71	28	17
75/25 DMF/MeOH	70	31	22
50/50 DMF/MeOH	61	30	20
25/75 DMF/MeOH	63	31	24
75/25 DMF/water	59	35	21
50/50 DMF/water	60	33	18
40/60 DMF/water	56	33	16
PSF-DMOA			
100/0 DMF/MeOH	72	32	15
75/25 DMF/MeOH	78	32	16
50/50 DMF/MeOH	58	28	16
45/55 DMF/MeOH	77	35	20
75/25 DMF/water	66	34	21
60/40 DMF/water	65	36	25

Table 11. Contact angle (°) of different liquids on films prepared from solutions of PSF-DMEA and PSF-DMOA in DMF/MeOH and DMF/water

These parameters are influenced by the solvent/nonsolvent composition from which the films are prepared (Albu et al., 2011; Ioan et al. 2011a). Some studies have reported that the chain shape of a polymer in solution could affect the morphology of the polymer in bulk. In this context, the conformations of both PSF-DMEA and PSF-DMOA are affected by the charged groups from different alkyl radicals of the studied quaternized samples, and also by the compositition of the solvent mixtures. For the PSF-DMEA in the DMF/MeOH system, one may observe that the polymer coil dimension decreases with increasing the DMF content in the 0.25–1 volume fraction domain; below a 0.25 volume fraction of DMF, the polymer precipitates. For the same polymer, but in a DMF/water solvent mixture, the dimensions increase with increasing the DMF content, starting from approximately the same volume fraction of DMF. The PSF-DMOA coil dimensions possess maximum values in DMF/MeOH and DMF/water, around 0.6 and 0.8 DMF volume fractions, respectively. For volume fractions of DMF below 0.25 in DMF/MeOH and 0.5 in DMF/water, the PSF-DMOA precipitates due to the nature of the alkyl radicals and content of nonsolvent in the system. Also, the values of intrinsic viscosity are higher for PSF-DMOA - with bulky carbon atoms in the alkyl side chain, compared with PSF-DMEA, where the alkyl side chain possesses two carbon atoms. Therefore, for a given composition of the DMF/MeOH and DMF/water solvent mixtures, one of the components is preferentially adsorbed by the quaternized polysulfone molecules in the direction of a thermodynamically most effective mixture (Filimon et al., 2010). These aspects influence the surface properties of the polymer. PSF-DMEA films possess low polar surface tension parameters, but slightly higher than those for PSF-DMOA. The hydrophobic character is given by the ethyl radical from the N-dimethylethylammonium chloride pendant group and by the octyl radical from the N-dimethyloctylammonium chloride pendant group, respectively.

Solvent mixtures	GM method			LW/AB method				
	γ_{sv}^{d}	γ_{sv}^{p}	γ_{sv}	γ_{sv}^{LW}	γ_{sv}^{+}	γ_{sv}^{-}	γ_{sv}^{AB}	$\gamma_{sv}^{LW/AB}$
PSF-DMEA, DMF/MeOH								
100/0	43.7	5.9	49.7	42.5	3.6	2.6	6.2	48.7
75/25	42.5	6.6	49.2	41.3	4.3	2.8	6.9	48.1
50/50	42.9	10.7	53.7	41.8	9.9	2.5	9.9	51.7
25/75	42.1	10.0	52.1	40.6	6.5	4.1	10.4	51.0
PSF-DMEA, DMF/water								
75/25	41.8	12.2	54.0	41.5	21.4	0.1	3.18	44.68
50/50	42.6	11.4	54.0	42.3	19.0	0.2	3.83	46.08
40/60	42.8	13.4	56.3	42.7	25.4	0.2	4.15	46.85
PSF-DMOA, DMF/MeOH								
100/0	40.9	5.1	46.0	42.4	0.4	1.4	1.5	43.9
75/25	39.6	3.0	42.6	42.7	0.2	0.8	0.8	43.5
50/50	43.6	1.9	45.2	43.1	1.5	2.0	1.8	44.9
45/55	39.1	3.6	42.7	41.3	1.0	0.5	1.4	42.7
PSF-DMOA, DMF/water								
75/25	38.4	7.5	46.0	41.4	0.5	1.2	1.5	42.9
60.40	38.5	8.6	47.1	40.2	0.8	3.5	3.3	43.6
50/50	41.4	5.9	47.4	42.8	0.5	1.3	1.6	44.4

Table 12. Surface tension parameters (mN/m) for quaternized polysulfone films PSF-DMEA and PSF-DMOA prepared from solutions in DMF/MeOH and DMF/water

Furthermore, the electron donor interactions, γ_{sv}^{-}, are smaller than the electron acceptor ones, γ_{sv}^{+}, for PSF-DMEA, and electron donor interactions, γ_{sv}^{-}, exceed the electron acceptor interactions, γ_{sv}^{+}, for PSF-DMOA, caused by the inductive phenomena from alkyl radical. The results reflect the capacity of the N-dimethylethylammonium chloride or N-dimethyloctylammonium chloride pendant groups to determine the acceptor or donor character of the polar terms, generated by these inductive phenomena.

In the same context, the substitution degrees of polysulfones with bulky phosphorus pendant groups enface the specific properties. Presence of substituted groups in position 1* for samples PS-DOPO-1 and PS-DOPO-2 and in positions 1* and 2* for sample PS-DOPO-3 influence *e.g.* the rheological properties, leading to modification of slopes as a function of viscosity *vs.* substitution degrees, concentration and temperature (Ioan et al., 2011b). This behavior is assigned to the modification of cohesive energy and molar volume for the studied samples. The Arrhenius equation evidenced an increased of activation energy and flow activation entropy, in the following order: PS-DOPO-1 < PS-DOPO-2 < PS-DOPO-3. This implies a higher energy barrier for the movement of an element of the fluid and a more rigid structure for the PS-DOPO3 polymer chain and, as well as, a strong variation with concentration of the activation energy and flow activation entropy. Thus, different forms of the entanglement in polymer solutions are generated by the specific molecular rearrangement of the polysulfone with bulky phosphorus pendant groups with different substitution degrees, at different concentrations and temperatures. Moreover, the viscoelastic characteristics showed that all samples exhibit the behavior characteristic of an elastic gel at high frequencies, when G′ > G″ and where both moduli are dependent on frequency. These features are characteristic for a physical gel network, in which the gelation process of phosphorus-modified polysulfones is influenced by intramolecular interactions, depending on the substitution degrees and by intermolecular attractions that depend on solution concentration.

Samples	W	F	EG	MI
PS-DOPO-1	77	56	59	43
PS-DOPO-2	78	52	49	21
PS-DOPO-3	82	46	31	≈ 0

Table 13. Contact angle of different probe liquids (water (W), formamide (F), ethylene glycol (EG) and methylene iodide (MI)) in (°) with polysulfones with bulky phosphorus pendant groups

Samples	Surface tension parameters							
	GM method			LW/AB method				
	γ_{sv}^{d}	γ_{sv}^{p}	γ_{sv}	γ_{sv}^{LW}	γ_{sv}^{AB}	γ_{sv}^{+}	γ_{sv}^{-}	$\gamma_{sv}^{LW/AB}$
PS-DOPO-1	33.18	4.80	37.98	34.23	3.99	0.557	7.15	37.22
PS-DOPO-2	44.62	2.34	46.96	39.20	1.508	0.112	5.063	40.710
PS-DOPO-3	52.85	1.14	53.99	48.85	1.378	0.358	1.327	50.228

Table 14. Surface tension parameters (mN/m) for membranes of PS-DOPO with different substitution degrees, prepared from solutions in DMF

On the other hand, according to Table 13 and, finally, Table 14, all phosphorus-modified polysulfones show low polar surface tension parameters, γ_{sv}^{p} or γ_{sv}^{AB}, which decrease with increasing the phosphorus pendant groups, and which manifest bring lower contributions to the total surface tension parameters. Furthermore, the electron donor interactions, γ_{sv}^{-} exceed the electro n acceptor interactions, γ_{sv}^{+}, considering electron-donor capacity of P=O group. These results show that the studied samples evidence high hydrophobicity, which increases by increasing the substitution degrees. The maximum hydrophobicity of the polysulfones with bulky phosphorus pendant groups would be advantageous, *e.g.*, for dielectric performance, since a low water absorption causes a significant decrease in the dielectric constant and, implicitly, low adhesion to different interfaces.

5. Surface and interfacial free energy of functionalized polysulfones

The effect of different radicals of functionalized polysulfones and of the history of the films formed from solutions on surface properties are analyzed by surface free energy, ΔG_w - expressing the balance between surface hydrophobicity and hydrophilicity (equation (3)) (Faibish et al., 2002; Rankl et al., 2003), by interfacial free energy between two particles of functionalized polysulfones in water phase, ΔG_{sws}^{GM} (equations (4) and (5) or (6)), and by the work of spreading of water, W_s (equation (7)).

$$\Delta G_w = -\gamma_{lv} \cdot (1 + \cos\theta_{water}) \tag{3}$$

where γ_{lv} is given in Tables 8 and θ_{water} is give in Table 9, 11 and 13 for functionalized polysulfones with N,N-dimethylethanolamine (PSFQ), N,N-dimethylethylamine and N,N-dimethyloctylamine (PSF-DMEA and PSF-DMOA), and polysulfones with bulky phosphorus pendant groups (DOPO), respectively.

$$\Delta G_{sws}^{GM} \text{ (or } \Delta G_{sws}^{LW/BAB}) = -2 \cdot \gamma_{sl} \text{ (or } \gamma_{sl}^{LW/AB}) \tag{4}$$

$$\gamma_{sl} = \left(\sqrt{\gamma_{lv}^{p}} - \sqrt{\gamma_{sv}^{p}}\right)^2 + \left(\sqrt{\gamma_{lv}^{d}} - \sqrt{\gamma_{sv}^{d}}\right)^2 \tag{5}$$

$$\gamma_{sl}^{LW/AB} = \left(\sqrt{\gamma_{sv}^{d}} - \sqrt{\gamma_{lv}^{d}}\right)^2 + 2\cdot\left(\sqrt{\gamma_{sv}^{+}\cdot\gamma_{sv}^{-}} + \sqrt{\gamma_{lv}^{+}\cdot\gamma_{lv}^{-}} - \sqrt{\gamma_{sv}^{+}\cdot\gamma_{lv}^{-}} - \sqrt{\gamma_{sv}^{-}\cdot\gamma_{lv}^{+}}\right) \tag{6}$$

$$W_s = W_a - W_c = 2\cdot[(\gamma_{sv}^{LW}\cdot\gamma_{lv}^{d})^{1/2} + (\gamma_{sv}^{+}\cdot\gamma_{lv}^{-})^{1/2} + (\gamma_{sv}^{-}\cdot\gamma_{lv}^{+})^{1/2}] - 2\cdot\gamma_{lv} \tag{7}$$

Generally, the literature (Faibish et al., 2002; van Oss, 1994) mentions that for $\Delta G_w < -113$ mJ/m^2, the polymer can be considered more hydrophilic while, when $\Delta G_w > -113$ mJ/m^2, it should be considered more hydrophobic. It is observed from Figure 9 that the surface free energy for PSF and CMPSF samples possesses low wettability . Moreover, Figures 9, 10 and Table 15 show that all functionalized polysulfones are characterized by small wettability, except for PSFQ3, where the hydrophylicity is higher. On the other hand, N,N-dimethyloctylamine groups determine a higher wettability as N,N-dimethylethylamine group (coresponding to PSF-DMOA and PSF-DMEA, respectively) and these wettability depends on the surphace morphology of samples prepared from different composition of DMF/MeOH and DMF/water solvent mixtures (see Table 6). Also, a small difference appears between PS-DOPO samples with different substitution degrees. Thus, one can say that increasing the substitution degrees lead to a slight increase in hydrophobicity.

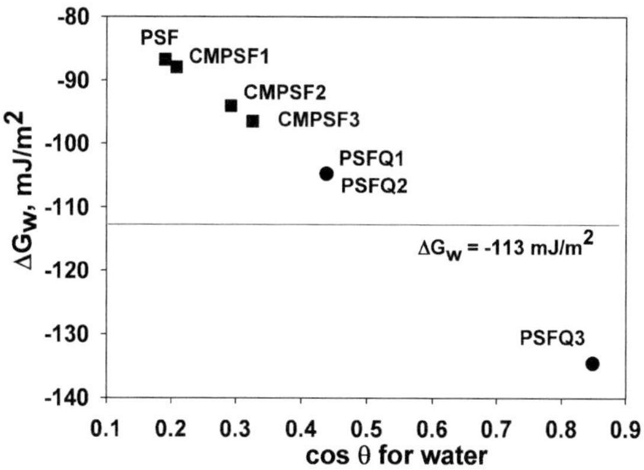

Fig. 9. Surface free energy *vs.* water contact angle for quaternized polysulfones with N,N-dimethylethanolamine

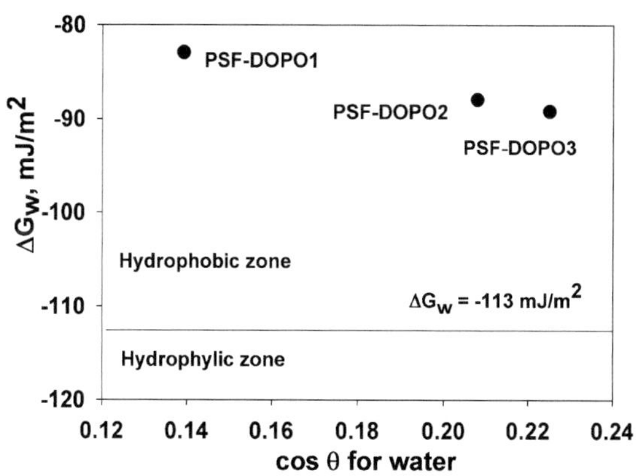

Fig. 10. Surface free energy vs. water contact angle for polysulfones with bulky phosphorus pendant groups

The interfacial free energy, ΔG_{sws}^{GM} evaluated from solid-liquid interfacial tension, γ_{sl}, has negative values (see Figure 11 - for PSFQ, Table 15 - for PSF-DMEA and PSF-DMOA, and Table 16 – for PSF-DOPO), indicating an attraction between the two polymer surfaces, s, immersed in water, w; hence, these materials are considered rather hydrophobic (Dourado et al., 1998; van Oss & Giese, 1995).

In addition, the hydrophobicity of these polymers is described by the work of spreading of water, W_s, over the surface, which represents the difference between the work of water adhesion, W_a, and the work of water cohesion, W_c (equation (7)).

Solvent mixtures	γ_{sl}		ΔG_w		ΔG_{sws}^{GM}	
	DMF/MeOH					
	PSF-DMEA	PSF-DMOA	PSF-DMEA	PSF-DMOA	PSF-DMEA	PSF-DMOA
100/0	25.95	26.92	-67	-95.30	-51.90	-53.84
75/25	24.26	32.06	-68	-87.94	-48.52	-64.12
50/50	18.47	37.15	-75	-111.38	-36.94	-74.30
45/55	-	29.91	-	-89.18	-	-59.82
25/75	19.15	-	-74	-	-38.30	-
	DMF/water					
75/25	16.50	21.66	-77	-102.41	-32.30	-43.32
60/40	-	20.08	-	-103.57	-	-40.16
50/50	17.60	25.30	-76	-97.70	-35.20	-50.60
40/60	15.57	-	-79	-	-31.14	-

Table 15. Water interfacial tensions and surface free energy for PSF-DMEA and PSF-DMOA films prepared in different DMF/MeOH and DMF/water, and interfacial free energy between two particles of quaternized polysulfones in water phase

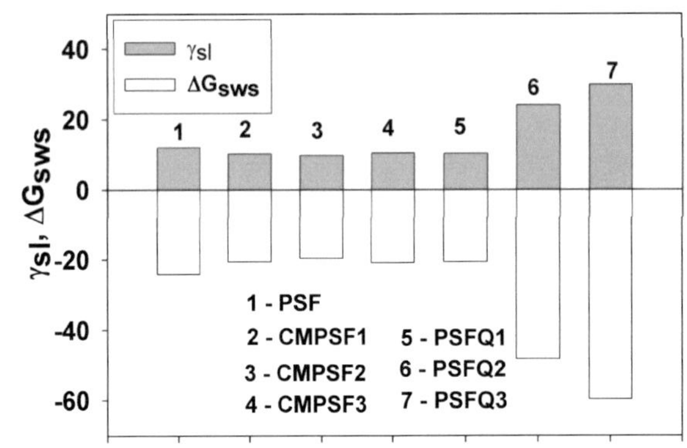

Fig. 11. Water interfacial tensions and surface free energy for polysulfone, chloromethylated polysulfones, and quaternized polysulfones films prepared in DMF/MeOH and DMF/water

Samples	γ_{sl}	$\gamma_{sl}^{LW/AB}$	ΔG_{sws}^{GM}	$\Delta G_{sws}^{LW/AB}$	W_s
PS-DOPO-1	25.69	21.84	-51.38	-43.69	-56.42
PS-DOPO-2	35.54	28.93	-71.08	-57.86	-61.03
PS-DOPO-3	43.65	40.09	-87.30	-80.17	-62.66

Table 16. Solid-water interfacial tensions, γ_{sl}, and interfacial free energy, ΔG_{sws}^{GM}, from the geometrical mean method (GM) and solid-water interfacial tensions, $\gamma_{sl}^{LW/AB}$ and interfacial free energy, $\Delta G_{sws}^{LW/AB}$, from the acid/base method (LW/AB), and the work of spreading of water, W_s, over the polymer surface

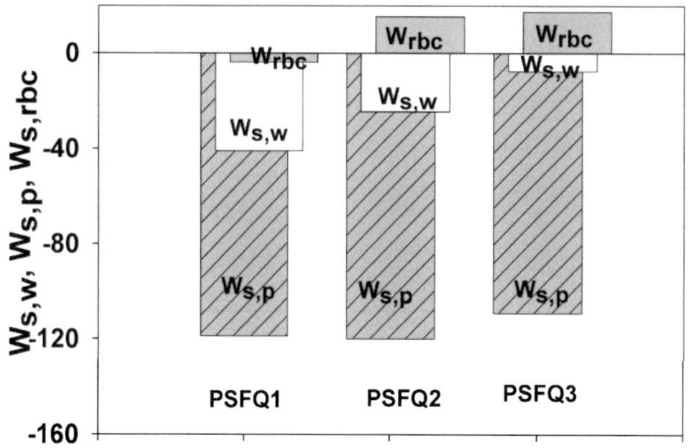

Fig. 12. Work of spreading of water, red blood cells and platelets over the surface of PSFQ1, PSFQ2 and PSFQ3 films

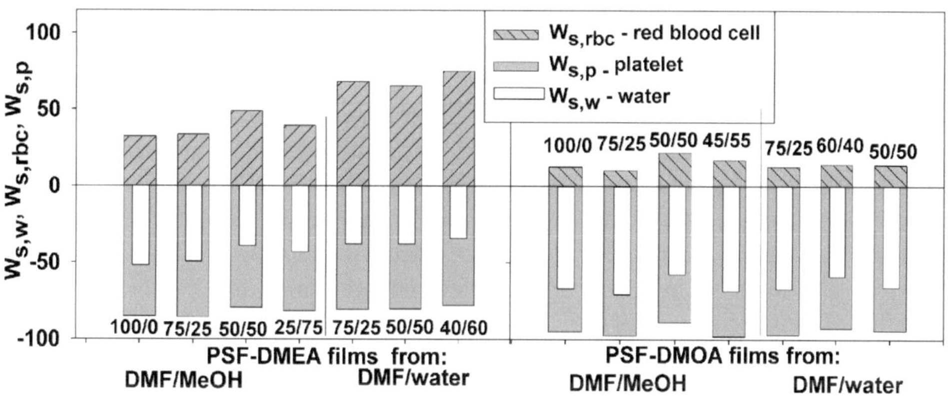

Fig. 13. Work of spreading of water, red blood cells and platelets over the surface of PSF-DMEA and PSF-DMOA films prepared in DMF/MeOH and DMF/water

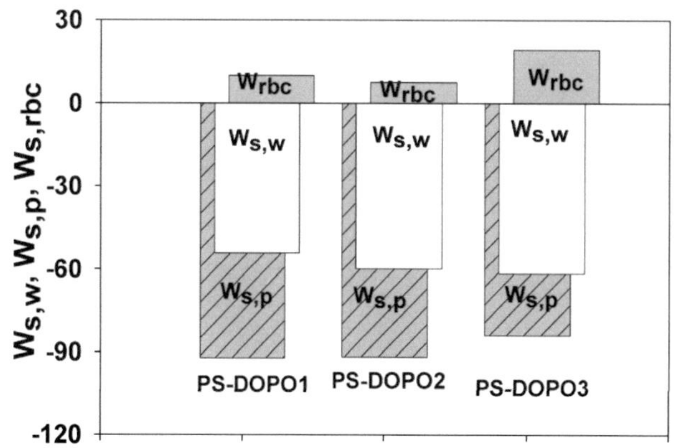

Fig. 14. Work of spreading of water, of red blood cells and of platelets over the surface of PSF-DOPO1, PSF-DOPO2 and PSF-DOPO3 films

According to the negative values of the interfacial free energy of all polysulfone samples, the work of spreading of water, $W_{s,w}$, takes negative values, caused by the hydrophobic surfaces, where the work of water adhesion is low, comparatively with the work of cohesion; at the same time, be noticed that adhesion and implicitly work of spreading of water, W_s, increases with increasing substitution degree for PSFQ samples ($W_{s,w,PSFQ1} < W_{s,w,PSFQ2} < W_{s,w,PSFQ3}$), is lower for PSF-DMOA than for PSF-DMEA, and decreases with increasing of substitution degrees for PSF-DOPO ($W_{s,w,PS-DOPO1} < W_{s,w,PS-DOPO2} < W_{s,w,PS-DOPO3}$).

6. Blood - functionalized polysulfone interactions

Blood compatibility is dictated by the manner in which their surfaces interact with blood constituents, like red blood cells and platelets. To analyze the possibilities of using the functionalized polysulfones in biomedical applications, and for establishing its compatibility with blood, equation (7) is used, where $W_{s,rbc}$ and $W_{s,p}$ describe the work of spreading of red blood cells and platelets (Vijayanand et al., 2005); when blood is exposed to a biomaterial surface, adhesion of cells occurs and the extent of adhesion decides the life of the implanted biomaterials; thus, cellular adhesion to biomaterial surfaces could activate coagulation and the immunological cascades. Therefore, cellular adhesion has a direct bearing on the thrombogenicity and immunogenicity of a biomaterial, and thus dictates its blood compatibility. The work of adhesion of the red blood cells can be considered as a parameter for characterizing biomaterials *versus* cell adhesion. The materials which exhibit a lower work of adhesion would lead to a lower extent of cell adhesion than those with a higher work of adhesion. Considering the surface energy parameters (γ_{lv}, γ_{lv}^{d}, γ_{lv}^{+}, γ_{lv}^{-}) given in Table 8 for red blood cells and platelets, the work of spreading of blood cells and platelets is estimated by equation (7), with surface tension parameters listed in Table 10 for PSFQ, Table 12 for PSF-DMEA and PSF-DMOA, and in Table 14 for PSF-DOPO.

Figures 12, 13 and 14 show positive values for the work of spreading of red blood cells, $W_{s,rbc}$, and negative values for the work of spreading of platelets, $W_{s,p}$, suggesting a higher work of adhesion comparatively with that of cohesion for the red blood cells, but a smaller work of adhesion comparatively with the one of cohesion for platelets. Among these polysulfones, the work of spreading of red blood cells, $W_{s,rbc}$ exhibits a negative value for the PSFQ1, generated by lower hydrophobicity, and a highest value for the PSF-DMEA, characterized by high disperse parameter of surface tension.

These results suggest that the exposure of platelets to functionalized polysulfone films determines an increase of platelets cohesion, and that a good hydrophobicity can be correlated with a good adhesion of the red blood cells on the surface of the polysulfone films.

In summary, both red blood cells and platelets are extremely important in deciding the blood compatibility of a material. Moreover, it is known that adhesion of the red blood cells onto a surface, *e.g.* modified polysulfones, requires knowledge of the interactions with the vascular components. Thus, endothelial glycocalyx along with the mucopolysaccharides adsorbed to the endothelial surface of the vascular endothelium reject clotting factors and platelets - which have a significant role in thrombus formation (Reitsma et al., 2007). In this context, adhesion of the red blood cells and cohesion of platelets to surface films must be discussed in correlation with future specific biomedical applications. These results seem to be applicable for evaluating bacterial adhesion to the surfaces, and could be subsequently employed for studying possible implanted induced infections, or for obtaining semipermeable membranes.

7. Antimicrobial activity assessments

The literature shows that quaternary ammonium salts represent one of the most popular types of antimicrobial agents (Hazziza-Laskar et al., 2005; Xu et al., 2006; Yu et al., 2007; Merianos, 1991). Their biological activity, depending on their structure and physico-chemical properties, affects the interaction with the cytoplasmic membrane of bacteria and influences cell metabolism.

The antimicrobial activity of quaternized polysulfones with various ionic chlorine contents, in DMSO, at different concentrations, is investigated against *Escherichia coli* (*E. coli*) and *Staphylococcus aureus* (*S. aureus*). As shown in Figure 15 and Table 17, these polymers inhibit the growth of microorganisms, the inhibition becoming stronger with increasing the polycationic nature of the modified polysulfone and with the polymer solution concentration. Also, for all solutions, the inhibition is intense compared to DMSO, which is used as a control sample.

Fig. 15. Antimicrobial screening tests: (a) PSFQ1 in DMSO, c = 1.08 g/L, against *Escherichia coli*; (b) PSFQ2 in DMSO, c = 1.10 g/L, against *Escherichia coli*; (c) PSFQ1 in DMSO, c = 1.08 g/L, against *Staphylococcus aureus*; (d) PSFQ2 in DMSO, c = 1.10 g/L, against *Staphylococcus aureus*. In each figure, the inhibition area on the right side is recorded for DMSO as a control sample

Microorganism	PSFQ1		PSFQ2		Control (DMSO)
	0.52 g/L	1.08 g/L	0.52 g/L	1.1 g/L	
E. coli	13	19	22	23	12
S. aureus	11	13	15	18	10

Table 17. Antimicrobial activity expressed by the diameter of the inhibition zone (mm) of PSFQ1 and PSFQ2 in DMSO at two concentrations and of DMSO used as a control sample against *Escherichia coli* and *Staphylococcus aureus*

The cationic modified polysulfones with quaternary ammonium group interfere with the bacterial metabolism by electrostatic stacking at the cell surface of the bacteria (Xu et al., 2006). This conclusion is evaluated in terms of the diameter of the inhibition zone presented in Table 17. The results show that the bacterial activity of the tested compounds is dependent on the microorganism nature. Thus, the *E. coli* is found be much more sensitive to the investigated polymers than the *S. aureus*.

On the other hand, the differences in the composition of the cell wall of Gram-negative (*E. coli*) and Gram-positive (*S. aureus*) bacteria cause different resistance to killing by antimicrobial agents. It is known that the component of Gram-positive bacteria cell walls is peptidoglycan, whereas the major constituents of Gram-negative bacteria cell walls are peptidoglycan, together with other membranes, such as lipopolysaccharides and proteins. These components of the cell walls generate the hydrophilic character of *E. coli* bacteria and the hydrophobic character of *S. aureus*. Thus, the different inhibiting effects of PSFQ1 and PSFQ2 on the growth of the tested *E. coli* and *S. aureus* bacteria may be due to different antimicrobial activities.

Therefore, all these aspects indicate that the antimicrobial activity depends not only on the substituent groups of the quaternized polysulfones but also on the hydrophilic or hydrophobic character of the bacteria, which generated different interactions of the quaternary ammonium salt groups with the bacterial cell membrane. In particular, the adhesion of the relatively hydrophilic *E. coli* to the hydrophilic quaternized polysulfones is higher than the adhesion of hydrophobic *S. aureus* cells.

8. Conclusion

New quaternized polysulfones, prepared by quaternization of chloromethylated polysulfone with N,N-dimethylethanolamine, N,N-dimethylethylamine, N,N-dimethyloctylamine and also polysulfones with bulky phosphorus pendant groups, obtained by chemical modification of the chloromethylated polysulfones by reacting the chloromethyl group with the P-H bond of 9,10-dihydro-oxa-10- phosphophenanthrene-10-oxide are investigated to obtain information on their hydrophilic/hydrophobic properties and blood compatibility. The history of the formed films, prepared by a dry-cast process in pure solvents or in different solvent/nonsolvent mixtures, influenced the surface tension parameters, surface and interfacial free energy and the work of spreading of water, maintaining the surfaces hydrophobic characteristics of polysulfones. On the other hand, the results reflect the capacity of N-dimethylethanolammonium chloride pendant group to determine the acceptor character of the polar terms, capacity of N-dimethylethylammonium or N-dimethyloctylammonium chloride pendant groups to determine the acceptor or donor character of the polar terms, and capacity of P-H bond of 9,10-dihydro-oxa-10-phosphophenanthrene-10-oxide to determine the donor character of the polar terms, caused by the inductive phenomena of different pendant groups.

The AFM images show that surface morphology is characterized by roughness and nodules formations, depending on the composition of solvent/nonsolvent mixtures, including the characteristics of polysulfones and the thermodynamic quality of the solvents. Moreover, the results suggest that:

- surface hydrophobicity and surface roughness are the parameters controlling the compatibility with the red blood cells and platelets: a good hydrophobicity can be correlated with a good adhesion of the red blood cells and with a good cohesions of the platelets on the surface of the polysulfone films;
- high work of adhesion comparatively with work of cohesion for the red blood cells, but a smaller work of adhesion comparatively with the one of cohesion for platelets is obtained. Among these polysulfones, the work of spreading of red blood cells, $W_{s,rbc}$ exhibits a negative value for the quaternized polysulfone with N,N-dimethylethanolamine pendant group and 2.15 ionic chlorine content (PSFQ1), generated by lower hydrophobicity, and a highest value for the quaternized polysulfone with N,N-dimethylethylamine pendant group (PSF-DMEA), characterized by high disperse parameter of surface tension.

On the other hand, the antimicrobial activity of polysulfones with quaternary ammonium groups is considered to be one of the important properties that are directly related to new possible applications. In this study, adhesion of *Escherichia coli* and *Staphylococcus aureus* cells to PSFQ is investigated. These polymers inhibit the growth of microorganisms, with inhibition becoming stronger with the increasing polycationic nature of the modified polysulfone and the concentration of the polymer solution. Moreover, the adhesion of the relatively hydrophilic *E. coli* to hydrophilic quaternized polysulfone is higher than the adhesion of the hydrophobic *S. aureus* cells.

These results are useful in investigations on specific biomedical applications, including evaluation of bacterial adhesion to the surfaces, and utilization of modified polysulfones as semipermeable membranes.

9. References

Albu, R.M.; Avram, E.; Stoica, I.; Ioanid, E.G.; Popovici, D. & Ioan, S. (2011). Surface Properties and Compatibility with Blood of New Quaternized Polysulfones. *Journal of Biomaterials and Nanobiotechnology*, Vol.2, No.2, (April 2011), pp. 114-123, ISSN 2158-7027

Barikani, M. & Mehdipour–Ataei, S. (2000). Synthesis, Characterization and Thermal Properties of Novel Arylene Sulfone Ether Polyimides and Polyamides. *Journal of Polymer Science, Part A: Polymer Chemistry*, Vol.38, No.9, (May 2000), pp. 1487-1492, ISSN 1099-0518

Della Volpe, C.; Maniglio, D.; Brugnara, M.; Siboni, S. & Morra, M. (2004). The Solid Surface Free Energy Calculation: in Defense of the Multicomponent Approach. *Journal of*

Colloid and Interface Science, Vol.271, No.2, (March 2004), pp. 434-453, ISSN 0021-9797

Dourado, F.; Gama, F.M.; Chibowski, E. & Mota, M. (1998). Characterization of Cellulose Surface Free Energy. *Journal of Adhesion Science and Technology*, Vol.12, No.10 (November 1998), pp. 1081-1090, ISSN 0169-4243

Faibish, R.S.; Yoshida, W. & Cohen, Y. (2002). Contact Angle Study on Polymer-Grafted Silicon Wafers. *Journal of Colloid and Interface Science*, Vol.256, No.2, (February 2003), pp. 341-350, ISSN 0021-9797

Filimon, A.; Albu, R.M.; Avram, E. & Ioan, S. (2010). Effect of Alkyl Side Chain on the Conformational Properties of Polysulfones with Quaternary Groups. *Journal of Macromolecular Science, Part B: Physics*, Vol.49, No.1, (January 2010), pp. 207-217, ISSN 0022-2348

Filimon, A.; Avram, E.; Dunca, S.; Stoica, I. & Ioan, S. (2009). Surface Properties and Antibacterial Activity of Quaternized Polysulfones. *Journal of Applied Polymer Science*, Vol.112, No.3, (February 2009), pp. 1808-1816, ISSN 1097-4628

Filimon, A.; Avram, E. & Ioan, S. (2007). Influence of Mixted Solvents and Temperature on the Solution Properties of Quaternized Polysulfones. *Journal of Macromolecular Science, Part B: Physics*, Vol.46, No.3 (March 2007), pp. 503-520, ISSN 0022-2348

Guan, R.; Zou, H.; Lu, D.; Gong, C. & Liu, Y. (2005). Polyetheresulfone Sulfonated by Chlorosulfonic Acid and its Membrane Characteristics. *European Polymer Journal*, Vol.41, No. 7, (July 2005), pp. 1554-1560, ISSN 0014-3057

Guiver, M.D.; Black, P.; Tam, C.M. & Deslandes, Y. (1993). Functionalized Polysulfone Membranes by Heterogeneous Lithiation. *Journal of Applied Polymer Science*, Vol.48, No.9, (June 1993), pp. 1597-1606, ISSN 1097-4628

Hazziza-Laskar, J.; Helary, G. & Sauvet, G. (2005). Biocidal Polymers Active by Contact. IV. Polyurethanes Based on Polysiloxanes with Pendant Primary Alcohols and Quaternary Ammonium Groups. *Journal of Applied Polymer Science*, Vol.58, No.1, (October 2005), pp. 77-84 , ISSN 1097-4628

Higuchi, A.; Iwata, N.; Tsubaki, M. & Nakagawa, T. (1988). Surface-Modified Polysulfone Hollow Fibers. *Journal of Applied Polymer Science*, Vol.36, No.8, (October 1988), pp. 1753-1767, ISSN 1097-4628

Higuchi, A.; Shirano, K.; Harashima, M.; Yoon, B.O.; Hara, M.; Hattori M. & Imamura, K. (2002). Chemically Modified Polysulfone Hollow Fibers with Vinylpyrrolidone Having Improved Blood Compatibility. *Biomaterials*, Vol.23, No.13, (July 2002), pp. 2659-2666, ISSN 0142-9612

Hopkins, A.R.; Rasmussen, P.G.; Basheer, R.A. (1996). Characterization of Solution and Solid State Properties of Undoped and Doped Polyanilines Processed from Hexafluoro-2-Propanol. *Macromolecules*, Vol.29, No.24, (November 1996), pp. 7838-7846, ISSN 0024-9297

Huang, D.H.; Ying, Y.M. & Zhuang, G.Q. (2000). Influence of Intermolecular Entanglements on the Glass Transition and Structural Relaxation Behaviors of Macromolecules. 2. Polystyrene and Phenolphthalein Poly(ether sulfone). *Macromolecules*, Vol. 33, No. 2, (January 2000), pp. 461-464, ISSN 0024-9297

Idris, A.; Zaina, N.M. & Noordinb, M.Y. (2007). Synthesis, Characterization and Performance of Asymmetric Polyethersulfone (PES) Ultrafiltration Membranes with Polyethylene Glycol of Different Molecular Weights as Additives. *Desalination*, Vol.207, No.1-3, (March 2007), pp. 324-339, ISSN 0011-9164

Ioan, S.; Albu, R.M.; Avram, E.; Stoica, I. & Ioanid, E.G. (2011a). Surface Characterization of Quaternized Polysulfone Films and Biocompatiblity Studies. *Journal of Applied Polymer Science*, Vol.121, No.1, (July 2011), pp. 127–137, ISSN 1097-4628

Ioan, S.; Buruiana, L.I.; Petreus, O.; Avram, E.; Stoica, I. & Ioanid, G.E. (2011b). Rheological and Morphological Properties of Phosphorus-Containing Polysulfones. *Polymer-Plastics Technology and Engineering*, Vol.50, No.1, (January 2011), pp. 36–46, ISSN 0360-2559

Ioan, S.; Filimon, A. & Avram, E. (2006a). Influence of the Degree of Substitution on the Solution Properties of Chloromethylated Polysulfone. *Journal of Applied Polymer Science*, Vol.101, No.1, (April 2006), pp. 524-531, ISSN 1097-4628

Ioan, S.; Filimon, A. & Avram, E. (2006b). Conformation and Viscometric Behavior of Quaternized Polysulfone in Dilute Solution. *Polymer Engineering Science*, Vol.46, No.7, (May 2006), pp. 827-836, ISSN 1548-2634

Ioan, S.; Filimon, A.; Avram, E. & Ioanid, G. (2007). Effect of Chemical Structure and Plasma Treatment on the Surface Properties of Polysulfones. *e-Polymers*, No.031, (March 2007), pp. 1- 13, ISSN 1618-7229

Johnson, R.N. (1969). Polysulfones. Plastics, Resins, Rubbers, Fibers, In: *Encyclopedia of Polymer Science and Technology*, F.M. Herman, G.G. Norman & M.N. Bikales, (Eds.), 447-463, John Wiley & Sons, New York, London, Sydney & Toronto

Kälble, D.H. (1969). Peel Adhesion: Influence of Surface Energies and Adhesive Rheology. *Journal Adheshion*, Vol.1, No.2, (April 1969), pp. 102-123, ISSN 0021-8464

Khang, G.; Lee H.B. & Park, J.B. (1995). Biocompatibility of Polysulfone. I. Surface Modifications and Characterizations. *Biomedical Materials and Engineering*, Vol.5, No.4, (April 1995), pp. 245-258, ISSN 0959-2989

Kesting, R.E. (1990). The Four Tiers of Structure in Integrally Skinned Phase Inversion Membranes and Their Relevance to the Various Separation Regimes. *Journal of Applied Polymer Science*, Vol.41, No.11-12, (December 1990), pp. 2739-2752 , ISSN 1097-4628

Kochkodan, V.; Tsarenko, S.; Potapchenko, N.; Kosinova, V. & Goncharuk, V. (2008). Adhesion of Microorganisms to Polymer Membranes: a Photobactericidal Effect of Surface Treatment with TiO_2. *Desalination*, Vol.220, No.1-3, (March 2008), pp. 380-385, ISSN 0011-9164

Kwok, D.Y.; Ng, H. & Neumann, A.W. (2000). Experimental Study on Contact Angle Patterns: Liquid Surface Tensions Less than Solid Surface Tensions. *Journal of Colloid and Interface Science*, Vol.225, No.2, (March 2004), pp. 323-328, ISSN 0021-9797

Luca, C.; Avram, E. & Petrariu, I. (1988). Quaternary Ammonium Polyelectrolytes. V. Amination Studies of Chloromethylated Polystyrene with N,N-Dimethylalkylamines. *Journal of Macromolecular Science, Part A: Pure Applied Chemistry*, Vol.25, No.4, (April 1988), pp. 345-361, ISSN 1520-5738

Mann, B.K. & West, J.L. (2001). Tissue Engineering in the Cardiovascular System: Progress Toward a Tissue Engineered Heart. *The Anatomical Record*, Vol.263, No.4, (August 2001), pp. 367-371, ISSN 1932-8494

Merianos, J.J. (1991). Quaternary Ammonium Antimicrobial Compounds, In: *Disinfection, Sterilization and Preservation*. Block, S.S., pp. 225, Lea & Febiger, Philadelphia

Owens, D.K. & Wend, R.C. (1969). Estimation of the Surface Free Energy of Polymers. *Journal of Applied Polymer Science*, Vol.13, No.8, (August 1969), pp. 1741-1747, ISSN 1097-4628

Petreus, O.; Avram, E. & Serbezeanu, D. (2010). Synthesis and Characterization of Phosphorus-Containing Polysulfone. *Polymer Engineering Science*, Vol.50, No.1, (January 2010), pp. 48-56, ISSN 1548-2634

Qian, J.W.; An, Q.F.; Wang, L.N.; Zhang, L. & Shen, L. (2005). Influence of the Dilute-Solution Properties of Cellulose Acetate in Solvent Mixtures on the Morphology and Pervaporation Performance of their Membranes. *Journal of Applied Polymer Science*, Vol.97, No.5, (September 2005), pp. 1891-1898, ISSN 1097-4628

Rankl, M.; Laib, S. & Seeger, S. (2003). Surface Tension Properties of Surface-Coatings for Application in Biodiagnostics Determined by Contact Angle Measurements. *Colloids and Surface B: Biointerfaces*, Vol.30, No.3, (July 2003), pp. 177-186, ISSN 0927-7765

Reitsma, S.; Slaaf, D.W.; Vink, H.; van Zandvoort, M.A.M.J. & oude Egbrink, M.G.A. (2007). The Endothelial Glycocalyx: Composition, Functions, and Visualization. *Pflügers Archiv - European Journal of Physiology*, Vol.454, No.3, (June 2007), pp. 345-359, ISSN 0031-6768

Ström, G.; Fredriksson, M. & Stenius, P. (1987). Contact Angles, Work of Adhesion, and Interfacial Tensions at a Dissolving Hydrocarbon Surface. *Journal of Colloid and Interface Science*, Vol.119, No. (2), (October 1987), pp 352-361, ISSN 0021-9797

Tomaszewska, M.; Jarosiewicz, A. & Karakulski, K. (2002). Physical and Chemical Characteristics of Polymer Coatings in CRF Formulation. *Desalination*, Vol.146, No.1-3, (September 2002), pp. 319-323, ISSN 0011-9164

van Oss, C.J.; Good R.J. & Chaudhury, M.K. (1988a). Additive and Nonadditive Surface Tension Components and the Interpretation of Contact Angles. *Langmuir*, Vol.4, No.4, (July 1988), pp. 884-891, ISSN 743-7463

van Oss, C.J.; Ju, L.; Chaudhury, M.K. & Good, R.J. (1988b). Interfacial Lifshitz-van der Waals and Polar Interactions in Macroscopic Systems. *Chemical Reviews*, Vol.88, No.6, (September 1988), pp. 927-941, ISSN 0009-2665

van Oss, C.J.; Ju, L.; Chaudhury, M.K. & Good, R.J. (1989). Estimation of the Polar Parameters of the Surface Tension of Liquids by Contact Angle Measurements on Gels. *Journal of Colloid and Interface Science,* Vol.128, No.2, (March 1989), pp. 313-319, ISSN 0021-9797

van Oss, C.J. & Giese, R.F. (1995). The Hydrophilicity and Hydrophobicity of Clay Minerals. *Clay Clay Miner*, Vol.43, No.4, (April 1995), pp. 474-477, ISSN 0009-8558

Vijayanand, K.; Deepak, K.; Pattanayak, D.K.; Rama Mohan, T.R. & Banerjee, R. (2005). Interpenetring Blood-Biomaterial Interactions from Surface Free Energy and Work of Adhesion. *Trends in Biomaterials and Artificial Organs*, Vol.18, No.2, (January 2005), pp. 73-83, ISSN 0971-1198

Xu, X. ; Li, S. ; Jia, F. & Liu, P. (2006). Synthesis and Antimicrobial Activity of Nano-fumed Silica Derivative with N,N-Dimethyl-n-Hexadecylamine. *Life Science Journal*, Vol.3, No.1, (January 2006), pp. 59-62, ISSN 1097-8135

Yildirim, E. (1997) Polymer Adsorption, In: *Handbook of Surface and Colloid Chemistry,* Birdi, K.S., pp. 266-307, CRC Press, ISBN 0-8493-945-7, Boca Raton

Yu, H.; Huang, Y.; Ying, H. & Xiao, C. (2007). Preparation and Characterization of a Quaternary Ammonium Derivative of Konjac Glucomannan. *Carbohydrate Polymers*, Vol.69, No.1, (May 2007), pp. 29-40, ISSN 0144-8617

14

The Design of Bacteria Strain Selective Antimicrobial Peptides Based on the Incorporation of Unnatural Amino Acids

Amanda L. Russell[1], David Klapper[2],
Antoine H. Srouji[3], Jayendra B. Bhonsl[4], Richard Borschel[5],
Allen Mueller[5] and Rickey P. Hicks[1,*]
[1]*Department of Chemistry, East Carolina University,*
Science and Technology Building, Greenville,
[2]*Peptide Core Facility in the School of Medicine*
of the University of North Carolina at Chapel Hill, Chapel Hill, North Carolina,
[3]*Synthetic Proteomics, Carlsbad, CA,*
[4]*Division of Experimental Therapeutics,*
Walter Reed Army Institute of Research, Silver Spring, Maryland,
[5]*Division of Bacterial and Rickettsial Diseases,*
Walter Reed Army Institute of Research, Silver Spring, Maryland,
USA

1. Introduction

An intensive international research effort is currently underway to develop new classes of compounds that exhibit novel mechanisms of antibacterial activity (Bush, 2004) (Klevens et al., 2007) due to the dramatic and continued evolution of drug resistant strains of bacteria (Hartl, 2000) (Shlaes, Projan, & Edwards, 2004) (Y. Huang, Huang, & Chen, 2010) (Godballe, Nilsson, Petersen, & Jenssen, 2011) (J. Song, 2008) (Findlay, Zhanel, & Schweizer, 2010). Both natural and synthetic antimicrobial peptides (AMPs) because of their novel mechanisms of antibiotic activity, coupled with the inherent difficulty for bacteria to develop resistance to them (Hartl, 2000) (Shlaes et al., 2004) (Y. Huang et al., 2010) (Godballe et al., 2011) (J. Song, 2008) (Findlay et al., 2010) exhibit a very high potential as new therapeutic agents. Several AMPs are currently in development as topical antibiotics (Zhang & Falla, 2009).

AMPs have evolved as a host defense mechanism against invading micro-organisms in most organisms (Hartl, 2000) (Shlaes et al., 2004) (Y. Huang et al., 2010) (Godballe et al., 2011) (J. Song, 2008) (Findlay et al., 2010) . In addition to antibacterial properties AMPs are believed to be key players in innate immunity (Devine & Hancock, 2002; Y. M. Song et al., 2005). AMPs are normally highly positively charged (+3 to +9) small peptides consisting of 5-50

* Corresponding Author

amino acid residues (Hancock, 1998). These peptides are amphipathic molecules with well-defined hydrophobic and hydrophilic regions (Toke, 2005; Zasloff, 2002). As of January 2009, more than 1330 (Wang, Li, & Wang, 2009) natural and synthetic antimicrobial peptides have been characterized. These peptides have exhibited, a diversity of biological activity (Khandelia, 2008).

The inherent selectivity of AMP's for prokaryotic vs eukaryotic cells is believed to be derived from the difference in the chemical compositions of their respective membranes (Dennison, Wallace, Harris, & Phoenix, 2005). Hancock and co-workers (Powers & Hancock, 2003) have extended this hypothesis to propose that the differences in membrane composition between different strains of bacteria are responsible for the diversity in the potency and selectivity exhibited by a particular AMP against different strains of bacteria. Bacterial cells contain a high percentage of negatively charged phospholipids while mammalian cells contain a much higher concentration of zwitterionic phospholipids (Papo & Shai, 2003) (Zhang & Falla, 2009) (Findlay et al., 2010) (Godballe et al., 2011) (Y. Huang et al., 2010) (Yeaman & Yount, 2003) (Leontiadou, Mark, & Marrink, 2006) (Y. Huang et al., 2010) (Hancock & Lehrer, 1998). Other differences also exist including membrane composition (sterols, lipopolysaccharide, peptidoglycan etc) (Yeaman & Yount, 2003), structure, and transmembrane potential and polarizability. These differences are in part responsible for the observed selectivity of some AMPs for bacterial vs. mammalian cells.(Hancock & Lehrer, 1998; Yeaman & Yount, 2003) Melo and co-workers have characterized the physical properties of bacterial membranes that are conducive for AMP interactions (Melo, Ferre, & Castanho, 2009).

The membranes surrounding different types of bacteria are also different. The lipid bilayer of Gram positive bacteria is covered by a porous layer of peptidoglycan, while the structure of Gram negative bacteria is more complex with two lipid membranes containing lipopolysaccharides and porins (Dennison et al., 2005; Giangaspero, Sandri, & Tossi, 2001). There is a developing preponderance of evidence in the literature supporting the concept that the selectivity and potency of a specific AMP is determined in a large measure by the chemical composition of the target membrane.(Powers & Hancock, 2003; Yeaman & Yount, 2003)

In our laboratory we have focused on developing antimicrobial peptides that contain unnatural amino acids to control the conformational and physicochemical properties (Hicks, Bhonsle, Venugopal, Koser, & Magill, 2007) of the resulting peptide. Our original skeletal design of an unnatural AMP incorporated placement of three L-Tic-L-Oic dipeptide units into the polypeptide backbone to induce an ordered structure onto the peptide (Hicks et al., 2007). The use of the conformationally restrained amino acids (Tic an Oic) reduces the local flexibility of the peptide backbone and thus reduces the total conformational space that may be sampled by the peptide during lipid binding. The AMPs developed in our laboratory, the Tic-Oic units are connected via an amino acid spacer with defined properties of charge and hydrophobicity resulting in peptides with well-defined physiochemical properties while maintaining sufficient conformational flexibility to allow the peptide to adopt different conformations on interacting with membranes of different chemical composition (Hicks et al., 2007; Russell et al., 2010; Venugopal et al., 2010). The basic skeleton of the AMPs under investigation is given in Fig. 1. The amino acid sequences for the AMPs used in this investigation are given in Table 1.

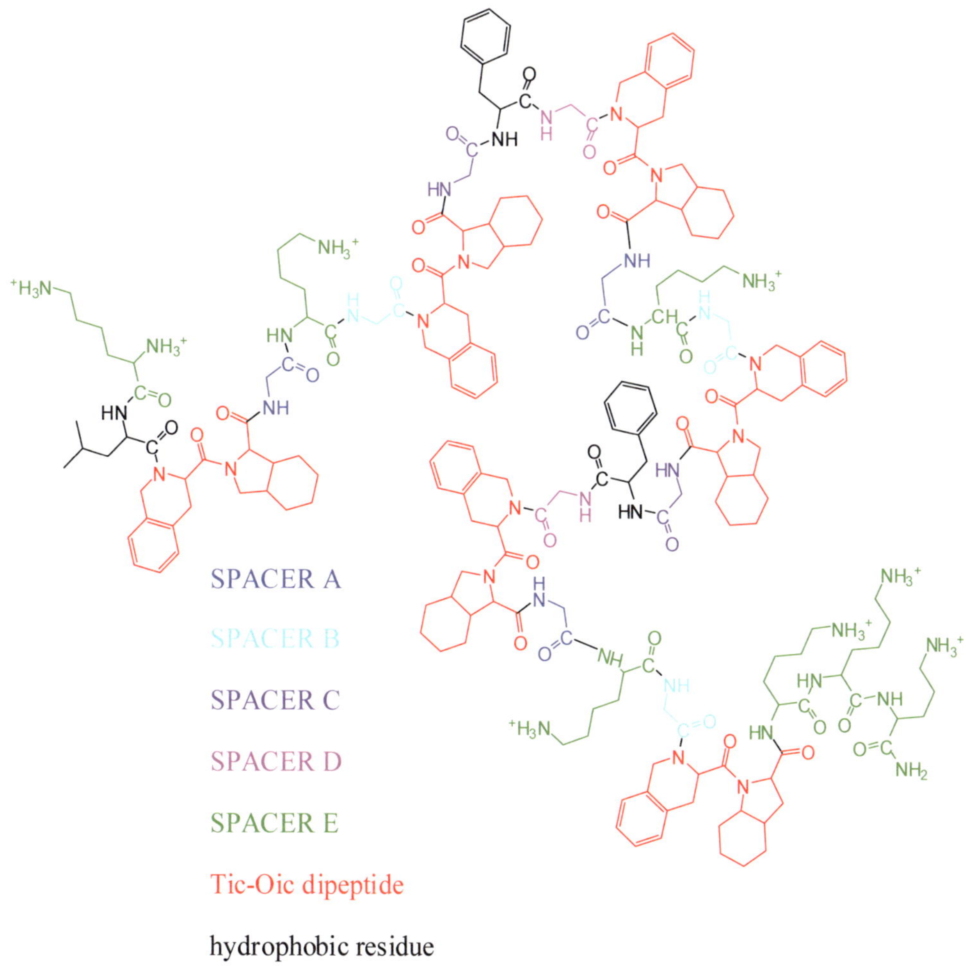

Fig. 1. The basic skeleton of the new analogs containing six Tic-Oic dipeptide units and the five SPACER residues.

In this report the in vitro antibacterial activity of a new series of AMPs incorporating six Tic-Oic dipeptide units as well as four additional Spacers, A, B, C and D on either side of the intervening hydrophobic and charged residues. These spacers define the overall conformational flexibility of the peptide backbone. A fifth Spacer, E, which defines the distance between the polypeptide backbone and the positively charged side chain amine group. Spacer E is involved in determining the overall surface charge density of the peptide as well as defining the distance between the membrane surface and the polypeptide backbone (Hicks et al., 2007). By varying the number of $-CH_2-$ groups in Spacer E from 1 to 4 in the side chain of the basic residues, the distance between the positive charge and the peptide backbone will decrease resulting in a) less side chain flexibility during binding -this flexibility is more important in the binding with zwitterionic lipids than with anionic lipids; and b) the positive charge density will reside closer to the peptide backbone. The additional Tic-Oic dipeptide units and spacers were incorporated in hopes of improving the organism selectivity of these AMPs.

Compound Number/ Amino Acid Sequence

22	H_2N-KL-Tic-Oic-K-Tic-Oic-F-Tic-Oic-K-Tic-Oic-F-Tic-Oic-K-Tic-Oic-KR-$CONH_2$
70	H_2N-KL-Tic-Oic-K-Tic-Oic-F-Tic-Oic-K-Tic-Oic-F-Tic-Oic-K-Tic-Oic-KKKK-$CONH_2$
71	H_2N-Orn-L-Tic-Oic-Orn-Tic-Oic-F-Tic-Oic-Orn-Tic-Oic-F-Tic-Oic-Orn-Tic-Oic-Orn-Orn-Orn-Orn-$CONH_2$
72	H_2N-Dpr-L-Tic-Oic-Dpr-Tic-Oic-F-Tic-Oic-Dpr-Tic-Oic-F-Tic-Oic-Dpr-Tic-Oic-Dpr-Dpr-Dpr-Dpr-$CONH_2$
73	H_2N-Dab-L-Tic-Oic-Dab-Tic-Oic-F-Tic-Oic-Dab-Tic-Oic-F-Tic-Oic-Dab-Tic-Oic-Dab-Dab-Dab-Dab-$CONH_2$
74	H_2N-KL-Tic-Oic-GK-Tic-Oic-F-Tic-Oic-GK-Tic-Oic-F-Tic-Oic-GK-Tic-Oic-KKKK-$CONH_2$
75	H_2N-KL-Tic-Oic-K-Tic-Oic-GF-Tic-Oic-K-Tic-Oic-GF-Tic-Oic-K-Tic-Oic-KKKK-$CONH_2$
76	H_2N-KL-Tic-Oic-GK-Tic-Oic-GF-Tic-Oic-GK-Tic-Oic-GF-Tic-Oic-GK-Tic-Oic-KKKK-$CONH_2$
77	H_2N-KL-Tic-Oic-K-Tic-Oic-FG-Tic-Oic-K-Tic-Oic-FG-Tic-Oic-K-Tic-Oic-KKKK-$CONH_2$
78	H_2N-KL-Tic-Oic-KG-Tic-Oic-F-Tic-Oic-KG-Tic-Oic-F-Tic-Oic-KG-Tic-Oic-KKKK-$CONH_2$
79	H_2N-KL-Tic-Oic-KG-Tic-Oic-FG-Tic-Oic-KG-Tic-Oic-FG-Tic-Oic-KG-Tic-Oic-KKKK-$CONH_2$

Table 1. Amino acid sequences of the compounds understudy

2. Biological activity

Antimicrobial peptides (AMP) have evolved in almost every class of living organism as a defense mechanism against invading micro-organisms (Yeaman & Yount, 2003) (Dennison et al., 2005). The exact mechanism of lipid-induced cytotoxicity of these peptides is currently debated in the literature (Powers & Hancock, 2003) with AMPs divided into two mechanistic classes, membrane-disruptors and non-membrane-disruptors (Powers & Hancock, 2003) (Brogden, 2005). However, there is a general agreement that the first target for either membrane-disruptor or non-membrane-disruptor AMPs is the negatively charged membranes of bacterial cells (Powers & Hancock, 2003). There are several different structural classes of membrane-disruptor AMPs but, for the purpose of this investigation we focused on linear amphipathic helical peptides. In many cases these peptides exhibit characteristics of a random coil conformation in aqueous or in organic solvents, however on binding to micelles or membranes they adopt an ordered amphipathic secondary structure (Hicks et al., 2003). This structural class of membrane-disruptors can be divided into two sub-classifications based on biological activity: 1) cell selective and 2) non-selective (Y. M. Song et al., 2005). As the name implies cell selective AMPs exhibit potent activity against bacterial cells while remaining inactive against mammalian cells, non-selective AMPs exhibit activity against bacterial as well as mammalian cells. We have previously reported the synthesis and biological evaluation of a series of potent (low μM to nm in vitro activity, and low acute toxicity 125 mg/kg in mice) bacteria selective membrane disruptors based on the incorporation of three Tic-Oic dipeptide units (Hicks et al., 2007) (Venugopal et al., 2010) (Bhonsle, Venugopal, Huddler, Magill, & Hicks, 2007). The in vitro antibacterial activity exhibited by the analogs containing six dipeptide units against seven strains of bacteria is given in Table 2. The in vitro antibacterial activity will be discussed as a function of the Spacers. The activity of Spacers A and C will be discussed together as will Spacers B and D. Spacer E will be discussed separately.

Bacteria Strain	Compound Number										
	22	70	71	72	73	74	75	76	77	78	79
Acinetobacter baumanii ATCC 19606	40.5	9.30	9.77	186	162	4.50	146.00	139	146	144	139
Acinetobacter baumanii WRAIR	162	150	156	186	162	144	146.00	139	146	144	139
Staphylococcus aureus ATCC 33591(MRSA)	162	150	156	186	162	17.90	146.00	139	146	144	139
Brucella melitensis 16M	162	150	156	186	162	144	73	139	146	144	139
Bacillus anthracis AMES	40.5	150	156	186	162	8.97	0.50	139	146	2.20	139
Francisella tularensis SCHUS4	81	150	156	186	162	144	146.00	139	146	144	139
Burkholderia mallei	162	150	156	84	81	144	146	139	146	144	139

Table 2. In Vitro MIC values (μM) for the compounds understudy against drug resistant strains and Gram Negative Select Agents.

2.1 Spacers A and C

In this series of analogs, Spacer A is the residue preceding each internal Lys residues (N-terminal side of the Lys) and Spacer C is the residue preceding each internal Phe residue (N-terminal side of the Phe) as listed in Table 3.

Compound **22** exhibited good in vitro activity against Acinetobacter baumanii ATCC 19606, and Bacillus anthracis AMES and moderate activity against Francisella tularensis SCHUS4 and poor activity against the other bacteria strains. Replacement of the Lys-Arg residues at the C-terminus with four Lys residues in compound **74** changes in vitro activity dramatically.

Compound number	Spacer A	Spacer C
22	Gly	none
70	none	none
74	Gly	none
75	none	Gly
76	Gly	Gly

Table 3. Spacers A and C

The activity against Acinetobacter baumanii ATCC 19606 and Staphylococcus aureus ATCC 33591(MRSA) are increased by a factor of 9, and the activity against Bacillus anthracis AMES, is increased by a factor of 5, compared to compound **22**. Deleting Spacer A (Gly residue) of compound **74** and introducing Spacer C (Gly residue) in Compound **75** again dramatically changes the activity compared to both compounds **22** and **74**. The activity against Acinetobacter baumanii ATCC 19606 is now very poor. However compound **75** exhibited the highest activity against Bacillus anthracis AMES (80 fold increase over compound **74**) and Brucella melitensis 16M. Activity against the other bacteria strains was very poor. Incorporating both Spacers B and C (both Gly residues) in compound **76** reduced the activity to a level that this compound is no longer therapeutically useful. Removing both Spacers in compound **70** resulted in a compound selective for Acinetobacter baumanii ATCC 19606 and inactive against the remaining bacteria strains. This data indicates that combined use of Spacer A and C coupled with increasing the density of the positive charge at the C-terminus of the peptide produces organism selectivity.

2.2 Spacers B and D

In this series of analogs Spacers B is the residue following each internal Lys residues (C-terminal side of the Lys) and Spacer D is the residue following each internal Phe residues (C-terminal side of the Phe) as listed in Table 4.

Compound Number	Spacer B	Spacer D
77	none	Gly
78	Gly	none
79	Gly	Gly

Table 4. Spacers D and E

Incorporation of Spacers D (Gly residues in both cases) in compounds **77** and **79** reduced to the activity of these two compounds against all seven strains of bacteria to poor making these compounds no longer therapeutically useful. Incorporation of Spacer B (Gly residue) only in compound **78** exhibited selective and very good activity against Bacillus anthracis AMES.

2.3 Spacer E

In this series of analogs, Spacer E replaces the charged Lys residues with charged residues with progressively shorter side chains as shown in Table 5. Replacement of the Lys residues with Orn residues (one less carbon atom in the side chain) in compound **71** dramatically reduces the activity against Staphylococcus aureus ATCC 33591(MRSA) and Bacillus anthracis AMES, while maintaining the activity against Acinetobacter baumanii ATCC 19606 as compared to compound **74**.

Compound Number	Spacer E
71	Orn
72	Dpr
73	Dab
74	Lys

Table 5. Spacer E

Replacement of the Lys residues with either Dpr residues (three carbon less) or Dab (two carbons less) residues in compounds **72** and **73** respectively dramatically reduces the activity against all of the bacteria strains except Burkholderia mallei. Clearly, the length of the side chain of Spacer E, and the resulting charge density, plays a major role in determining organism potency and selectivity against these seven strains of bacteria.

3. Characterization of peptide lipid interactions

The observed differences in bacteria strain selectivity for these compounds are believed to be derived directly from the conformational and physicochemical properties presented by these AMPs to the surface of the bacteria membrane. Circular Dichroism (CD), isothermal calorimetry (ITC) and calcein fluorescence leakage experiments were conducted to provide insight into the mechanisms of binding of a series of antimicrobial peptides to simple membrane models for bacterial and mammalian cells. LUVs and SUVs consisting of 1-Palmitoyl-2-Oleoyl-sn-Glycero-3-Phosphocholine (POPC) were selected as a simple model for the zwitterionic membranes of mammalian cells and membrane models consisting of (4:1) 1-Palmitoyl-2-Oleoyl-sn-Glycero-3-Phosphocholine (POPC) / **1-**Palmitoyl-2-Oleoyl-sn-Glycero-3-[Phospho-rac-(1-glycerol)] (Sodium Salt)) (POPG) were selected as a simple model for the anionic membranes of bacteria cells (Bringezu et al., 2007) Hicks, 2007 #19}.

3.1 Calcein leakage studies

It is well documented that AMPs containing three Tic-Oic dipeptide units exhibited potent and selective antibacterial activity (Hicks et al., 2007; Venugopal et al., 2010) **and t**hese AMPs were designed to be members of the general class of AMPs known as membrane disruptors (Shlaes et al., 2004) (Kamysz, 2005) (Zhang, Harris, & Falla, 2005) (Toke, 2005) (Godballe et al., 2011) (Ryge, Frimodt-Moller, & Hansen, 2008). Calcein leakage studies clearly indicated that the analogs containing three Tic-Oic dipeptides units are in fact membrane disruptors (Russell et al., 2010). The first step in this investigation was to confirm that the analogs containing six Tic-Oic dipeptide units are also membrane disruptors. Peptide induced calcein leakage from LUVs monitored through fluorescence, is a well documented technique for probing AMP activity to confirm membrane disruption (Andrushchenko, Aarabi, Nquyen, Prenner, & Vogel, 2008) (Medina, Chapman, Bolender, & Plesniak, 2002). Peptides at a concentration of 4μM were introduced to solutions of either POPC or 4:1 POPC/POPG LUVs containing encapsulated 70 mM calcein (Figure 2) and the resulting induced fluorescence due to leakage of calcein was monitored over a 90 minute time period.

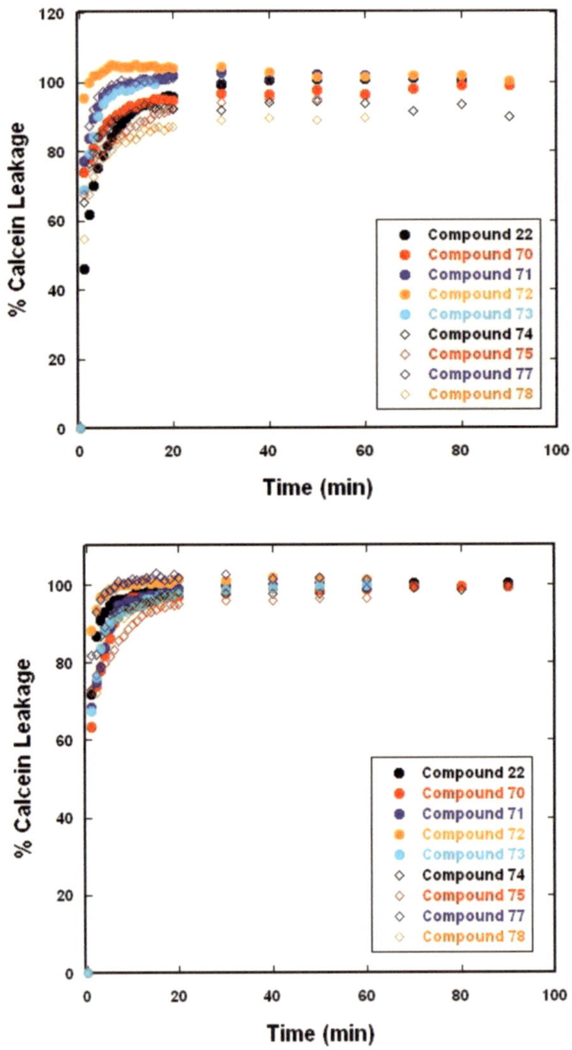

Fig. 2. Induced leakage of calcein from TOP) POPC LUVs and BOTTOM) 4:1 POPC/POPG.

The induced calcein leakage data indicates that all AMPs in this investigation interact in a very similar fashion with both zwitterionic (POPC) and anionic (4:1 POPC/POPG) LUVs. This data also strongly suggests that the mechanism of action of these AMP involves some type of membrane disruption. However, this data doesn't explain the observed organism selectivity. The observed induced calcein induce leakage behavior of the analogs containing six Tic-Oic dipeptide units is very different from that observed for the corresponding three Tic-Oic dipeptide containing analogs. The analog containing three Tic-Oic dipeptide units corresponding to compound **74** has the amino acid sequence: Ac-GF-Tic-Oic-GK-Tic-Oic-GF-Tic-Oic-GK-Tic-KKKK-$CONH_2$ (compound **23**) (Russell et al., 2010). For compound **23** different concentrations of peptide (4 - 20 µM) were introduced to solutions of either POPC or 4:1 POPC/POPG LUVs and their induced fluorescence, due to leakage of calcein was monitored over a 90 minute time period. It can be seen in Fig. 3, compound **23** interacts with

POPC LUVs, inducing calcein leakage, in a non linear concentration dependent manner (Russell et al., 2010). However, compound **23** induces calcein leakage from 4:1 POPC/POPG LUVs in a concentration dependent manner. This data indicates the analogs containing three Tic-Oic dipeptide units interact with zwitterionic POPC LUV and anionic 4:1 POPC/POPG LUVs via different mechanisms (Russell et al., 2010). This data also indicates that the incorporation of three additional Tic-Oic dipeptide units, and thus increasing the overall length of the peptide, changes the mechanism of binding with both types of LUVs.

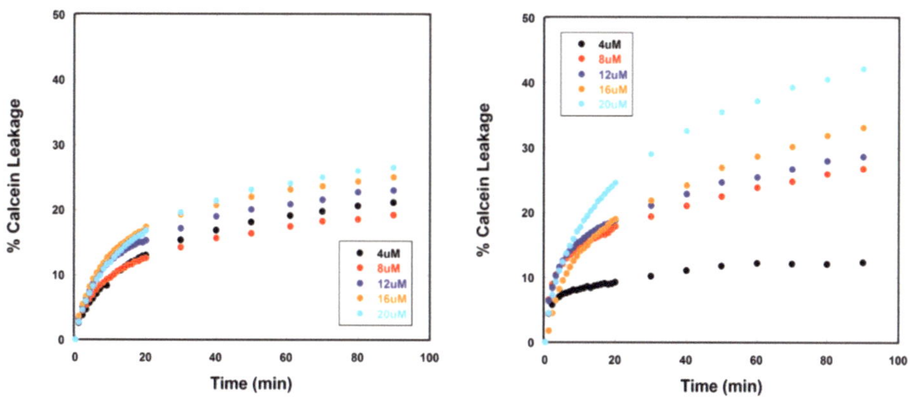

Fig. 3. The calcein leakage induced by compound **23** as a function of peptide concentration from LEFT) POPC LUVs and RIGHT) 4:1 POPC/POPG LUVs

The corresponding experiment using compound **74** were conducted. Compound 74 at various concentrations (4 - 20 μM) was introduced to solutions of either POPC or 4:1 POPC/POPG LUVs and their induced fluorescence, due to leakage of calcein was monitored over a 90 minute time period. It can be seen in Fig. 4, that compound **74** interacts with POPC LUVs, and 4:1 POPC/POPG LUVs in a totally concentration independent manner. In fact almost 100% calcien leakage was observed at all concentrations. This suggests that the analogs containing six Tic-Oic dipeptide units bind to the surface of the LUV and cause the LUV to lysis or break down in some fashion. After causing this disruption of the surface of the LUVs these peptides most likely then separate from the "disrupted" LUV and bind to another LUV and repeat the process.

3.2 Circular dichroism studies

Circular dichroism (CD) spectroscopy is very sensitive and its use to monitor conformational changes in peptides and proteins is well documented (Glattli, Daura, Seebach, & van Gunsteren, 2002), (Ladokhin, Selsted, & White, 1999) (Ladokhin, Vidal, & White, 2010). Traditionally, SUVs have been employed almost exclusively to investigate the binding of peptides and proteins with lipids in CD studies in order to minimize the contribution of light scattering on the spectra (Ladokhin et al., 2010) (C. H. Huang, 1969).

In the analysis of CD data, it must be noted that the spectrum of a peptide represents the linear combination of a number of different conformers (Glattli et al., 2002)-(Berova, Nakanishi, & Woody, 2000). This is particularly true when liposomes are used as there are several different peptide-liposome interactions possible depending on the lipid to peptide ratio.

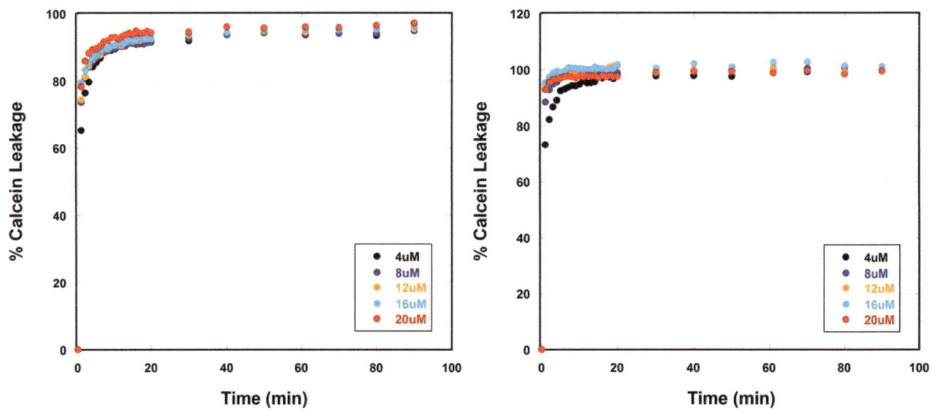

Fig. 4. The calcein leakage induced by compound **74** as a function of peptide concentration from LEFT) POPC LUVs and RIGHT) 4:1 POPC/POPG LUVs

Changes in the intensity or shape of the CD spectrum of a peptide in the presence of a liposome indicate that the peptide is adopting different conformations on interacting with that particular liposome as compared to another environment such as a buffer. (Berova et al., 2000; Glattli et al., 2002). Due to the high percentage of unnatural amino acids incorporated into the peptide under investigation here, no quantitative estimation of secondary structural features are possible. Therefore, analysis of CD spectra will be limited, at best, to qualitative comparisons highlighting differences in the spectra of these compounds with references to possible secondary structural features.

3.2.1 CD spectra in buffer

The CD spectra of 100 μM solutions of compounds **22, 70, 71, 72, 73, 74, 75, 76, 77, 78** and **79** in 40 mM phosphate buffer, pH = 6.8 are given in Fig. 5. As can be seen in Fig. 5, these peptide adopt a variety of different conformations varying from random coil to possible β-sheet like. However, overall these peptides are unordered in buffer only.

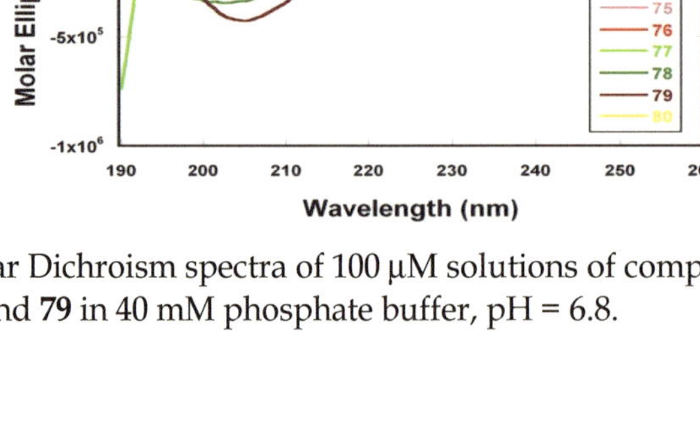

Fig. 5. Far-UV Circular Dichroism spectra of 100 μM solutions of compounds **22, 70, 71, 72, 73, 74, 75, 76, 77, 78** and **79** in 40 mM phosphate buffer, pH = 6.8.

3.2.2 CD spectra in the presence of POPC SUVs

The CD spectra of compounds **70, 71, 72, 73, 74, 75, 76, 77, 78** and **79** in the presence of 1.75 mM POPC SUVs are shown in Fig. 6. This figure illustrates the diversity of conformations adopted by these compounds on binding to POPC SUVs. The CD spectra of these compounds in the presence of POPC SUVs are different in most cases from the corresponding spectra in buffer. This indicates that these compounds are interacting with the POPC SUVs via some mechanism. This is consistent with the very strong interactions observed with POPC LUVs in the induced calcein leakage studies. From this figure it is difficult to characterize the conformational difference associated with the variations in the five Spacers. To accomplish analysis, these CD spectra will be divided into smaller sub sets of spectra.

The first sub set of CD spectra to be discussed will be those of compounds **70**, **74**, **75**, and **76** which contain Spacers A or C, or both. The CD spectra of these four compounds, given in Fig. 7, are different indicating that these compounds adopt different conformations on binding to POPC SUVs. This observation is critical in explaining possible organism selectivity. Since this POPC SUVs are a model for mammalian cells, this data suggests that these compounds would adopt different conformations on binding to red blood cells. From a therapeutic perspective this information may allow for the development of analogs with reduced toxicity. The CD spectrum of compound **75** exhibits a maxima at approximately 195 nm and double minima at approximately 220 and 205 nm which would suggest the presence of α-helical conformers (Ladokhin et al., 1999) (Fuchs et al., 2006). While the CD spectra of compounds **74** and **76** exhibit double minima at approximately 220 and 205 nm without the maxima at 195 nm suggesting these compound adopt a mixture of β-turn and β-sheet like conformers on binding to POPC SUVs (Ladokhin et al., 1999) (Fuchs et al., 2006). The CD spectrum of compound **70** exhibits a maxima at approximately 195 nm, however no negative minima is observed. It is very difficult to qualitatively characterize the possible secondary structures adopted by compound **70**.

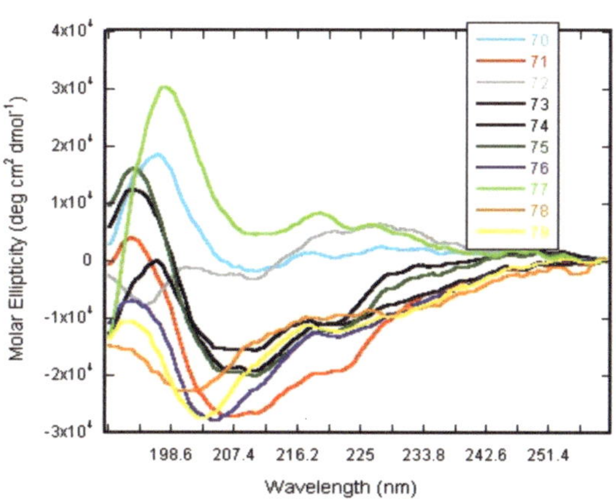

Fig. 6. Far-UV Circular Dichroism spectra of 100 μM solutions of compounds **70, 71, 72, 73, 74, 75, 76, 77, 78** and **79** in the presence of 1.75 mM POPC SUV in 40 mM phosphate buffer, pH = 6.8.

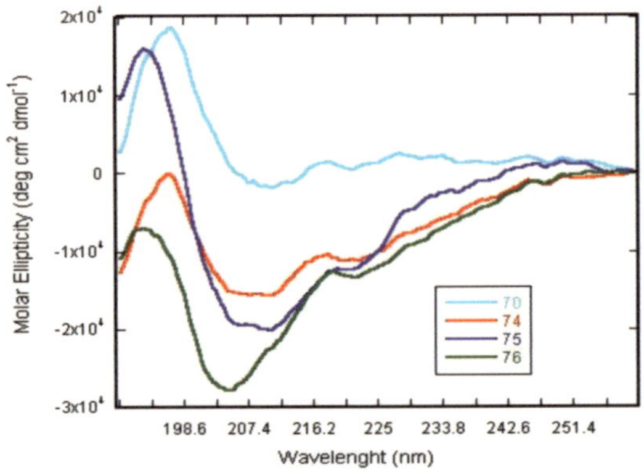

Fig. 7. Far-UV Circular Dichroism spectra of 100 μM solutions of compounds **70, 74, 75** and **76** in the presence of 1.75 mM POPC SUV in 40 mM phosphate buffer, pH = 6.8

The next sub set of CD spectra to be discussed in the presence of 1.75 mM POPC SUVs are of compounds **77**, **78** and **79** (Fig. 8) which contain Spacers B and D. The CD spectrum of compound **77** is similar in shape and intensity to the CD spectrum of compound **70,** and therefore it is very difficult to qualitatively characterize the possible secondary structures adopted by compound **77**. While compounds **78** and **79** exhibited similar CD spectra to each other, both with minim at approximately 202 to 205 nm, it is very difficult to qualitatively characterize the possible secondary structures adopted by compounds **78** and **79,** however, this suggests these compound adopt a mixture of β-turn and β-sheet like conformers on binding to POPC SUVs (Ladokhin et al., 1999) (Fuchs et al., 2006) . What is clear from these three CD spectra is Spacer D (compound **77**) doesn't play a major role in stabilizing a particular secondary structure on binding to POPC SUVs. Spacer B (contained in compounds **78** and **79**) seems to play a more significant role in stabilizing a β-turn/sheet conformation on binding to POPC SUVs.

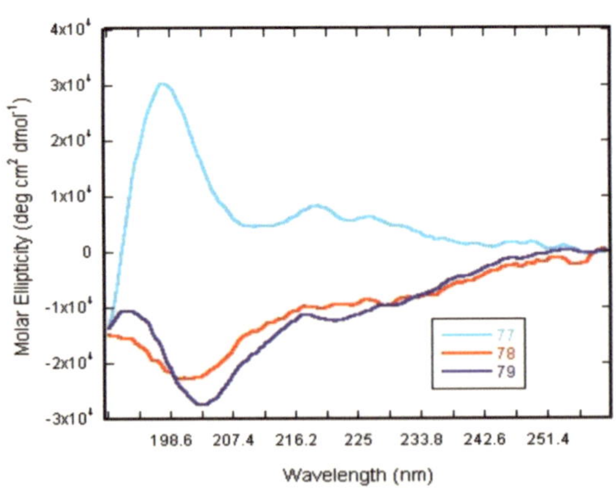

Fig. 8. Far-UV Circular Dichroism spectra of 100 μM solutions of compounds **77, 78** and **79** in the presence of 1.75 mM POPC SUV in 40 mM phosphate buffer, pH = 6.8.

The last sub set of CD spectra to be discussed in the presence of 1.75 mM POPC SUVs are those of compounds **71**, **72**, **73**, and **74** (Fig. 9) which contain Spacer E. Compounds **71**, and **74** exhibit double minima at approximately 220 and 205 nm without the maxima at 195 nm suggesting these compound adopt a mixture of β-turn and β-sheet like conformers on binding to POPC SUVs (Ladokhin et al., 1999) (Fuchs et al., 2006). While the CD spectrum of compound **72** exhibits characteristics of a random coil conformation. The CD spectrum of compound **73** exhibited a maxima at approximately 195 nm and a double minima at approximately 220 and 205 nm which would suggest the presence of α-helical conformers (Ladokhin et al., 1999) (Fuchs et al., 2006). This data clearly indicates that the length of Spacer E and the resulting location of the positive charge density relative to the peptide's backbone plays a major role in determining the conformations adopted by these peptides on binding to POPC SUVs.

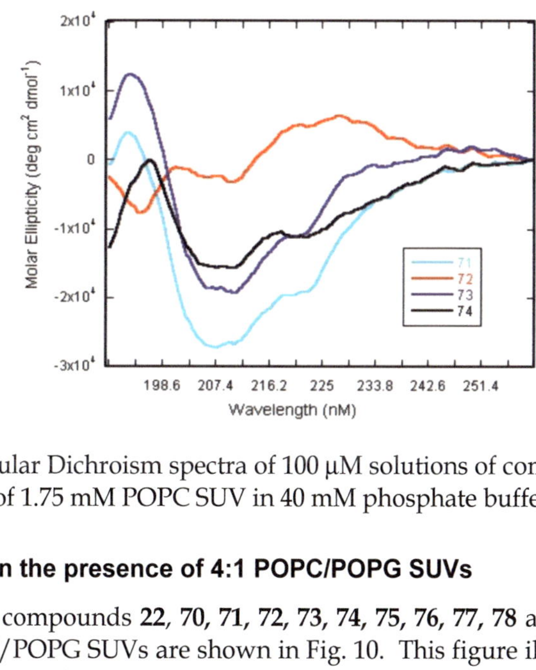

Fig. 9. Far-UV Circular Dichroism spectra of 100 μM solutions of compounds **71, 72, 73,** and **74** in the presence of 1.75 mM POPC SUV in 40 mM phosphate buffer, pH = 6.8.

3.2.3 CD spectra in the presence of 4:1 POPC/POPG SUVs

The CD spectra of compounds **22**, **70**, **71**, **72**, **73**, **74**, **75**, **76**, **77**, **78** and **79** in the presence of 1.75 mM 4:1 POPC/POPG SUVs are shown in Fig. 10. This figure illustrates the diversity of conformations adopted by these compounds binding to 4:1 POPC/POPG SUVs. The CD spectra of these compounds in the presence of 4:1 POPC/POPG SUVs are different in most cases from the corresponding spectra in buffer. This indicates that these compounds are interacting with the 4:1 POPC/POPG SUVs. From this figure it is difficult to characterize the conformational difference associated with the variations in the five Spacers. To accomplish this analysis, these CD spectra will be divided into smaller sub sets of spectra. As will become evident from the analysis of the sub sets of spectra, many of these compounds adopt different conformation on binding to POPC and 4:1 POPC/POPG SUVs.

The first sub set of spectra to be discussed will be focused on Spacers A and C, consisting of compounds **70**, **74**, **75**, and **76**. The CD spectra (Fig. 11) of these four compounds are different indicating that these compounds adopt different conformations on binding to 4:1 POPC/POPG SUVs. The differences in the CD spectra for these compounds binding to 4:1

POPC/POPG SUVs is critical in that it helps to explain the observed organism selectivity between the different bacteria strains. The CD spectrum of compound **75** exhibits a maxima at approximately 195 nm and double minima at approximately 220 and 205 nm which would suggest the presence of α-helical conformers just as it did in the presence of POPC SUVs (Ladokhin et al., 1999) (Fuchs et al., 2006).

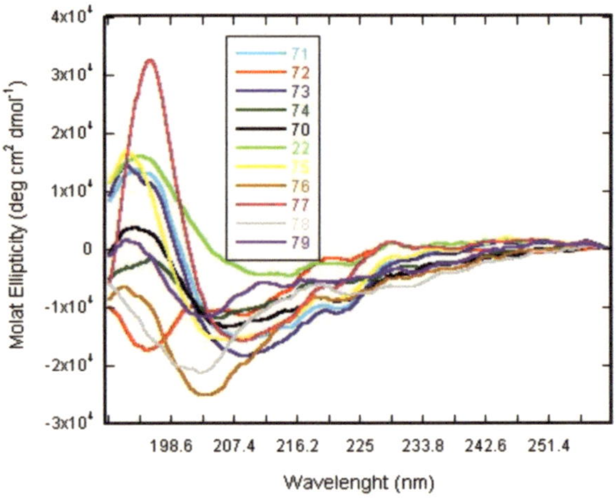

Fig. 10. Far-UV Circular Dichroism spectra of 100 μM solutions of compounds **22, 70, 71, 72, 73, 74, 75, 76, 77, 78** and **79** in the presence of 1.75 mM 4:1 POPC/POPG SUV in 40 mM phosphate buffer, pH = 6.8.

The CD spectra of compounds **76** exhibited a double minima at approximately 220 and 205 nm without the maxima at 195 nm suggesting these compound adopt a mixture of β-turn and β-sheet like conformers on binding to 4:1 POPC/POPG SUVs just as it did with POPC SUVs (Ladokhin et al., 1999) (Fuchs et al., 2006). The CD spectrum of compound **70** exhibits a weak maxima at approximately 195 nm and a weak double minima at approximately 220 and 205 nm is also observed which would suggest the presence of α-helical conformers just as it did in the presence of POPC SUVs (Ladokhin et al., 1999) (Fuchs et al., 2006). This CD spectrum is different from the one observed in the presence of POPC SUVs. The CD spectrum of compound **74** was very interesting as it exhibits only a single minima at approximately 205 nm. Also the overall spectra intensity is less than that observed in the presence of POPC SUVs.

The second set of CD spectra to be discussed are of compounds **77**, **78** and **79** (Fig. 12) (Spacers B and D) in the presence of 4:1 POPC/POPG SUVs. The CD spectrum of compound **77**, unlike in the presence of POPC SUVs, is very different in shape and intensity from the CD spectrum of compound **70**. Compound **77** in the presence of 4:1 POPC/POPG SUVs exhibits a strong and a weak minima at approximately 207 nm. Again compounds **78** and **79** exhibited similar CD spectra to each other both with minima at approximately 202 to 205 nm. These spectra are similar to those observed in the presence of POPC SUVs. These CD spectra are different in shape, thus implying a different conformation of binding to the SUVs, from the CD spectra of the active analogs, which from a therapeutic respective is a critical observation.

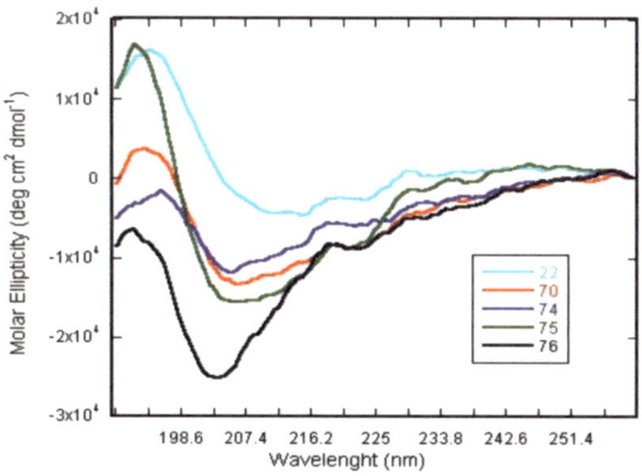

Fig. 11. Far-UV Circular Dichroism spectra of 100 μM solutions of compounds **22, 70, 74, 75,** and **76** in the presence of 1.75 mM 4:1 POPC/POPG SUV in 40 mM phosphate buffer, pH = 6.8.

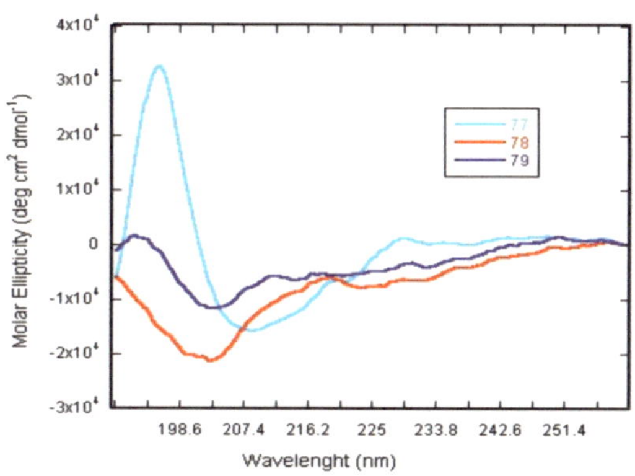

Fig. 12. Far-UV Circular Dichroism spectra of 100 μM solutions of compounds **77, 78** and **79** in the presence of 1.75 mM 4:1 POPC/POPG SUV in 40 mM phosphate buffer, pH = 6.8.

The fact that the CD spectra of the inactive compounds **77**, **78** and **79** in the presence of 4:1 POPC/POPG SUVs are very different from the CD spectra of the active compounds supports our hypothesis that the incorporation of multiple Tic-Oic dipeptide units that are connected via an amino acid spacer with defined properties of charge and hydrophobicity will result in peptides with well-defined physiochemical properties while maintaining sufficient conformational flexibility to allow the peptide to adopt different conformations on interacting with membranes of different chemical compositions, and thus exhibit organism selectivity.

The last sub set of CD spectra to be discussed in the presence of 1.75 mM 4:1 POPC/POPG SUVs are those of compounds **71**, **72**, **73**, and **74** (Fig. 13) which contain Spacer E. The CD spectra of compounds **71**, and **73** exhibited a maxima at approximately 195 nm and a double minima at approximately 220 and 205 nm which would suggest the presence of α-helical conformers (Ladokhin et al., 1999) (Fuchs et al., 2006). In the presence of POPC SUVs the CD spectrum of compound **71** exhibit double minima at approximately 220 and 205 nm without the maxima at 195 nm suggesting that this compound adopts a mixture of β-turn and β-sheet like (Ladokhin et al., 1999) (Fuchs et al., 2006). While the CD spectrum of compound **72** exhibits a minima at approximately 195 nm. This spectrum was very different from the CD spectrum observed in the presence of POPC SUVs. The CD spectrum of compound **74** exhibited a double minima at approximately 220 and 205 nm which would suggest the presence of β–turn or β- sheet conformers (Ladokhin et al., 1999) (Fuchs et al., 2006). Again the observed difference in the CD spectra in the presence of 4:1 POPC/POPG SUVs are consistent with the observed differences in bacteria strain selectivity and potency.

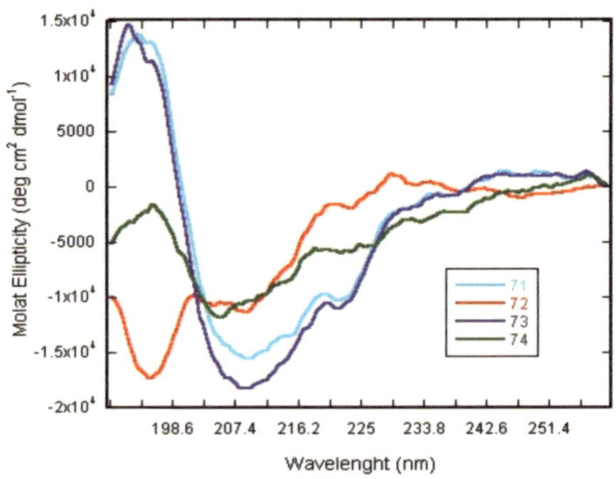

Fig. 13. Far-UV Circular Dichroism spectra of 100 μM solutions of compounds **71, 72, 73,** and **74** in the presence of 1.75 mM 4:1 POPC/POPG SUV in 40 mM phosphate buffer, pH = 6.8.

3.3 Isothermal titration calorimetry studies

The isothermal titration calorimetry (ITC) binding enthalpies for peptides with SUVs, generally, are more exothermic than those in the presence of LUVs. Therefore, SUVs are not ideal membrane models for thermodynamic measurements. (Beschiaschvili & Seelig, 1992; Seelig & Ganz, 1991; Wieprecht, Apostolov, & Seelig, 2000). LUVs formed by extrusion methods are equilibrium structures and are more appropriate to use as a membrane model in thermodynamic studies (Enoch & Strittmatter, 1979). ITC studies were conducted on the most active analogs in hopes of gaining insight into the possible explanations for the observed organism selectivity.

3.3.1 ITC studies of POPC LUVs

The application of isothermal titration calorimetry (ITC) to the study of peptide-membrane interactions is well documented in the literature (Hunter et al., 2005) (Wieprecht, Beyermann, & Seelig, 2002) (Meier & Seelig, 2007). The binding interactions between these AMPs and 35 mM POPC LUVs was investigated using ITC. In an ideal case, ITC would allow for the complete analysis of the thermodynamic parameters (free energy, enthalpy, entropy, binding constants and heat capacity changes) associated with the binding interactions between peptides and LUVs (Wieprecht et al., 2002) (Thomas, Surolia, &

Surolia, 2001). The interaction of peptides with lipid vesicles are driven by three forces: the hydrophobic effect, the coil-helix transition and non-classical hydrophobic effect (Wieprecht, Apostolov, Beyermann, & Seelig, 2000) (Wieprecht, Beyermann, & Seelig, 1999). Further, Wieprecht and co-workers (Wieprecht, Apostolov, Beyermann, et al., 2000) proposed that the binding of a peptide to a LUV is dependent on the global structural physicochemical properties including the overall charge, hydrophobicity, and amphipathicity of the AMP.

As seen in Fig. 14 the titration of 35 mM POPC LUVs into compounds **22**, **70**, **71**, **72**, **74**, **75**, resulted in different thermograms, suggesting that the process that contributes to the binding interactions vary in magnitude with each peptide. All six peptides begin their thermograms with an endothermic phase (Wieprecht, Apostolov, Beyermann, et al., 2000). Previous studies reported in the literature (Abraham, Lewis, Hodges, & McElhaney, 2005) of the interaction between peptides and lipids using ITC have noted that an endothermic phase can be attributed to a combination of several events. These events may include electrostatic interactions between the peptide and the membrane surface, disruption of polar head groups accompanied by reorganization of lipids on the surface of the membrane, disruption of the solvation spheres on both the peptide and the membrane surfaces, and other less understood phenomena (Abraham et al., 2005).

Compounds **22**, **74** and **75** all produced single endothermic phases. With these compounds, titration with POPC resulted in the continuous decrease of heat as the experiment progressed until only the heat of dilution was observed. This behavior is similar to that previously observed for compound 23. While compounds **70**, **71** and **72** all produced two phased thermograms. These thermograms begin with an endothermic phase of varying intensity and then transition into a second exothermic phase of varying intensity.

This behavior is rare, but not unknown. Jelokhani-Niaraki et. al. reported a similar observation for the interaction of aromatic amino acid analogues of gramicidin S with phospholipid membranes (Jelokhani-Niaraki, Hodges, Meissner, Hassenstein, & Wheaton, 2008). Andrushchenko and co-workers reported similar thermograms for the binding of the tryptophan-rich cathelicidin antimicrobial peptides tritrpo4 and tritrpo6 with 7:3 POPE/POPG LUVs (Andrushchenko et al., 2008). This type of two phased thermogram was not observed for the titration of the corresponding three Tic-Oic didpeptide containing analogs into POPC LUVs, only single phased endothermic thermograms were observed.

Several factors may contribute to the observed enthalpies of binding to LUVs such as attractive and repulsive electrostatic interactions, conformational changes of the peptide, changes in Van der Waals contacts due to the dehydration of both the peptide and membrane surfaces and disruption of polar head groups on the surface of the LUV (Seelig, 1997) (Andrushchenko et al., 2008) (Seelig, 2004) (Abraham et al., 2005).The endothermic heat event occurring at low lipid to peptide ratios has been interpreted as the summation of several simultaneous processes. These processes include, but are not limited to, pore formation (Wieprecht, Apostolov, Beyermann, et al., 2000), peptide aggregation (Andrushchenko et al., 2008), as well as changes in the phase properties of the lipids at different points during the titration (Epand, Segrest, & Anantharamaiah, 1990) (Seelig, 2004). The exothermic heat event at high lipid-to-peptide ratios has been attributed to the binding of the peptide to lipid surfaces (Seelig, 1997).

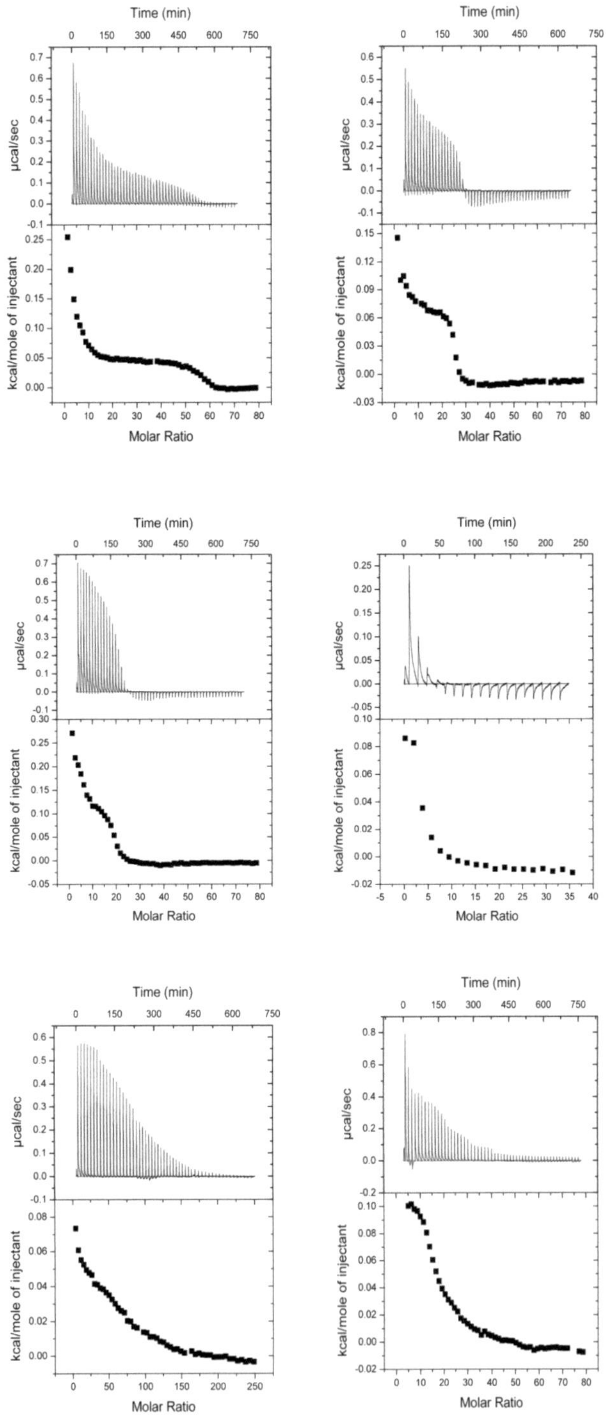

Fig. 14. The ITC thermograms of POPC LUVs titrated into compounds **22**, **70**, **71**, **72**, **74**, and **75**.

3.3.2 ITC studies of 4:1 POPC/POPG LUVs

Titration of 35 mM 4:1 POPC/POPG LUVs into compounds **22**, **70**, **71**, **74** and **75** all resulted in two phased thermograms beginning (Fig. 15) with an endothermic phase of varying intensity transitioning into an exothermic phase of varying intensity. This behavior is similar to that previously observed for compound **23** (Russell et al., 2010). This behavior is very similar to the thermograms observed for the three Tic-Oic dipeptide containing analogs. Not only does the magnitude of the endothermic and exothermic components vary, also the transition point between the endothermic and exothermic phase vary at different lipid to peptide molar ratios. These ratios are given in Table 6. Replacement of the Lys residues in compound **74** with Orn residues in compound **71** reduce the lipid to peptide ratio for the transition from the endothermic to the exothermic phase from 30 to 12. This reduction in the lipid to peptide ratio for the transition from the endothermic to the exothermic phase may possibly help explain the reduction in the overall antibacterial activity of compound **71** compared to compound **74**. Also compound **22** exhibits the most intense exothermic phase of the five compounds investigated, at a lipid to peptide ratio for the transition from the endothermic to the exothermic phase of 18. This observation suggests that a combination of factors such as the lipid to peptide ratio for the transition from the endothermic to the exothermic phase, and the magnitude of the exothermic and endothermic phase all play major roles in defining antibacterial activity, with no single factor correlating directly with the observed antibacterial activity.

Compound	Lipid to Peptide Ratio for transition
22	18
70	18
71	12
74	30
75	10

Table 6. Lipid to Peptide molar ratios for the transition from the endothermic to the exothermic phase of the thermogram for the titration of 4:1 POPC/POPG LUVs into the following peptides

As previously stated there is a growing preponderance of evidence in the literature, indicating that the selectivity and potency of a specific AMP for bacterial cells is determined in a large measure by the chemical composition of the target cell's membrane (Powers & Hancock, 2003; Yeaman & Yount, 2003). Thus we and other researchers have postulated that the bacteria cell membrane's physicochemical surface interactions with the physicochemical surface of the AMP defines organism selectivity.(Dennison et al., 2005; Giangaspero et al., 2001; Glukhov, Stark, Burrows, & Deber, 2005; Powers & Hancock, 2003) In addition, it is our hypothesis that the physicochemical properties of the target cell's membrane interact with the physicochemical properties of the approaching AMP defining its selectivity and potency against that particular cell. During this process conformational changes are induced onto the AMP that will maximize the attractive interactions between the two to facilitate AMP-membrane binding. The observation of different thermograms for these compounds on interacting with POPC and 4:1 POPC/POPG LUVs supports our hypothesis.

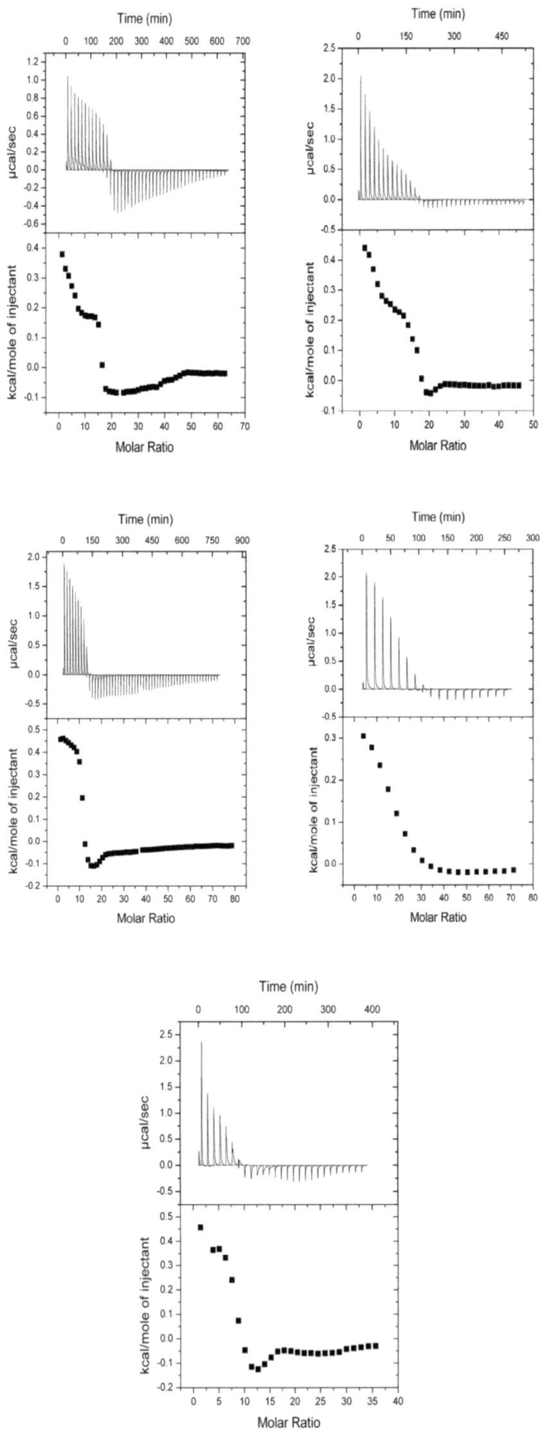

Fig. 15. ITC thermogram of the titration of 4:1 POPC/POPG into compounds **22**, **70**, **71**, **74**, **75**.

4. Material and methods

Sodium dodecyl sulfate (SDS) and Bis-Tris buffer were purchased from Sigma-Aldrich. Monobasic and dibasic sodium phosphate, EDTA and NaCl were purchased from Fischer Scientific. POPC (1-Palmitoyl-2-Oleoyl-sn-Glycero-3-Phosphocholine), POPG (1-Palmitoyl-2-Oleoyl-sn-Glycero-3-[Phospho-rac-(1-glycerol)] (Sodium Salt)) and Dodecylphosphocholine (DPC) were purchased from Avanti Polar Lipids. High purity calcein was purchased from Invitrogen. All chemicals were used without further purification.

4.1 Peptide synthesis

Peptide synthesis was performed either manually using tBOC chemistry or with an automated peptide synthesizer using FMOC chemistry (Grant, 2002) as previously reported (Hicks et al., 2007) (Venugopal et al., 2010) (Russell et al., 2010). All peptides were purified by Reverse Phase HPLC using an Agilent 1100 Series Preparative Instrument and a Vydac C18 Reverse Phase Preparative HPLC Column as previously reported (Venugopal et al., 2010) (Russell et al., 2010). All purified peptides were analyzed again by HPLC and Mass-Spec. Mass Spectral analyses were carried out using a Finnigan LTQ ESI-MS instrument running Xcalibur 1.4SR-1 or a Kratos PC Axima CFR Plus instrument (MALDI) running Kompact V2.4.1. ESI-MS showed multiply charged ions and the accurate mass was calculated. MALDI analyses were performed in reflectron mode (Venugopal et al., 2010) (Russell et al., 2010).

4.2 Preparation of POPC and 4:1 POPC/POPG SUVs

The appropriate amount of dried POPC or 4:1 POPC/POPG (mol to mol) lipid was weighed out to yield a final lipid concentration of 35 mM. The lipid was dissolved in chloroform and vortexed for 3 minutes. The sample was dried under a stream of nitrogen gas for four hours and under high vacuum overnight. The lipid was then hydrated with 2 mL of buffer (40 mM sodium phosphate, pH = 6.8) and vortexed extensively. SUVs were prepared by sonification of the milky lipid suspension using a titanium tip ultrasonicator (Qsonica Sonicators model Q55) for approximately 40 minutes in an ice bath until the solution became transparent. The titanium debris was removed by centrifugation at 14,000 rev/min for 10 minutes using an Eppendorf table top centrifuge (Wieprecht, Apostolov, & Seelig, 2000). It has been reported that the mean diameter for SUVs formed by sonication is approximately 30 nm (C. H. Huang, 1969; Ladokhin et al., 2010). Final concentration used for CD was 1.75 mM or as otherwise stated in the text.

4.3 Preparation of POPC and 4:1 POPC/POPG LUVs

A defined amount of dried POPC or 4:1 POPC/POPG (mol to mol) was weighed, suspended in buffer (40 mM sodium phosphate, pH = 6.8) and spun for 30 minutes, resulting in a total lipid stock solution concentration of 35 mM. Other concentrations of LUVs used are noted in the text and resulted from the dilution of the stock solution. Large unilamellar vesicles (LUVs) were prepared by extrusion using a Mini-Extruder (Avanti Polar Lipid Inc) (Wei, 2006) (Hunter et al., 2005). The solution was extruded through a 100 nm pore size polycarbonate membrane 21 times. After extrusion the LUVs were allowed to "rest" for at least two hours before use to allow for equilibration to occur. The final lipid concentration was calculated based on the weight of the dried lipid (Wieprecht, Apostolov, & Seelig, 2000) (Wieprecht, Apostolov, Beyermann, et al., 2000) (Russell et al., 2010) (Wieprecht et al., 2002) (Wieprecht & Seelig, 2002) (Wen, 2007). Kennedy and co-workers previously reported the preparation of LUVs using this procedure resulting in a homogeneous population of LUVs with >95% of the particles falling into the particle size range of 70-100 nm (Kennedy et al., 2002). We have also previously shown by ^{31}P NMR that these LUVs are unilamellar (Kennedy et al., 2002).

4.4 LUVs for calcein release experiments

A defined amount of dried POPC or 4:1 POPC/POPG LUVs (mol to mol) was weighed and suspended in a calcein-containing buffer (70mM calcein, 10mM Bis-Tris, 150mM NaCl, 1mM EDTA, pH=7.1, the pH was corrected using 3mM NaOH and the final calcein concentration was calculated based on dilution). The resulting solution was vortexed for one minute (5 times). The calcein-encapsulated LUVs were extruded using the same technique as described above. Following extrusion the unencapsulated calcein was removed by gel filtration on a Sephadex G50 column (eluent: buffer containing 10 mM Bis-Tris, 150 mM NaCl, 1 mM EDTA, pH =7.1) The fraction containing calcein-encapsulated liposomes was collected and retained for fluorescence studies (Wieprecht, Apostolov, & Seelig, 2000) (Wieprecht, Apostolov, Beyermann, et al., 2000) (Russell et al., 2010) (Wieprecht et al., 2002) (Wieprecht & Seelig, 2002) (Wen, 2007) (Wieprecht et al., 1996) (Dathe, 1996) (Tamba & Yamazaki, 2005). Prior to use each batch of calcein encapsulated LUVs was subjected to a self-quenching efficiency test. The self-quenching efficiency (Q) for each lipid suspension was set at a minimum value of 80% before it could be used in these investigations. The Q value was calculated using the following equation: $Q = (1 - (F_0/F_T)) \times 100\%$ where F_0 and F_T are the background fluorescence of the lipid suspension and the total fluorescence after the addition of a solution of 10% Triton X, respectively (Jing, Hunter, Hagel, & Vogel, 2003) (Benachir & Lafleur, 1995) (Tachi T, Epand RF, Epand RM, & K., 2002) (Mason, Marquette, & Bechinger, 2007).

4.5 Circular dichroism studies

Peptide stock solutions were dissolved in 40mM phosphate buffer (pH = 6.8). Binding studies were conducted using SUV preparations consisting of 1.75 mM POPC or 4:1 POPC/POPG in 40 mM phosphate buffer (pH = 6.8) with peptide concentrations of 100 µM. All CD spectra were obtained by acquiring 8 scans on a Jasco J-815 CD Spectrometer using a 0.1 mm cylindrical quartz cell (Starna Cells, Atascadero, CA) from 260 to 178 nm at 20 nm/min, with a 1 nm bandwidth, a data pitch of 0.2 nm, a response time of 2.0 sec and a sensitivity of 5 mdeg at room temperature (~25°C). Contributions due SUVs were eliminated by subtracting the lipid spectra of the corresponding peptide-free solutions. All analysis of CD spectra was conducted after smoothing (with a means-movement function) and conversion to molar ellipticity using the JASCO Spectra Analysis program (Russell et al., 2010) (Wei, 2006) (Bringezu et al., 2007). CD spectra that exhibited HT values greater than 400 were not used due to excessive light scattering and / or absorption.

4.6 Isothermal titration calorimetry studies

Data was acquired using a Microcal VP-ITC calorimeter (Microcal, Northampton, MA). All experiments were run in 40 mM sodium phosphate buffer, pH = 6.8, at 25°C. All solutions were degassed for approximately 10 minutes under vacuum before loading the reaction cell and syringe. For full titration experiments, 15 μL aliquots of 35 mM lipid solutions in buffer were titrated into peptide (100 - 200 μM). For binding single injection experiments, 15 μL of dilute samples of peptide (100-200 μM) were titrated into excess lipid (15-20 mM). A stirring speed of 220 rpm and injection duration of 30 sec were chosen to ensure sufficient mixing while keeping the baseline noise to a minimum. To ensure complete equilibration between injections, a delay of 700 sec between injections was used. The background heat of dilution was obtained by titrating LUVs into the reaction cell containing only buffer and was subtracted prior to analysis. Data was analyzed with Origin® software (version 7.0). ITC data collection was obtained in duplicate in an effort to ensure reproducibility (Wieprecht, Apostolov, & Seelig, 2000) (Wieprecht, Apostolov, Beyermann, et al., 2000) (Russell et al., 2010) (Wei, 2006) (Wieprecht et al., 2002) (Wieprecht & Seelig, 2002) (Wen, 2007).

4.7 Calcein leakage assays

Peptide induced calcein leakage studies were conducted using an ISS PC1 photon counting spectrofluorometer (ILC Technology) at an excitation wavelength of 494 nm and an emission wavelength of 518 nm. An aliquot of peptide (4 - 20 μM) in buffer (10 mM Bis-Tris, 150 mM NaCl, 1 mM EDTA, pH = 7.1) was added to the cell containing calcein-encapsulated liposomes (36.6 μM lipid concentration). Measurements were taken every minute for the first 20 minutes of the experiment and every 10 minutes after until no further changes in the emission intensity occurred (approximately 90 minutes). To determine the maximum fluorescence intensity that corresponded to hundred percent leakage, an aliquot of 10 μL of a 10% Triton X solution was added to the sample at the end of each experiment. The apparent percent leakage was calculated using the following equation: % leakage = $[(F_I-F_0)/F_T)] \times 100$ %, where F_0 and F_T are the initial fluorescence before introduction of peptide and after the addition of Triton X, respectively (Dathe, 1996; Russell et al., 2010) (Wieprecht, Apostolov, Beyermann, et al., 2000) (Tamba & Yamazaki, 2005) (Wieprecht et al., 1996) (Wei, 2006).

4.8 Preparation of bacteria samples

A small amount of the organism was streaked onto the appropriate agar plates and the plates incubated at either 30° or 37° C (depending on organism to be tested - Plague and Anthrax were incubated at 30°, Brucella, Francisella, and Burkholderia were incubated at 37°C) overnight. At the end of the incubation period the organisms were harvested using a sterile loop and suspended into a 15 mL centrifuge tube containing 5 mL Mueller Hinton Broth + IsoVitaleX (Bekton Dikinson, Inc.) and thoroughly mixed. A standardized suspension of 1.0 OD_{600} was prepared from a suspension using sterile saline as the diluent and the absorbance was read against a tube containing saline only as a blank using either a spectrophotometer (Spectronix 20, Bausch & Lomb, Inc.) or by a microplate reader (Spectramax Plus384, Molecular Devices, Inc.) at a wavelength of 600nm. The suspension was adjusted to obtain a volume of 5 mL containing a 1.0 OD solution. One 50 mL disposable Erlenmeyer flask containing 20mL of broth with 1 mL of the organism dilution was then inoculated for 24-48 hours (dependent on growth characteristics of the organism in question) in a shaking incubator rotating at 200 rpm (Standards, 2000) (Standard, 2002) (Venugopal et al., 2010) . After 24-48 hours, the suspension was thoroughly mixed, the tubes centrifuged at high speed (2,000 x g) for 15 minutes, the supernatant removed, and the pelleted organisms resuspended to yield a total volume of 45 mL. This was repeated for a total of 3 washes. After the final wash the pelleted organisms were resuspended in 15 mL of broth and a 0.1 OD_{600} standardized suspension was then prepared by making dilutions in sterile cuvettes using broth as the diluent and the absorbance read blanked against a tube containing broth. The suspensions were then diluted to a final concentration of $1x10^5$ cfu/ml with sufficient volume for the number of plates being used for the MIC determination (Standards, 2000) (Standard, 2002; Venugopal et al., 2010) .

4.9 Minimum Inhibitory Concentration

Minimum Inhibitory Concentration (MIC) was determined for the following organisms; Plague (Yersinia pestis), (Pohanka & Skladal, 2009; Revazishvili et al., 2008) Brucella, (Brucella melitensis, suis, or abortus),[64, 65] Anthrax (Bacillus anthracis), (Hicks et al., 2005; Koehler, 2009; Pohanka & Skladal, 2009) Glanders (Burkholderia mallei),(Harley, Dance, Drasar, & Tovey, 1998; Lehavi, Aizenstien, Katz, & Hourvitz, 2002; Manzeniuk et al., 1999) Melioidosis (Burkholderia pseudomallei), [71, 72] Staphylococcus aureus -MRSA [73, 74] and Tularemia (Francisella tularensis) (Pohanka & Skladal, 2009; Santic, Al-Khodor, & Abu Kwaik, 2009) Acinetobacter baumanii and a drug resistant clinical isolate strain of Acinetobacter baumanii (Walter Reed Army Institute of Research) using the following protocol. All protocols used were approved by the appropriate review committee at the Walter Reed Army Institute of Research and are summarized below. Modifications of the NCCLS methods were employed for these analyses. [76, 77] A solution of Mueller Hinton Broth + IsoVitaleX (Bekton Dikinson, Inc.) was prepared containing1% Dimethylsulfoxide (DMSO Sigma Scientific) and referred to herein as HIBCD broth. Frozen antibiotic solutions were thawed and diluted to a final concentration of 500 μg/mL in HIBCD broth (Standards, 2000) (Standard, 2002). Using a robotic sample processor (Precision XS, Biotek Instruments, Inc.), 15μL of HIBCD broth was transferred into all wells of a 384 well plate and the plate was then inoculated using 15μL of the organism suspension into all 384 wells using one organism per plate. The plates were incubated overnight and the optical density was read using the microplate reader (Spectromax Plus384) at a wavelength of 600nm and repeated at 24 hour intervals until the control wells reached an optical density of 1.0OD. The MIC values correspond to the highest compound dilution with no measurable growth as determined from OD readings compared to both negative and control measurements (Standards, 2000) (Standard, 2002) (Venugopal et al., 2010).

5. Conclusion

Our laboratory has previously reported the development of a series of novel antimicrobial peptides (AMPs) that incorporate unnatural amino acids into their primary sequence to impart specific three-dimensional physicochemical properties onto the peptide (Hicks et al., 2007,) (Bhonsle et al., 2007) (Venugopal et al., 2010) (Russell et al., 2010). Those AMPs exhibited low μM to nM in-vitro MIC antibacterial activity against several strains of Gram positive, Gram negative and mycobacterium. Many also exhibited very low hemolytic activity and low acute in vivo toxicity (125 mg/kg) in mice; these AMPs have also exhibited excellent metabolic stability in pooled human liver microsomes for up to 60 minutes (unpublished results). (Hicks et al., 2007)(Venugopal et al., 2010). The first generation AMPs focused on the incorporation of three Tic-Oic dipeptide units separated by an amino acid spacer (Spacer #1) and either a charged residue (Spacer 2) or a hydrophobic residue to control the conformational physicochemical properties of the AMP. Unnatural amino acids were selected for incorporation into the primary sequence of the AMP to provide peptide chemists with a "toolbox" of new functionality. This "toolbox" would allow for the development of novel peptides with specific physicochemical properties that will interact with cell membranes in novel ways thus providing entry to the synthesis of organism specific AMPs. The incorporation of unnatural amino acids into the peptide sequence of an AMP offers several advantages over an AMP consisting exclusively of naturally occurring amino acids. 1) Unnatural amino acids inherently exhibit greater metabolic stability. Therefore, incorporation of unnatural amino acids into the primary sequence generally results in an increase in metabolic stability compared to peptides consisting of only natural amino acids (Toke, 2005, Hancock and Lehrer, 1998). 2) The use of the conformationally restrained amino acids (Tic an Oic) reduces the local flexibility of the peptide backbone and thus reduces the total conformational space that may be sampled by the peptide during lipid binding.

Here in the in vitro antibacterial activity and physical characterization of a new series of AMPs incorporating six Tic-Oic dipeptide units as well as four spacers, A, B, C and D on either side of the intervening hydrophobic and charged residues as well as a fifth spacer, E, which defines the distance between the polypeptide backbone and the positively charged side chain amine group is reported. This new series of AMPs is overall less active than the series containing only three Tic-Oic dipeptide units and two spacers. However, the new analogs exhibit far greater bacteria strain selectivity than the previously reported analogs. Induced calcein leakage studies, CD and ITC investigations indicate that the analogs containing six Tic-Oic dipeptide units and the additional spacers interact with POPC and 4:1 POPC/POPG LUVs and SUVs somewhat differently than the previous analogs. These differences could account for the observed differences in bacteria strain selectivity. Induced calcein leakage studies, CD and ITC studies also indicate that the five spacers, A, B, C, D, and E contribute differently to organism selectivity and potency. For example Spacer D completely eliminates any activity, while Spacer A and E seem to play more important roles in defining organism selectivity and potency.

What is clear from this study is that correct positioning of amino acids that influence the conformational and physicochemical properties that are favorable to interactions with a membrane with a specific chemical composition is critical to obtain organism selectivity and potency.

6. Acknowledgements

The authors would like to acknowledge funding from the Bacterial Therapeutics Program 2.1 of the Defense Threat Reduction Agency. Contract # W81XWH-08-2-0095. The authors would also like to acknowledge funding from the North Carolina Biotechnology Center grant number 2006-FRG-1015 and from East Carolina University.

7. References

Abraham, T., Lewis, R. N. A. H., Hodges, R. S., & McElhaney, R. N. (2005). Isothermal titration calorimetry studies of the binding of the antimicrobial peptide gramicidin S to phospholipid bilayer membranes. *Biochemistry*, 44, 11279-11285.

Andrushchenko, V. V., Aarabi, M. H., Nquyen, L. T., Prenner, E. J., & Vogel, H. J. (2008). Thermodynamics of the interaction of tryptophan-rich cathelicidin antimicrobial peptides with model and natural membranes. *Biochimica et Biophysica Acta*, 1778, 1004-1014.

Benachir, T. Lafleur, M. (1995). Study of vesicle leakage induced by melittin. *Biochimica et Biophysica Acta*, 1235, 452-460.

Berova, N., Nakanishi, K., Woody, R. W. (2000). *Circular Dichroism Principles and Applications*. New York: Wiley-VCH.

Beschiaschvili, G., Seelig, J. (1992). Peptide binding to lipid bilayers: nonclassical hydrophobic effect and memrane-induced pK shifts. *Biochemistry*, 31, 10044-10053.

Bhonsle, J. B., Venugopal, D., Huddler, D. P., Magill, A. J., Hicks, R. P. (2007). Application of 3D-QSAR for identification of descriptors defining bioactivity of antimicrobial peptides. *J. Med. Chem.*, 50(26), 6545-6553.

Bringezu, F., Wen, S., Dante, S., Hauss, T., Majerowicz, M., & Waring, A. (2007). The insertion of the antimicrobial peptide dicynthaurin monomer in model membranes: thermodynamic and structural characterization. *Biochemistry*, 46, 5678-5686.

Brogden, K. A. (2005). Antimicrobial peptides: pore formers or metabolic inhibitors in bacteria? *Nature Reviews Microbiology*, 3, 238-250.

Bush, K. (2004). Why it is important to continue antibacterial drug discovery. *ASM News*, 70, 282-287.

Dathe, M. (1996). Peptide Helicity and Membrane Surface Charge Modulate the Balance of Electrostatic and Hydrophobic Interactions with Lipid Bilayers and Biological Membranes. *Biochemistry*, 35, 12612-12622.

Dennison, S. R., Wallace, J., Harris, F., Phoenix, D. A. (2005). Amphiphilic a-helical antimicrobial peptides and their structure/function relationships. *Protein and Peptide Letters*, 12, 31-39.

Devine, D. A., Hancock, R. E. W. (2002). Cationic Peptides: Distribution and Mechanisms of Resistance. *Current Pharmaceutical Design*, 8, 703-714.

Enoch, H. G., Strittmatter, P. (1979). Formation and properties of 1000-A- diameter, single-bilayer phospolipid vesicles. *PNAS*, 76, 145-149.

Epand, R. M., Segrest, J. P., Anantharamaiah, G. M. (1990). Thermodynamics of the binding of human apolipoprotein A-1 to dimyristoylphosphatidylglycerol. *J. Biol. Chem.*, 265, 20829-20832.

Findlay, B., Zhanel, G. G., Schweizer, F. (2010). Cationic amphipiles, a new generation of antimicrobials inspired by natural antimicrobial peptide scaffold. *Antimicrobial Agents and Chemotherapy*, 54, 4049-4058.

Fuchs, P. F. J., Bonvin, A. M. J. J., Bochicchio, B., Pepe, A., Alix, A. J. P., Tamburro, A. M. (2006). Kinetics and thermodynamics of tyoe VIII β-turn formation: A CD, NMR and microsecond explicit molecular dynamics study of the GDNP tetrapeptide. *Biophysical J.*, 90, 2745-2759.

Giangaspero, A., Sandri, L., Tossi, A. (2001). Amphipathic α-helical antimicrobial peptides. *Eur. J. Biochem.*, 268, 5589-5600.

Glattli, A., Daura, X., Seebach, D., van Gunsteren, W. F. (2002). Can one derive the conformational preference of a β-peptide from its CD spectrum. *J. Am. Chem. Soc.*, 124, 12972-12978.

Glukhov, E., Stark, M., Burrows, L. L., Deber, C. M. (2005). Basis for selectivity of cationic antimicrobial peptides for bacterial versus mammalian membranes. *J. Biol. Chem.*, 280, 33960-33967.

Godballe, T., Nilsson, L. L., Petersen, P. D., Jenssen, H. (2011). Antimicrobial β-peptides and α-peptoids. *Chem Biol Drug Des*, 77, 107-116.

Grant, G. A. (2002). *Synthetic Peptides, A user's guide* (2nd ed.). New York, NY: Oxford University Press.

Hancock, R. E. W. (1998). The therapeutic potential of cationic peptides. *Exp. Opin. Invest. Drugs*, 7, 167-174.

Hancock, R. E. W., Lehrer, R. (1998). Cationic peptides: a new source of antibiotics. *Trends Biotechnol*, 16, 82-88.

Harley, V. S., Dance, D. A., Drasar, B. S., Tovey, G. (1998). Effects of Burkholderia pseudomallei and other Burkholderia species on eukaryotic cells in tissue culture. *Microbios*, 96(384), 71-93.

Hartl, G. (2000). Drug resistance threatens to reverse medical progress.: World Health Organization.

Hicks, R. P., Bhonsle, J. B., Venugopal, D., Koser, B. W., Magill, A. J. (2007). De Novo Design of Selective Antibiotic Peptides by Incorporation of Un-natural Amino Acids, *J. Med. Chem.* , 50(13), 3026-3036.

Hicks, R. P., Hartell, M. G., Nichols, D. A., Bhattacharjee, A. K., van Hamont, J. E., Skillman, D. R. (2005). The medicinal chemistry of botulinum, ricin and anthrax toxins. *Curr Med Chem*, 12(6), 667-690.

Hicks, R. P., Mones, E., Kim, H., Koser, B. W., Nichols, D. A., Bhattacharjee, A. K. (2003). Comparison of the conformation and electrostatic surface properties of magainin peptides bound to SDS and DPC micelles: Insight into possible modes on antimicrobial activity. *Biopolymers*, 68, 459-470.

Huang, C. H. (1969). Studies on phosphatidylcholine vesicles. Formation and physical characteristics. *Biochemistry*, 8, 344-352.

Huang, Y., Huang, J., Chen, Y. (2010). Alpha-helical cationic antimicrobial peptides:relationships of structure and function. *Protein Cell*, 1, 143-152.

Hunter, H. N., Jing, W., Schibli, D. J., Trinh, T., Park, I. Y., Kim, S. C., et al. (2005). The interactions of antimicrobial peptides derived from lysozyme with model membrane systems. *Biochimica et Biophysica Acta*, , 1668, 175-189.

Jelokhani-Niaraki, M., Hodges, R. S., Meissner, J. E., Hassenstein, U. E., Wheaton, L. (2008). Interaction of gramicidin S and its aromatic amino acid analoques with phospholipid membranes. *Biophys J.*, 95(7), 3306-3321.

Jing, W., Hunter, H. N., Hagel, J., Vogel, H. J. (2003). The structure of the antimicrobial peptide Ac-RRWWRF-NH2 bound to micelles and its interactions with phospholipid bilayers. *J. Pept. Res.*, 61, 219-229.

Kamysz, W. (2005). Are antimicrobial peptides an alternative for conventional antibiotics. *Nuclear Med. Reviews*, 8(1), 78-86.

Kennedy, A., Hmel, P. J., Seelbaugh, J., Quiles, J. Q., Hicks, R. P., Reid, T. J. (2002). Characterization of the main phase transition in 1,2-

dipalmitoylphosphateidylcholine LUV's by 1H NMR. *J. Liposome Research*, 12(3), 221-237.

Khandelia, H. I., J. H.; Mouritsen, O. G. . (2008). The impact of peptides on lipid membranes. *Biochim. Biophys. Acta* 1778, 1528-1536.

Klevens, R. M., Edwards, J. R., Richards, C. L., Horan, T. C., Gaynes, R. P., Pollack, D. A., et al. (2007). Estimating Health Care-Associated Infections and Deaths in U.S. Hospitals, 2002. Public Health Reports, 122, 160-166.

Koehler, T. M. (2009). Bacillus anthracis physiology and genetics. *Mol Aspects Med*, 30(6), 386-396.

Ladokhin, A. S., Selsted, M. E., White, S. H. (1999). CD spectra of indolicidin antimicrobial peptides suggest turns, not polyproline helix. *Biochemistry*, 38, 12313-12319.

Ladokhin, A. S., Vidal, M. F., White, S. H. (2010). CD spectroscopy of peptides and proteins bound to large unilamellar vesicles. *J. Membrane Biol*, 236, 247-253.

Lehavi, O., Aizenstien, O., Katz, L. H., Hourvitz, A. (2002). Glanders--a potential disease for biological warfare in humans and animals. Harefuah, 141 Spec No, 88-91, 119.

Leontiadou, H., Mark, A. E., Marrink, S. J. (2006). Antimicrobial peptides in action. *J Am Chem Soc*, 128(37), 12156-12161.

Manzeniuk, I. N., Galina, E. A., Dorokhin, V. V., Kalachev, I., Borzenkov, V. N., Svetoch, E. A. (1999).[Burkholderia mallei and Burkholderia pseudomallei. Study of immuno- and pathogenesis of glanders and melioidosis. Heterologous vaccines. *Antibiot Khimioter*, 44(6), 21-26.

Mason, A. J., Marquette, A., Bechinger, B. (2007). Zwitterionic phospholipids and sterols modulate the antimicrobial peptide-induced membrane destabilization. *Biophysical J.*, 93, 4289-4299.

Medina, M. L., Chapman, B. S., Bolender, J. P., Plesniak, L. A. (2002). transient vesicle leakage initiated by synthetic apoptotic peptides derived from the death domain of neurotrophin receptor p75NTR. *J. Peptide Science*, 59, 149-158.

Meier, M., Seelig, J. (2007). Thermodynamics of the Coil β-Sheet Transition in a Membrane Environment. *J. Mol. Biol.* , 369, 277-289. .

Melo, M. N., Ferre, R., Castanho, M. A. (2009). Antimicrobial peptides: linking partition, activity and high membrane-bound concentrations. *Nat Rev Microbiol*, 7(3), 245-250.

Papo, N., Shai, Y. (2003). New lytic peptides based on the D,L amphipathic helix motif preferentially kill tumor cells compared to normal cells. *Biochemistry*, 42, 9346-9354.

Pohanka, M., Skladal, P. (2009). Bacillus anthracis, Francisella tularensis and Yersinia pestis. The most important bacterial warfare agents - review. *Folia Microbiol* (Praha), 54(4), 263-272.

Powers, J.-P. S., & Hancock, R. E. W. (2003). The relationship between peptide structure and antibacterial activity. *Peptides*, 24, 1681-1691.

Revazishvili, T., Rajanna, C., Bakanidze, L., Tsertsvadze, N., Imnadze, P., O'Connell, K., et al. (2008). Characterisation of Yersinia pestis isolates from natural foci of plague in the Republic of Georgia, and their relationship to Y. pestis isolates from other countries. *Clin Microbiol Infect*, 14(5), 429-436.

Russell, A. L., Kennedy, A. M., Spuches, A., Venugopal, D., Bhonsle, J. B., Hicks, R. P. (2010). Spectroscopic and thermodynamic evidence for antimicrobial peptide membrane selectivity. *Chem. Phys. Lipids*, 163, 488-497.

Ryge, T. S., Frimodt-Moller, N., Hansen, P. R. (2008). Antimicrobial activities of twenty lysine-peptoid hybrids against clinically relevant bacteria and fungi. *Chemotherapy,* 54, 152-156.

Santic, M., Al-Khodor, S., Abu Kwaik, Y. (2009). Cell biology and molecular ecology of Francisella tularensis. *Cell Microbiol.*

Seelig, J. (1997). Titration calorimetry of lipid-peptide interactions. *Biochimica et Biophysica Acta,* 1331, 103-116.

Seelig, J. (2004). Therodynamics of lipid-peptide interactions. *Biochimica et Biophysica Acta,* 1666, 40-50.

Seelig, J., Ganz, P. (1991). Nonclassical hydrophobic effect in membrane binding equlibria. *Biochemistry,* 30, 9354-9359.

Shlaes, D. M., Projan, S. J., & Edwards, J. E. (2004). Antibiotic discovery: state of the state. *ASM News,* 70, 275-281.

Song, J. (2008). What's new on the antimicrobial horizon? *Int. J. Antimicrob. Agents,* 32, S207-S213.

Song, Y. M., Park, Y., Lim, S. S., Yang, S.-T., Woo, E.-R., Park, S., et al. (2005). Cell selectivity and mechanism of action of antimicrobial model peptides containing peptoid residues. *Biochemistry,* 44, 12094-12106.

Standard, N. C. f. C. L. (2002). Performance Standards for Antimicrobial Susceptibility Testing: . In N. C. f. C. L. S. (2002). & P. S. f. A. S. T. . (Eds.) (Vol. Twelfth Informational Supplement M100-S12.): NCCLS, Wayne, PA, USA

Standards, N. C. f. C. L. (2000). Methods for Dilution Antimicrobial Susceptibility Tests for Bacteria That Grow Aerobically—Fifth Edition:. In N. C. f. C. L. Standards. (Ed.) (Vol. Approved Standard M7-A5.): NCCLS, Wayne, PA, USA

Tachi T, Epand RF, Epand RM, K., M. (2002). Position-dependent hydrophobicity of the antimicrobial magainin peptide affects the mode of Peptideâ^'Lipid interactions and selective toxicity. *Biochemistry* 41(34), 10723-10731.

Tamba, Y., & Yamazaki, M. (2005). Single Giant Unilamellar Vesicle Method Reveals Effect of Antimicrobial Peptide Magainin 2 on Membrane Permeability. Biochemistry, 44, 15823-15833.

Thomas, C. J., Surolia, N., Surolia, A. (2001). Kinetic and thermodynamic analysis of the interactions of 23-residue peptides with endotoxin. *J. Biological Chemistry,* 276, 35701-35706.

Toke, O. (2005). Antimicrobial peptides;new candidates in the fight against bacterial infections. *Biopolymers,* 80, 717-735.

Venugopal, D., Klapper, D., Srouji, A., Bhonsle, J. B., Borschel, R., Mueller, A., et al. (2010). Novel antimicribial peptides that exhibit activity against select agents and other drug resistant bacteria. *Bioorganic & Medicinal Chemistry,* 18, 5137-5147.

Wang, G., Li, X., Wang, Z. (2009). APD2: the updated antimicrobial peptide database and its application in peptide design. Nucleic Acids Res., 37, D933-937.

Wei, S.-T. (2006). Solution Structure of a Novel Tryptophan-Rich Peptide with Bidirectional Antimicrobial Activity. *Journal of Bacteriology,* 188, 328-334.

Wen, S. (2007). Dicynthaurin (ala) Monomer Interaction with Phospholipid Bilayers Studied by Fluorescence Leakages and Isothermal Titration Calorimetry. . *Journal of Physical Chemistry,* 111, 6280-6287.

Wieprecht, T., Apostolov, O., Beyermann, M., Seelig, J. (2000). Membrane binding and pore formation of the antibacterial paprtide PGLa: thermodynamic and mechanistic aspects. *Biochemistry* (39), 442-452.

Wieprecht, T., Apostolov, O., Seelig, J. (2000). Binding of the antibacterial peptide magainin 2 amide to small and large unilamellar vesicles. *Biophysical Chem.*, 85, 187-198.

Wieprecht, T., Beyermann, M., Seelig, J. (1999). Binding of Antibacterial Magainin Peptides to Electrically Neutral Membranes: Thermodynamics and Structure. *Biochemistry*, 38, 10377-10387.

Wieprecht, T., Beyermann, M., Seelig, J. (2002). Thermodynamics of the coil a-helix transition of amphipathic peptides in a membrane environment: role of vesicle curvature. *Biophysical Chem.*, 96, 191-201.

Wieprecht, T., Dathe, M., Schumann, M., Krause, E., Beyermann, M., Bienert, M. (1996). Conformation and functional study of magainin 2 in model membrane environment using the new approach of systematic double D-amino acid replacement. *Biochemistry* (35), 10844-10853.

Wieprecht, T., Seelig, J. (2002). Isothermal titration calorimetry for studying interactions between peptides and lipid membranes. In S. A. Simon & T. J. McIntosh (Eds.), *Peptide-Lipid Interactions* inVol. *Current Topics in Membranes*, ,32-58). San Diego: Academic Press.

Yeaman, M. R., Yount, N. Y. (2003). Mechanisms of antimicrobial pepitde action and resistance. *Pharmacological Reviews*, 55(1), 27-55.

Zasloff, M. (2002). Antimicrobial peptides of multicellular organisms. *Nature* 415, 389-395.

Zhang, L., & Falla, T. J. (2009). Host defense peptides for use as potential therapeutics. *Curr Opin Investig Drugs*, 10(2), 164-171.

Zhang, L., Harris, S. C., Falla, T. J. (2005). *Therapeutic application of innate immunity peptides.* San Diego: Horizon Bioscience

15

Synthesis, Spectral, Magnetic, Thermal and Antimicrobial Studies on Symmetrically Substituted 2, 9, 16, 23-tetra-phenyliminophthalocyanine Complexes

M. H. Moinuddin Khan[1,*], K. R. Venugopala Reddy[2] and J. Keshavayya[3]
[1]Department of Chemistry, Jawaharlal Nehru National College of Engineering, Shimoga, Karnataka,
[2]Department of Studies in Industrial Chemistry, Sahyadri Science College (Autonomous) Kuvempu University, Shimoga, Karnataka,
[3]Department of Studies in Chemistry, School of Chemical Sciences, Kuvempu University, Jnanasahyadri, Shankaraghatta, Shimoga, Karnataka, India

1. Introduction

The Phthalocyanine (or tetraazatetrabenzporphyrin) ligand has a heteroaromatic π-system and readily forms complexes with many transition metals. The aza-nitrogen and peripheral fused benzene ring imparts chemical and thermal stability to the phthalocyanine macromolecule. The optical properties of metal phthalocyanine complexes have been studied extensively for several decades [1-3]. The low solubility of these complexes combined with the intense π–π* bands associated with phthalocyanine ligand led to the industrial applications as pigments in paints and dyestuffs [4]. In recent decades, there has been renewed interest in the use of metal phthalocyanine complexes in number of high technological applications, including those based upon the close structural relationships of the phthalocyanine to porphyrin complexes. Mimicking the natural energy cycle of chlorophyll, the oxygen binding capacity and activation properties of the heme proteins has been a key role in phthalocyanine research [5-9]. New applications include as photosensitizers in PDT and in anti-scrapie treatments [10-11], as power leads and as molecular switches in nanotechnology [12] and as potential industrial catalysts [13].

Eventhough the information on synthesis and structural investigations of metal (II) 2, 9, 16, 23-tetraamino phthalocyanines were documented [14] in the literature, no evidences are available on synthesis and structural studies on metal (II) 2, 9, 16, 23-tetraimino phthalocyanines starting from the respective amino phthalocyanine complexes.

In the present chapter we report the synthesis, characterization and antimicrobial studies of 2, 9, 16, 23-tetra-phenyliminophthalocyanine complexes of copper (II), cobalt (II), nickel (II)

* Correspondig Author

and zinc (II). The procedure available from the literature [14-16] was suitably modified and used for the synthesis of title complexes.

2. Materials and methods

4-nitrophthalic acid was synthesized by using phthalic anhydride adopting the procedure reported elsewhere [14]. All other chemicals were of analytical grade and were used as such. Metal (II) 2, 9, 16, 23-tetra-phenyliminophthalocyanines (M-PhImPc) were prepared as per the scheme 1.

2.1 Preparation of cobalt (II) 2, 9, 16, 23-tetra-phenyliminophthalocyanine complex

The procedure adopted for the synthesis of cobalt (II) 2, 9, 16, 23-tetra-nitro phthalocyanine (M-PcTN) was reported elsewhere [15]. The nitro derivative of the aforesaid complex was

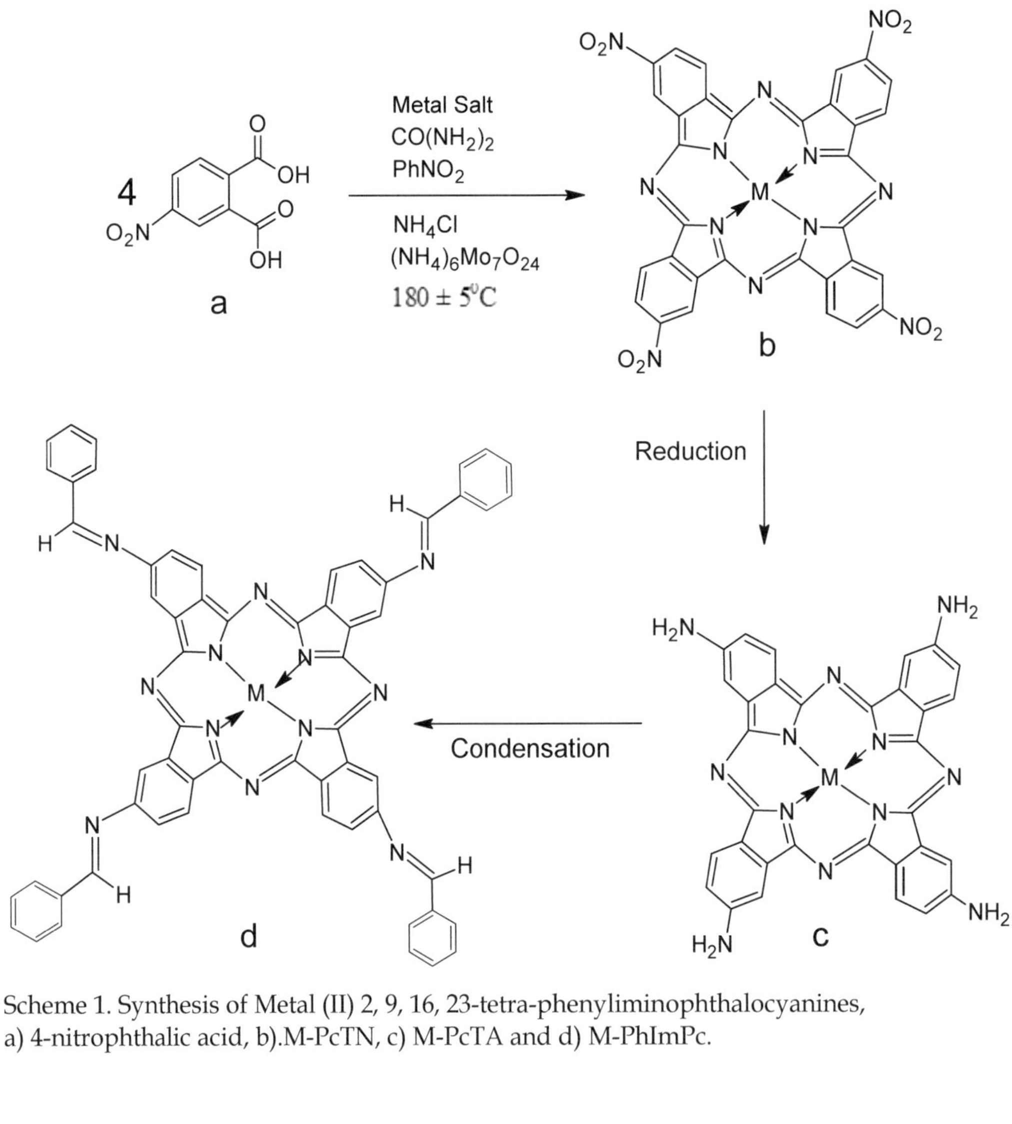

Scheme 1. Synthesis of Metal (II) 2, 9, 16, 23-tetra-phenyliminophthalocyanines, a) 4-nitrophthalic acid, b).M-PcTN, c) M-PcTA and d) M-PhImPc.

converted into amino derivative quantitatively by reduction using sodium sulfide nonahydrate in aqueous medium [14]. The finely powdered cobalt (II) 2, 9, 16, 23-tetra-amino phthalocyanine (M-PcTA) (6.30 g 0.01 mole) was dissolved in 4.2 mL of DMSO and stirred with benzaldehyde. Above mixture was refluxed for 5 hours in the presence of catalytic quantity of concentrated sulphuric acid and contents were poured into ice cold water. The settled bluish green coloured condensed phenylimino phthalocyanine complex was washed with alcohol several times till free from aldehyde.

The pigment form of the above complex was obtained by the acid pasting process, in which 1 part of powdered sample was dissolved in 6-10 parts of concentrated sulphuric acid. The mixture was allowed to stand for 1-2 hour and then poured on to 45-50 parts of crushed ice and stirred thoroughly. The pigment thus obtained was filtered off and washed with hot water. Finally it was washed with distilled water and dried in vacuum over phosphorous pentoxide.

Metal (II) 2, 9, 16, 23-tetra-phenyliminophthalocyanines of Cu (II), Ni (II) and Zn (II) were prepared by the above procedure using the respective metal amino phthalocyanines C, H and N elemental analyses were done at STIC, Kerala, India. Magnetic susceptibility studies were made at room temperature (29° C) using Gouy magnetic balance consisting of NP-53 type electromagnets with a DC power supply unit and a semi microbalance. Pascal's constants were used to calculate the diamagnetic corrections. A $Hg[Co(SCN)_4]$ complex was used as calibrant [17]. Varian Cary 5000 with 1 cm width silica cell was used for electronic absorption spectral studies. IR spectra were recorded using Nicolet MX-FT IR spectrometer. Philips analytical PW-1710 X-ray diffractrometer was used to study the diffraction pattern of the complexes. The spectra were recorded using Cu Kα at a voltage of 40 kv, a current of 20 mA, a time constant of 4, a channel width of 7mm and a chart speed of 10 mm/min. TGA studies were carried out by using a Perkin Elmer, Diamond TG/DTA thermal analyzer at a heating rate of 10°/min both in the air and nitrogen atmosphere.

3. Results and discussion

The procedure adopted for the synthesis of M-PhImPcs yield pure complexes. The title complexes are dark bluish green in color. These complexes give clear solution in concentrated sulfuric acid, dimethyl sulphoxide (DMSO), dimethyl formamide (DMF) and pyridine and are sparingly soluble in alcohol. The results of elemental analysis for carbon, hydrogen and nitrogen and gravimetric methods for metals (Table II) are in good agreement with the calculated values and are consistent with the proposed structure.

3.1 Electronic spectra

The electronic spectra of M-PhImPcs were recorded in DMSO in the concentration range of 1.0-1.5 x 10^{-4} M and the results were summarized in Table 1 and the graphs in Fig. 2. The observed deep bluish green color of the complexes may be due to $a_{2u} \rightarrow e_g$ and $b_{2u} \rightarrow e_g$ transitions [17]. For all the complexes absorption bands were observed in the wavelength range 732-770 nm, which are considerably higher than the corresponding parent metal phthalocyanines [17]. This observed red shift was attributed to the increase in conjugation of π-electrons of the phthalocyanine molecule with that of peripheral substituted aromatic imino groups. The splitting of the Q-band was observed in all the complexes in the wavelength range 531-572 nm. The origin of the Q-band was attributed to the $a_{1u} \rightarrow e_g$ transition of the phthalocyanine molecule. A sharp and intense B-band was observed in all the complexes in the range of 331-355 nm. A weak L-band was observed for all metal-imino phthalocyanines at 252-266 nm.

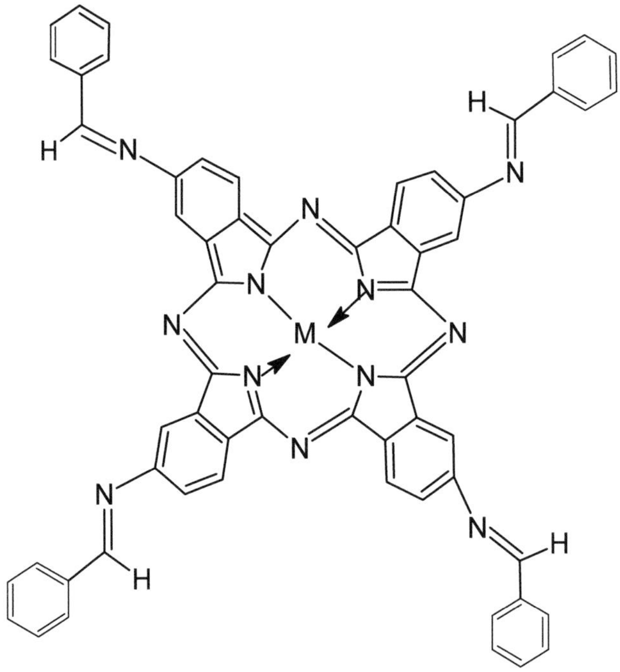

Fig. 1. Suggested structure of symmetrically tetra substituted phenyliminophthalocyanine, where M = Co, Cu, Ni and Zn.

Name of the Complex	UV-Visible wavelength λ nm (log ε)	IR Spectral Data (cm^{-1})	Powder XRD Data 2θ angle (d A°)	Relative Intensity %
Co-PhImPc	256 (4.13) 336 (3.98) 572 (3.53) 769 (4.19)	607,752,1103, 1316,1491,1626	25.96 (3.42) 25.52 (3.48) 43.71 (2.10)	100.00 64.12 24.25
Cu-PhImPc	266 (3.62) 335 (3.99) 569 (3.00) 770 (4.09)	690,747,1098, 1310,1497,1615	28.95 (3.08) 42.96 (2.10) 25.51 (3.48)	100.00 59.59 24.00
Ni-PhImPc	260 (4.61) 331 (4.60) 531 (4.10) 765 (4.53)	648,752,1098, 1316,1491,1624	26.54 (3.35) 29.39 (4.27) 40.31 (2.10)	100.00 29.39 25.45
Zn-PhImPc	252 (4.02) 355 (4.89) 578 (4.45) 732 (4.82)	700,742,1093, 1341,1486,1634	26.87 (3.56) 28.58 (3.21) 32.75 (2.54)	100.00 86.26 85.98

Table 1. Spectral data of metal (II) 2, 9, 16, 23-tetra-phenylimino phthalocyanines.

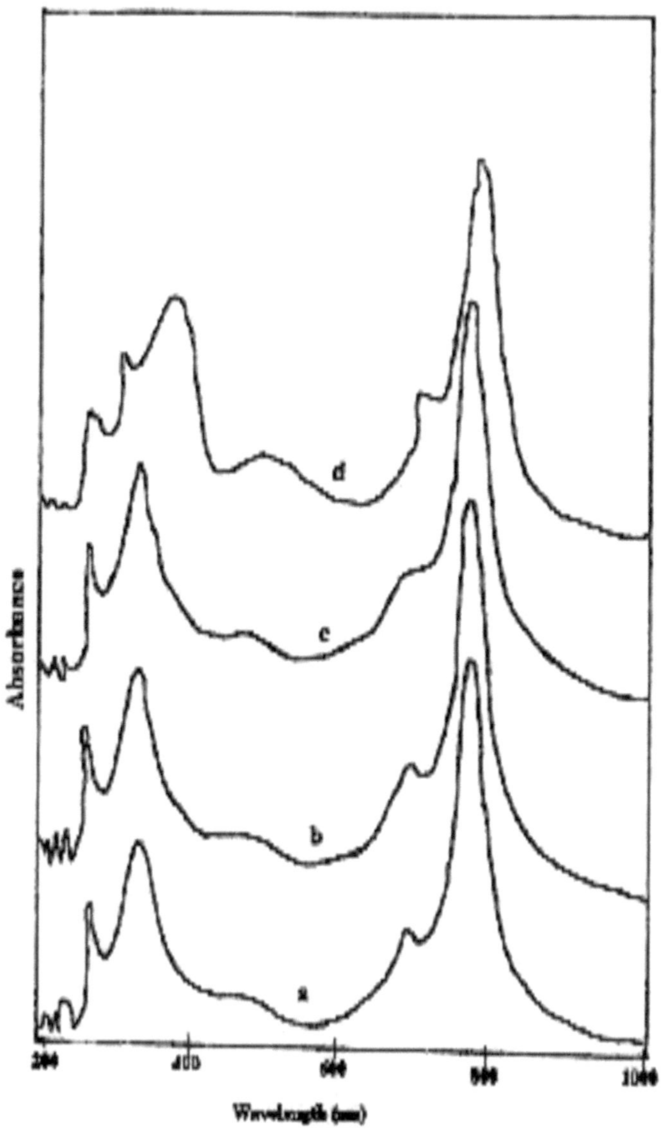

Fig. 2. Electronic spectra of (a) Co–PhImPc, (b) Cu–PhImPc, (c) Ni–PhImPc and (d) Zn–PhImPc.

3.2 IR spectra

IR spectral data were recorded in KBr pellets and the results were presented in Table 1 and the graphs in Fig. 3. The sharp peaks at the range of 1615-1634 cm^{-1} were attributed to C=N of imine group and peaks at the range of 1468-1497 cm^{-1} were due to C-N aromatic stretching. The peaks observed in the range of 1310-1341 cm^{-1} were due to C-H symmetric bending. All the remaining bands observed in the range 742-752 and 607-700 cm^{-1} may be assigned to various skeletal vibrations of the phthalocyanine ring.

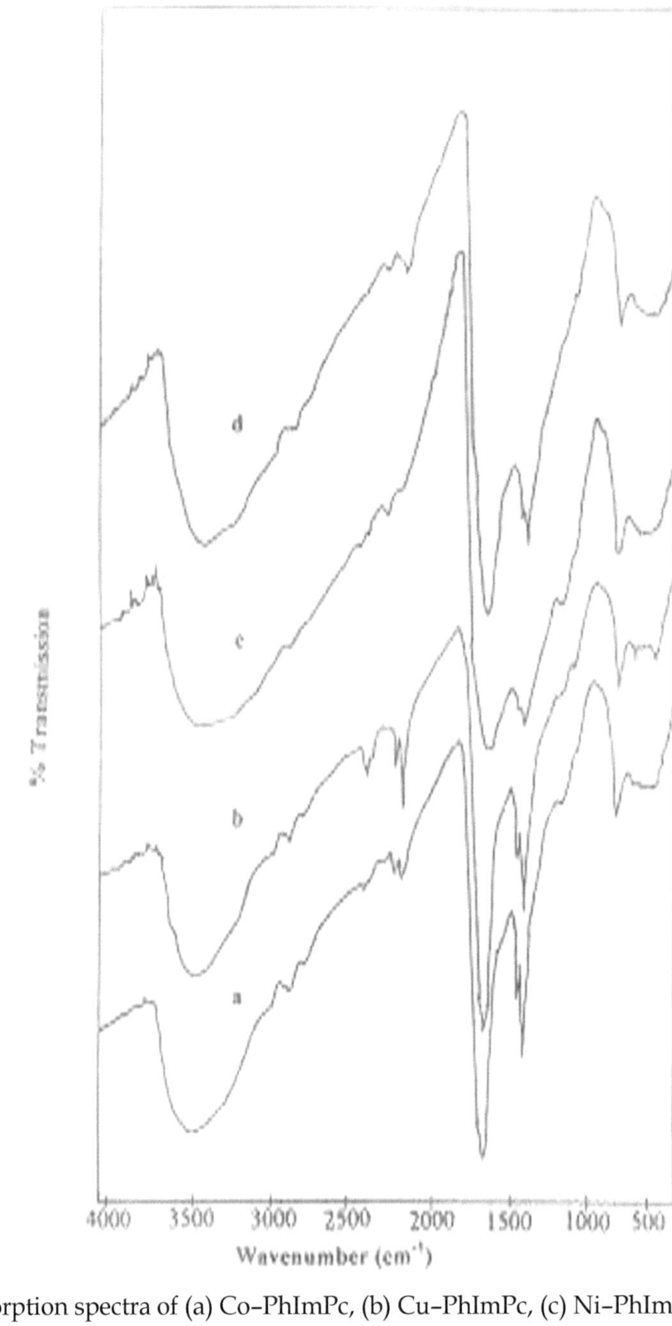

Fig. 3. IR absorption spectra of (a) Co–PhImPc, (b) Cu–PhImPc, (c) Ni–PhImPc and (d) Zn–PhImPc.

3.3 Powder XRD

Powder X-ray diffraction patterns of M-PhImPcs were taken in the range of 2θ angles 6-70⁰ showed identical peaks with relatively poor crystallinity (Table 1, Fig. 4). The observed patterns were very much similar to that of unsubstituted parent phthalocyanines except for the broadening of the peaks with diffused intensity. The broadening may be due to the presence of substitutents at the periphery of the molecule, which seems to provide hindrance for the effective stacking of the molecule and thus the poor crystallinity of the complexes.

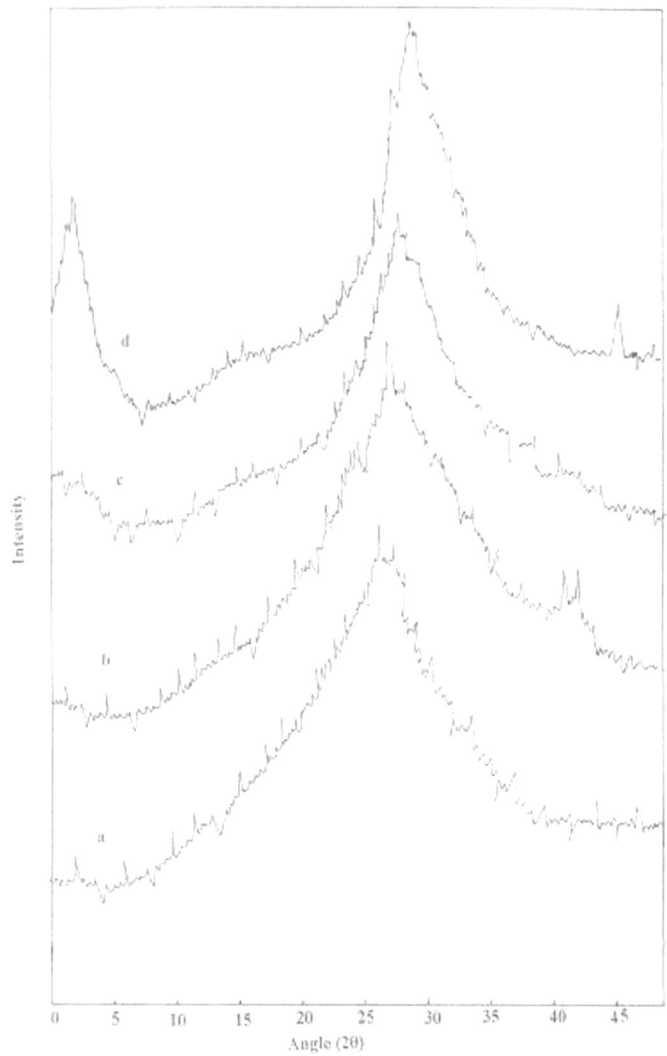

Fig. 4. Powder X-ray diffraction patterns of (a) Co–PhImPc, (b) Cu–PhImPc, (c) Ni–PhImPc and (d) Zn–PhImPc.

3.4 Magnetic susceptibility

Magnetic susceptibility studies were carried out at ambient temperature and summary of the results were in Table 2 and the magnetic moment values reported in the table were the average of three independent determinations. The measured magnetic moment of complexes was higher than the spin only values due to orbital contributions, and these values were higher than those for the corresponding parent phthalocyanines. The magnetic susceptibility studies revealed that Co-PhImPc and Cu-PhImPc are paramagnetic, whereas Ni-PhImPc and Zn-PhImPc are diamagnetic. The measured magnetic moments for Co-PhImPc and Cu-PhImPc are higher than the spin only value corresponding to the one unpaired electron (1.73 BM), due to the mixing of ground state orbitals with higher energy degenerate states and intermolecular co-operative effect [18]. This effect decreases with the increase in field strength and μ_{eff} value approaches spin only value at higher field strength. The observed higher μ_{eff} value at lower field strength is attributed to intermolecular magnetic interaction coupled with magnetic anisotropy of phthalocyanine π-current [19].

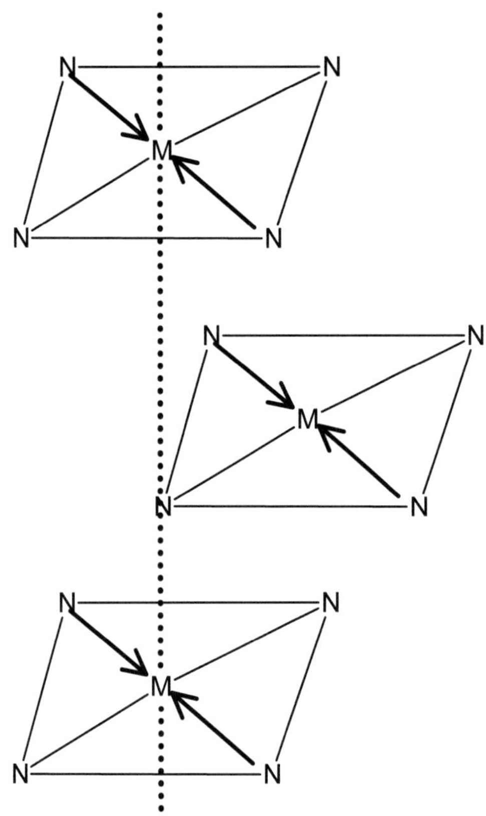

Fig. 5. Probable molecular stacking of metal phthalocyanines. M = Cu, Co, Ni and Zn, N = azamethine atom of phthalocyanine.

The crystallographic studies revealed that the metal phthalocyanines of Co, Cu, Ni and Zn have square planar structure with D_{4h} symmetry and are isomorphous [20]. The molecular plane is approximately normal to a-b plane and molecules are stacked along the short b-axis. The molecular planes are inclined to the a-c axis at an angle of 45°. Thus the complexes stacked in columns with N-atom above and below on every metal atom (Fig. 4) and hence the nearest neighboring molecule along the b-axis contributes a nitrogen atom at the interplanar distance 3.4 Å [21].

Complex (Colour) (Yield)	Empirical Formulae (Molecular weight)	Field strength K Guass	Magnetic susceptibility (χ_m X 10^{-6} cgs unit)	Magnetic moments μ_{eff} (B.M)	Elemental analysis Found (Calc.)
Co-PhImPc (Dark green) (45%)	$C_{60}H_{36}N_{12}Co$ (982.93)	1.02 1.30 1.50 1.82 2.09	+3754.20 +3439.03 +3127.39 +2983.82 +2778.52	3.02 2.91 2.75 2.68 2.59	C: 72.68; (73.25) H: 3.48; (3.66) N:16.96; (17.09) Co: 5.82; (5.99)
Cu-PhImPc (Dark green) (43%)	$C_{60}H_{36}N_{12}Cu$ (987.53)	1.02 1.30 1.50 1.82 2.09	+3574.75 +3390.05 +3135.61 +2975.12 +2582.05	2.94 2.88 2.76 2.69 2.51	C: 72.59; (72.91) H: 3.65; (3.64) N: 16.98; (17.01) Cu: 6.38; (6.43)
Ni-PhImPc (Dark green) (40%)	$C_{60}H_{36}N_{12}Ni$ (982.69)	1.50	-580.15	---	C: 72.98; (73.24) H: 3.67; (3.66) N: 16.69; (16.09) Ni: 5.90; (5.97)
Zn-PhImPc (Dark green) (32%)	$C_{60}H_{36}N_{12}Zn$ (989.39)	1.50	-675.23	---	C: 72.22; (72.77) H: 3.63; (3.63) N: 16.76; (16.98) Zn: 6.66; (6.60)

Table 2. Elemental analysis and magnetic susceptibility data of metal (II) 2, 9, 16, 23-tetra-phenyliminophthalocyanine.

3.5 Thermogravimetric and kinetic studies

Thermogravimetric analytical data of M-PhImPc pigments were summarized in the Table 3. It was observed that the decomposition of the title complexes occur generally in two steps. The first step of degradation in air, which takes place in the temperature region of 100-360°C may be accounted for the loss of four substituted imino groups. The major weight loss was observed for all the complexes in the last step in the temperature range of 360-580°C correspond to the oxidative degradation of phthalocyanine moiety. The residues remained after the thermal decompositions were found to be the corresponding metal oxides [22]. The thermal decomposition of these imino substituted complexes in the nitrogen atmosphere appears to be very slow. For Co-PhImPc, 54% of the complex was found to be decomposed at 700°C. For Cu-PhImPc, Ni-PhImPc and Zn-PhImPc about 48%, 58% and 55% loss of the mass was observed even at 700°C. Above trend confirms the relatively higher stability of these complexes in nitrogen atmosphere than in air. Eventhough, all the four functional groups seem to be lost in first step, a dimer or a polymer was suspected to be formed via the nitrogen atoms of the peripheral end groups before the second step of decomposition starts [23]. DTA results revealed that all degradation steps were exothermic in nature. Kinetic and thermodynamic parameters of the title complexes have been evaluated by Broido's method [24]. Plots of ln (ln1/y) versus 1/T (where y is the fraction of the complex undecomposed) were developed for the decomposition segment where loss of functional groups occur. From the plots, energy of activation (Ea), frequency factor (ln A), enthalpy (ΔH), entropy (ΔS) and free energy (ΔG) of the title complexes have been computed by using standard equations (Table 3).

Name of the Compound	Activation energy Ea KJ mole^{-1}	Frequency factor Ln A,	ΔH KJ mole^{-1}	ΔS J k^{-1}	ΔG KJ mole^{-1}
Co-PhImPc	1.059 (0.733) 3.038 (59.86)	4.314 (3.557) 5.118 (22.48)	0.021 (0.513) 3.030 (55.708)	-159.87 (- 173.6) -162.19 (322.09)	20.80 (25.53) 115.37 (105.3)
Cu-PhImPc	0.943 (0.333) 3.829 (71.77)	4.096 (2.446) 5.462 (26.59)	0.136 (0.914) 2.156 (67.86)	-156.11 (- 184.9) -157.61 (457.92)	20.43 (26.82) 111.32 (147.2)
Ni-PhImPc	0.857 (0.542) 3.168 (65.40)	3.913 (3.544) 5.137 (22.58)	0.223 (0.516) 2.983 (56.20)	-153.38 (- 174.4) -161.82 (326.56)	20.163 (24.94) 116.76 (107.3)
Zn-PhImPc	0.426 (0.435) 4.352 (48.687)	2.841 (2.734) 5.545 (18.04)	0.575 (1.060) 2.631 (43.865)	-141.35 (- 182.5) -157.14 (174.37)	17.53 (31.79) 129.36 (57.27)

The values in the bracket corresponds to the nitrogen atmosphere

Table 3. TGA and Kinetic parameters of metal (II) 2, 9, 16, 23-tetra phenyliminophthalocyanine.

3.6 Antimicrobial studies

The ligand and all complexes synthesized in the present investigation and the respective metal salts were evaluated for their antimicrobial studies like antibacterial and antifungal.

3.7 Antibacterial activity

Bacterial strains: Bacterial strains of *Xanthomonas* were procured from department of Biotechnology, Sahyadri Science College, Kuvempu University, Shimoga.

Method: The above said imino phthalocyanine complexes were tested against pathogenic bacteria *Xanthomonas citri* and *Xanthomonas Compstris.*

The agar diffusion cup plate method was followed for antibacterial assay as described in Indian pharmacopoeia [25]. Inoculum was prepared from 24 hr old culture in nutrient broth. The M-PhImPc complexes of 500 ppm were prepared by dissolving the required quantity of complexes in DMSO. It was further diluted with DMSO for the preparations of 200 ppm, 100 ppm and 50 ppm solution of M-PhImPc. With the help of stainless steel well cutter (6mm) cups were cut out and into each of these cups 100 μl of each of the complexes of different concentrations were placed separately under aseptic conditions with the help of a sterile micropipette. Ciprofloxacin was used as standard drug. The plates were then maintained at room temperature (26°) for 1 hr to allow the diffusion of the solutions into medium and incubated at 37° C for *Xanthomonas citri* and *Xanthomonas Compstris*. Inhibition was recorded by measuring the diameter of the inhibition zone at the end of 24 hr and compared with standard drug [26-27]. As per the observations made, the maximum inhibition effect was observed in Zn-PhImPc with tested organism and the least inhibition effect was observed in Co-PhImPc. The datas for zone of inhibition was presented in Table 4.

Name of the Compound	Conc. (in ppm)	*Xanthomonas compstris* Zone of inhibition (in mm)	*Xanthomonas citri* Zone of inhibition (in mm)
Ciprofloxacin (Standard drug)	50	04	05
	100	06	07
	200	09	10
	500	14	15
Co-PhImPc	50	06(03)	07(03)
	100	09(05)	11(05)
	200	13(06)	15(08)
	500	16(08)	17(10)
Cu-PhImPc	50	06(04)	08(02)
	100	11(07)	13(05)
	200	16(09)	16(07)
	500	19(10)	19(11)
Ni-PhImPc	50	07(04)	06(03)
	100	11(06)	09(05)
	200	14(08)	13(07)
	500	18(11)	18(10)
Zn-PhImPc	50	06(04)	08(04)
	100	11(05)	13(07)
	200	16(08)	16(09)
	500	19(11)	19(11)

The values in the bracket correspond to the parent metal (II) amino phthalocyanines.

Table 4. Zone of inhibition of metal (II) 2, 9, 16, 23-tetra-phenylimino phthalocyanines.

3.8 Antifungal activity

The *Aspergilus Niger* and *Aspergilus Flavous were* studied for its growth, color and sporulation characteristics in the presence of the selected metal phthalocyanine complexes. The M-PhImPc complexes of 500 ppm were prepared by dissolving the required quantity of complexes in DMSO. It was further diluted with DMSO for the preparations of 200 ppm, 100 ppm and 50 ppm solution of M-PhImPc. Ketaconazole was used as a standard drug. Potato Dextrose Agar (PDA) media with the above preparations were sealed with aluminum foils and sterilized in an autoclave at a temperature of 120° C and 15 psi. The hot sterilized media was poured into petriplates in an aseptic chamber and then to room temperature (26°). The *Aspergilus Niger* and *Aspergilus Flavous* were inoculated as a point at the center of the plate and are incubated at 23±1° C for one week and the observations were made everyday [28].

The summary of the observations were presented in Table 5. It was found that all the complexes inhibit the radial growth of the fungi. After 2 days of inoculation, the fungi exhibited minimal growth. It was observed that the inhibiting effects of M-PhImPcs were more for *Aspergilus Niger* compared to *Aspergilus Flavous*. After 5 days, all the complexes show distinct inhibiting effect. However, Zn-PhImPc induced maximum effect. The inhibition of growth effect is in the order of Zn-PhImPc > Co-PhImPc > Cu-PhImPc > Ni-PhImPc.

Name of the Compound	Conc. (in ppm)	*Aspergilus Niger* Radial growth in cm		*Aspergilus Flavous* Radial growth in cm	
		2 days	5 days	2 days	5 days
Ketaconazole (Standard drug)	50	1.50	4.15	1.45	3.80
	100	1.35	3.90	1.35	3.65
	200	1.25	3.60	1.20	3.45
	500	1.00	3.35	1.05	3.10
Co-PhImPc	50	1.20 (1.45)	3.60 (3.85)	1.25 (1.35)	3.35 (3.40)
	100	1.00 (1.35)	3.30 (3.75)	1.15 (1.30)	3.15 (3.30)
	200	0.95 (1.20)	3.15 (3.55)	1.00 (1.20)	3.00 (3.25)
	500	0.75 (1.05)	3.00 (3.40)	0.90 (1.10)	2.85 (3.15)
Cu-PhImPc	50	1.35 (1.40)	3.50 (3.75)	1.20 (1.35)	3.30 (3.50)
	100	1.00 (1.35)	3.10 (3.70)	1.00 (1.30)	3.15 (3.40)
	200	0.80 (1.20)	2.70 (3.45)	0.90 (1.15)	3.05 (3.25)
	500	0.50 (1.05)	2.20 (3.25)	0.80 (1.05)	1.95 (3.10)
Ni-PhImPc	50	1.30 (1.50)	3.60 (3.80)	1.25 (1.30)	3.30 (3.45)
	100	1.20 (1.45)	3.10 (3.65)	0.90 (1.25)	2.95 (3.20)
	200	1.00 (1.35)	2.85 (3.55)	0.80 (1.15)	2.65 (3.05)
	500	0.85 (1.05)	2.35 (3.15)	0.70 (1.05)	2.00 (2.85)
Zn-PhImPc	50	1.45 (1.45)	3.75 (3.60)	1.20 (1.25)	3.45 (3.45)
	100	1.10 (1.40)	1.80 (3.45)	0.95 (1.40)	1.93 (3.05)
	200	0.85 (1.25)	1.50 (3.20)	0.70 (1.15)	1.65 (2.85)
	500	0.50 (1.00)	1.30 (2.85)	0.55 (1.05)	1.45 (2.55)

The values in the bracket correspond to the parent metal (II) amino phthalocyanines.

Table 5. Antifungal data of metal (II) 2, 9, 16, 23-tetra-phenylimino phthalocyanines.

The interesting observation made during the investigation was the change in the colour of the fungus. *Aspergilus Niger* was known for its black colour and *Aspergilus Flavous* for its green colour. However, in the presence of metal complexes the matured colonies of the fungus were pale brown and the new colonies were pale green [29]. It was confirmed by the parallel experiment with and with out the addition of 2 ml DMSO in the media that the change in colour of the fungi was not due to the presence of DMSO in the medium. The change of colour of the fungi may be due to the effect of complexes as the pigmentation properties of the growing fungi.

4. Conclusions

The synthetic route adopted was very simple and give good yield. The measured magnetic moment of complexes was higher than the spin only values due to orbital contributions, and these values were higher than those for the corresponding parent phthalocyanines. The variations of magnetic moment as a function of field strength indicated the presence of inter molecular co-operative interactions. The magnetic susceptibility studies clearly revealed the structural information of the complexes. The red shift of the complexes compared to the parent phthalocyanine is due to increase in conjugation of π-electron with the π-electron cloud of peripheral substituted imino groups. Metal phthalocyanines are biologically active and the complexes show enhanced antimicrobial activity against one or more strains compared to conventional standard drugs. Imino substituents tend to make the complexes act as more powerful and potent bactericidal agent.

5. Acknowledgements

The authors are thankful to the Principal, J. N. N. College of Engineering, Shimoga. The thanks are also to Prof. A. M. A. Khader, Mangalore University and Prof. M. A. Pasha, Bangalore University, for their help in recording spectra.

6. References

[1] CE Dent; RP Linstead; AR Lowe; *J. Chem. Soc.*, 1934, 1033.
[2] D Whorle, *Adv. polym. Sci.* Review Article., 1983, 50.
[3] W Kobel; M Hanack; *Inorg. chem.*, 1986, 25, 103.
[4] AE Dandridge; HA Drescher; J Thomas; (To Scottish Dyes Ltd) *British Patent* 822, 169, (1929).
[5] RB Eltis; JE Lyons; *Coord. Chem. Rev.*, 1990, 105.
[6] JF Bartoli; P Brigoud; P Battioni; D Mansuy; *J. Chem. Soc, Chem. Commun.*, 1994, 440 .
[7] MW Grinstaff; MG Hill; I Labinger; HB Gray; *Science.*, 1994, 264, 311.
[8] RA Sheldon; Metalloporphyrin catalytic oxidation, Marcel Decker, New York, 1194.
[9] B Paquette; H Ali; EJ Van Lier; *J. Phys-Chim. Biol.*, 1991, 88, 1113.
[10] SA Priola; A Raines; WS Caughey; *Science.*, 2000, 287, 1503 .
[11] MC Harsans; NP Guisinger; JW Lyding; *Nanotechnology.*, 2000, 11, 70.
[12] D Scblewein; E Karmann; T Ockermann; H Yanagi; *Electrochim. Acta.*, 2000, 45, 4697.
[13] R Jasinski; *Nature.*, 1964, 201, 1212.
[14] MP Somashekarappa; J Keshavayya; *Synth. React. Inorg. Met-org. chem.*, 1999, 29 (5) 767.
[15] MP Somashekarappa; J Keshavayya; *J. Soudi. Chem. Soc.*, 1999, 3(2) 113.
[16] BN Achar; GM Fohlen; JA Parker; J Keshavayya; *Polyhedron.*, 1989, 6(6), 1463 .
[17] CC Leznoff; ABP Lever; Phthalocyanines, properties and applications: VCH publications, Inc., New York, 1989, 1, 173.
[18] P Selwood; Magnetochemistry, New York, 1956, Interscience.
[19] MP Somashakarappa; J Keshavayya; *Spectrochemica Act.*, Part-A, 2003, 59, 767.
[20] BN Achar; GM Fohlen; JA Parker; J Keshavayya; *Polyhedron.*, 1999, 6(6), 1463.
[21] BN Achar; JM Bhandari; *Trans. Metal. Chem.*, 1993, 18, 123.
[22] Arthur I. Vogel, Quantitative Inorganic Analysis, 3rd Ed., Longmans Publishers, London (1964).
[23] MP Somashekarappa; KR Venugopala Reddy; MNK Harish; J Keshavayya; *J. T. R. chem.*, 2004, 11(1) 1.
[24] A. Broido, *J. Polym. Sci.*, Part A-2, 1969, 7 1761.

[25] Indian pharmacopoeia, 1985, Government of India 3rd Edition, New Delhi,Appendix IV p. 90

[26] L Ahmed; Z Mohammed; F Mohammed; *J. Ethnopharmacol.*, 1998, 62, 183.

[27] SN Padhy; SBMahato; NL Dutta; *Phytochem.*, 1973, 12, 217-221.

[28] NN Shah, AS Biradar, Seema I Habib, JA Dhole, MA Baseer, PA Kulkarni, *Der Pharma Chemica,* 2011, 3(1): 167-171.

[29] MH Moinuddin Khan; Fasiulla; J Keshavayya; KR Venugopala Reddy; *Russ. J. Inorg. chem.*, 2008, 53(1), p. 66-77.

16

Antisense Antibacterials: From Proof-Of-Concept to Therapeutic Perspectives

Hui Bai[1,2] and Xiaoxing Luo[1]
[1]*Department of Pharmacology, School of Pharmacy, Fourth Military Medical University, Xi'an,*
[2]*Department of Biotechnology, Institute of Radiation Medicine, Academy of Military Medical Sciences, Beijing, China*

1. Introduction

Recent years have witnessed several gram-negative bacteria (GNB) species and a few gram-positive bacteria (especially the *Staphylococcus aureus*) posing overwhelming threats to the healthcare-associated infections as a series of frightening superbugs (Engel, 2010; Peleg & Hooper, 2010). It is primarily due to the fact that incidence of multidrug resistance (MDR) or pan-drug resistance (PDR) bacteria have been escalating in a manner of global dimension, frequent prevalence and alarming magnitude. The predominate resistance issues are those related to GNB species, including *Enterobacteriaceae* (Deshpande & et al, 2010), *Klebsiella pneumonia, Pseudomonas aeruginosa* and *Acinetobacter baumannii*. Theses circulating isolates have created big problems for treatment of nosocomial infection because they carry highly transmissible elements encoding multiple resistance genes, e.g. extended-spectrum beta-lactamases (ESBLs) that inactivates different classes of first-line antibiotics (Bush, 2010; Engel, 2010), metallo-beta lactamase that hydrolyzes penicillins, cephalosporins and carbapenems, efflux pumps that decrease bacterial transporting ability to almost all antibiotics and natural antimicrobial products (Pages & et al, 2010), and promoters that ensure the transcription of these genes.

Traditional antimicrobial drugs target only a few cellular processes and are derived from a few distinct chemical classes. Despite that genetic screens to identify new drug targets and classic searching for new chemical leads with diverse structures (Moellering, Jr., 2011), the constant need of new broad-spectrum antimicrobial agents has rarely been met (Cattoir & Daurel, 2010). Meanwhile, antibacterial strategies that favor in offering timely therapeutic countermeasures are urgently required for possible outbreaks of new super bug infections. One promising strategy is antisense antibacterial, which can contribute to both aspects of the problem. It is generally described as RNA silencing in bacteria using synthetic nucleic acid oligomer mimetics to specifically inhibit essential gene expression and achieve gene-specific antibacterial effects. First proposed in 1991, RNA targeting in bacterial has been made more flexible by 20 years of technology refinement, circumventing major problems of target selection/validation and efficient delivery (Bai & et al, 2010). Antisense antibacterials have been developed by constructing sequence-designed synthetic RNA silencers using new chemical classes, e.g. nucleic acid mimics peptide nucleic acid (PNA) and phosphorodiamidate morpholino (PMO), that conjugated with cell penetrating peptide (CPP) in multiple functional ways (Geller, 2005; Hatamoto & et al, 2010). And their potent bactericidal effects have been displayed in a variety of pathogens by targeting several growth essential genes *in vitro* and *in vivo* (Bai & et al, 2010). Advantage of RNA silencing is unique in having the potential to selectively kill target pathogens with species and even strain specificity. Of particular interest are possibilities to tailor the antibacterial spectrum, aid the use of conventional antibiotics by potentiating their activity, and reverse resistance. Further, antisense antibacteirals may present an unusual opportunity for developing broad-spectrum therapeutics against upgrading infections caused by multi-drug or pan-drug resistant pathogenic species, where many successful compounds have failed. This review will describe the characteristics of the antisense antibacteiral strategy (including antisense mechanism, basic chemistry involved in nucleic acid analogs, their anti-infection applications *in vitro* and *in vivo*, and preliminary studies on pharmacokinetics and toxicity), and focus on the major determinants of target accessibility and CPP-mediated delivery in the general context of antisense antibacterials. We will also highlight the promising targets and delivery strategies that favor the possible development of broad-spectrum nucleic acid-based therapeutic molecules and provide overall information of their potentials as functional component of systemic broad-spectrum antisense antibacterial agents.

2. Antisense antibacterials: 20 years of technology refinement

Antisense antibacterial strategy is revolutionary for silencing essential genes at mRNA level by antisense oligodeoxyribonucleotides (AS-ODNs) for realization of bacterial cell death or restoration of susceptibility. Significant technology advances in aspects of microbial genomics (Monaghan & Barrett, 2006), structural modification of oligonucleotides and efficient delivery systems have fundamentally promoted the transformation of antisense antibacterials from concept to future therapeutic "antisense antibiotics".

2.1 Mechanism of action and chemistry

AS-ODNs are designed to bind the target mRNA to prevent translation or bind DNA to prevent gene transcription respectively. And once bound to the target, AS-ODNs modulate its function through a variety of post binding events. Meanwhile, AS-ODNs based on the three generations of modified structures, have overcome the biological disadvantages of RNA and DNA, and shown great potency in gene expression inhibition with apparently high degree of fidelity and exquisite specificity both *in vitro* and *in vivo*.

2.1.1 Antisense mechanism

Most of the reported work on antisense drugs has been accomplished in eukaryotic systems and their mechanisms have been well explored (Houseley & Tollervey, 2009). AS-ODNs bind to the target RNA by well-characterized Watson-Crick base pairing mechanism. The effect of gene silencing or "knock down" that happens after the binding can be broadly categorized as cleavage-dependent or occupancy-only mechanism (Figure 1). Cleavage-dependent mechanism includes degradation of RNA:mRNA duplexes by double-strand RNA (dsRNA)-specific RNAases (a natural means of transcriptional regulation), degradation of stable DNA:RNA heterodimers through the activity of RNAse H, and degradation via the action of RNase P (only if external guide sequences are coupled to the oligonucleotide). Occupancy-only mechanism, also known as translation arrest, features as that AS-ODN:RNA heteroduplexes inhibit translation by steric blocking the ribosomal maturation and polypeptide elongation process. Antisense antibacterials function on the base of above antisense mechanisms, whereas the specific mechanism is dependent on the structural chemistry and design of the modified oligonucleotides (Bennett & Swayze, 2010).

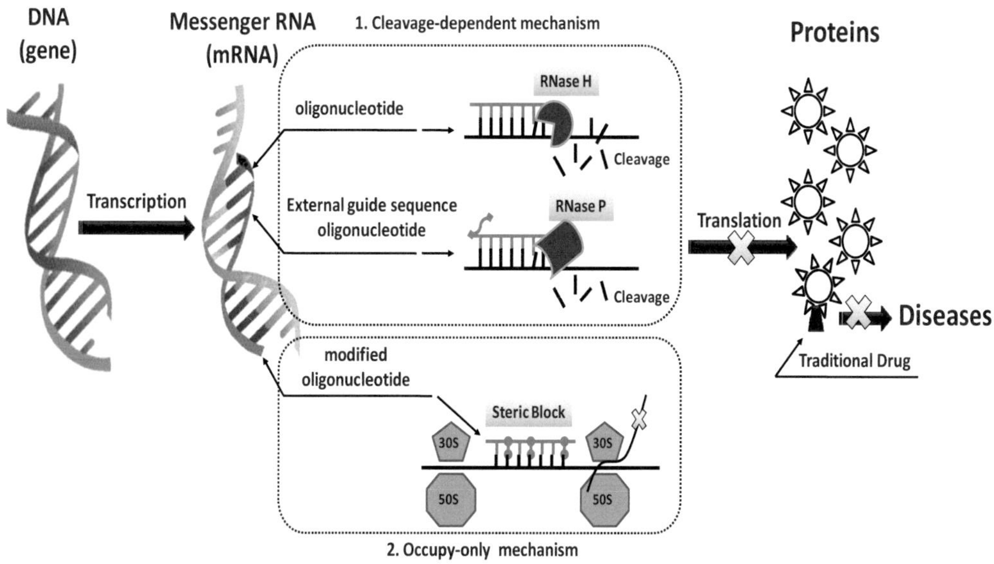

Fig. 1. Different antisense mechanisms: antisense oligodeoxyribonucleotides (AS-ODNs) are known to interact and block the function of the mRNA. Different antisense mechanisms shown include the nondegradative mechanisms (e.g., inhibition of translation) and mechanisms that promote degradation of the RNA (e.g., RNase H mediated cleavage and external guided sequence mediated RNase P cleavage).

2.1.2 Nucleic acid chemistry: structure and binding

Unmodified DNA/RNA is susceptible to nucleases attack and degradation. Furthermore, their poor pharmacokinetics properties (including weak binding to plasma proteins, rapid filter by kidney and excretion into urine, and et al) make them undesirable and unacceptable therapeutic agents for systemic administration. In order to increase their nuclease stability and intrinsic affinity to complementary target RNAs, many efforts have been made to the structural modification of DNA or RNA (Kurreck, 2003). Key modifications concentrate on the backbone, phosphodiester bond, and sugar ring, giving births to three generations of nucleic acid anologs. Representative oligonucleotide derivatives include phosphorothioate oligodeoxyribonucleotides (PS-ODNs), 2′-O-methyloligoribonucleotides (2′-OMes), 2′-O-methoxyethyl oligonucleotides (2′-MOE), locked nucleic acids (LNAs), phosphorodiamidate morpholino oligonucleotides (PMOs), thiophosphoroamidate oligonucleotides and peptide nucleic acids (PNAs) (Figure 2).

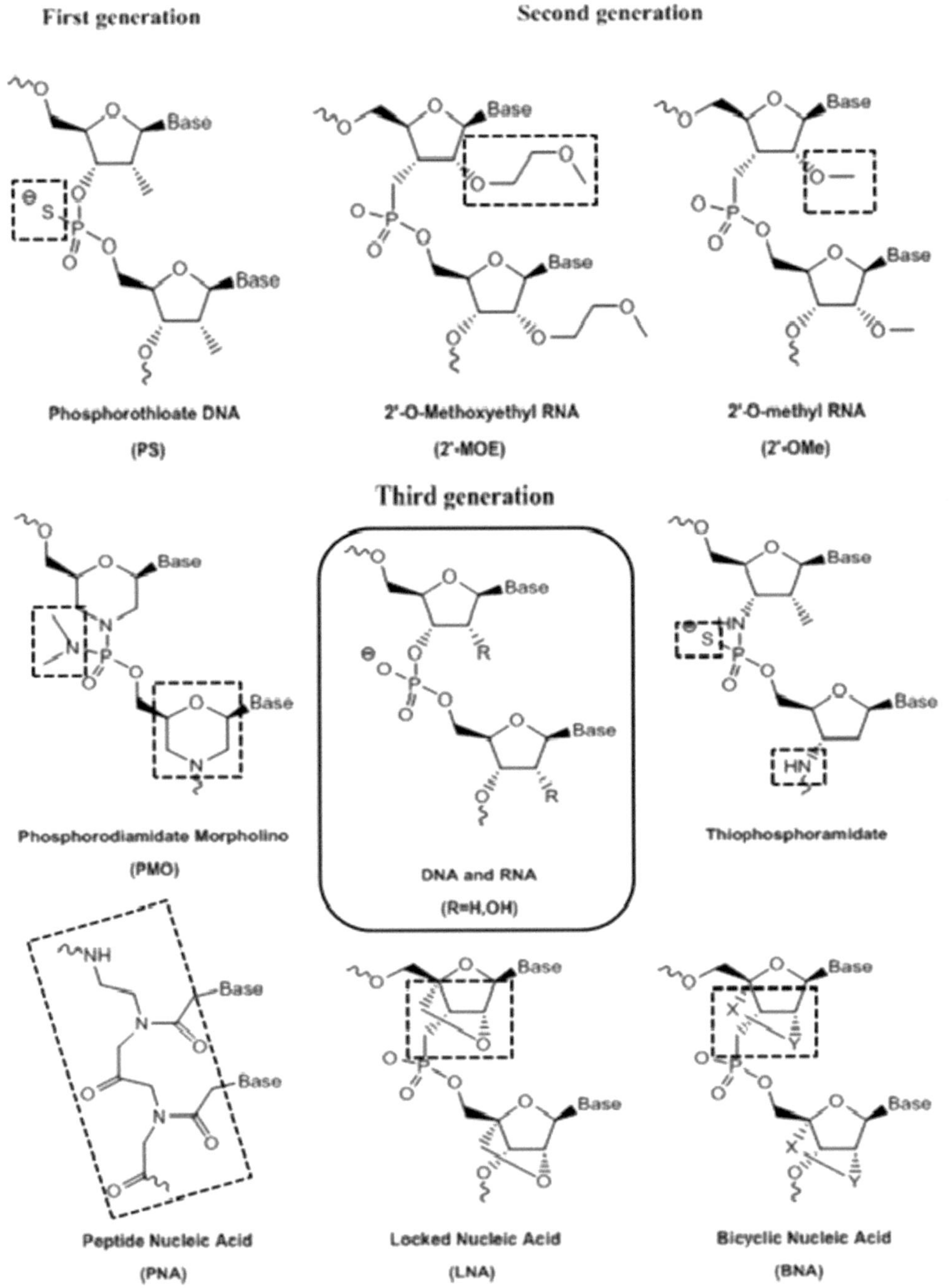

Fig. 2. Representative modified antisense oligodeoxyribonuleotides. Replacement in structure compared with DNA/RNA is highlighted by dashed rectangle. First generation of modified form shown includes only PS-ODNs. Second generation mainly includes 2′-OMes and 2′-MOE. Third generation includes a series of DNA/RNA analogs, e.g., LNA, PNA, BNA, PMO and thiophosphoroamidate oligonucleotides.

Like DNA or RNA, PS-ODNs, LNA and thiophosphoroamidate oligonucleotides are negatively charged. Other modified oligonucleotides like PNA and PMO are electric neutral, showing little repulsion during hybridization to target DNA or RNA. 2'-OMes, 2'-MOE, LNAs, PMOs and PNAs all bind to RNA more tightly than unmodified oligonucleotides or PS-ODNs. Therefore, they can be used at shorter lengths and lower concentrations for exerting specific and potent RNA silencing effect. Meanwhile, PNA and PMO have provided substantially better specificity to the same target sequence than DNA, phosphorothioate DNA, and 2'-O-methyl RNA, either at low concentration of 50 nM or at high concentration of 3.5 μM (Deere & et al, 2005). Furthermore, it is acknowledged that only PS-ODNs activate RNase H to degrade mRNA in eukaryotic cells, whereas the other modified oligonucleotides show direct translation arrest effect. The same results have also been confirmed for gene manipulation by antisense strategy in bacteria.

2.2 Antisense antibacterial strategy

The hypothesis that any gene can be antisense inhibited is quite tantalizing. Therefore, antisense oligomers have been studied as bacterial growth inhibitors for developing new types of antibiotics. In 1991, Rahman et al firstly observed the inhibited protein synthesis and colony formation in normal *E. coli* by using PEG 1000 attached methylcarbamate DNAs targeting the start codon sequence of prokaryotic 16S rRNA (Rahman & et al, 1991). Ever since, the potential of gene specific modified AS-ODNs as biomedically useful antibiotics has been well accepted and further explored. The present-day antisense antibacterial strategy has overcome major obstacles that hampered this innovative approach developing into clinically applicable therapeutics: (i) target validation and (ii) efficient delivery system. Meanwhile, modified AS-ODNs, e.g. PNA and PMO, have accepted thorough preclinical and clinical evaluation on the aspects pharmaceutical properties as promising antisense antibiotics.

2.2.1 Inherent advantages

Compared to human genome, bacterial genome is much less complicated and homogenous. Unlike eukaryotic cells, the double-strand DNA (dsDNA) of bacterial cells locates in nucleoid. And in this low electron density zone, there is no nucleic membrane to strictly separate biochemical reactions into different time and space level. DNA replication, RNA transcription and protein synthesis in bacteria are processed in cytoplasm, which allows exogenous AS-ODNs to interfere with genes and/or RNAs more readily. Meanwhile, RNA interfering (RNAi) mechanism found in eukaryotic cells has not been reported so far in bacteria. Bacteria themselves use antisense as a natural mechanism to inhibit specific gene expression, therefore, antisense technology suits better as an effective gene modulating tool in bacteria.

2.2.2 Target identification and validation

A key objective for discovery of new antisene antimicrobial agents is to determine the genes essential for survival of the pathogenic organisms. In paticular, the main critireas for measuring the quality of a candidate gene as a good target include vitality of the gene and its targeting accessability for antisense oligomers. Compared to gene knockout technique, antisense approach itself has been proved to be an effective tool for target validation in bacteria, with controllable sensitivity, larger breadth of applicability and more realistically mimic effect of a therapeutic inhibitor (Wright, 2009).

2.2.2.1 Target site selection and design of AS-ODNs

Theoretically, antisense antibacterials as modulators of bacterial essential genes can be used to alter biological state or behavior in potentially any pathogenic species. Their growth inhibitory activity relies on sequence-specific inhibition of gene expression, which offers the potential for high specificity in immediate bacteriocidal or bacteriostatic therapeutic consequences (Rasmussen & et al, 2007). However, the fundamental requirements for potent antisense activity include sufficient concentration of antisense agent at the most sensitive targeting site, an ability to hybridize to the target mRNA sequence, the capacity of the ODN/mRNA duplex to interfere with gene expression, and sufficient biological stability of the antisense agent.

Antisense inhibitors must bind accessible regions of the target mRNA so that stable ODN:RNA(DNA) heteroduplexes or triplex (as for PNA) can be formed to elicit antisense effect. In order to obtain the antisense sequence with best potency and efficacy, researchers normally follow a comparatively fixed procedure in design (Shao & et al, 2006). Generally, possible targeting regions are those nucleotide sequences free of any double strand (e.g. hairpin) in secondary structure, which are determined by RNA secondary structure softwares (Ding & Lawrence, 2003) within full sequences. Notably, most previous studies have demonstrated that the start codon region of the mRNA (see Table 1&2) is the most effective region for RNase H independent antisense inhibition, because this region initiates the translation and includes the Shine-Dalgarno (SD) sequence (Dryselius & et al, 2003). However, a few studies also have confirmed that specific AS-ODNs complementary to sites beyond the start codon region receive equal positive results in *in vitro* efficacy test (e.g. antisense targeting of *rpoD* by PNA in methicillin resistant *Staphloccous aureus*, Bai & et al, 2012a). Then, bioinformatic algorithms are used to calculate the DNA: ODN binding parameters (e.g. minimal free energy and melting temperature, et al) with setted lengths for AS-ODN. According to the combind data, rational analysis was performed to confirm 3-10 different targeting sites/sequences with highest binding affinity and stability for sequence-specific antisense inactivation of target genes. AS-ODNs complementary to the chosen target sites are synthesized. The length of AS-ODN is predominantly determined by their chemical properties. In principle, nucleic acid analogs with stronger affinity to target RNA require shorter lengths. Customarily, policy that 14-30 monomers for PS-ODNs and 8-16 monomers for PNA, LNA and PMO have been adopted for potent inhibition and achieving better hits. In order to improve the uptake of AS-ODNs by the cell, AS-ODNs with attached membrane permeabilizing peptides have been developed (elaborated in 2.2.3).

Presently, *in vitro* modified minimal inhibitory concentration (MIC) and minimal bactericidal inhibitory concentration (MBC) tests of peptide-ODN conjugates are well-acknowledged methods to preliminarily confirm the antibacterial effect of AS-ODNs. Targeted gene vitality and accessability are determined by comparing MIC and MBC values. AS-ODN that shows the lowest MIC value indicates the most sensitive targeting site for antisense inhibition, whereas the MBC value suggests if the antisense antibacteial effect is bacteriocidal. Target specificity is generally evlauated at the same time by testing the antibacterial activity of designed control AS-ODNs (e.g. AS-ODN with mismacted or scrambled nucleotide sequences) and pepides. Further, RT-PCR and western blotting can be used to observe the reduction of mRNA and protein product of the particularly targeted gene. Collectively, with regard to target selection, researcher are dedicated to identify an ideal essential gene that is with small nucleotide content and upmost stringency but effective region of coding sequences for potent antisense ihibition (Goh & et al, 2009). Meanwhile, the antisene property of AS-ODN itself should also be taken into consideration.

2.2.2.2 Validated targets in bacteria

2.2.2.2.1 Targeting essential genes

Essential genes that regulate or control bacterial growth, proliferation, virulence and synthesis of important living-dependent substances are candidates attracting the majority of research enthusiasms. Validated essential genes among different bacterial species include *fbpA/ fbpB/fbpC* (Harth & et al, 2002, 2007) and *glnA1* (Harth & et al, 2000) in *Mycobacterium tuberculosis, gyrA/ompA* in *Klebsiella pneumonia* (Kurupati & et al, 2007), *inhA* in *Mycobacterium smegmatis* (Kulyte & et al, 2005), *oxyR/ahpC* in *Mycobacterium avium* (Shimizu & et al, 2003), *NPT/EhErd2* in *Entamoeba histolytica* (Stock & et al, 2000, 2001), *gtfB* in *Streptococcus mutans* (Guo & et al, 2006), *fmhB/ gyrA/hmrB* (Nekhotiaeva & et al, 2004a) and *fabI* (Ji & et al, 2004) in *Staphylococcus aureus, 23S rRNA* (Xue-Wen & et al, 2007), 16S rRNA plus *lacZ/bla* (Good & Nielsen, 1998), and RNAse P (Gruegelsiepe & et al, 2006) in *Escherichia coli,* and *acpP* in *Burkholderia cepacia* (Greenberg & et al, 2010), *Escherichia coli* (Deere & et al, 2005b; Geller & et al, 2003a, 2003b, 2005; Mellbye & et al, 2009, 2010; Tan & et al, 2005; Tilley & et al, 2007) as well as *Salmonella enterica serovar Typhimurium* (Mitev & et al, 2009; Tilley & et al, 2006).

Target[a]		AS-ODN[b]	Test organism[c]	Efficacy identified[d]	Delivery Method[e]	References
23S rRNA	P.T. center	PNA	*E. coli* AS19 (permeable membrane)	*in vitro*/ $IC_{50} > 20\ \mu M$ (duplex)	–	Good & Nielsen, 1998
				in vitro/ $IC_{50} = 5\ \mu M$ (triplex)		
	domain II		*E. coli* Dh5α	*in vitro*/MIC = 10 μM	CPP=$(KFF)_3K$	Xue-Wen et al,2007
	α-sarcin loop		*E. coli* AS19	*in vitro*/$IC_{50} = 2\ \mu M$	–	Good & Nielsen, 1998
			E. coli K12 (wild-type)	*in vitro*/MIC* = 5 μM	–	Good et al, 2001
				in vitro/MIC* = 0.7 μM	CPP=$(KFF)_3K$	
				in vitro/MIC = 3 μM		
16S rRNA	preceding the start codon region	MDNA	*E. coli* lacking outer cell wall	*in vitro* / inhibit protein synthesis and colony formation	–	Rahman MA et al, 1991
			normal *E. coli*		PEG attached	
	mRNA binding site	PNA	*E. coli* AS19	*in vitro* / $IC_{50} > 20\ \mu M$	–	Good & Nielsen, 1998
			E. coli K12	*in vitro*/MIC=10 μM	CPP=$(KFF)_3K$	Hatamoto et al, 2010

Table 1. Continued

Target[a]		AS-ODN[b]	Test organism[c]	Efficacy identified[d]	Delivery Method[e]	References
acpP (start codon region)	SD site -24 to -13 nt	PNA	*E. coli* K12	*in vitro* / MIC= 1.5 µM	CPP=$(KFF)_3K$	Dryselius et al, 2003
	-9 to 3 nt					
	-5 to 5 nt		*E. coli*	*in vitro* / MIC=0.2* or 1 µM		Good et al, 2001
			E. coli SM101 (defective membrane)	*in vivo*/100% rescued mice at a single i.p. dose of > 5 nmol		Tan et al, 2005
			E. coli K12	*in vivo*/60% rescued mice at a single i.v. dose of 100 nmol		
			E. coli K12	*in vivo*/ MIC=0.8µM, post antibiotic effect duration 11.7h		Nikravesh et al, 2007
	6 to16 nt	PMO	*E. coli* AS19	*in vitro* luciferase system/ most potent inhibition	–	Deere et al, 2005
			E. coli SM105 (normal membrane)	*in vitro* / EC = 20µM *in vivo*/sustanied post-infectin reduction in cfu at single i.p. dose of 76 nmol	–	Geller et al, 2005
			E. coli W3110 (ATCC27325)	*in vitro* / IC_{50} CPP_1 = 9.5µM IC_{50} CPP_2 = 10.8µM IC_{50} CPP_3 = 3.6µM	CPP_1=$(KFF)_3KXB$ CPP_2=RTRTRFLR RTXB CPP_3=$(RFF)_3XB$ CPP_4=$(RXX)_3B$	Tilley et al, 2006
			EPEC (E. coli E2348.69)	Ex vivo cocultured Caco-2 culture / IC_{50} CPP_2= 5.3µM IC_{50} CPP_3= 0.5µM		
			S. enterica (ATCC29629)	Ex vivo cocultured Caco-2 culture / IC_{50} CPP_2= IC_{50} CPP_3= 0.5µM		
			E. coli W3110 (ATCC27325)	*in vivo* /100% 48h-after survival in mice at i.p. injection of 2 treatments with 30µg or 300µg conjugate	CPP_3=$(RFF)_3XB$	Tilley et al, 2007
			E. coli W3110 (ATCC27325)	*in vitro* /MIC from 0.625 to > 80µM	19 synthetic CPPs	Mellbye, 2009
				in vivo /100% 48h-after survival in mice at i.p. injection of 2 treatments with 30µg CPP_2-PMO	CPP_1=$(RX)_6B$ CPP_2=$(RXR)_4XB$ CPP_3=$(RFR)_4XB$	
		PMO	*S. enterica* LT1	*in vivo* /MIC = 1.25µM	CPP=$(RXR)_4XB$	Mitev et al, 2009
		3+Pip-PMO		*in vivo* /MIC = 0.625µM intracellular infected macrophage/99% decrease in intracellular bacteria at 3µM		
		Pip-PMO	*E. coli* W3110 (ATCC27325)	*in vivo* /MIC3+ = 0.3 µM		Mellbye et al, 2010
				in vivo /100% 48h-after survival in mice at i.p. injection of 2 treatments with 5 or 15 mg/Kg CPP-PMO		
		Gux-PMO		*in vivo* /MIC5+ = 0.6 µM		
	4 to 14 nt	PMO	14 *B. cepacia strains* (5 clinical isolates+9 from ATCC)	*in vitro*/lowest MIC = 2.5 µM	CPP= $(RFF)_4XB$	Greenberg et al, 2010
	-5 to 6 nt			*in vitro*/lowest MIC = 2.5 µM *in vivo* /55% 30d-after survival in mice at i.p. injection of single dose of 200 µg CPP-PMO		

Table 1. Continued

Target[a]	AS-ODN[b]	Test organism[c]	Efficacy identified[d]	Delivery Method[e]	References
P15 loop of RNase P	LNA	*E. coli*	*in vitro* / bingding affinity value only	–	Gruegelsiepe et al, 2006
	PNA		*in vitro*/MIC = 5 μM	CPP=$(KFF)_3K$	
floA	PNA	*E. coli* AS19	*in vitro*/MIC = 2.5 μM	CPP=$(KFF)_3K$	Hatamoto et al, 2010
floP			*in vitro*/MIC = 6.5 μM		
glnA1	PS-ODN	*M. tuberculosis*	*in vitro* / combination of 3 PS-ODNs for transcript mRNA, EC = 10 μM	ethambutol or polymyxin B nonapeptide	Harth et al, 2000
fbpA,fbpB, fbpC			*in vitro* / combination of 4 PS-ODNs for each transcript mRNA, EC = 10 μM	–	Harth et al, 2002
	5'-, 3'-HP PS-ODN		*in vitro* / combination of 3 PS-ODNs for each transcript mRNA, EC = 10 μM	–	Harth et al, 2007
inhA	PNA	*M. smegmatis*	*in vitro* / MIC < 6.5 μM	CPP=$(KFF)_3K$	Kulyté A et al, 2005
adk	PNA	*S. aureus* RN4220	*in vitro* / MIC = 15 μM	CPP=$(KFF)_3K$	Hatamoto et al, 2010
fmhB	PNA	*S. aureus* RN4220	*in vitro* / MIC = 10 μM	CPP=$(KFF)_3K$	Nekhotiaeva et al. 2004
gyrA			*in vitro* / MIC = 20 μM		
hmrB			*in vitro* / MIC = 12 μM		
fabI	UM	*S. aureus*	*in vitro* / MIC = 15 μM	–	Ji et al, 2004
	PNA	*E. coli* K12	*in vitro*/ MIC = 3 μM	CPP=$(KFF)_3K$	Hatamoto et al, 2010
fabD	PNA	*E. coli* K12	*in vitro* / MIC = 2.5 μM	CPP=$(KFF)_3K$	
gyrA	PNA	*K. pneumoniae*	*in vitro* / MIC = 20 μM	CPP=$(KFF)_3K$	Kurupati et al, 2007
ompA			*in vitro* / MIC = 40 μM		
gtfB	PS-ODN	*S. mutans*	*in vitro* / reduce biomass	–	Guo et al, 2006
oxyR/ahpC	UM	*M. avium complex*	*in vitro* / ineffective	–	Shimizu T, 2003
NPT/ EhErd2	UM	*E. histolytica*	*in vitro* / inhibited cell growth	–	Stock et al, 2001, 2000

[a] The essential genes that were targeted encode the following proteins: *acpP*, acyl carrier protein; *fabI*, enoyl-acyl carrier protein reductase; *fabD*, malonyl coenzyme A acyl carrier protein transacylase; *folP*, dihydropteroate synthase; *fmhB*, protein involved in the attachment of the first glycine to the pentaglycine interpeptide; *gyrA*, DNA gyrase subunit A; *hmrB*, ortholog of the *E. coli* acpP gene; *adk*, adenylate kinase; inhA, enoyl-acyl carrier protein reductase; *ompA*, outer membrane protein A; *gtfB*, synthesis of water-insoluble glucans; *inhA*, enoyl-(acyl carrier protein) reductase; RNase P, P15 loop of RNase P; *gyrA*, DNA gyrase subunit A; *oxyR*, oxidative stress regulatory protein; *ahpC*, alkyl hydroperoxide reductase subunit C; *glnA1*, glutamine synthetase; *fbpA,fbpB, fbpC*, 30/32-kDa mycolyl transferase protein complex; *NPT*, Neomycin phosphorotransferase; *EhErd2*, marker of the Golgi system; *LacZ/bla*, beta-galactosidase/beta-lactamase; P.T. indicates peptidyl transferase; SD, Shine-Dalgarno; nt, nucleotide.

[b] UM, unmodfied; MDNA, ethylcarbamate DNA; Pip-PMO and n+Gux-PMO, cations (piperazine or N-(6-guanidinohexanoyl)piperazine) attached to the phosphorodiamidate linkages;

[c] EPEC, enteropathogenic *E. coli*.

[d] Minimal inhibitory concentrations (MIC) were tested in Mueller–Hinton broth, except in case marked with an asterisk(*), in which the MIC values were determined in LB broth at 10% of the normal strength; IC_{50} values are the concentrations that caused a 50% inhibition of cell growth relative to control cultures that lacked AS-ODN; EC values are the concentrations that caused significant decrease in cell grwoth relative to control cultures that lacked AS-ODN; PAE, post antibiotic effect; i.p. indicates intraperitoneal and i.v. indicates intravenous.

[e] CPP, cell penetrating peptide; PEG, Polyethylene glycol; "–", no delivery method used. For synthetic peptides, X is 6-aminohexanoic acid and B is beta-alanine.

Table 1. Examples of AS-ODNs targeting essential genes in antibacterial therapy.

2.2.2.2.2 Targeting resistance mechanism

Developing resistance inhibitors in traditional antibiotic industry is a sound, well-validated strategy for tackling resistance problems, because they postpone the "expire date" of on-market antibiotcs and expand their application. The economic and clinical value of this rationale is well recognized and demonstrated by offering new combinations to clincal practice. Thus, a few studies focused on interrupting the expression of genes involved in resistant mechanism by antisense approach, aiming to restore bacterial susceptibility to key antibiotics in clinical practice.

Target	Encoding Proteins	AS-ODN[a]	Test organism	Efficacy identified	Delivery Method[b]	References
oprM	outer membrane efflux protein	PS-ODN	*P. aeruginosa*	*in vitro*	liposome	Wang et al, 2010
mecA	penicillin-binding protein 2 prime	PS-ODN	*S. aureus*	*in vitro* & *in vivo*	liposome	Meng J et al, 2006, 2009
cmeA	CmeABC multidrug efflux transporter	PNA	*C. jejun*	*in vitro*	CPP	Jeon B et al, 2009
aac(6')-Ib	aminoglycoside 6'-N-acetyltransferase type Ib, mediate amikacin resistance	UM	*E. coli*	EGS mediated RNaseP leavage / *in vitro*	–	Soler Bistué AJ et al, 2007
				in vitro	EP	Sarno R et al, 2003
metS/ murB	methionyl-tRNA synthetase /UDP-N-acetylenolpyruvoylglucosamine reductase	UM	*B. anthracis*	*in vitro*	–	Kedar GC et al, 2007
act	chloromycetin acetyl transferase	UM	*E. coli*	EGS mediated RNaseP leavage / *in vitro*	–	Gao MY et al, 2005
				in vitro	–	Chen H et al, 1997
vanA	class A (VanA) glycopeptide-resistant related protein	UM	*E. faecalis*	*in vitro*	–	Torres VC et al, 2001
marOR AB	multiple antibiotic resistance operon	PS-ODN	*E. coli*	*in vitro*	HS/EP	White DG et al, 1997
LacZ/bla	β-galactosidaze/ β-l actamase	PNA	*E. coli* AS19 (permeable membrane)	*in vitro*	–	Good & Nielsen, 1997

[a] UM, unmodfied.
[b] CPP, cell penetrating peptide; EP, electroporation; HS, heat shock; EGS, external guide sequences; "–", no delivery method used.

Table 2. Examples of AS-ODNs targeting resistance mechanism in antibacterial therapy.

First proof-of-principle evidence was given by White et al in 1997 for successful increasing the bactericidal activity of norfloxacin by antisense inhibiting the *marRAB* operon in *Escherichia coli* (White & et al, 1997). Hitherto, limited but successful trials have extended to dominating resistant genes and bacterial species with highest incidence of resistance (Table 2), e.g., antisense targeting *aac(6')-Ib* (Sarno & et al, 2003; Soler Bistue & et al, 2007), *act* (Chen & et al, 1997; Gao & et al, 2005) in *Escherichia coli*, *vanA* in *Enterococcus faecalis* (Torres & et al, 2001), *cmeA* in *Campylobacter jejun* (Jeon & Zhang, 2009), *mecA* in *Staphylococcus aureus* (Meng & et al, 2006; Meng & et al, 2009), *metS/murB* in *Bacillus anthracis* (Kedar & et al, 2007) and *oprM* in *Pseudomonas aeruginossa* (Wang & et al, 2010).

2.2.3 Efficient delivery systems

Virtually, any microbial gene could be targeted and highly organism-specific drugs could be envisioned in the development of antisense therapeutic agents. The obvious obstacle is stringent bacterial cell membrane for penetration or poor cellular uptake of AS-ODNs. An unmodified 10-mer oligonucleotide is 2-3 kDa, and various chemical modifications outlined above add further to this size. In short, AS-ODNs are likely to be considerably larger than vancomycin, therefore require efficient delivery systems. A variety of strategies exist to deliver compounds to bacterial cells in the laboratory, including electroporation, permeablilizing solvents, cationic lipid formulations (e.g. liposome), and pore-forming peptides (see Table 1&2). Although what exactly will work for AS-ODNs remains to be determined, the cell penetrating peptide (CPP) mediated delivery of AS-ODNs (especially peptide-PNA and peptide-PMO conjugates) outperformed other delivery systems in way of reaching future therapeutic applications.

2.2.3.1 Limitation in cellular uptake

Many barriers exist for the efficient transfer of genes/oligonucleotide anologs into cells, including the extracellular matrix, the endosomal/lysosomal environment, the endosomal membrane, and the nuclear envelope. Many delivery systems have been proved to serve suitably for antisense approach in eukaryotic cells regardless of their types (non-viral or viral) *vs* cell types. However, like most oligonucleotide-based strategies, the major limitation of antisense antibacterials is their poor cellular uptake due to low permeability of bacterial cell membrane to modified nucleic acids (Nekhotiaeva & et al, 2004). In particular, lipopolysaccharide outer membrane of gram-negative bacteria is a major barrier to molecule uptake (Good & et al, 2000). Meanwhile, decreased membrane permeability has been permanently observed for originally antibiotic-susceptible bacterial species after frequent exposure to multiple antibiotics present in commensal environments. Alternatively, membrane-associated energy-driven efflux in bacteria is of extremely broad substrate specificity, preventing intracellular drugs to release sustained effects.

2.2.3.2 Cell penetrating peptide (CPP) mediated delivery

Further, antisense antibacteirals may require development of delivery conditions for each bacterial species. Several strategies have been developed to improve delivery of oligonucleotides both in cultured cells and *in vivo*. So far, there is no universally applicable method for their delivery into different gram-positive and gram-negative specie, as they all present several limitations. Peptide-based strategies, representing a new and innovative concept to bypass the problem of bioavailability, have been demonstrated to improve the cellular uptake of nucleic acids both in cultured cells and *in vivo*.

Cell-penetrating peptides (CPPs) are short peptides of less than 30 amino acids that are able to penetrate cell membranes and translocate different cargoes into cells. The only common feature of these peptides appears to be that they are amphipathic and net positively charged. CPPs constitute very promising tools and have been successfully applied *in vivo* (Crombez & et al, 2008; Morris & et al, 2008). Two CPP strategies have been described to date; the first one requires chemical linkage between the drug and the carrier peptide for cellular drug internalization, and the second is based on the formation of stable complexes with drugs, depending on their chemical nature. Recently, the second strategy has the tendency to replace the first strategy for convenient delivery of DNA or AS-ODNs into eukaryotic cells, especially considering the synthesis and cost issues. However, CPP-conjugated method is now extensively applied for antisense antimicrobial ODNs.

In order to improve cellular uptake of PNA into bacterial cells, Good L and Nielsen PE, who first established CPP conjugating to the end of PNA in chemical synthesis, have realized efficient delivery of PNA through bacterial out membrane by observing its potent bacteriocidal antisense effects at micromolar ratio (Eriksson & et al, 2002). Further evidence demonstrates that introducing spacers or linkers between PNA and CPP in the direct covalent conjugate may increase its antisense efficacy and antibacterial potency. It also has been demonstrated that the release property of the chemical bond between PNA and CPP (e.g. the more stable amide bond or the less stable disulfide bond) has no influence on the antisense efficacy of PNAs. Later chemistry inventions make possible the conjugation CPP to other oligonucleotide analogs (e.g., PS-ODNs, LNAs, PMOs) for imparting them into bacterial cells and specific intracellular targets.

Regarding CPP itself, the mechanism of cell wall penetration is controversial and still under exploration. Nonetheless, the improvement in bacterial uptake of AS-ODNs with the aid of CPP has been well-recognized, making it an indispensable delivery system before no advanced system is developed (Lebleu & et al, 2008). The "carrier peptide" KFFKFFKFFK, originally reported for efficient penetrating ability through brain blood barrier, is the first also the most extensively applied peptide sequence verified by Nielsen PE and Good L et al for successful delivery of PNA into *Escherichia coli*. Previous studies suggested that the repeated amphipathic motif with cationic residues followed by hydrophobic regions is an important structure for carrier efficiency of CPP. The more efficient peptide sequences RFFRFFRFFRXB and RXRRXRRXRRXRXB (X is 6-aminohexanoic acid and B is β-alanine), have been recently reported for improved delivery efficacy of CPP-attached PMOs. Many efficient and simple penetrating efficacy test models for CPPs have been established in eukaryotic cells, whereas standard qualification method has been developed for only a few bacterial species. Evidence shows that the efficacy of CPPs differs according to bacterial species, and the underlying mechanisms are still unclear. Inadequate information has been accumulated from sporadic studies, e.g., $(KFF)_3K$ facilitates delivery of PNAs and PMOs into *Escherichia coli, Salmonella enterica serovar Typhimurium, Klebsiella pneumoniae* (however much less potent) and *Staphylococcus aureus*. But it is not working for *Pseudomonas aeruginosa* membrane even at higher concentrations of conjugated PNAs. $(RFF)_3RXB$ and $(RXR)_4XB$ enables more efficient transporting of PNAs and PMOs across the membrane of *Escherichia coli, Salmonella enterica serovar Typhimurium, Klebsiella pneumonia, Staphylococcus aureus* and *Pseudomonas aeruginosa*. Notably, the membrane of two gram-negative species *Acinetobacter baumanni* and *Shigella flexneri* show highest sensitivity to $(RXR)_4XB$ mediated PNA-CPP conjugates (unpublished results, and Bai & et al, 2012b). Following studies demonstrated that gram positive bacteria *Bacillus subtilis* and *Corynebacterium efficiens* exhibit increased susceptibility to CPP-PNAs, but in contrast, the gram-negative bacterium Ralstonia eutropha was not affected by addition of CPP-PNA.

2.2.3.3 Nano-material based delivery system

Nanomedicine is a growing field with a great potential for introducing new generation of targeted and personalized drugs. Membranes of eukaryotic cells and organelles, as well as the cell wall and membrane of pathogenic microorganisms, constitute a serious barrier for the access of hydrophilic drugs to their target molecules inside the cell structures. To overcome gene delivery problems of macro-molecule like AS-ODNs, various nano-mateiral based delivery techniques, including linear polymers, dendrimers and carbon nano tubes, have been developed and further studied as delivery tool for gene therapy purposes. And some of them are definitely worthy of extended trials in the antisense antibacterial aspect, with regard to delivery efficiency and other important pharmaceutical properties (e.g. 3D size, large scale synthesis and chemical modification, solubility, bioavailability, biocompatibility, toxicity, and pharmacokinetics, et al).

2.2.3.3.1 Dendrimers as vectors

Dendrimers are new class of synthetic polymeric materials characterized by well-defined and extensively branched 3D structure (Figure 3A). They have narrow polydispersity, nanometer size range, which can allow easier passage across biological barriers (e.g. small enough to undergo extravasations through vascular endothelial tissues). Notably, affordable commercialization of different types of size-controllable and surface-functionalized dendrimers is now available, providing a high degree of versatility. Besides, the unique properties of funtinonalized dendrimers, such as uniform size, high degree of branching, water solubility, multivalency, well-defined molecular weight and available internal cavities, have made them promising biological and drug-delivery systems for traditional drug (i.e. classical organic types) and gene therapy (e.g. DNA, small interfering RNAs, AS-ODNs, IgG antibodies, etc.) applications (Ravina & et al, 2010). And their excellent pharmacological properties, such as cytotoxicity, bacteriocidal and virucidal effect, biodistribution and biopermeability, may be modulated to fit specific medicinal purposes.

The wide range of applications reported for the use of dendrimers as delivery vectors for versatile cargos in the patent and literature demonstrates the general applicability of these molecules as carrier candidate for antisense antibacterials. There have been reports on the use of poly(amidomide) based dendrimers (e.g. PAMAM) for the development of antibacterial drugs mainly by destroying the cell walls of pathogenic organisms with their cationic surface groups, leading to direct cell death. However, our research suggests that lower generation of polyamide dendrimers, such as G1.0 and G2.0 PAMAM, showed no cell-wall impairment to many bacterial species (unpublished data, Xue et al, 2010), becoming a highly active vector for bacteria-specific oligonucleotides. Thus, dendrimers are highly suitable tools in antisense drug discovery to a wide variety of bacterial receptors.

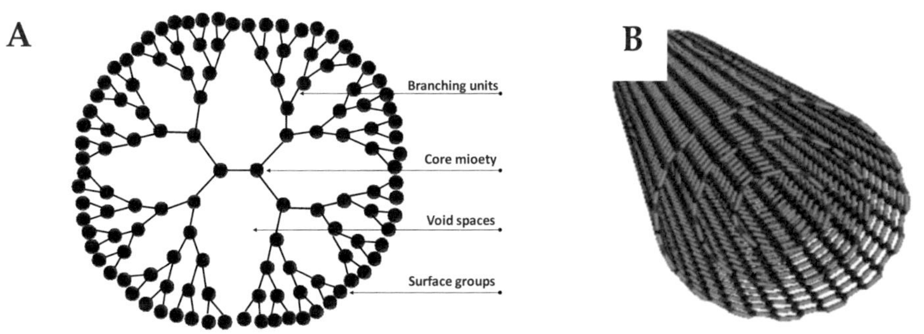

Fig. 3. (A) Schematic representation of generation 4 (G4.0) dendrimer. (B) Molecular structures of a multi-walled carbon nanotube (MWNT).

2.2.3.3.2 Funtionalized multi-walled carbon nanotubes (MWCNTs) as vectors

Synthetic inorganic gene nanocarriers have received limited attention in the transformation of bacterial cells. Amongst new generation of nano-vectors are carbon nanotubes (CNTs), a new form of carbon made-up of graphene layers rolled-up into a cylindrical from which can be produced as single or multi-walled (Figure 3B). The physico-chemical features of CNTs, such as needle-like shape, nanorange size, surface modification flexibility, and electronic properties, make them unique materials in nanoscience and nanotechnology. Multi-walled carbon nanotubes (MWCNTs) can be fabricated as biocompatible nanostructures (cylindrical bulky tubes), forming supramolecular complexes with proteins, polysaccharides and nucleic acids (Kateb & et al, 2010). These structures have been under investigation in the biomedical domain and in nanomedicine as viable and safe nanovectors for gene and drug delivery.

Research work based on nanobiotechnologies has allowed us to develop complex antigenic systems and novel delivery routes for peptides, nucleic acids and drugs covalently linked or simply adsorbed onto carbon nanotubes. In particular, Rojas-Chapana J and et al have presented a plasmid delivery system based on water dispersible multi-walled carbon nanotubes (CNTs) that can simultaneously target the bacterial surface and deliver the plasmids into *E. coli* cells via temporary nanochannels across the cell envelope (Rojas-Chapana & et al, 2005). It is the first experimental evidence that shows high potential of CNTs for nanoscale cell electroporation in bacteria. However, the study of metabolism, the toxicity and the mechanism of elimination of water-soluble carbon nanotubes in order to evaluate their impact on the health and validate the concept of CNT as new delivery system still arouse concerns in many critial ways. Recently initiated researches on hybriding the dendrimers with MWCNT has offered us new hopes (Qin & et al, 2011; Zhang & et al, 2011). Besides increased dispersility, solubility, biocompatiability and stability, (MWCNTs)-polyamidoamine (PAMAM) hybrid prepared by covalent linkage has possessed good plasmid DNA immobilization ability and efficiently delivered GFP gene into cultured HeLa cells. The surface modification of MWCNTs with PAMAM improved the transfection efficiency and simultaneously decreased cytotoxicity by about 38%, as compared with mixed acid-treated MWCNTs and pure PAMAM dendrimers. The MWCNT-PAMAM hybrid can be considered as a new carrier for the delivery of biomolecules into both mammalian and bacterial cells.

2.2.4 Other pharmaceutical properties

As far as the other properties in therapeutic application of antisense antibacterials are concerned, there should always be a systematic overview of cargo (i.e. oligonucleotides) and vector (especially the CPP, Heitz & et al, 2009). With regard to AS-ODNs, the electric neutral PNA and PMO calsses emerge and show desirable properties (especially their non-ionic backbones) as better therapeutic alternative to other antisense agents. And with regard to delivery strategies, alternative nonviral methods, such as electroporation and the use of liposomes, have been developed for delivery of antibacterial AS-ODNs. These methods have been proved to be effective *in vitro* and for research purposes, but showed limited potential for delivery *in vivo* due to toxicity, cell damage, and immunogenicity. They are also technically demanding in their application, lack tissue and cell specificity, and can deliver material to only a limited amount of cells. In view of these considerations, peptide conjugated AS-ODNs (i.e. peptide-PNAs and peptide-PMOs) offer a promising noninvasive version of gene scilencers with potent antisense antibacterial activity, the pharmaceutical properties of which this part will mainly focus on.

2.2.4.1 PNAs, PMOs and their peptide conjugates

PNAs and PMOs (also known as morpholino) are novel classes of antisense agents that offer a better therapeutic alternative to other antisense antibacterial oligomers. They both possess a non-ionic backbone, but differed in ribose sugar replacement. For PMO, the backbone of DNA is replaced by a 6-membered morpholine moiety and the phosphorodiester intersubunit bonds with phosphorodiamidate linkages. PNA has a pseudopeptide backbone composed of (2-aminoethyl)glycine units, in which the geometry and the spacing of the bases is nearly identical to that found in a native DNA or RNA strand. The polyamide backbone of PNA has no phosphate groups, having an amino (NH_2) to carboxyl (CO_2H) orientation instead of 5' to 3' orientation as do phosphodiester backbones. Specifically, PNAs can bind to either single-stranded DNA or RNA, in which the resulting hybrid resembles the B-form of DNA, or double-stranded DNA, in which the PNA invades the DNA double stranded helix and hybridizes to the target sequence, thus displacing the second DNA strand into a 'D'loop. Although departing significantly from the sugar-phosphate backbone found in regular DNA, oligomers of both types independently (i.e. with or without delivery strategies) have been found to be remarkable steric-block ODNs for inhibiting translation and blocking mRNA activity, as demonstrated in embryos, cells and animals. Now PMOs have been taken to pre-clinical studies for treatment of cardiovascular diseases, viral diseases and genetic disorders, such as Duchenne muscular dystrophy (DMD).

Since the conjugation of CPP to negatively charged ODNs (e.g. PS-ODNs) did not result in a level of delivery into cells sufficient for biological activity, PNA- and PMO- CPP conjugates (covalently linked with or without spacers) confer on these compounds more desirable properties over the original ODN forms, as well as ribozyme and siRNA counterparts (Thompson & Patel, 2009), especially with respect to antisense antibacterials. Several CPPs have been developed for bacterial-specific transformation purposes (as mentioned in 2.2.3.2), and they can be coupled to PNA or PMO by flexible linker types (Venkatesan & Kim, 2006). No general rules have yet emerged as to optimal linkage types, since the factors affecting biological activity are often complex. Early popular labile linkers for PNA and CPP include AEEA (8-amino-3,5-dioxo-octanoic acid, a polyether spacer also known as an O-linker), and disulfide bond linkage, which were proposed to be cleavable within the reducing environment of the cell. Stable linkage such as glycine linkage, thioether linkage and thiol-maleimide linkage have also been reported for improved *in vivo* stability. The conjugation of CPPs and PMOs through a thioether (maleimide), disulfide or amide linker have previously been described. The nuclear antisense activities of the CPP–PMOs with the three linkage types were similar (Lebleu & et al, 2008). But, the amide linkage is advantageous with regard to synthetic procedures (e.g. greater yield and less steps) and *in vivo* stability.

2.2.4.2 Tissue distribution, pharmacokinetics and stability

The modified chemistry of PNAs and PMOs provides excellent resistance to nuclease and protease activity, which is the basis for the enhanced stability in plasma, tissues, cerebrospinal fluid and urine. Independently, the non-ionic character of the PNA/PMO portion of the conjugates avoids potential non-specific drug interactions with bacteiral cellular components (except for the target RNA sequence) observed with PS-ODNs. In addition, the neutral character of PNAs and PMO chemistry not only guarantees a high safety profile but also sufficient tissue concentrations required for effective PNA or PMO oligonucleotide:RNA duplex formation, thus enhancing their affinity for the target RNA sequence and hence increasing efficacy. Rational design of conjugates (i.e. the optimal linker type and position of CPP) may eliminate CPP's stereospecific blockade that might significantly influence the antisense effect of PNA or PMO in target recognization, base matching and binging affinity. However, stability of CPPs coupled to antisense PNA or PMO may be a matter of concern. This is partially due to the fact that degradation of CPP in solution and plasma has been observed for systematic delivery of CPP-2'MOE in mouse model (Henke & et al, 2008). Another concern is the non-specificity of CPP mediated delivery of PNA or PMO when eukaryotic cells and prokaryotic cells exist in commensal environment. Although little lethal damage to cells would be done by conjugated CPP at equal molar ratio of PNA or PMO used at the highest concentrations *in vitro* and *ex vivo*, the consequences of its non-specific physical disruption to normal human cell membranes *in vivo* have not been thoroughly evaluated. Systematic study on CPP mediated antisense antibacterial therapy needs to be done if any possible candidate for clinical development is ever recommended (Zorko & Langel, 2005).

The application of unmodified PNAs as antisense therapeutics has been limited by their low solubility under physiological conditions, insufficient cellular uptake, and poor biodistribution due to rapid plasma clearance and excretion. The excelent stability of PNA-CPP conjugates has been confirmed for a 48h period at 37°C in rat's plasma (Bai & et al, 2012b). However, there has been no report of *in vivo* tissue distribution and pharmacokinetics properties of CPP-PNA conjugates targeting bacterial genes. Nontheless, limited information from PNA-CPP conjugates targeting genes in eukaryotic cells can be refered. Jia et al have determined that PNA-CPP conjugates targeting bcl-2 mRNA showed specific tumor uptake, low uptake in blood and organs (e.g. liver and slpeen) except for kidney, as well as slower urinary clearance in Mec-1-bearing severe combined immunodeficiency (SCID) mice. Recently, Wancewicz et al have reported that conjugation of PNA (targeting murine phosphatase and tensin homolog) to short basic peptides (serve as solubility enhancers and delivery vehicles) allowed for rapidly distribution and accumulation of conjugates in liver, kidney and adipose tissue, while their rates of elimination via excretion were dramatically reduced compared to unmodified PNA.

Unlike CPP-PNA conjugates, the pharmaceutical properties of CPP-PMO conjugates have been evaluated in an extensive scope besides specific gene modulators (Amantana & et al, 2007). In general, the conjugation of CPP to PMO enhances the PMO pharmacokinetic profile, tissue uptake, and subsequent retention. Amantana et al have reported that conjugation of a PMO to the $(RXR)_4XB$ peptide increased the tissue uptake (in all organs except in brain, with greater increase being seen in liver, spleen and lungs) and retention time in these organs, while efflux of the conjugated PMO from tissues to the vascular space was slow. They have also confirmed that peptide conjugation also improved the kinetic behaviour of PMO as demonstrated by increased volume of distribution, estimated elimination half life, and area under the plasma concentration versus time curve. Youngblood et al have determined that the stabilities of CPP–PMO conjugates in cells and in human serum varied according to CPP sequences, amino acid compositions and/or linkers.

The stability of a $(RXR)_4XB$ peptide in the conjugate exhibited time- and tissue-dependent degradation, with biological stability ranked in the order of liver>heart>kidney>plasma. Meanwhile, the PMO portion of the conjugates was completely stable in cells, serum, plasma and tissues.

2.2.4.3 Toxicity

Large amount of data concerning toxicity of CPP-PNA conjugates have been collected from *ex vivo* studies (Alksne & Projan, 2000; Kurupati & et al, 2007), in which antisense peptide-PNAs cured cell cultures that were infected with bacteria in a dose-dependent manner without any noticeable toxicity to the human cells. *In vivo* inhibition of gene expression and growth have been observed for anti-*acpP* CPP-PNA conjugates in mouse intraperitoneal *E. coli* infection. However, none toxicity issues have yet been seriously addressed.

CPP-PMO conjugates targeting bacterial essential genes have been evaluated to confirm their bacteriocidal antisense effect in several animal bacteremia models, and have proven to be efficacious with an excellent safety profile (Amantana & et al, 2007) within doeses for 100% survival 48h after treatment. They have also found that survival was significantly reduced for mice treated with 2×300 mg and 2×1 mg of the 11-base AcpP peptide-PMO, indicating toxicity at these high doses. Generally, the toxicity of $(RXR)_4XB$ -PMOs is caused by $(RXR)_4XB$ while the PMO portions of the conjugates are essentially non-toxic. In particular, data from CPP-PMO conjugates targeting genes in eukaryotic cells have demonstrated that the degree of toxicity depends on the dose, dose frequency and route of administration. Collectively, mice tolerated $(RXR)_4XB$-PMOs well with repeated intraperitoneal (i.p.) or intravenous (i.v.) injection doses of ≤15 mg/kg at diverse time intervals, showing no changes in behaviour, weight and serum chemistry, and no histopathological abnormalities were detected in major organs. However, at higher doses and dosing frequency, animals experienced weight loss, despite maintaining their normal organ weights and appearances. Rats treated with a single 150 mg/kg dose appeared lethargic immediately after the injection and proceeded to lose weight, accompanied by affected kidney function. The LD_{50} of a $(RXR)_4XB$-PMO in rats was around 220–250 mg/kg.

3. Broad-spectrum antisense antibacterials

A range of functional genes in bacteria have been validated as potential targets by using unmodified PNAs or CPP conjugated ODNs. Collectively, consistent efforts on antisense targeting of a small bacterial gene *acpP* (encoding the essential fatty acid biosynthesis protein) have passed the proof-of-principle phase and have gathered plenty of positive results, especially from recent studies focusing on *in vivo* confirmation of anti-*rpoD* peptide-PMO's bactericidal effect in mice infected with several pathogenic bacteria (i.e., *Escherichia coli, Salmonella enterica serovar Typhimurium*, and *Burkholderia cepacia*. However, few reports describe promising gene targets that have potential for broad-spectrum antisense growth inhibition among different bacterial species. Indeed, the validated targets in different bacterial species show discouragingly low similarity in gene sequence and homology. Thus, identification of gene targets for broad-spectrum antisense inhibition would aid the development of new antimicrobial agents that could relieve the exacerbating therapeutic consequences caused by MDR/PDR infections.

3.1 Target accessibility

A challenging aspect of identifying essential genes in bacteria for broad-sepctrum antisense inhibition mainly involves efforts to locate the exact targeting site within a specific gene for realization of the most potent and specific antisense inhibitory effect of complementary AS-ODNs against different species (e.g. among gram-negatives, gram-positives, or both). Naturally, prequisities in target selection require searching for genes with high similarity and identidy amongst as many bacterial species as possible. As time comsuming as it is, an economical way of identifying genes that acturally fit this critea should focus on the validated targets for both traditional antibiotics and antisense antibacteirals, certainly because massive open data of their gene sequencing are available.

Antisense suppression of the above mentioned essential genes (e.g. 16S rRNA, *acpP*, *gyr*, and et al) in single bacterial species has showed potent growth inhibitory and cell death effect in a sequence-specific and dose-dependent manner. However, the issue of target accessability among different species still needs investigation and validation. (1) The ribosome has a complex structure involving rRNAs and ribosomal proteins, and therefore, inaccessibility of the target site could be one of the reasons for the ineffectiveness of the antiribosomal ODN. (2) Systematic researches *in vitro* and in animal models have demonstrated that one potential target *acpP* (encoding acyl carrier protein *AcpP*) opens limitless possibility for recommending the very first "antisense antibiotic" into market. Besides, *acpP* gene in pathogenic gram-negative species share highly homology in sequences, making itself an ideal candidate for antisense antibacterials with broad anti-gram-negative spectrum, although more candidate bacterial species are needed to confirm the accessibility of an already-validated 11-nucleotide targeting site in its start codon region of mRNA. (3) Meanwhile, newly discovered gene targets for new types of protein-targeting antibiotics, i.e. the bacterial cell division inhibitor (targeting bacterial cell division protein *FtsZ for* terminating bacterial proliferation (Boberek & et al, 2010)) and virulence inhibitor (targeting quorum sensing sensor protein *QseC* without affecting bacterial growth (Alksne & Projan, 2000)), also show promises and potential for developing specific or broad-spectrum antisense antibacterials based on their homology assessment. (4) our researches focus on validating the known target in broad-spectrum antibiotic development by antisene strategy, in which the DNA-dependent RNA polymerase (RNAP) is a candidate of great interest for it distinct advantanges (see 3.3).

3.2 Universally applicable delivery systems

Furthermore, the term "broad-spectrum" also qualifies the delivery systems for AS-ODNs. Specifically, rational design of peptide-ODN conjugates could optimize the effective AS-DONs in way of enhancing antibacterial potency and expanding antibacterial spectrum, in which CPP choice is of equal importance. The range of sensitivities observed for different bacterial species to CPPs largely determines their application potential. To our knowledge, the synthetic CPP $(RXR)_4XB$ has shown by far the most broad cell pernetrating range as an effective tool for intracellular AS-ODN delivery into many major gram-negative and gram-positive pathogens. And the transfection efficiency of the widely used CPPs $(RXR)_4XB$ and $(KFF)_3K$ appears to reflect with the features of bacterial cell walls of clinical isolates, where less potent effects were observed for $(KFF)_3K$ against species with stringent cell barrier (e.g., *Klebsiella pneumoniae* and *Pseudomonas aeruginosa*). Collectively, our results suggested that the peptide component of peptide-PNA conjugates may be developed for a wide range of indications to realize broad antisense antibacterial spectrum.

3.3 proof-of principle studies

Bacterial DNA-dependent RNA polymerase (RNAP) is a key enzyme in transcription regulation and gene expression. Its function requires coordination of a core enzyme (comprising five subunits α_2, β, β' and ω) and an independent σ subunit that is reversibly recruited by core enzyme. The RNAP core enzyme is responsible for transcription elongation, and different σs are in charge of transcription initiations from promoters that express genes in diverse function. Deactivation of RNAP by any possible means leads to direct cell death, attracting much exploration for developing specific RNAP inhibitors, the most representative class of broad-spectrum antibiotics (e.g. the rifamycins) with fundamental clinical significance. The most developed σ^{70} family of σs, especially the primary σ^{70}, is essential for initiating transcription of multiple genes in exponentially growth cells , which to our knowledge has not previously been demonstrated for target validation. And most importantly, gene *rpoD* (encoding the primary σ^{70} of RNAP) is highly conserved in identity and homologous in sequence among different pathogenic gram-negative species. Such features are distinct advantages for developing broad-spectrum antisense antibacterial agents (Bai & et al, 2011).

Results from our lab (unpublished and Bai & et al, 2012b) gives the first proof-of-principle evidence for exploring and identifying bacterial RNAP σ^{70} as an antibacterial target by antisense strategy. We identified a conserved target sequence within the native *rpoD* mRNA start codon region, and a cell penetrating peptide $(RXR)_4XB$ conjugated 10-mer peptide nucleic acid was developed for potent sequence-selective bacteriocidal antisense effect against six pathogenic gram-negative species, including *Escherichia coli, Salmonella enterica, Klebsiella pneumoniae, Shigella flexneri, Citrobacter freundii, and Enterobacter cloacae.* It cured endothelial cell cultures from lethal infection with single or triple GNB without showing any apparent toxicity. It specifically interferes with *rpoD* mRNA, and inhibited the expression of σ^{70} in a concentration-dependent manner. Its *in vivo* antibacterial activity has also been confirmed by increased survival in bacterial infected mice.

4. Conclusion

New classes of antisense antibacterial agents (bactericidal agents or resistance inhibitors), represent an evolutionary inevitability in antibiotic industry. In the past 20 years, many essential genes have been studied as potential targets for developing bactericidal antisense agents or resistance inhibitors against clinically pathogenic bacteria. Nonetheless, much investment needs to be infused for converting this concept into real drugs. Identified targets with application potentiality or new targets under investigation should be further evaluated for posing the least risk for selection of resistant variants. Nucleic acid monomers with simple synthesis and cheap source of starting materials are more viable as antisense drugs. Versatile of delivery systems developed for eukaryotic cells, e.g., polymers, dendrimers, nanotubes, should be given considerations for AS-ODNs delivery in bacteria, if no more innovative systems than the non-invasive CPP-mediated delivery system is created. Furthermore, it is more practical and economical to develop antisense agent that targets only multiple-resistant or pan-resistant bacteria, particularly when it allows co-administration of a narrow-spectrum antibiotic. Most importantly, "broad-spectrum" antisense antimicrobials should also be developed to meet future clinical requirements, in which target selection and validation address more attention.

We have already lagged behind our therapeutic initiatives to meet the challenges of increasing isolation of new antibiotic-resistant bacterial strains. A great many functional genes discovered in the past decade represent themselves as potential targets for developing antibacterial therapeutic agents with whole new mechanisms. Thus, innovative approaches must become a priority in antibitoitc discovery, in which antisense antibacterial strategy is absolutely a leap in our ability to effectively treat human pathogens of great concern (Woodford & Wareham, 2009). The theoretical advantage of antisense antibacterials is obvious and has been well-acknowledged by strong evidence in the long process of conquering major technological obstacles. It is our continous efforts that will make ultimate success in this glorious field.

5. References

Alksne, Lefa E. & Projan, Steven J. (2000). Bacterial virulence as a target for antimicrobial chemotherapy. *Current Opinion in Biotechnology,* Vol11, No6, (December 2000), pp.625-636, ISSN0958-1669

Amantana, A., Moulton, H. M., Cate, M. L., Reddy, M. T., Whitehead, T., Hassinger, J. N., Youngblood, D. S., & Iversen, P. L. (July 2007). Pharmacokinetics, biodistribution, stability and toxicity of a cell-penetrating peptide-morpholino oligomer conjugate. *Bioconjugate Chemistry,* Vol18, No4, (July 2007), pp.1325-1331, ISSN1043-1802

Bai, H., Sang, G., You, Y., Xue, X., Zhou, Y., Hou, Z., Meng, J., & Luo, X. (2012a). Targeting RNA Polymerase Primary sigma as a Therapeutic Strategy against Methicillin-Resistant Staphylococcus aureus by Antisense Peptide Nucleic Acid. *PLoS One,* Vol7, No1, (Janurary 2012), pp.e29886, ISSN1932-6203

Bai, H., Xue, X., Hou, Z., Zhou, Y., Meng, J., & Luo, X. (2010). Antisense antibiotics: a brief review of novel target discovery and delivery. *Curr Drug Discovery Technology,* Vol7, No2, (June 2010), pp.76-85, ISSN1570-1638

Bai, H., You, Y., Yan, H., Meng, J., Xue, X., Hou, Z., Zhou, Y., Ma, X., Sang, G., & Luo, X. (2012b). Antisense inhibition of gene expression and growth in gram-negative bacteria by cell-penetrating peptide conjugates of peptide nucleic acids targeted to rpoD gene. *Biomaterials,* Vol33, No2, (Janurary 2012), pp.659-667, ISSN0142-9612

Bai, H., Zhou, Y., Hou, Z., Xue, X., Meng, J., & Luo, X. (2011). Targeting bacterial RNA polymerase: promises for future antisense antibiotics development. *Infectious Disorder Drug Targets,* Vol11, No2, (April 2011), pp.175-187, ISSN1871-5265

Bennett, C. F. & Swayze, E. E. (2010). RNA targeting therapeutics: molecular mechanisms of antisense oligonucleotides as a therapeutic platform. *Annual Review of Pharmacology Toxicology,* Vol50, (February 2010), pp.259-293, ISSN0362-1642

Boberek, J. M., Stach, J., & Good, L. (2010). Genetic Evidence for Inhibition of Bacterial Division Protein FtsZ by Berberine. *PLoS One,* Vol5, No10, (October 2010), pp.e13745-, ISSN1932-6203

Bush, K. (2010). Alarming beta-lactamase-mediated resistance in multidrug-resistant Enterobacteriaceae. *Current Opinion in Microbiology,* Vol13, No5, (October 2010), pp.558-564, ISSN 1369-5274

Cattoir, V. & Daurel, C. (2010). Update on antimicrobial chemotherapy. *Medecine et Maladies Infectieuses,* Vol40, No3, (March 2010), pp.135-154, ISSN0399-077X

Chen, H., Ferbeyre, G., & Cedergren, R. (1997). Efficient hammerhead ribozyme and antisense RNA targeting in a slow ribosome Escherichia coli mutant. *Nature Biotechnology,* Vol15, No5, (May 1997), pp.432-435, ISSN1087-0156

Crombez, L., Morris, M. C., Deshayes, S., Heitz, F., & Divita, G. (2008). Peptide-based nanoparticle for ex vivo and in vivo drug delivery. *Current Pharmaxeutical Design,* Vol14, No34, (December 2008), pp.3656-3665, ISSN1381-6128

Deere, J., Iversen, P., & Geller, B. L. (2005). Antisense phosphorodiamidate morpholino oligomer length and target position effects on gene-specific inhibition in Escherichia coli. *Antimicrobial Agents and Chemotherary,* Vol49, No1, (Janurary 2005), pp.249-255, ISSN0066-4804

Deshpande, P., Rodrigues, C., Shetty, A., Kapadia, F., Hedge, A., & Soman, R. (2010). New Delhi Metallo-beta lactamase (NDM-1) in Enterobacteriaceae: treatment options with carbapenems compromised. *The Journal of the Association of Physicians India,* Vol58, (March 2010), pp.147-149, ISSN0004-5772

Ding, Y. & Lawrence, C. E. (2003). A statistical sampling algorithm for RNA secondary structure prediction. *Nucleic Acids Research,* Vol31, No24, (December 2003), pp.7280-7301, ISSN0305-1048

Dryselius, R., Aswasti, S. K., Rajarao, G. K., Nielsen, P. E., & Good, L. (2003). The translation start codon region is sensitive to antisense PNA inhibition in *Escherichia coli. Oligonucleotides,* Vol13, No6, (December 2003),pp.427-433, ISSN1545-4576

Engel, L. S. (2010). The dilemma of multidrug-resistant gram-negative bacteria. *The American journal of the Medical Sciences,* Vol340, No3, (September 2009), pp.232-237, ISSN0002-9629

Eriksson, M., Nielsen, P. E., & Good, L. (2002). Cell permeabilization and uptake of antisense peptide-peptide nucleic acid (PNA) into *Escherichia coli. The Journal of Biological Chemistry,* Vol277, No9, (March 2002), pp.7144-7147, ISSN0021-9258

Gao, M. Y., Xu, C. R., Chen, R., Liu, S. G., & Feng, J. N. (2005). Chloromycetin resistance of clinically isolated E coli is conversed by using EGS technique to repress the chloromycetin acetyl transferase. *World Journal of Gastroenterology,* Vol11, No46, (December 2005), pp.7368-7373, ISSN1007-9327

Geller, B. L. (2005). Antibacterial antisense. *Current Opinion in Molecular Therapy,* Vol7, No2, (April 2005), pp.109-113, ISSN1464-8431

Geller, B. L., Deere, J., Tilley, L., & Iversen, P. L. (2005). Antisense phosphorodiamidate morpholino oligomer inhibits viability of Escherichia coli in pure culture and in mouse peritonitis. *The Journal of Antimicrobial Chemotherapy,* Vol55, No6, (June 2005), pp.983-988, ISSN0305-7453

Geller, B. L., Deere, J. D., Stein, D. A., Kroeker, A. D., Moulton, H. M., & Iversen, P. L. (2003). Inhibition of gene expression in Escherichia coli by antisense phosphorodiamidate morpholino oligomers. *Antimicrobial Agents and Chemotherary,* Vol47, No10, pp.3233-3239, ISSN0066-4804

Geller, B. L., Deere, J. D., Stein, D. A., Kroeker, A. D., Moulton, H. M., & Iversen, P. L. (2003). Inhibition of gene expression in *Escherichia coli* by antisense phosphorodiamidate morpholino oligomers. *Antimicrobial Agents and Chemotherary,* Vol47, No10, (October 2003), pp.3233-3239, ISSN0066-4804

Goh, S., Boberek, J. M., Nakashima, N., Stach, J., & Good, L. (2009). Concurrent growth rate and transcript analyses reveal essential gene stringency in *Escherichia coli. PLoS One,* Vol4, No6, (June 2009) pp.e6061, ISSN1932-6203

Good, L. & Nielsen, P. E. (1998). Inhibition of translation and bacterial growth by peptide nucleic acid targeted to ribosomal RNA. *Proc Natl Acad Sci U S A,* Vol95, No5, pp.2073-2076, ISSN

Good, L., Sandberg, R., Larsson, O., Nielsen, P. E., & Wahlestedt, C. (2000). Antisense PNA effects in *Escherichia coli* are limited by the outer-membrane LPS layer. *Microbiology,* Vol146 (Pt 10), (Octoer 2000), pp.2665-2670, ISSN1350-0872

Greenberg, D. E., Marshall-Batty, K. R., Brinster, L. R., Zarember, K. A., Shaw, P. A., Mellbye, B. L., Iversen, P. L., Holland, S. M., & Geller, B. L. (2010). Antisense phosphorodiamidate morpholino oligomers targeted to an essential gene inhibit Burkholderia cepacia complex. J Infect Dis, Vol201, No12, pp.1822-1830, ISSN

Gruegelsiepe, H., Brandt, O., & Hartmann, R. K. (2006). Antisense inhibition of RNase P: mechanistic aspects and application to live bacteria. *The Journal of Biological Chemistry,* Vol281, No41, (October 2006), pp.30613-30620, ISSN0021-9258

Guo, Q. Y., Xiao, G., Li, R., Guan, S. M., Zhu, X. L., & Wu, J. Z. (2006). Treatment of Streptococcus mutans with antisense oligodeoxyribonucleotides to *gtfB* mRNA inhibits GtfB expression and function. *FEMS Microbiology Letters,* Vol264, No1, (November 2006), pp.8-14, ISSN0378-1097

Harth, G., Horwitz, M. A., Tabatadze, D., & Zamecnik, P. C. (2002). Targeting the Mycobacterium tuberculosis 30/32-kDa mycolyl transferase complex as a therapeutic strategy against tuberculosis: Proof of principle by using antisense technology. *Proc Natl Acad Sci U S A,* Vol99, No24, (November 2002), pp.15614-15619, ISSN0027-8424

Harth, G., Zamecnik, P. C., Tabatadze, D., Pierson, K., & Horwitz, M. A. (2007). Hairpin extensions enhance the efficacy of mycolyl transferase-specific antisense oligonucleotides targeting Mycobacterium tuberculosis. *Proc Natl Acad Sci U S A,* Vol104, No17, (April 2007), pp.7199-7204, ISSN0027-8424

Harth, G., Zamecnik, P. C., Tang, J. Y., Tabatadze, D., & Horwitz, M. A. (2000). Treatment of Mycobacterium tuberculosis with antisense oligonucleotides to glutamine synthetase mRNA inhibits glutamine synthetase activity, formation of the poly-L-glutamate/glutamine cell wall structure, and bacterial replication. *Proc Natl Acad Sci U S A,* Vol97, No1, (Janurary 2000), pp.418-423, ISSN0027-8424

Hatamoto, M., Ohashi, A., & Imachi, H. (2010). Peptide nucleic acids (PNAs) antisense effect to bacterial growth and their application potentiality in biotechnology. *Applied Microbiology and Biotechnology,* Vol86, No2, (March 2010), pp.397-402, ISSN0175-7589

Heitz, F., Morris, M. C., & Divita, G. (2009). Twenty years of cell-penetrating peptides: from molecular mechanisms to therapeutics. *British Journal of Pharmacology,* Vol157, No2, (May 2009), pp.195-206, ISSN0007-1188

Henke, E., Perk, J., Vider, J., de, C. P., Chin, Y., Solit, D. B., Ponomarev, V., Cartegni, L., Manova, K., Rosen, N., & Benezra, R. (2008). Peptide-conjugated antisense oligonucleotides for targeted inhibition of a transcriptional regulator in vivo. *Nature Biotechnology,* Vol26, No1, (Janurary 2008), pp.91-100, ISSN1087-0156

Houseley, J. & Tollervey, D. (2009). The many pathways of RNA degradation. *Cell,* Vol136, No4, (February 2009), pp.763-776, ISSN0092-8674

Jeon, B. & Zhang, Q. (2009). Sensitization of Campylobacter jejuni to fluoroquinolone and macrolide antibiotics by antisense inhibition of the CmeABC multidrug efflux transporter. *The Journal of Antimicrobial Chemotherapy,* Vol63, No5, (May 2009), pp.946-948, ISSN 0305-7453

Ji, Y., Yin, D., Fox, B., Holmes, D. J., Payne, D., & Rosenberg, M. (2004). Validation of antibacterial mechanism of action using regulated antisense RNA expression in *Staphylococcus aureus. FEMS Microbiology Letters,* Vol231, No2, (February 2004), pp.177-184, 0378-1097 ISSN

Kateb, B., Yamamoto, V., Alizadeh, D., Zhang, L., Manohara, H. M., Bronikowski, M. J., & Badie, B. (2010). Multi-walled carbon nanotube (MWCNT) synthesis, preparation, labeling, and functionalization. *Methods in Molecular Biology,* Vol651, (September 2010), pp.307-317, ISSN1064-3745

Kedar, G. C., Brown-Driver, V., Reyes, D. R., Hilgers, M. T., Stidham, M. A., Shaw, K. J., Finn, J., & Haselbeck, R. J. (2007). Evaluation of the *metS* and *murB* loci for antibiotic discovery using targeted antisense RNA expression analysis in *Bacillus anthracis. Antimicrobial Agents and Chemotherary,* Vol51, No5, (May 2007), pp.1708-1718, ISSN0066-4804

Kulyte, A., Nekhotiaeva, N., Awasthi, S. K., & Good, L. (2005). Inhibition of *Mycobacterium smegmatis* gene expression and growth using antisense peptide nucleic acids. *Journal of Molecular Microbiology Biotechnology,* Vol9, No2, (November 2005), pp.101-109, ISSN1464-1801

Kurreck, J. (2003). Antisense technologies. Improvement through novel chemical modifications. European Journal Biochemistry, Vol270, No8, (April 2003), pp.1628-1644, ISSN0014-2956

Kurupati, P., Tan, K. S., Kumarasinghe, G., & Poh, C. L. (2007). Inhibition of gene expression and growth by antisense peptide nucleic acids in a multiresistant beta-lactamase-producing *Klebsiella pneumoniae* strain. *Antimicrobial Agents and Chemotherary,* Vol51, No3, (March 2007)pp.805-811, ISSN0066-4804

Lebleu, B., Moulton, H. M., Abes, R., Ivanova, G. D., Abes, S., Stein, D. A., Iversen, P. L., Arzumanov, A. A., & Gait, M. J. (2008). Cell penetrating peptide conjugates of steric block oligonucleotides. *Advanced Drug Delivery Reviews,* Vol60, No4-5, (March 2008), pp.517-529, ISSN0169-409X

Mellbye, B. L., Puckett, S. E., Tilley, L. D., Iversen, P. L., & Geller, B. L. (2009). Variations in amino acid composition of antisense peptide-phosphorodiamidate morpholino oligomer affect potency against *Escherichia coli* in vitro and in vivo. *Antimicrobial Agents and Chemotherary,* Vol53, No2, (February 2009), pp.525-530, ISSN0066-4804

Mellbye, B. L., Weller, D. D., Hassinger, J. N., Reeves, M. D., Lovejoy, C. E., Iversen, P. L., & Geller, B. L. (2010). Cationic phosphorodiamidate morpholino oligomers efficiently prevent growth of Escherichia coli in vitro and in vivo. *The Journal of Antimicrobial Chemotherapy,* Vol65, No1, (Janurary 2010), pp.98-106, ISSN0305-7453

Meng, J., Hu, B., Liu, J., Hou, Z., Meng, J., Jia, M., & Luo, X. (2006). Restoration of oxacillin susceptibility in methicillin-resistant Staphylococcus aureus by blocking the MecR1-mediated signaling pathway. *Journal of Chemotherapy,* Vol18, No4, (August 2006), pp.360-365, ISSN1120-009X

Meng, J., Wang, H., Hou, Z., Chen, T., Fu, J., Ma, X., He, G., Xue, X., Jia, M., & Luo, X. (2009). Novel anion liposome-encapsulated antisense oligonucleotide restores susceptibility of methicillin-resistant *Staphylococcus aureus* and rescues mice from lethal sepsis by targeting mecA. *Antimicrobial Agents and Chemotherary,* Vol53, No7, (July 2009), pp.2871-2878, ISSN0066-4804

Mitev, G. M., Mellbye, B. L., Iversen, P. L., & Geller, B. L. (2009). Inhibition of intracellular growth of *Salmonella enterica serovar Typhimurium* in tissue culture by antisense peptide-phosphorodiamidate morpholino oligomer. *Antimicrobial Agents and Chemotherary,* Vol53, No9, (September 2009), pp.3700-3704, ISSN0066-4804

Moellering, R. C., Jr. (2011). Discovering new antimicrobial agents. International *Journal of Antimicrobial Agents,* Vol37, No1, (Janurary 2011), pp.2-9, ISSN0924-8579

Monaghan, R. L. & Barrett, J. F. (2006). Antibacterial drug discovery--Then, now and the genomics future. *Biochemical Pharmacology,* Vol71, No7, (March 2006), pp.901-909, 0006-2952

Morris, M. C., Deshayes, S., Heitz, F., & Divita, G. (2008). Cell-penetrating peptides: from molecular mechanisms to therapeutics. *Biology of Cell,* Vol100, No4, (April 2008), pp.201-217, ISSN0248-4900

Nekhotiaeva, N., Awasthi, S. K., Nielsen, P. E., & Good, L. (2004). Inhibition of *Staphylococcus aureus* gene expression and growth using antisense peptide nucleic acids. *Molecular Therapy,* Vol10, No4, (October 2004), pp.652-659, ISSN1525-0016

Nekhotiaeva, N., Elmquist, A., Rajarao, G. K., Hallbrink, M., Langel, U., & Good, L. (2004). Cell entry and antimicrobial properties of eukaryotic cell-penetrating peptides. *FASEB J,* Vol18, No2, (February 2004), pp.394-396, ISSN0892-6638

Pages, J. M., Alibert-Franco, S., Mahamoud, A., Bolla, J. M., Davin-Regli, A., Chevalier, J., & Garnotel, E. (2010). Efflux Pumps of Gram-Negative Bacteria, a New Target for New Molecules. *Current Topics in Medicinal Chemistry,* Vol10, No18, (December 2010), pp.1848-1857, ISSN1568-0266

Peleg, A. Y. & Hooper, D. C. (2010). Hospital-acquired infections due to gram-negative bacteria. The *New England Journal of Medicine,* Vol362, No19, (May 2010), pp.1804-1813, ISSN0028-4793

Qin, W., Yang, K., Tang, H., Tan, L., Xie, Q., Ma, M., Zhang, Y., & Yao, S. (2011). Improved GFP gene transfection mediated by polyamidoamine dendrimer-functionalized multi-walled carbon nanotubes with high biocompatibility. *Colloids and Surfaces. B Biointerfaces,* Vol84, No1, (May 2011), pp.206-213, ISSN0927-7765

Rahman, M. A., Summerton, J., Foster, E., Cunningham, K., Stirchak, E., Weller, D., & Schaup, H. W. (1991). Antibacterial activity and inhibition of protein synthesis in Escherichia coli by antisense DNA analogs. *Antisense Research Development,* Vol1, No4, (winter 1991), pp.319-327, ISSN1050-5261

Rasmussen, L. C., Sperling-Petersen, H. U., & Mortensen, K. K. (2007). Hitting bacteria at the heart of the central dogma: sequence-specific inhibition. *Microbial Cell Factories,* Vol6, (August 2007), pp.24, ISSN1475-2859

Ravina, M., Paolicelli, P., Seijo, B., & Sanchez, A. (2010). Knocking down gene expression with dendritic vectors. *Mini Review in Medicinal Chemistry,* Vol10, No1, (Janurary 2010), pp.73-86, ISSN1389-5575

Rojas-Chapana, J., Troszczynska, J., Firkowska, I., Morsczeck, C., & Giersig, M. (2005). Multi-walled carbon nanotubes for plasmid delivery into *Escherichia coli* cells. *Lab on a Chip,* Vol5, No5, (May 2005), pp.536-539, ISSN1473-0197

Sarno, R., Ha, H., Weinsetel, N., & Tolmasky, M. E. (2003). Inhibition of aminoglycoside 6'-N-acetyltransferase type Ib-mediated amikacin resistance by antisense oligodeoxynucleotides. *Antimicrobial Agents and Chemotherary,* Vol47, No10, (October 2003), pp.3296-3304, ISSN0066-4804

Shao, Y., Wu, Y., Chan, C. Y., McDonough, K., & Ding, Y. (2006). Rational design and rapid screening of antisense oligonucleotides for prokaryotic gene modulation. *Nucleic Acids Research,* Vol34, No19, (October 2006), pp.5660-5669, ISSN0305-1048

Shimizu, T., Sato, K., Sano, C., Sano, K., & Tomioka, H. (2003). Effects of antisense oligo DNA on the antimicrobial activity of reactive oxygen intermediates and antimycobacterial agents against *Mycobacterium avium* complex. *Kekkaku,* Vol78, No1, (Janurary 2003), pp.33-35, ISSN0022-9776

Soler Bistue, A. J., Ha, H., Sarno, R., Don, M., Zorreguieta, A., & Tolmasky, M. E. (2007). External guide sequences targeting the aac(6')-Ib mRNA induce inhibition of amikacin resistance. *Antimicrobial Agents and Chemotherary,* Vol51, No6, (June 2007), pp.1918-1925, ISSN0066-4804

Stock, R. P., Olvera, A., Sanchez, R., Saralegui, A., Scarfi, S., Sanchez-Lopez, R., Ramos, M. A., Boffa, L. C., Benatti, U., & Alagon, A. (2001). Inhibition of gene expression in *Entamoeba histolytica* with antisense peptide nucleic acid oligomers. *Nature Biotechnology,* Vol19, No3, (March 2001), pp.231-234, ISSN1087-0156

Stock, R. P., Olvera, A., Scarfi, S., Sanchez, R., Ramos, M. A., Boffa, L. C., Benatti, U., Landt, O., & Alagon, A. (2000). Inhibition of neomycin phosphorotransferase expression in *Entamoeba histolytica* with antisense peptide nucleic acid (PNA) oligomers. *Archives of Medical Research,* Vol31, No4 Suppl, (July-August 2000), pp.S271-S272, ISSN0188-4409

Tan, X. X., Actor, J. K., & Chen, Y. (2005). Peptide nucleic acid antisense oligomer as a therapeutic strategy against bacterial infection: proof of principle using mouse intraperitoneal infection. *Antimicrobial Agents and Chemotherary,* Vol49, No8, (August 2005), pp.3203-3207, ISSN0066-4804

Thompson, A. J. V. & Patel, K. (2009). Antisense Inhibitors, Ribozymes, and siRNAs. *Clinics in Liver Disease,* Vol13, No3, (August 2009), pp.375-390, ISSN 1089-3261

Tilley, L. D., Hine, O. S., Kellogg, J. A., Hassinger, J. N., Weller, D. D., Iversen, P. L., & Geller, B. L. (2006). Gene-specific effects of antisense phosphorodiamidate morpholino oligomer-peptide conjugates on Escherichia coli and Salmonella enterica serovar typhimurium in pure culture and in tissue culture. *Antimicrobial Agents and Chemotherary,* Vol50, No8, (August 2006), pp.2789-2796, ISSN0066-4804

Tilley, L. D., Mellbye, B. L., Puckett, S. E., Iversen, P. L., & Geller, B. L. (2007). Antisense peptide-phosphorodiamidate morpholino oligomer conjugate: dose-response in mice infected with *Escherichia coli. The Journal of Antimicrobial Chemotherapy,* Vol59, No1, (Janurary 2007), pp.66-73, ISSN0305-7453

Torres, V. C., Tsiodras, S., Gold, H. S., Coakley, E. P., Wennersten, C., Eliopoulos, G. M., Moellering, R. C., Jr., & Inouye, R. T. (2001). Restoration of vancomycin susceptibility in Enterococcus faecalis by antiresistance determinant gene transfer. *Antimicrobial Agents and Chemotherary,* Vol45, No3, (March 2001), pp.973-975, ISSN0066-4804

Venkatesan, N. & Kim, B. H. (2006). Peptide conjugates of oligonucleotides: synthesis and applications. *Chemical Reviews,* Vol106, No9, (September 2006), pp.3712-3761, ISSN0009-2665

Wang, H., Meng, J., Jia, M., Ma, X., He, G., Yu, J., Wang, R., Bai, H., Hou, Z., & Luo, X. (2010). oprM as a new target for reversion of multidrug resistance in Pseudomonas aeruginosa by antisense phosphorothioate oligodeoxynucleotides. *FEMS Immunology Medical Microbiology,* Vol60, No3, (December 2010), pp.275-282, ISSN0924-8244

White, D. G., Maneewannakul, K., von, H. E., Zillman, M., Eisenberg, W., Field, A. K., & Levy, S. B. (Decenber 1997). Inhibition of the multiple antibiotic resistance (mar) operon in *Escherichia coli* by antisense DNA analogs. *Antimicrobial Agents and Chemotherary,* Vol41, No12, pp.2699-2704, ISSN0066-4804

Woodford, N. & Wareham, D. W. (2009). Tackling antibiotic resistance: a dose of common antisense? *The Journal of Antimicrobial Chemotherapy,* Vol63, No2, (2009), (February 2009),pp.225-229, ISSN0305-7453

Wright, G. D. (2009). Making sense of antisense in antibiotic drug discovery. *Cell Host a Microbe,* Vol6, No3, (September 2009), pp.197-198, ISSN1931-3128

Xue-Wen, H., Jie, P., Xian-Yuan, A., & Hong-Xiang, Z. (2007). Inhibition of bacterial translation and growth by peptide nucleic acids targeted to domain II of 23S rRNA. *Journal of Peptide Science,* Vol13, No4, (April 2007), pp.220-226, ISSN1075-2617

Zhang, Y., Qin, W., Tang, H., Yan, F., Tan, L., Xie, Q., Ma, M., Zhang, Y., & Yao, S. (2011). Efficient assembly of multi-walled carbon nanotube-CdSe/ZnS quantum dot hybrids with high biocompatibility and fluorescence property. *Colloids and Surfaces. B Biointerfaces,* Vol87, No2, (Octiber 2011), pp.346-352, ISSN0927-7765

Zorko, M. & Langel, U. (2005). Cell-penetrating peptides: mechanism and kinetics of cargo delivery. *Advanced Drug Delivery Review,* Vol57, No4, (February 2005), pp.529-545, ISSN0169-409X